Lecture Notes in Mathematics   1555

Editors:
A. Dold, Heidelberg
B. Eckmann, Zürich
F. Takens, Groningen

T0215666

Otto Liess

Conical Refraction and Higher Microlocalization

Springer-Verlag

Berlin Heidelberg New York
London Paris Tokyo
Hong Kong Barcelona
Budapest

Author

Otto Liess
Department of Mathematics
University of Bologna
Piazza Porta di San Donato 5
Bologna, Italy

Mathematics Subject Classification (1991): 35-02, 35A27, 58G17

ISBN 3-540-57105-1 Springer-Verlag Berlin Heidelberg New York
ISBN 0-387-57105-1 Springer-Verlag New York Berlin Heidelberg

© Springer-Verlag Berlin Heidelberg 1993
Printed in Germany

2146/3140-543210 - Printed on acid-free paper

Preface

These notes focus on some results concerning propagation of analytic microlocal singularities for solutions of partial differential equations with characteristics of variable multiplicity, and on the tools from the theory of higher involutive microlocalization needed in the proofs.

The simplest model to which the results apply is Maxwell's system for homogeneous anisotropic optical media (typical examples of which are crystals); then the underlying physical phenomenon is that of conical refraction.

The main difficulty in the study of operators with characteristics of variable multiplicity stems from the fact that the characteristic variety of such operators is not smooth. Indeed, near a singular point, a number of constructions usually performed in the study of the propagation of singularities will degenerate or break down. In the analytic category, these difficulties can be best investigated from the point of view of higher microlocalization.

Unfortunately, none of the theories on higher analytic microlocalization in use nowadays completely covers the situation that we encounter later in the notes. Rather than adapting or extending the existing theories to the present needs, we have chosen to build up a new theory from a uniform point of view. Actually the results on higher microlocalization are sufficiently well delimited from the other results of the text and could in principle be read independently of the rest. Special emphasis is put, on the other hand, upon the relation and interplay between the results on propagation of microlocal singularities and similar results and constructions in geometrical optics.

All microlocalization processes seem to follow some underlying common pattern. Therefore some overlap with other articles and books on higher microlocalization has been inevitable. It is also clear in this situation that we have been greatly influenced by the published literature. However, the point of view on higher microlocalization taken here is different from both that of Kashiwara-Laurent on second microlocalization, and from that of Sjöstrand-Lebeau on higher microlocalization. In particular, the intersection with the Astérisque volume of Sjöstrand and the Birkhäuser volume of Laurent (see references) is reasonably small. Otherwise, the text is based to a large extent on results that were obtained by the author in the last few years and have not been published in detail before.

Acknowledgements. This volume could not have been written without the help of my friends and family. This applies as much to a number of discussions on parts of the mathematical content as to practical matters connected with the writing, or typing, of these notes. My thanks go to H.D.Alber, E.Bernardi, A.Bove, F.John, S.Hansen, K.H.Hofmann, L.Liess, E.Meister, C.Parenti, L.Rodino, N.Tose and M.Uchida who all helped me in some way. In practical terms, the notes were written while I was working at, or visiting the, Universities of Bologna, Darmstadt, Palermo, Turin and the Courant Institute. I am indebted to all these institutions for their generous hospitality. The permission of the Editel Publishing Company to use material from a seminar report on part of these notes (see references) is gratefully acknowledged.

Contents

Chapter 1

Introduction

1.1 Statement of the main results

1. In the present notes we study propagation of analytic microlocal singularities for operators with characteristics of variable multiplicity: the classes of operators which we consider are described in the subsections 6 and 9 below and our main results are the theorems 1.1.3 and 1.1.6. The simplest case to which these theorems apply, and for which in fact the results they predict are wellknown, is the case of the equations of crystal optics. As in that case, or for that matter, as for a large class of operators which have characteristics of variable multiplicity, singularities will not propagate along uniquely defined curves or leaves associated with the characteristic surface but will rather split along families of such curves or leaves. In crystal optics this phenomenon is closely related to the physically observed phenomenon of conical refraction of light in transmission problems. It has therefore become common practice to refer to such results as to results in "conical refraction". While we shall try to discuss the relation between "conical refraction in the interior" and "conical refraction at the boundary" later on, we should point out that no special effort is made to develop specific methods to study boundary or transmission problems for their own sake. In particular our results refer basically to propagation of singularities in free space and boundary or transmission problems are considered only marginally. (Cf. here however chapter VIII.) This explains the first part of the title of these notes.

As for the second part of the title, it refers to the methods of proof. In fact, the main difficulties in the study of operators with characterstics of variable multiplicity, come from the fact that the characteristic variety associated with them is not smooth. Indeed, near a singular point a number of constructions which one usually performs to study propagation of singularities will degenerate or break down. A typical situation how this happens is when the directions of propagation of microlocal singularities are given by some Hamiltonian vector field in the regular part of the characteristic variety and when this vector field degenerates at singular points. If then one looks at the limits of directions of propagation associated with regular points nearby, one will be left in general with a family of possible directions or subspaces of propagation: the appearance of this kind of families is in fact the true reason why conical refraction occurs. It is then natural to ask if one could not possibly simplify the picture by pushing the process of microlocalization further and localize around individual such limit points. (Also cf. here section 1.2.) This is often possible with the aid of so called "second microlocalization" and indeed it has turned out that in a number of cases in which conical refraction appears, the propagation of the second wave front set is governed by the same kind of simple laws which have been established for propagation of the first wave front set for operators with characteristics of constant multiplicity: cf. Esser [1], Laubin-Esser [1], Laubin [1,2], Tose [1,2,3,4,5]. Actually, it may be safely said at the present moment that many results on propagation of analytic singularities and conical refraction can be understood best from the point of view of higher microlocalization. (The first mathematicians to use second microlocalization in a systematic way in in the analytic category in this context have been P.Laubin and N.Tose.) Higher microlocalization will also play a central role in the present notes. It is then perhaps worth mentioning that since we consider also cases of characteristics of "highly" variable multiplicity, a real simplification will occur sometimes only at the level of rather high order microlocalization.

Since several theories of second or higher microlocalization have been elaborated, we would like to say a few words on how our present theory relates to the ones already in print. In fact, second analytic microlocalization has been introduced by Kashiwara, in what we shall call the involutive case: cf. Kashiwara- Kawai [1], Kashiwara-Laurent [1] and Laurent [1] for a description and elaboration of Kashiwara's theory. A related definition has been considered in a Lagrangian setting by Sjöstrand [2,3], who also considered microlocalization of arbitrary order. Here we should note that while Kashiwara argued in terms of relative cohomology, Sjöstrand's approach is based on the Fourier- Bros-Iagolnitzer (FBI) transformation. Finally Lebeau [1], working in the spirit of Sjöstrand's theory, has introduced second microlocalization with respect to isotropic submanifolds. (Cf. also the expository notes of Esser- Laubin [1].) A survey on related results in the C^∞-category is Bony [2]. (Also cf. Delort [1] and comment 2.1.14 below.)

The situation which we shall encounter here is essentially an involutive one. When we resort to microlocalization of order higher than two, however none of the theories in use nowadays seems to cover the present situation and rather than adapting one of the existing theories, we have chosen to build up a theory of higher microlocalization from scratch. In particular we obtain for the case of second microlocalization a theory which is parallel to the theory of Kashiwara-Laurent, but our construction is based on completely different ideas: while, as mentioned above, Kashiwara's definitions are based on relative cohomology, we remain very close to Hörmander's first definition of the (standard) analytic wave front set. (Cf. Hörmander [2].) In fact, one of the main objectives in our study of second microlocalization is to analyze how Fourier integral operators (F.I.O.'s, henceforth), defined following the prescription from Hörmander [1], act (for suitable symbols and phases) on the second analytic wave front set. Unfortunately, in the case of microlocalization of order greater than two, things become very technical. While we have analyzed pseudodifferential operators also in that case, we have not tried to build up a reasonable calculus of F.I.O.'s. As a consequence, we have not analyzed the invariant meaning of the notions which we consider and our theory of higher microlocalization of order higher than two remains essentially a theory in special coordinates.

2. To state our first result, let us now assume that $p(x, D)$ is a linear partial differential operator of order m with analytic coefficients defined on an open set X in R^n, or, alternatively, a classical analytic pseudodifferential operator of order m defined on a set of form $X \times G$, where G is a cone in the space of phase variables. (For a definition of analytic pseudodifferential operators defined on conic subsets of T^*X, cf. Sato-Kawai-Kashiwara [1], Treves [1] or chapter III below.) We denote by W the characteristic variety of $p(x, D)$ and consider some point $(x^0, \xi^0) \in W$. When the principal part p_m of p vanishes of constant multiplicity, no conical refraction occurs, so we shall assume later on that p_m vanishes of variable multiplicity at (x^0, ξ^0). To describe our assumptions, let us assume that p_m vanishes of order s at (x^0, ξ^0) and denote by Σ the set

$$\Sigma = \{(x, \xi); \partial_{x,\xi}^\alpha p_m(x, \xi) = 0 \text{ if } |\alpha| < s , \sum_{|\alpha|=s} |\partial_{x,\xi}^\alpha p_m(x, \xi)| \neq 0\}. \qquad (1.1.1)$$

(Multiindex notation is used in a standard fashion.)
We make the following assumptions:

i) Σ is a homogeneous real-analytic manifold of codimension d .

ii) In a conic neighborhood of (x^0, ξ^0), Σ is regular involutive.

By this we mean that we can find a conic neighborhood Γ of (x^0, ξ^0) and positively homogeneous real-analytic functions r_i, $i = 1, ..., d$, on Γ with the following properties:

$\Sigma \cap \Gamma = \{(x, \xi) \in \Gamma; r_i(x, \xi) = 0, \ i \leq d\},$

$\{r_i, r_j\} = 0$, where $\{f, g\}$ is the Poisson bracket of f and g,

the differentials dr_i, $i \leq d$, and the canonical one-form $\sum \xi_i dx_i$ are linearly independent at all points from Σ which lie in Γ.

An important notion in this context is that of the bicharacteristic leaves of Σ. By definition these are the d−dimensional submanifolds in Σ which have as tangent vectors the Hamiltonian vectorfields associated with the r_i. As is standard, it follows from the assumptions on Σ that we can find homogeneous real-analytic canonical coordinates such that in a conic neighborhood of (x^0, ξ^0),

$$\Sigma = \{(x, \xi); \xi_i = 0 \text{ for } i \leq d\}. \tag{1.1.2}$$

In these coordinates, the bicharacteristic leaves are of course just of form $x_i = constant$, $\xi_i = constant$, $i = d + 1, ..., n$. We shall also often set in this situation $\xi' = (\xi_1, ..., \xi_d)$, so that Σ becomes $\xi' = 0$ in some suitable local coordinate patch. Correspondingly, we set

$$x' = (x_1, \ldots, x_d), x'' = (x_{d+1}, \ldots, x_n), \xi'' = (\xi_{d+1}, \ldots, \xi_n).$$

(Related notations shall also be considered later on.)

An interesting case for results on propagation of singularities is here when p_m is transversally elliptic to Σ. By this we mean that if we fix $(y, \eta) \in \Sigma$, then we can find a conic neighborhood Γ of (y, η) and some c_1, c_2 so that

$$d_\Sigma^s(x, \xi)|\xi|^{m-s} \leq c_1 |p_m(x, \xi)| \leq c_2 d_\Sigma^s(x, \xi)|\xi|^{m-s} \text{ if } (x, \xi) \in \Gamma, \tag{1.1.3}$$

where d_Σ is some homogeneous distance function to Σ. Actually, Σ is then precisely the characteristic variety of p, so p_m vanishes of constant multiplicity on its characteristic variety. We have then the following classical result of Bony-Schapira [1]:

Theorem 1.1.1. *Let u be a solution of $p(x, D)u = 0$ and denote by $WF_A u$ the analytic wave front set of u. Then $WF_A u$ is a union of bicharacteristic leaves of Σ. (More explicitly, if L is a connected bicharacteristic leaf of Σ and if $(x^0, \xi^0) \in WF_A u$, then $L \subset WF_A u$.)*

The case when p_m is real-valued and $s = 1$ in (1.1.1), i.e. when p is of so-called real principal type, theorem 1.1.1 had of course been considered already by Hörmander [2] and Kashiwara (cf. Sato-Kawai-Kashiwara [1]) and has been the prototype of all results on propagation of microlocal singularities ever since. Note that in this case the bicharacteristic leaves are just the null bicharacteristic curves of p_m. (For related results for propagation of C^∞ singularities, we mention here Hörmander-Duistermaat [1], Bony [3] and Sjöstrand [1]. Theorem 1.1.1 has also been studied in the Gevrey category. Cf. Kessab [1] and Arisumi [1].) A more general result in the analytic category is the one from Grigis-Schapira-Sjöstrand [1].

3. The theorem of Bony-Schapira settles the case when p_m vanishes of constant multiplicity on W near (x^0, ξ^0). When p_m does not vanish of constant multiplicity, we must in general expect that the situation becomes more complicated. A trivial example of what can happen is when $n = 3$, $p(D) = D_1 D_2$, $D_j = -\sqrt{-1}\partial/\partial x_j$, and $\zeta^0 = (0, 0, 1)$. In this case the general solution of $p(D)u = 0$ is of form $u = u_1 + u_2$, where u_1 is constant in x_1 and u_2 is constant in x_2. The microlocal singularities at $(0, \xi^0)$ may therefore propagate in the x_1-direction or in the x_2-direction or in both and it cannot be predicted without additional information on u which of these possibilities will actually occur. A slightly more general situation of the same kind is when we start from a constant coefficient partial differential operator $p(D)$ in d variables $x = (x_1, x_2, \ldots, x_d) \in R_x^d$ and regard it as an operator in the variables $(x, y) \in R^n$, $y = (y_1, y_2, \ldots, y_{d'}) \in R_y^{d'}$, $n = d + d'$, $d' \geq 1$. We also assume that p is hyperbolic in the direction $\theta = (1, 0, \ldots, 0) \in R^n$ and consider the vector $\lambda^0 = (0, \ldots, 0, 1) \in R^n$. We want to see what information we can obtain from $(0, \lambda^0) \in WF_A u$ if u is a solution of $p(D)u = 0$, say on R^n. Let Γ be the hyperbolicity cone of p_m with respect to θ, i.e., let Γ be the connected component of $\{\lambda \in R^n; p_m(\lambda) \neq 0\}$ which contains θ. If Γ^0 denotes the polar of Γ, then we can find a fundamental solution E of $p(D)E = \delta$ concentrated in Γ^0. (Cf. e.g. Hörmander [5].) Of course, $\Gamma^0 \subset \{\lambda; \lambda_{d'+1} = \cdots = \lambda_n = 0\}$. It is easy to conclude from this (e.g.) that if $x_1 > 0$ is fixed, then there is $x' = (x_2, \ldots, x_d)$ so that $z = (x_1, x', 0) \in \Gamma^0$ and so that $(z, \lambda^0) \in WF_A u$: the singularity has spread inside the cone Γ^0 from $(0, \lambda^0)$ to (z, λ^0). It is a phenomenon of this kind which we have in mind when we say that conical refraction has occured. It should be noted that Γ^0 is a cone of dimension strictly lower than the full dimension n. Let us also mention that an equivalent way of formulating the preceding result is the following: if for some fixed $x_1^0 > 0$ it is known that $(z, \lambda^0) \notin WF_A u$ for any z of form $(x_1^0, x', 0) \in \Gamma^0$, then it follows that $(0, \lambda^0) \notin WF_A u$. It is in this way that results will be formulated later on.

4. To study the structure of p_m near $\lambda^0 = (x^0, \xi^0)$ when (1.1.3) is not satisfied, it is a natural idea to consider the Taylor polynomial p_m^T of order s of p_m at points $\lambda \in \Sigma$. For constant coefficient equations one obtains then what was called the "localization polynomial" of p_m in Atiyah-Bott-Gårding [1] in a related context (-more general forms of "localizations" had been considered in Hörmander [4] before-) but the idea of working

with this kind of Taylor expansions is of course implicit or explicit in most papers in which higher microlocalization is used. Here we recall that the homogeneous part of order s of the Taylor expansion of p_m is, under the assumptions above, an invariantly defined function on $T_\Sigma \Gamma$, which is the tangent space to Γ considered over points in Σ. When $\lambda \in \Sigma$ and $v \in T_\lambda \Gamma$ we may in fact just set

$$p_m^T(\lambda)(v) = (V)^s p_m(\lambda), \tag{1.1.4}$$

where V is any C^∞-vectorfield on Γ such that $V(\lambda) = v$. Note that the right hand side of (1.1.4) depends (in view of (1.1.1)) only on v. Moreover, when $v' \in T\Sigma$ (which we identify with a subspace in $T\Gamma$, using $\Sigma \subset \Gamma$), then

$$p_m^T(\lambda)(v) = p_m^T(\lambda)(v + v') \tag{1.1.5}$$

since we can conclude from (1.1.1) that

$$V_1 V_2 \cdots V_s p_m(\lambda) = 0 \text{ for } \lambda \in \Sigma, \tag{1.1.6}$$

if the V_j are vectorfields of which at least one is tangent to Σ. It follows that p_m^T is in fact a function defined on the normal bundle $N\Sigma = T_\Sigma\Gamma/T\Sigma$ of Σ. (Thus for each λ in Σ, the fiber over λ of $N\Sigma$ is $T_\lambda\Gamma/T_\lambda\Sigma$.) Furthermore, we denote by $\dot{N}\Sigma$ the normal bundle to Σ with the zero-section removed. It is on $\dot{N}\Sigma$ that our further analysis takes place.

Definition 1.1.2. *Let $(\lambda^0, v^0) \in \dot{N}\Sigma$ be given. We say that (λ^0, v^0) is microcharacteristic for p_m relative to Σ (or for (p_m, Σ), for short) if $p_m^T(\lambda^0)(v^0) = 0$.*

Related notions have been introduced in Bony [1], respectively Hanges-Sjöstrand [2]. When the set of microcharacteristic vectors relative to Σ is void, we are in the situation from (1.1.3). (This is checked best by assuming that Σ has been brought to the form (1.1.2).) The localization polynomial is then $p_m^T(\lambda)(v) = \sum_{|\alpha|=s} \partial_\xi^\alpha p_m(\lambda) v^\alpha$ and theorem 1.1.1 applies. In the first result of the present notes we shall assume (and this is in a rather complete analogy with the assumptions in the theorem of Bony-Schapira above,) that the set Λ of microcharacteristic vectors is regular involutive with respect to a natural bi-symplectic structure on $\dot{N}\Sigma$ and that p_m^T vanishes precisely of some order s' on Λ. What this means shall be explained in detail later on, but we can describe the situation briefly well enough working in the special coordinates from (1.1.2). Let us assume in fact that real-analytic homogeneous canonical coordinates have been chosen so that $\Sigma = \{(x, \xi); \xi_i = 0 \text{ for } i \leq d\}$ in a conic neighborhood of λ^0. In these coordinates the bicharacteristic foliation of Σ is given by the leaves $\{(x, \xi), x_i = constant, \xi_i = constant \text{ for } i \geq d+1, \xi_i = 0 \text{ for } i \leq d\}$, and the coordinates in $\dot{N}\Sigma$ may be chosen to be (x, ξ'', v), with $v \in R^d$. Next we observe that p_m^T is homogeneous of order s in the fiber variable v of $\dot{N}\Sigma$ and homogeneous of order $m - s$ in the variable ξ of Σ. This is closely related to the fact that $\dot{N}\Sigma$ itself has a bi-homogeneous structure: on one hand we have the multiplication in the fiber variables of $\dot{N}\Sigma$ and on the other hand the multiplication in Σ induces another multiplication in $\dot{N}\Sigma$ for which, in our special coordinates, (x, ξ'', v) is mapped to $(x, t\xi'', tv)$. It is now clear that $(x, \xi'', v) \in \Lambda$ implies $(x, t\xi'', tv)$ and $(x, \xi'', tv) \in \Lambda$ for all $t > 0$. In the special coordinates considered here, the assumptions on Λ and p_m^T are now that for each fixed (λ^0, v^0) in Λ we can find a set G in $\dot{N}\Sigma$, a natural number d' and real-analytic functions $f_j : G \to R$, $j \leq d'$, with the following properties :

a) G is open, contains $(x^0, \xi^{0''}, v^0)$ and from (x, ξ'', v) in G and $t > 0$ it follows that $(x, t\xi'', tv)$ is in G. Moreover we also require that $(x, \xi'', v) \in G$ and $0 < t < 1$ implies $(x, \xi'', tv) \in G$.

b) The f_i are homogeneous of degree one in v, of degree zero in ξ'', and
$$\Lambda \cap G = \{(x, \xi'', v) \in G; f_i(x, \xi'', v) = 0, \forall i \leq d'\}.$$

c) We then also assume that the functions f_i are for each fixed (x'', ξ'') in involution in the variables (x', v).

By this we mean (and this is a definition) that
$$\sum_{k=1}^{d} [(\partial f_i / \partial x_k)(\partial f_j / \partial v_k) - (\partial f_i / \partial v_k)(\partial f_j / \partial x_k)] = 0, \forall i, j \leq d'.$$

The left hand side of the preceding expression will be called, following the terminology introduced in Laurent [1], the "relative" Poisson bracket of f_i and f_j.

Our next assumptions are:

d) for each fixed (x'', ξ'') the differentials of the functions $(x', v) \to f_i(x', x'', \xi'', v)$ and the one-form $\sum_{k \leq d} v_k dx_k$ are linearly independent.

e) rank $(\partial f_i / \partial v_j)(\lambda^0, v^0)_{i \leq d', j \leq d} = d'$.

f) Finally we assume that
$$|p_m^T(x, \xi)(v)| \sim (dist_\Lambda(x, \xi'', v))^{s'} |v|^{s-s'} |\xi''|^{m-s}, \text{ for some } s'$$

where $dist_\Lambda(x, \xi'', v)$ is the distance from (x, ξ'', v) to Λ. Practically this means that p_m^T is transversally elliptic to Λ.

Note that all these conditions have been defined in special coordinates, so we must see
later on which is their invariant meaning. This is essentially a consequence of the results
in Laurent [1], but we shall study the "relative" symplectic structure on $\dot{N}\Sigma$ associated
with the relative Poisson brackets above in detail in the sections 4.5, 4.6, 4.7. We also
observe that the conditions on Λ are just meant to make Λ an involutive regular manifold
in the relative symplectic structure; we are thus in a completely involutive setting.

5. We may then consider for each fixed (x, ξ, v) in Λ the bicharacteristic leaf in the variables (x', v) which contains (x, ξ, v) for the simplectic structure induced by the relative
Poisson brackets. We call such leaves "relative" in what follows. We can now state the
following result:

Theorem 1.1.3. *Let $(x^0, \xi^0), \Sigma, \Lambda$ be as before and assume that all the assumptions
above on them are satisfied. Also consider a (distribution) solution u of $p(x, D)u = 0$ in
a conic neighborhood Γ of (x^0, ξ^0). (By this we mean that $WF_A\, p(x, D)\, u \cap \Gamma = \emptyset$. When
$p(x, D)$ is a "true" differential operator it is perhaps more natural to consider solutions
u defined in a full neighborhood U of x^0.) Let further a subset Y in $R^d_{x'}$, the space of the
x' variables, be given. We assume that we can find a connected open set $W \subset R^d_{x'}$ with
the following properties:*

$x^{0'} \in W, Y \subset W,$

u is defined in a neighborhood of $W \times \{x^{0''}\}$,

*for each $x' \in W$ and each v for which $((x', x^{0''}), \xi^0, v)$ is microcharacter-
istic for (p_m, Σ), the relative bicharacteristic leaf L of Λ which contains
$(x', x^{0''}, \xi^{0''}, v)$ has a nontrivial intersection with $Y \times \{x^{0''}\} \times \{\xi^{0''}\} \times R^d$.*

Finally assume that

$$((x', x^{0''}), \xi^0) \notin WF_A\, u, \forall x' \in Y.$$

Then it follows that $(x^0, \xi^0) \notin WF_A\, u.$

Despite its technical character, the meaning of the conditions in the statement will
become quite clear at the level of second microlocalization. Actually, these assumptions
are such that the second analytic wave front set will propagate along the leaves of the
relative bicharacteristic foliation. This is again in a rather strong analogy with the result
of Bony-Schapira above, where the first analytic wave front set was propagated along
the leaves of the initial bicharacteriastic foliation of Σ. In the conditions from theorem
1.1.3 the conclusion there will then follow from the fact that the first analytic wave front

set must propagate in regions where there is no second analytic wave front set. (That this is so is the content of the micro-Holmgren theorem below.)

The most interesting particular case in theorem 1.1.3 is perhaps when $d' = 1$. When p_m^T is second order and strictly hyperbolic, the result has been proved by Laubin [1] and Tose [4], independently of each other and using different methods. The same result in the C^∞-setup had been proved earlier by Melrose-Uhlmann [1]. Also cf. Bernardi [1] and Wakabayashi [3,5]. In an oral communication I was told by N.Tose that it is also possible to obtain the general case in theorem 1.1.3 as a consequence of the theory of "two-microfunctions with holomorphic parameters". (Details of the argument have not been published.) Actually, the result of Melrose-Uhlmann and Laubin is stated in a somewhat different way, so we shall establish the relation between the statement of Melrose, Uhlmann, Laubin and theorem 1.1.3 in section 6.2. Also note that the equations of crystal optics are essentially of the type $d' = 1$ with p_m^T second order and of real principal type and that the bicharactersitic leaves reduce to bicharacteristic curves in this case.

6. To get a feeling of what theorem 1.1.3 says, let us return to our example in nr. 4 above in a slightly more general situation already mentioned briefly in Melrose-Uhlmann [1]. We shall assume in fact that M is a real analytic manifold of product type

$$M = M_1 \times M_2,$$

and let p be a partial differential operator of order $m > 1$ with analytic coefficients on M_1. We extend p to an operator on M by the rule

$$p(z, D_z)f(x, y) = p(x, D_x)f(x, y),$$

where the local variables from M_1 are denoted by x, those from M_2 by y and $z = (x, y)$ are local variables in M. (In other words, p acts effectively only in the variables from M_1.) We assume that p, when regarded as an operator on M_1, is non-elliptic, but of real principal type, and let P be an operator on M of form

$$P = p + q,$$

where q is some partial differential operator with analytic coefficients on M which is of lower order than p. No additional conditions will be put on q and all our statements and results are microlocal, so it is clear that we may always assume in local coordinates that p is homogeneous. Next we consider some vector $(x^0, \xi^0) \in T^*M_1$, $\xi^0 \neq 0$, which is characteristic for p. Finally fix $(y^0, \eta^0) \in T^*M_2$ such that $\eta^0 \neq 0$ and denote (x^0, y^0)

by z^0 and $(0, \eta^0)$ by λ^0. The vector (z^0, λ^0) is then characteristic for P and p vanishes precisely of order m at (z^0, λ^0). In particular the bicharacteristic curve of p starting at (z^0, λ^0) stays forever at (z^0, λ^0) and no propagation phenomena are directly related to it. It is also clear that the multiplicity of the characteristic variety of P is changing at (z^0, λ^0). In local coordinates near (z^0, λ^0) the set of characteristic vectors at which p vanishes of order m is

$$\Sigma = \{(x, y, \xi, \eta); \xi = 0, \eta \neq 0\}$$

and the leaves of Σ are $y = \tilde{y}$, $\eta = \tilde{\eta}$ for some $(\tilde{y}, \tilde{\eta})$. Moreover, $p^T(x, y, \lambda, v) = p(x, v)$ and the set of microcharacteristic vectors is $\{(x, y, \lambda, v); p(x, v) = 0, \lambda = (0, \eta), \eta \neq 0, v \neq 0\}$, which is a set of co-dimension one. In particular, the relative bicharacteristic leaves are just curves. Further, the relative Poisson bracket of f and g with respect to Σ is

$$\{f, g\}_r(x, y, \eta, v) = \sum_{k=1}^{d} [(\partial f/\partial x_k)(\partial g/\partial v_k) - (\partial f/\partial v_k)(\partial g/\partial x_k)]$$

where d is the dimension of M_1. It is thus immediate that, as expected, the relative bicharacteristics are of form $L \times (y^0, \eta^0)$, where L is some bicharacteristic curve for p on M_1.

Here we recall that the microlocal singularities of solutions of $p(x, D_x)v = 0$ on M_1, propagate along the bicharacteristic curves of p in T^*M_1. It is then natural to expect that a microlocal singularity at (z^0, λ^0) of a solution of $P(z, D_z)u = 0$ may propagate along any of the relative bicharacteristic curves $L \times (y^0, \eta^0)$. The theorem 1.1.3 thus relates, under appropriate geometric conditions, propagation of singularities for the operator p to conical refraction for P. Of course, all this works also when we only assume that p is a differential operator on M_1 with coefficients which depend parametrically on the variables from M_2. Unfortunately, operators in such product type situations are not general enough to be reasonable models for the situations covered by the general theorems.

To describe another elementary situation to which theorem 1.1.3 applies, recall that if f is a function which depends only on the phase variables ξ and if $f(\xi^0) = 0$, then the null bicharacteristic associated with f which starts at (x^0, ξ^0) is $(x^0 + s \operatorname{grad}_\xi f(\xi^0), \xi^0)$, $s \in R$. More generally, if $\Lambda = \{(x, \xi); f_1(\xi) = 0, \ldots, f_{d'}(\xi) = 0\}$, then the leaf of the bicharacteristic foliation of Λ which contains (x^0, ξ^0) is given by

$$\{(x, \xi^0); x = x^0 + \sum_{j=1}^{d'} s_j \operatorname{grad}_\xi f_j(\xi^0), s_j \in R\}.$$

Assume then that p is an operator as in theorem 1.1.3, with the additional assumption that p_m has constant coefficients. Since we do not necessarily assume that p is a differential operator, it is no loss of generality to assume that $\Sigma = \{(x, \xi) \in X \times G; \xi' = 0\}$ so that

$$p_m(\xi) = \sum_{|\alpha|=s} a_\alpha(\xi)\xi'^\alpha$$

for some real analytic functions a_α which are defined in a conic neighborhood of ξ^0 and are homogeneous of order $m - s$ there. It is then easy to see that

$$p_m^T(\lambda)(v) = \sum_{|\alpha|=s} a_\alpha(0, \xi'')v^\alpha,$$

so $(\tilde{x}, \tilde{\xi}, \tilde{v})$, $\tilde{\xi}' = 0$, is micro-characteristic precisely if $\sum_{|\alpha|=s} a_\alpha(0, \tilde{\xi}'')\tilde{v}^\alpha = 0$. Let us now fix $(\tilde{x}, \tilde{\xi}, \tilde{v})$ microcharacteristic. The assumption in the theorem says that we can write the set Λ of micro-characteristic vectors near $(\tilde{x}, \tilde{\xi}, \tilde{v})$ in the form

$$\{(x, \xi, v); \ \xi' = 0, f_i(\xi'', v) = 0, \ i = 1, ..., d'\},$$

for some real-analytic functions f_i which satisfy the conditions b), d) and e) above. The conditions $\{f_i, f_j\}_r = 0$ are here of course automatically fulfilled. The bicharacteristic leaf of Λ containing the point $(\tilde{x}, \tilde{\xi}'', \tilde{v})$ is then, (as seen in our introductory remark) of form $\{(x, \xi'', v); x \in \tilde{x} + A, \xi'' = \tilde{\xi}'', v = \tilde{v}\}$, where

$$A = \text{ span } (\text{ grad}_v f_1(\tilde{\xi}'', \tilde{v}), ..., \text{ grad}_v f_{d'}(\tilde{\xi}'', \tilde{v})).$$

7. To describe the second class of operators which we consider, assume once more that X is open in R^n and consider $(x^0, \xi^0) \in \dot{T}^*X$. (Which is the cotangent space of X with the zero section removed.) Further we consider a classical analytic pseudodifferential operator of order m defined in a conic neighborhood Γ of (x^0, ξ^0). Our first assumption on p is

I) the principal symbol p_m of p does not depend on x, i.e. has constant coefficients.

(We have assumed that p is classical, so the principal symbol is well-defined by the requirement that it is positively homogeneous of order m and that $p - p_m$ is of order $m - 1$.)

Actually we should say that we do not use the assumption I) in the proofs in the strong form in which we have stated it here. A particularly interesting condition which could replace I) is to assume that

"the localization p_m^T of p_m to $\Sigma = \{(x, \xi); \xi = t\zeta^0, t \in \dot{R}\}$ has constant coefficients."

Moreover, given the fact that we are studying microlocal phenomena, we could in principle assume that

I)' p_m (or, for that matter, p_m^T) has constant coefficients in a conic neighborhood of (x^0, ξ^0), provided p has been conjugated previously with a suitable elliptic analytic F.I.O. and has been multiplied with a convenient elliptic analytic pseudodifferential operator.

That the passage from I) to I)' can be interesting, will be shown in an example from section 1.3. However, we should say that we have not studied the invariant meaning of the constructions which we perform starting from $p(D)$ and in particular we cannot give (except for some special cases) an invariant statement of the conclusions one would obtain.

8. Let us now assume that p satisfies I). Our aim, as in theorem 1.1.3, is to estimate $WF_A u$ where u is a solution of $p(x, D)u = 0$ on Γ, for which we assume that possibly $(x^0, \xi^0) \in WF_A u$. Of course it follows from this in particular that $p_m(\xi^0) = 0$. Next we denote by s the multiplicity with which p_m vanishes at ξ^0, i.e., s is defined by the condition:

$$D^\alpha p_m(\xi^0) = 0 \text{ for } |\alpha| < s, \quad \sum_{|\alpha|=s} |D^\alpha p_m(\xi^0)| \neq 0. \tag{1.1.7}$$

When $s = 1$, then p_m is of principal type at ξ^0. In this case (and also when p_m vanishes of constant multiplicity at ξ^0 on its characteristic variety) propagation of singularities is well understood. (Cf. theorem 1.1.1, Hörmander [2], Hanges [1], Hanges-Sjöstrand [1], among others.) In the sequel we shall therefore almost always assume that $s > 1$. As opposed to the case from the preceding theorem, we do not have here however a distinguished involutive manifold with respect to which we could consider localizations other than $\Sigma = \{(x, \xi); \xi \text{ is proportional to } \xi^0\}$. We shall therefore define a localization polynomial $p_{m,1}$ by

$$p_{m,1}(\xi) = \sum_{|\alpha|=s} \partial_\xi^\alpha p_m(\xi^0)(\xi - \xi^0)^\alpha / \alpha!. \tag{1.1.8}$$

(For homogeneity reasons it suffices to consider $p_{m,1}$ at ξ^0 rather than at all points $t\xi^0$.) The relation which corresponds here to (1.1.5) is

$$p_{m,1}(\xi + \sigma \xi^0) = p_{m,1}(\xi), \forall \sigma \in R, \tag{1.1.9}$$

a property which is also explicitly stated in Atiyah-Bott-Gårding [1]. (Let us briefly show, for the convenience of the reader, how (1.1.9) is proved. We have, with $\nu = \sigma - 1$, that

$$p_{m,1}(\xi + \sigma \xi^0) = \sum_{|\alpha|=s} \partial_\xi^\alpha p_m(\xi^0)(\xi + \nu \xi^0)^\alpha / \alpha! =$$

$$\sum_{|\alpha|=s} \partial_\xi^\alpha p_m(\xi^0) \sum_{\gamma \leq \alpha} (\alpha!/(\gamma!(\alpha - \gamma)!))(\xi - \xi^0)^\gamma (\xi^0)^{\alpha-\gamma} / \alpha! =$$

$$\sum_{l=0}^{s}\sum_{|\gamma|=l}(\xi-\xi^0)^\gamma\sum_{|\delta|=s-l}[(\xi^0)^\delta\partial_\xi^{\gamma+\delta}p_m(\xi^0)/\delta!]/\gamma! = \sum_{|\gamma|=s}\partial_\xi^\gamma p_m(\xi^0)(\xi-\xi^0)^\gamma/\gamma!\,.$$

Here we have used that $\sum_{|\delta|=s-l}(\xi^0)^\delta\partial_\xi^{\gamma+\delta}p_m(\xi^0)/\delta! = 0$ for $s-l>0$, as a consequence of Euler's relation for homogeneous functions.) We also note that another convenient way to write (1.1.9) is sometimes

$$\langle\xi^0,\partial/\partial\xi\rangle^j p_{m,1}(\xi) = 0,\,\forall\xi\,,\forall j. \tag{1.1.10}$$

It follows once more that we may regard $p_{m,1}$ as a function on $\dot{N}\Sigma$, but since $p_{m,1}$ is constant coefficient we shall prefer to work in special coordinates and in fact regard it as a polynomial in the variables orthogonal fo ξ^0. We shall not yet make any explicit hypothesis on $p_{m,1}$ and must therefore hope that $p_{m,1}$ itself already corresponds to an operator which is simpler to understand than p, and that one can read off properties for p from the corresponding properties of $p_{m,1}$. As far as the first point is concerned, we observe that (1.1.9) implies in particular that $p_{m,1}$ is homogeneous, the degree of homogeneity being of course s. Moreover, in view of (1.1.9), it is natural to choose coordinates for which $\xi^0 = (0,...,0,1)$, so $p_{m,1}$ may be regarded as a polynomial on R^{n-1}. In some sense, we have then indeed a simpler operator, since we have lowered the dimension by one. However, it may well be that the simplification so achieved does not quite suffice to make $p_{m,1}$ as simple as desired.

To give an example, assume that coordinates have been chosen so that $\xi^0 = (0,...,0,1)$ and let $k \le n-2$ be given. Let us then also assume that p_m depends effectively only on $\xi_1,...,\xi_k$, in the sense that

$$p_m(\xi_1,...,\xi_k,\sigma_{k+1},...,\sigma_n) = p_m(\xi_1,...,\xi_k,0,...,0),\,\forall(\sigma_{k+1},...,\sigma_n)\in R^{n-k}.$$

It is then obvious that $p_{m,1}(\xi) = p_m(\xi)$ for all $\xi \in R^n$. If we now assume that the restriction of p_m to $\hat{R}^k = \{\xi \in R^n; \xi_j = 0 \text{ for } j > k\}$ is non-elliptic, then p_m vanishes of variable multiplicity at ξ^0.

We pause here for a moment for a brief comment on notations: in the sequel we shall often have to regard R^d for $d \le n$ as a subspace in R^n. Unless otherwise specified, we shall identify R^d with the subspace $\{\xi \in R^n; \xi_j = 0 \text{ for } j > d\}$ in the canonic way. Occasionally we shall write in such situations $\hat{R}^d$ rather than R^d, but often we shall just consider some vector $\mu \in R^d$ and also regard it as a vector in R^n by identifying it with $(\mu, 0,...,0)$.

Not much seems to have been gained by passing from p_m to $p_{m,1}$, since $p_{m,1}$ will have, even when regarded as a polynomial on $\hat{R}^{n-1} = \{\xi; \xi_n = 0\}$ zeros of variable multiplicity, e.g. at $(0, ..., 0, 1, 0) \in \hat{R}^{n-1}$. The main point however is that the dimension of the space on which $p_{m,1}$ lives is $n - 1$. Iteration of the localization process will then lead in the end to operators which are easy to understand. To see how this comes about, let us regard $p_{m,1}$ as a polynomial on R^{n-1} and let us denote by W^1 the characteristic variety of $p_{m,1}$ as a polynomial on R^{n-1}. It is clear that W^1 is nontrivial, since otherwise $p_{m,1}$ would be elliptic on R^{n-1} and ξ^0 would be the only zero of p_m in a conic neighborhood of ξ^0, contrary to our assumption. Moreover, it is natural to expect that the difficulties in our further study come from the zeros of $p_{m,1}$. In fact at this step, we can now localize $p_{m,1}$ as a polynomial in $n - 1$ variables at the points ξ^1 from W^1. We shall then arrive at some polynomials $p_{m,2}$ which live essentially on a space complementary to the space generated by $\xi^0, \xi^1, \xi^1 \in W^1$, and which are either elliptic on the space of variables on which they effectively depend (what we mean by this, will be explained in a moment) or have a nontrivial characteristic variety on that space. In the first case we stop our process of localizations for the $p_{m,2}$ under consideration, but in the second case we continue it. Arguing in this way, it is now clear that we shall arrive at families of chains of vectors $(\xi^0, \xi^1, \xi^2, ..., \xi^{k-1})$ and of polynomials $(p_{m,0}, p_{m,1}, ..., p_{m,k})$ which are related to p by the following properties:

a) $p_{m,0} = p_m$ and p_m vanishes of multiplicity $s(0) = s \geq 1$ at ξ^0,

b) $p_{m,j-1}$ vanishes of (some) multiplicity $s(j - 1)$ at ξ^{j-1} and $p_{m,j}$ is the localization of $p_{m,j-1}$ at ξ^{j-1}, $j = 1, ..., k$,

c)   the vectors $\xi^0, \xi^1, ..., \xi^{k-1}$ are linearly independent,

d) $p_{m,k}$ is elliptic as a polynomial on the space of variables on which it effectively depends, i.e. as a polynomial on any space Λ' which is complementary to its lineality Λ,

$$\Lambda = \{\xi \in R^n ; p_{m,k}(\eta + t\xi) = p_{m,k}(\eta), \forall \eta \in R^n, \forall t \in R\}.$$

Note that by relation (1.1.9) the lineality of $p_{m,1}$ contains ξ^0 and that if Λ' is the lineality of $p_{m,j-1}$ and Λ'' that of $p_{m,j}$, then Λ'' contains both ξ^{j-1} and Λ'. In particular, the lineality of $p_{m,k}$ will always contain ξ^0. To understand the meaning of theorem 1.1.6 below, we observe further that when $s(0) = s = 1$, then $p_{m,1}$ is a linear form, i.e., $p_{m,1} = \langle a, \xi \rangle$ for some $a \in C^n$. The lineality of $p_{m,1}$ is the space $\{\xi \in R^n ; \langle \operatorname{Re} a, \xi \rangle = 0, \langle \operatorname{Im} a, \xi \rangle = 0\}$. Also note that the space orthogonal to the lineality of $p_{m,1}$ is generated by $\operatorname{Re} a$ and $\operatorname{Im} a$ and that in fact $a = \operatorname{grad} p_{m,1}$ is proportional to $\operatorname{grad} p_m(\xi^0)$. More generally, we shall show in section 1.3 that if p_m vanishes of constant multiplicity at

$W = \{\xi; p_m(\xi) = 0\}$, and if W is given by the equations $\{\xi; f_i(\xi) = 0, \ i = 1, \ldots, d\}$, where the f_i have linearly independent differentials in a conic neighborhood of ξ^0, then the lineality of $p_{m,1}$ is $\{\xi; \langle d\, f_i(\xi^0), \xi \rangle = 0\}$ and a vector is orthogonal to the lineality precisely if it is in the span of the vectors grad $f_i(\xi^0)$. Finally we mention, and this is what is needed in order to understand the example in ur.3 above from the point of view of theorem 1.1.6 that if $p(D)$ is a linear constant coefficient partial differential operator which is strictly hyperbolic with respect to $\theta = (1, 0, \ldots, 0)$ and if Γ is the hyperbolicity cone of $p(D)$ with respect to θ, then we will have grad $p_m(\xi^0) \in (\Gamma^0 \cup -\Gamma^0)$ for any $\xi^0 \neq 0$ which satisfies $p_m(\xi^0) = 0$. (This is a consequence of the fact that we can construct a fundamental solution E for $p(D)$ which is supported by Γ^0. In fact, it is also known that there are solutions of $p(D)u = 0$ for which the singularities effectively propagate in the direction grad $p_m(\xi^0)$, which can only be if grad $p_m(\xi^0) \in (\Gamma^0 \cup -\Gamma^0)$. Also cf. section 1.3.)

Remark 1.1.4. *If $\xi^0, \xi^1, \ldots, \xi^{k'-1}$, $p_{m,0}, \ldots, p_{m,k'}$, are constructed according to the prescriptions a), b) and c), then $p_{m,k'}$ does not vanish identically. Since the number of variables on which some $p_{m,j}$ effectively depends is at most $n - j$, it follows that the process a),b),c),d) stops, at the latest, when we compute $p_{m,n-1}$. In fact, if $p_{m,n-1}$ is computed at all (for some particular chain of ξ-vectors), then it is a nontrivial homogeneous polynomial in essentially one variable and is therefore elliptic in that variable.*

Remark 1.1.5. *Actually, as is clear from the construction, $p_{m,j}$ depends on the choice of ξ^{j-1} and therefore, implicitly, also on the choices of the $\xi^0, \xi^1, \ldots, \xi^{j-2}$. This is not made visible in the notation.*

Theorem 1.1.6. *Let G be some open cone in R^n which contains ξ^0 and let $p(x, D)$ be a classical analytic pseudodifferential operator on $X \times G$, where X is some open set in R^n which contains the point $0 \in R^n$. Assume that the principal symbol p_m of p does not depend on x (but also see the discussion above). Let further B be a closed convex set in R^n_x such that $0 \notin B$ and assume that for any chains $(\xi^0, \xi^1, \ldots, \xi^{k-1}), (p_{m,0}, p_{m,1}, \ldots, p_{m,k})$ constructed according to the prescriptions a),b),c),d) above we can find some point $x \in B$ which is orthogonal to the lineality Λ of $p_{m,k}$.*

Finally, consider $u \in D'(X)$ a solution of $p(x, D)u = 0$ on $X \times G$ and assume that there is $\tilde{t} > 0$ so that $\tilde{t}B \subset X$ and

$$\tilde{t}B \times \{\xi^0\} \cap WF_A u = \phi.$$

Then it follows that $(0, \xi^0) \notin WF_A u$.

Here we should remark that if B has the properties from the assumptions in this theorem, then so has its intersection with the space $X = \{x; (x, \xi^0) = 0\}$. It is then natural to assume from the very beginning that $B \subset X$. A number of additional comments on the geometric situation in this result will be made in section 1.3.

9. It seems worthwhile here to consider a variant of definition 1.1.2 adapted to the situation from theorem 1.1.6.

Definition 1.1.7. *Consider* $(\xi^0, \xi^1, ..., \xi^{k-2})$, $(p_{m,0}, p_{m,1}, ..., p_{m,k-1})$ *as in theorem 1.1.6 and let* $\sigma \in \dot{R}^n$ *be orthogonal to all vectors* ξ^j. σ *will be called (k-1)-microcharacteristic for* p_m *at* $(\xi^0, \xi^1, ..., \xi^{k-2})$ *if and only if* $p_{m,k-1}(\sigma) = 0$. *A vector* $\sigma \in \dot{R}^n$, *which is orthogonal to the* $\xi^0, \xi^1, ..., \xi^{k-2}$, *but for which* $p_{m,k-1}(\sigma) \neq 0$ *shall be called (k-1)-micro-noncharacteristic.*

Remark 1.1.8 *The present terminology is chosen so that "1-microcharacteristic" corresponds to what was called "microcharacteristic" in definition 1.1.2. Here we recall that classical characteristic vectors are in fact objects of first microlocalization and that the "microcharacteristic vectors" of definition 1.1.2 are associated with second microlocalization. Correspondingly, (k-1)-microcharacteristic vectors are objects of k-microlocalization.*

Let us now also state a result which explains the relation between this notion and the notion of "micro-noncharacteristicity" considered (for two-microlocalization) by Hanges-Sjöstrand [1]. In doing so it is convenient to choose coordinates so that

$$\xi^j = (0, ..., 0, 1, 0, ..., 0), \text{ with the "1" on position } n - j$$

and to denote by Π_j the projection from R^n to $M_j = \{\xi; \xi_i = 0, i \geq n - j\}$, respectively from C^n to $X_j = \{\zeta \in C^n; \zeta_i = 0, i \geq n - j\}$.

Proposition 1.1.9. *Let* $(\xi^0, \xi^1, ..., \xi^{k-2})$, $(p_{m,0}, p_{m,1}, ..., p_{m,k-1})$ *be as in theorem 1.1.6 and consider* $\sigma = (0, ..., 0, 1, 0, ..., 0) \in M_{k-1}$. *Assume in addition that* $p_{m,0}$ *is a polynomial. If* $\beta > 0$ *is small enough, there are equivalent:*

a) σ *is (k-1)-micrononcharacteristic for* p_m *at* $(\xi^0, \xi^1, ..., \xi^{k-2})$.

b) *We can find open cones* $G^j \subset X_j$, $j \leq k - 2$, *which contain* $\Pi_j(\xi^j)$, *an open cone* $G^{k-1} \subset X_{k-1}$ *which contains* σ, *and a constant* $c > 0$ *so that* $p_m(\zeta) \neq 0$ *for any* $\zeta \in C^n$ *which satisfies* $\Pi_j(\zeta) \in G^j$, $j \leq k - 1$ *and for which*

$$|\Pi_i(\zeta)| \geq c|\Pi_{i-1}(\zeta)|^{1+\beta}/|\Pi_{i-2}(\zeta)|^\beta \text{ for } 2 \leq i \leq k - 1.$$

Remark 1.1.10. *The condition*

$$|\Pi_i(\xi)| \geq c|\Pi_{i-1}(\xi)|^{1+\beta}/|\Pi_{i-2}(\xi)|^{\beta} \, , \, 2 \leq i \leq k - 1$$

is of course to be considered void when k=2. We also mention that we shall encounter conditions of this type in a number of definitions and results later on. The reasons why such conditions appear at all will become clear in section 3.5. (Also cf. the introduction to chapter III.)

The proof of proposition 1.1.9 will be given in section 3.12. We also mention explicitly the following converse of the preceding proposition, since this gives rise to a useful reformulation of condition b) in that result.

Proposition 1.1.11. *In the conditions from proposition 1.1.9 the following two assertions are equivalent if $\beta > 0$ is small enough:*

i) $p_{m,k-1}(\sigma) = 0$,

ii) there is a sequence $\zeta^r \in C^n$ so that

$$p_m(\zeta^r) = 0,$$

$$\Pi_j(\zeta^r)/|\Pi_j(\zeta^r)| \to \Pi_j(\xi^j)/|\Pi_j(\xi^j)| \, , \, \forall j \leq k - 2,$$

$$\Pi_{k-1}(\zeta^r)/|\Pi_{k-1}(\zeta^r)| \to \Pi_{k-1}(\sigma)/|\Pi_{k-1}(\sigma)|,$$

$$|\Pi_{i-1}(\zeta^r)|^{1+\beta}|\Pi_i(\zeta^r)|^{-1}|\Pi_{i-2}(\zeta^r)|^{-\beta} \to 0, \, \forall 2 \leq i \leq k - 1.$$

(The last condition is once again void for $k = 2$.)

10. Proposition 1.1.11 gives a geometric interpretation in C^n of higher order characteristic vectors. We can obtain a similar characterization for the directions involved in conical refraction. This is based on the following remark, which for simplicity, we state in the case $k = 2$.

Proposition 1.1.12. *Assume that σ is orthogonal to ξ^0, that $p_{m,1}(\sigma) = 0$, but that grad $p_{m,1}(\sigma) \neq 0$. Also let $\zeta^r \in C^n$ be a sequence of vectors such that*

$$p_m(\zeta^r) = 0, \, \zeta^r/|\zeta^r| \to \xi^0/|\xi^0|, \, \Pi_1(\zeta^r) \neq 0, \, \Pi_1(\zeta^r)/|\Pi_1(\zeta^r)| \to \Pi_1(\sigma)/|\Pi_1(\sigma)|.$$

It follows that

$$\lim_{r \to \infty} \frac{grad \, p_m(\zeta^r)}{|grad \, p_m(\zeta^r)|} = \lambda \, \, exists$$

and that grad $p_{m,1}(\sigma)$ is proportional to λ.

Also this result will be proved in section 3.12. (The geometric interpretation of proposition 1.1.12 is seen best if one assumes that $\{\xi \in R^n; p_m(\xi) = 0\}$ has codimension one in R^n.)

11. In view of the preceding result it is an interesting question to establish when in proposition 1.1.11 one can choose the vectors ζ^r to be real. This is true, at least for k=2, when p_m is hyperbolic.

Proposition 1.1.13. *Assume that p_m is hyperbolic with respect to $N \in \dot{R}^n$ and assume that $\sigma \in \dot{R}^n$ is orthogonal to ξ^0 and satisfies $p_{m,1}(\sigma) = 0$. Then we can find a sequence $\{\xi^r\}_r \subset R^n$ such that*

a) $\xi^r \to \xi^0$,

b) $p_m(\xi^r) = 0$,

c) $\Pi_1(\xi^r)/|\Pi_1(\xi^r)| \to \Pi_1(\sigma)/|\Pi_1(\sigma)|$,

where Π_1 is the orthogonal projection on the subspace $\{\xi; \langle \xi, \xi^0 \rangle = 0\}$.

The proof of this result will be given in section 3.13.

12. We continue this introduction with some remarks related to the history of conical refraction and to the connection of our results with the classical theory. In fact, as a physical phenomenon, conical refraction is known since 1833, when it was discovered experimentally by H. Lloyd after a spectacular prediction of R.W. Hamilton from 1832. Indeed, the experimental verification of Hamilton's prediction was of big impact on the development of the theory of light, since it seemed to favor Fresnel's theory over Newton's theory. The simpler, but related, phenomenon of double refraction had been discovered in the 17-th century by E.Bartolinus and had been explained by Ch. Huygens. Correspondingly, more complicated phenomena of conical refraction have been discovered in later times and other contexts, e.g. crystal-elasticity. (Cf. Musgrave [1].) Actually, results are stated in crystal optics or elasticity theory rather as results on the geometry of what is usually called (following Hamilton) the "slowness" surface and of the "wave" surface of the system under consideration. In the case of the system of crystal optics both these surfaces are of the same type and are called Fresnel surfaces; from a differential-geometric point of view they have received considerable attention in the last century. (See e.g. the Mathematische Enzyklopädie [1], the book of Loria [1] or that of Darboux [1].) Closer to our times and from the point of view of partial differential equations, we should mention the contributions of Herglotz, who studied the fundamental solutions

of the system of crystal optics (cf. Herglotz [1]). The results of Herglotz were then simplified and improved by F.John : cf. e.g. Courant-Hilbert [1], and, for the history, Gårding [1]. The first paper in which conical refraction has been studied in a spirit close to the general theory of linear partial differential equations, seems to have been that of Ludwig, [1]. (Also cf. Alber [1].) The microlocal point of view was introduced in conical refraction by Ivrii [1,2,3]. It was then further developed in Melrose-Uhlmann [1], Uhlmann [1], Wakabayashi [1,2,3,4,5], Bernardi [1], Bernardi-Bove-Parenti [1], Bernardi-Bove [1], Petrini-Sordoni [1,2], in the other papers mentioned in the first part of this introduction and, of course, others. We should also mention that many papers on hyperbolic operators with characteristics of variable multiplicity have developed methods similar to those used in conical refraction. O.Oleinik, V.Ivrii, V.Petkov, L.Hörmander, R.Melrose, T.Nishitani, N.Iwasaki, E.Bernardi, A.Bove, C.Parenti and others have made important contributions, but since the scope of many of these papers is quite different we have chosen not to mention such papers explicitly.

13. To understand the relation between the results in the classical theory, stated in terms of the geometry of the slowness surface, and the statements in the microlocal theory, which are in terms of localizations of the principal symbol, we shall recall here for the convenience of the reader the definition of the slowness surface and make some comments. (Additional comments will be made in chapter VI.) Let us assume then that $p(D_x, D_t)$ is a constant coefficient linear partial differential operator in the (x,t) space, $x \in R^n$, $t \in R$. We assume that $p(D_x, D_t)$ is hyperbolic in the direction t and that for the principal symbol p_m we have that $p_m(\xi, \tau) = 0$ together with $\xi \neq 0$ implies $\tau \neq 0$. Here $\lambda = (\xi, \tau)$ are of course the variables Fourier-dual to x and t. The slowness surface is then by definition the surface

$$S = \{\xi \in R^n ; p_m(\xi, 1) = 0\}.$$

(Here we should add perhaps the points ξ which satisfy $p_m(\xi, -1) = 0$. For the cases of interest to us this will not change S since $p_m(\xi, 1) = p_m(\xi, -1)$ for them.)

The main remark here is now that conditions stated in terms of p_m can be immediately reformulated in terms of $q(\xi) = p_m(\xi, 1)$, as a consequence of $p_m(\lambda) = \tau^m q(\xi/\tau)$, if $\tau \neq 0$. Thus for example we have the following

Remark 1.1.14. *If $\lambda^0 = (\xi^0, 1)$ is such that*

$$\partial_\lambda^\alpha p_m(\lambda^0) = 0 \text{ for } |\alpha| < s , \sum_{|\beta|=s} |\partial_\lambda^\beta p_m(\lambda^0)| \neq 0 \qquad (1.1.11)$$

and if we set

$$p_{\lambda^0}^T(v, w) = \sum_{|\beta|+j=s} \partial_\xi^\beta \partial_\tau^j p_m(\lambda^0)(v^\beta/\beta!)(w^j/j!),$$

$$q_{\xi^0}^T(\theta) = \sum_{|\beta|=s} \partial_\xi^\beta q(\xi^0)\theta^\beta/\beta!$$

then it follows that we have

$$p_{\lambda^0}^T(v, w) = q_{\xi^0}^T(v - w\xi^0). \tag{1.1.12}$$

This is based on the fact that for $|\beta| + j = s$,

$$\partial_\xi^\beta \partial_\tau^j p_m(\lambda^0) = \sum_{|\gamma|=j} \partial_\xi^{\beta+\gamma} q(\xi^0)(j!/\gamma!)(-\xi^0)^\gamma.$$

(Here we use of course the first part of (1.1.11).) It follows that

$$\begin{aligned}
p_{\lambda^0}^T(v, w) &= \sum_{|\beta|+j=s} \sum_{|\gamma|=j} \partial_\xi^{\beta+\gamma} q(\xi^0)(j!/\gamma!)(-\xi^0)^\gamma v^\beta w^j (\beta! j!)^{-1} \\
&= \sum_{|\beta|+|\gamma|=s} \partial_\xi^{\beta+\gamma} q(\xi^0)(\beta! \gamma!)^{-1} v^\beta(-w\xi^0)^\gamma \\
&= \sum_{|\alpha|=s} \partial_\xi^\alpha q(\xi^0)(v - w\xi^0)^\alpha/\alpha! = q_{\xi^0}^T(v - w\xi^0).
\end{aligned}$$

We have thus proved (1.1.12), which gives an interesting relation between microcharacteristic vectors and solutions of $q_{\xi^0}^T(\theta) = 0$. In fact, if $p_{\lambda^0}^T(v, w) = 0$, then

$$\theta = v - w\xi^0 \tag{1.1.13}$$

will satisfy $q_{\xi^0}^T(\theta) = 0$ and conversely it is clear that we can always solve $\theta = v - w\xi^0$ for v and w. At a first glance, the relation $\theta \to (v - w\xi^0)$ is not uniquely defined, but here we should recall that a natural domain of definition of $p_{\lambda^0}^T$ was the space of variables orthogonal to λ^0. This means that to the equation (1.1.13) we should add the relation $\langle v, \xi^0 \rangle + w = 0$. It follows immediately that

$$w = -\langle \theta, \xi^0 \rangle/(1 + |\xi^0|^2),$$

so that then also

$$v = \theta - \xi^0 \langle \theta, \xi^0 \rangle/(1 + |\xi^0|^2).$$

To obtain geometric interpretations from all this, we must now also note that if $\theta \neq 0$ satisfies $q_{\xi^0}^T(\theta) = 0$, then the corresponding (v, w) is of type $v \neq 0$. To prove this, we have to use that p is hyperbolic in the direction t. It is well-known that the same is then also true for $p_{\lambda^0}^T$ (cf. e.g. Atiyah-Bott-Gårding [1] or Hörmander [5]), so in particular $p_{\lambda^0}^T(0, w) = 0$ will imply $w = 0$. If now we had $v = 0$, it would follow from $p_{\lambda^0}^T(v, w) = 0$ that also $w = 0$, but then we would have $\theta = v = 0$, contrary to our assumption. Conversely, if $v \neq 0$ and (v, w) is orthogonal to λ^0, then we will have $\theta \neq 0$, as is easy to see.

Finally, our last remark before coming to the interpretations in terms of the slowness surface, is

Remark 1.1.15. *Let (v^1, w^1) be a solution of $p_{\lambda^0}^T(v^1, w^1) = 0$ and denote by $\theta^1 = v^1 - w^1\xi^0$ the associated solution of $q_{\xi^0}^T(\theta) = 0$. Then we have*

$$grad_{v,w} p_{\lambda^0}^T(v^1, w^1) = (grad_\theta q_{\xi^0}^T(\theta^1), \langle grad_\theta q_{\xi^0}^T(\theta^1), \xi^0 \rangle). \qquad (1.1.14)$$

In particular,

$$grad_{v,w} p_{\lambda^0}^T(v^1, w^1) \neq 0 \Leftrightarrow grad_v p_{\lambda^0}^T(v^1, w^1) \neq 0 \Leftrightarrow grad_\theta q_{\xi^0}^T(\theta^1) \neq 0.$$

The geometric interpretation of the condition $grad_\theta q_{\xi^0}^T(\theta^1) \neq 0$ itself, starts from the observation that for $s > 1$ (in (1.1.11)) the surface $S = \{\xi; q(\xi) = 0\}$ will in general be singular at ξ^0. It is nevertheless possible to consider "tangent vectors" to S at ξ^0, which in fact are the directions given by nontrivial solutions of the equation $q_{\xi^0}^T(\theta) = 0$. (Precise definitions will be recalled for the convenience of the reader in chapter VI.) If θ^1 is such a solution and if $grad_\theta q_{\xi^0}^T(\theta^1) \neq 0$, then the set $q_{\xi^0}^T(\theta) = 0$ is a smooth surface in a conic neighborhood in the θ-space of θ^1, to which the direction $grad_\theta q_{\xi^0}^T(\theta^1)$ is normal. All this becomes very simple when, e.g., $s = 2$. Then $q_{\xi^0}^T$ is a quadratic form, $\{\theta; q_{\xi^0}^T(\theta) = 0\}$ is a cone and the directions $grad_\theta q_{\xi^0}^T(\theta^1)$ belong to the polar or the antipolar of this cone. (Details are discussed in section 1.3.)

14. We can now finally turn to the relation of, for example, the statement of theorem 1.1.6, and classical results on conical refraction. Of course, we have to consider now some hyperbolic equation $p(D)$ which has a physical interpretation and, as before, we have replaced R_x^n by $R_{x,t}^{n+1}$. We shall apply the arguments above for p_m, but shall use the notation $p_{m,1}$ rather than $p_{\lambda^0}^T$ considered previously. We will then also assume that $\lambda^0 = (\xi^0, 1)$ and consider a solution u, e.g. on R^{n+1}, of $p(D)u = 0$ for which $(0, \lambda^0) \in WF_A u$. Finally we assume that $grad_\lambda p_{m,1}(\lambda^1) \neq 0$ for any nontrivial solution λ^1 of $p_{m,1}(\lambda^1) = 0$ which is orthogonal to λ^0. Here λ^1 will correspond to the (v, w) in the discussion above. Under these conditions, theorem 1.1.6 states, roughly speaking, that $WF_A u$ must propagate at least in one of the directions of type $grad\, p_{m,1}(\lambda^1)$. An observer in the physical x-space could perceive such a "propagation" in principle then as a movement of

the singularity on the line through 0 of direction $\mathrm{grad}_\xi\, p_{m,1}(\lambda^1)$. According to the above, this direction is also equal to the direction $\mathrm{grad}_\theta\, q_{\xi^0}^T(\theta^1)$ where $q(\xi) = p_m(\xi, 1)$ and θ^1 is associated with $(v, w) = \lambda^1$ as before. The speed of the movement of the singularity will be

$$|\mathrm{grad}_\xi\, p_{m,1}(\lambda^1)/(\partial/\partial\tau)p_{m,1}(\lambda^1)|,$$

and is possibly very high. Instead of seeing an effective movement along our line, the observer will then see some halfray $\{s\ \mathrm{grad}\, q_{\xi^0}^T(\xi^1 - \tau^1\xi^0), s \geq c\}$. The fact that this ray has precisely the direction of a normal to the tangent cone to the singular point is then in agreement with what is predicted by conical refraction in the classical theory. (For the case of crystal optics, cf. Courant-Hilbert [1].) More details are given in chapter VI and applications to transmission problems are considered in chapter VIII.

15. Let us, finally, say a few words about the plan of these notes. As explained above, our main aim is to proof the theorems 1.1.3, 1.1.6 and the tools from the theory of higher microlocalization needed to do so. This is more or less what is done in the first five chapters. Actually, most of the content of the chapters II, III, IV, V and IX is just on higher microlocalization, and the theorems 1.1.3, 1.1.6 are either immediate consequences (this is the case of theorem 1.1.3) or in some sense illustrations (in case of theorem 1.1.6) of the general theory, so we hope that these chapters could also serve as an introduction to the theory of higher analytic microlocalization in the involutive case. The exposition is not completely selfcontained however. The relation of the results in theorem 1.1.3 and 1.1.6 with a number of earlier results of the same kind, respectively with the classical theory of conical refraction in crystal optics, forms, on the other hand, the object of the chapters VI, VII, VIII (and of part of chapter I).

1.2   An heuristic justification

1. All notations in this section are as in the statement of theorem 1.1.3. We want to give a heuristic argument which shows why we must expect p_m^T, Λ and relative bicharacteristic leaves to play an important role in our theory. The discussion is also related to proposition 1.1.11. Actually we shall perform our discussion only under the additional assumption that

$$\mathrm{grad}_v p_m^T(\lambda^0)(v^0) \neq 0$$

and that p_m is real valued. Recall here that $\lambda^0 = (x^0, 0, \xi^{0''})$. It can then be proved that we can find $\delta > 0$ and a real analytic curve

$$\gamma : (-\delta, \delta) \to T^*X$$

such that:

i) $p_m(\gamma(t)) \equiv 0$,

ii) $\gamma(0) = \lambda^0$,

iii) $\gamma_{(x)}(t) = x^0$, $\gamma_{(\xi'')}(t) = \xi^{0''}$, ($\gamma_{(x)}$ and $\gamma_{(\xi'')}$ are respectively the x- and the ξ''-component of γ. Similar notations shall also be used later on.)

iv) $\dot\gamma(0) = (0, v^0, 0)$.

In fact, we regard here p_m for fixed λ^0 as a function in ξ', so it vanishes of order s at $\xi' = 0$. The polynomial $v \to p_m^T(\lambda^0)(v)$ is thus the Taylor expansion of order s at $\xi' = 0$ of this function. The existence of a curve γ with the desired properties will then follow from proposition 6.4.2 below. It is also convenient here to regard p_m^T as a function of $(\lambda, v) \in \dot N\Sigma$: $(\lambda, v) \to p_m^T(\lambda)(v)$. Since $\Sigma = \{(x, \xi); \xi' = 0\}$, this makes p_m^T a function of (x, ξ'', v). In addition to the properties i), ii), iii), iv) we now also have

v) $(d/dt)^i \operatorname{grad}_{x,\xi} p_m(\gamma(0)) = 0$ for $i \leq s - 2$,

vi) $(d/dt)^{s-1} \operatorname{grad}_{x,\xi''} p_m(\gamma(0)) = 0$ and

$$(d/dt)^{s-1} \operatorname{grad}_{\xi'} p_m(\gamma(0)) = (\operatorname{grad}_{\xi'} p_m)^T(\lambda^0)(v^0) = \operatorname{grad}_v p_m^T(\lambda^0)(v^0),$$

where f^T is (for given f) the Taylor expansion of order $s - 1$ of f in the ξ'-variables,

vii) $\operatorname{grad} p_m(\gamma(t)) \neq 0$, if $t \neq 0$ is small,

viii) $\lim_{t \to 0} \dfrac{\operatorname{grad} p_m(\gamma(t))}{|\operatorname{grad} p_m(\gamma(t))|} = \dfrac{(0, \operatorname{grad}_v p_m^T(\lambda^0)(v^0), 0)}{|\operatorname{grad}_v p_m^T(\lambda^0)(v^0)|}$

To prove v), we note that $|\gamma_{(\xi')}(t)| \sim |t|$, that

$$|\operatorname{grad}_{(x,\xi'')} p_m(\gamma(t))| \leq c|\gamma_{(\xi')}(t)|^s$$

and that

$$|\operatorname{grad}_v p_m(\gamma(t))| \leq c|\gamma_{(\xi')}(t)|^{s-1}.$$

The relations vi) and vii) are easy consequences. As for viii), we denote $\operatorname{grad} p_m(\gamma(t))$ by $f(t)$ and have

$$\lim_{t \to 0} \frac{f(t)}{|f(t)|} = \lim_{t \to 0} \frac{t^{s-1}(d/dt)^{s-1}f(0) + O(t^s)}{|t^{s-1}(d/dt)^{s-1}f(0) + O(t^s)|} = \frac{(d/dt)^{s-1}f(0)}{|(d/dt)^{s-1}f(0)|} \ .$$

We have now discussed the relations i) to viii) and return to our heuristic discussion above. In doing so, we shall denote for some given function f by H_f the Hamiltonian field $\Sigma(\partial f/\partial\xi_j)(\partial/\partial x_j) - (\partial f/\partial x_j)(\partial/\partial\xi_j)$ of f. We observe at first that it follows from viii) that

ix)
$$\lim_{t\to 0} \frac{H_{p_m}(\gamma(t))}{|H_{p_m}(\gamma(t))|} = \frac{(w,0)}{|(w,0)|},$$

where $w = \mathrm{grad}_v\, p_m^T(\lambda^0)(v^0)$. Of course, w is parallel to the $\delta x'$ component of the direction of the relative bicharacteristic of p_m^T which passes through (λ^0, v^0). On the other hand, $\gamma(t)$ lies for small t in the smooth part of the characteristic variety and the analytic wave front set at $\gamma(t)$ propagates in the direction $H_{p_m}(\gamma(t))/|H_{p_m}(\gamma(t))|$. The relation ix) now tells us that $(w,0)/|(w,0)|$ is the limit of these directions of propagation. We may therefore expect that the analytic wave front set at the limit point λ^0 will have a "tendency" to propagate in the direction $(w,0)$ determined from the relative bicharacteristic curve. The extent to which this is true is expressed in the theorems, but note anyway that we may have many microcharacteristic vectors v^0, so that no unique direction of propagation is associated with λ^0. For this reason it will be convenient to localize around microcharacteristic directions in a way similar to which one localizes in standard microlocalization around characteristic directions: this is a first justification for the need of second microlocalization. Moreover, in the computations above all information on $\mathrm{grad}_{x'}p_m^T(\lambda^0)(v^0)$ is lost. Technically this corresponds to the fact that in the propagation results above propagation will occur inside the bicharacteristic leaves of Σ, so in the special coordinates which we have chosen, the ξ-component of WF_Au will remain unchanged. We shall see later on that computations are more natural at the level of second microlocalization, in that there we shall take into account all the information associated with the relative bicharacteristic curves. It is for this reason mainly that we have used such curves (or leaves) already in our statements.

1.3   Discussion of theorem 1.1.6

1. Our criterium to stop in d), section 1.1, was that $p_{m,k}$ had to be elliptic in the space of variables on which it effectively depended. Other criteria to stop could be taken into consideration: we could for example assume that $p_{m,k}$ were hyperbolic with respect to some direction. (Cf. here the result in section 6.3). The important point is that propagation of singularities for $p_{m,k}$ should be easy to understand, although it would be hard to make this more precise. In any case, we think that in many cases one will loose nothing in the quality of the results if one does not stop the process of

localization at the first possible moment. We give an example to this effect. Let us in fact assume, to simplify the situation, that the characteristic variety W of p_m is a real analytic homogeneous manifold of codimension d in a conic neighborhood of ξ^0 and that p_m is transversally elliptic to W, so that actually it would not be necessary to start the process of localizations at all. It follows indeed in this case from theorem 1.1.1 that we have propagation of the analytic wave front set in the directions of the normals to W at ξ^0, which is to say, in the terminology of theorem 1.1.6, that the conclusion from theorem 1.1.6 is satisfied if B is any fixed point in the set of normals to W at ξ^0. We want to show that the same conclusion can also be obtained from theorem 1.1.6, in that the assumptions from that theorem are valid for such a set B. To do this, let us assume that W is defined in a conic neighborhood of ξ^0 by the equations $f_i = 0$, $i = 1, ..., d$, where the f_i are real-valued real analytic functions which are positively homogeneous in a conic neighborhood of ξ^0 and have linearly independent differentials there. Assume further that p_m vanishes of order s on W and that it is transversally elliptic to W. By this we mean that if we write

$$p_m = \sum_{|\alpha|=s} a_\alpha f^\alpha$$

for some functions a_α which are real-valued, real analytic and positively homogeneous of order $m - s$, then

$$\left| \sum_{|\alpha|=s} a_\alpha(\xi) u^\alpha \right| \geq c|\xi|^{m-s}|u|^s,$$

for all ξ in a small conic neighborhood of ξ^0 and all u in R^d.

Also assume that $\xi^0 = (0, ..., 0, 1)$ is in W. It is then easy to see that the localization polynomial $p_{m,1}$ at ξ^0 is given by

$$p_{m,1} = \sum a_\alpha(\xi^0) g^\alpha,$$

where $g_i = \langle (df_i)(\xi^0), \xi \rangle$ for $i = 1, ..., d$ and df_i is the differential of f_i. (We use here that $f_i(\xi^0) = 0$.) It follows that $p_{m,1}$ is transversally elliptic to

$$W^1 = \{\xi; \sum a_\alpha(\xi^0) g^\alpha(\xi) = 0\}.$$

Of course W^1 is just the set $\{\xi; g_i(\xi) = 0, i = 1, ..., d\}$, and it is also clear that the lineality of $p_{m,1}$ is equal to W^1. Further linearizations will not change the form of $p_{m,1}$, but only restrict the possibilities to choose ξ^j. To conclude our remark, it then suffices to note that the set of normals to W at ξ^0 is equal to the set of normals to W^1 at ξ^0.

2. To illustrate theorem 1.1.6, let us first consider the case when p_m vanishes of order two precisely at ξ^0. In this case $p_{m,1}$ is a quadratic form in the variables orthogonal to ξ^0

and is therefore at least of principal type in the variables on which it effectively depends. We shall assume for simplicity that p_m is real valued. An interesting particular case in which $p_{m,1}$ is elliptic in the variables on which it effectively depends, is when

$$p_{m,1}(\xi) = \langle x^0, \xi \rangle^2 \text{ for some } x^0 \in \dot{R}^n, \tag{1.3.1}$$

in which case $\Lambda = \{\xi; \langle x^0, \xi \rangle = 0\}$. Thus $\dim \Lambda = n - 1$ and $p_{m,1}$ only depends on one variable effectively. We may take here $B = \{x^0\}$. The conclusion of the theorem 1.1.6 is then that $WF_A u \cap \{(x,\xi); \xi = \xi^0\}$ propagates for solutions of $p(x, D)u = 0$ in the direction x^0, so no splitting of singularities occurs. This is an interesting phenomenon, in that the characteristic variety V of p is not smooth at ξ^0 and $p_{m,1}$ is not transversally elliptic to the variety $\{\theta \xi^0; \theta \in R\}$. An example of a case when we are in a situation of the type just described occurs for the system of elasticity for cubic crystals. (We thus refer here explicitly to the case of a system, whereas in the main body of these notes our results are stated for scalar equations. It should be noted that in many cases the passage from scalar equations to systems is, by the nature of the arguments, very easy and only leeds to notational complications. In any case, the system of crystal elasticity is a determined constant coefficient system, so the components of the solutions will automatically satisfy the scalar equation associated with the determinant of the symbol of the system.) Explicitly, the characteristic surface V of the system of crystal elasticity for cubic crystals is if coordinates are suitable (cf. here (6.8.2))

$$\frac{b\xi_1^2}{\tau^2 - c|\xi|^2 + (b-a)\xi_1^2} + \frac{b\xi_2^2}{\tau^2 - c|\xi|^2 + (b-a)\xi_2^2} + \frac{b\xi_3^2}{\tau^2 - c|\xi|^2 + (b-a)\xi_3^2} = 1\,.$$

Here $n = 4$ and the variables in R^4 are of course written as (ξ, τ), $\xi = (\xi_1, \xi_2, \xi_3)$. It is not difficult to see that -under suitable assumptions on the constants a, b and c (cf. section 6.8) - there are fourteen singular directions on V: eight of them are determined by the condition $\xi_1^2 = \xi_2^2 = \xi_3^2$ and the remaining six are determined by the condition that the ξ-component lies on one of the coordinate axes. It is an easy matter to check that the last six directions are of the type considered in relation (1.3.1), so no conical refraction is associated with them. Actually this is true already at the level of geometrical optics and in both situations the true reason is just the same: the singularities of the characteristic surface at these directions are just too weak to generate conical refraction . A detailed discussion of all this is made in section 6.8.

3. Next we observe that the situation remains essentially the same when $p_{m,1}$ is just elliptic in the space of variables on which it effectively depends, in that, even in the case when the dimension of this space is bigger than one, no "true" conical refraction occurs.

However, the situation will be different if $p_{m,1}$ is not elliptic on the space of variables on which it effectively depends. After an orthogonal change of variables in the ξ-space, we may assume that $p_{m,1}$ is of form

$$p_{m,1}(\xi) = \sum_{i \leq k'} a_i^2 \xi_i^2 - \sum_{k' < i \leq k} a_i^2 \xi_i^2$$

for some positive numbers a_i and suitable $k' \geq 1, k \leq n - 1, 1 \leq k' < k \leq n - 1$. The zeros of $p_{m,1}$ lie then on the cone

$$\Gamma = \{\sigma \in R^n; \sum_{i \leq k'} a_i^2 \sigma_i^2 - \sum_{k' < i \leq k} a_i^2 \sigma_i^2 = 0\},$$

and if $\sigma \in R^n$ with $\sigma' = (\sigma_1, ..., \sigma_k) \neq 0$ satisfies $p_{m,1}(\sigma) = 0$, then the localization $p_{m,2}$ of $p_{m,1}$ at σ is

$$p_{m,2}(\xi) = 2 \left[\sum_{i \leq k'} a_i^2 \sigma_i \xi_i - \sum_{k' < i \leq k} a_i^2 \sigma_i \xi_i \right].$$

Of course, a true cone will here appear only when $k \geq 3$, in which case we will have $n \geq 4$, which is therefore the case of greatest interest to us. (In physical applications, one variable is the time, so we need three more variables in the space of the x-variables. This is in fact precisely the situation in crystal optics or crystal acoustics.) The sets B from the statement in theorem 1.1.6 are then precisely those closed convex sets which contain for each $\sigma \in \Gamma$ some point of direction

$$(a_1^2 \sigma_1, ..., a_{k'}^2 \sigma_{k'}, -a_{k'+1}^2 \sigma_{k'+1}, ..., -a_k^2 \sigma_k, 0, ..., 0).$$

(We have here a case when we can essentially stop our process of successive localizations at step 2. This does not formally follow from theorem 1.1.6, although one could reformulate the theorem in such a way as to take care of such situations. We have not done so in order to avoid additional notational complications; it is anyway clear when $k = n - 1$.) Of particular interest is here the case when $k' = 1$. If σ and $\tilde{\sigma}$ in R^n satisfy $p_{m,1}(\sigma) = 0$, $p_{m,1}(\tilde{\sigma}) = 0$, if $\sigma_1 > 0$, $\tilde{\sigma}_1 > 0$, and if $\nu \in R^k$ is of form

$$\nu = \lambda(a_1^2 \tilde{\sigma}_1, -a_2^2 \tilde{\sigma}_2, ..., -a_k^2 \tilde{\sigma}_k),$$

for some $\lambda > 0$, then we will have that $\sum_{i \leq k} \sigma_i \nu_i \geq 0$. Indeed, we may take $\lambda = 1$ and have that

$$\sum_{i \leq k} \sigma_i \nu_i = a_1^2 \sigma_1 \tilde{\sigma}_1 - \sum_{i \geq 2} a_i^2 \sigma_i \tilde{\sigma}_i \geq a_1^2 \sigma_1 \tilde{\sigma}_1 - (\sum_{i \geq 2} a_i^2 \sigma_i^2)^{1/2} (\sum_{i \geq 2} a_i^2 \tilde{\sigma}_i)^{1/2}$$

in view of the Cauchy-Schwarz inequality. It remains then to use that $a_1^2 \sigma_1^2 = \sum_{i \geq 2} a_i^2 \sigma_i^2$ and that $a_1^2 \tilde{\sigma}_1^2 = \sum_{i \geq 2} a_i^2 \tilde{\sigma}_i^2$. Let us next denote by

$$\Gamma^+ = \{\mu \in R^k; a_1^2 \mu_1^2 \geq \sum_{i=2}^{k} a_i^2 \mu_i^2, \mu_1 \geq 0\},$$

and by

$$\Gamma^{+0} = \{y \in R^k; \langle y; \mu \rangle \geq 0, \forall \mu \in \Gamma^+\}$$

the polar of Γ^+. The computation just made shows that if $\mu \neq 0$ satisfies $Q(\mu) = 0$ for $Q(\mu) = a_1^2\mu_1^2 - \sum_{i=2}^k a_i^2\mu_i^2$, then

$$\text{grad } Q(\mu) \in \Gamma^{+0},$$

a fact which is also clear from the geometrical interpretation of the polar of a smooth convex cone. Also denote by

$$\tilde{Q}(y) = y_1^2/a_1^2 - \sum_{i=2}^k y_i^2/a_i^2$$

the quadratic form dual to Q. It is then standard to observe that we have the following

Lemma 1.3.1.
$$\Gamma^{+0} = \{y; y_1 \geq 0, \tilde{Q}(y) \geq 0\}.$$

Proof. Denote $K^+ = \{y; y_1 \geq 0, \tilde{Q}(y) \geq 0\}$. It suffices then to show that

$$K^+ \subset \Gamma^{+0}. \tag{1.3.2}$$

In fact, once we have checked that (1.3.2) is valid, we also have by the same argument that

$$\Gamma^+ \subset K^{+0},$$

which gives $K^+ = K^{+00} \supset \Gamma^{+0}$ by duality.

We have thus reduced ourselves to the proof of (1.3.2), which comes to the fact that $y_1 \geq (\sum_{i \geq 2} y_i^2/a_i^2)^{1/2}$, together with $\mu_1 \geq (\sum_{i \geq 2} \mu_i^2 a_i^2)^{1/2}$, implies $\langle y, \mu \rangle \geq 0$. That this is so is once more a consequence of the Cauchy-Schwarz inequality since we may write

$$|\sum_{i=2}^k y_i \mu_i| = |\sum_{i=2}^k (y_i/a_i)(\mu_i a_i)| \leq (\sum_{i=2}^k y_i^2/a_i^2)^{1/2} (\sum_{i=2}^k \mu_i^2 a_i^2)^{1/2}.$$

It follows from the preceding lemma that we can take B to be the intersection of Γ^{+0} with some transversal hyperplane and theorem 1.1.6 says then, roughly speaking, that we have conical refraction along the polar cone of Γ^+. This is of course wellknown, at least when $k = n - 1$. Also note that this polar lives in a space of dimension k and is therefore relatively thin if $k < n - 1$. It is worthwhile to mention that the eight singular directions for the system of crystal elasticity for cubic crystals which were not of the type of (1.3.1) are all of the type discussed here (with $k = n - 1$), so microlocal singularities associated with such directions will in general split along cones. We obtain in this way a complete discussion of all singular directions for the system of crystal elasticity for cubic crystals. (Cf. section 6.8.) Moreover, we think that phenomena of conical refraction for the system of elasticity can be discussed completely for all crystal classes with the aid of theorem 1.1.6, but the computations to check this seem at least unpleasant to perform. Also note that the situation is somewhat simpler for the system of crystal optics. The interesting case is here that of optically biaxial crystals. The relevant part of the characteristic variety is here given (if coordinates are suitable) by

$$p(\lambda) = \tau^4 - \psi(\xi)\tau^2 + \varphi(\xi)|\xi|^2 = 0$$

where

$$\varphi(\xi) = d_2 d_3 \xi_1^2 + d_3 d_1 \xi_2^2 + d_1 d_2 \xi_3^2,$$

$$\psi(\xi) = (d_2 + d_3)\xi_1^2 + (d_3 + d_1)\xi_2^2 + (d_1 + d_2)\xi_3^2,$$

and where d_1, d_2, d_3 are constants which are computed in terms of the dielectric and magnetic permeability tensor of the crystal under consideration. (Cf. section 6.7 for more details. We are of course once more in $R_\xi^3 \times R_\tau$.) There are four singular directions in this case and when numerotation is such that $d_1 < d_2 < d_3$ then these directions are determined by the conditions $\xi_2 = 0$,

$$\frac{\xi_1}{\xi_3} = \pm\sqrt{\frac{d_2 - d_1}{d_3 - d_2}}.$$

All four of these directions are of the type considered in this subsection, so they will lead to conical refraction at the level of the first wave front set. (For details we refer to section 6.7 and chapter VIII.)

4. The fact that in the computations above we relied so much on the euclidean structure of R^n (e.g. when we computed Γ^{+0}) might seem strange and is indeed misleading in that it seems to preclude any invariant meaning of our conclusions. It should therefore be noted that actually Γ^{+0} is the symplectic polar of Γ^+. This is well-known in the present context (cf. e.g. Wakabayashi [3] or Parenti [1]), and the reason why we formulate our conditions in the euclidean setting are largely notational. Let us explain however the situation here in some detail. We denote then, as above, by

$$\Sigma = \{(x, \xi); \xi \text{ is of form } t\xi^0 \text{ for some } t \in R\}$$

and by $p^T : T\Sigma \to R$ the localization polynomial of p_m at the points from Σ. To simplify notations we shall choose coordinates such that $\Sigma = \{(x, \xi); \xi_1 = \cdots = \xi_{n-1} = 0\}$ and denote tangent vectors to Σ at $\lambda \in \Sigma$ by $(\delta x, \delta \xi)$. It is then clear that p_m^T at a point of form (x, ξ^0) is just $p_{m,1}$. Let further F_λ be

$$F_\lambda = \{(\delta x, \delta \xi) \in T_\lambda \Sigma; \ p_{m,1}(\delta \xi) > 0, (\delta \xi)_1 > 0\}.$$

(For some vector δz we denote by $(\delta z)_i$ its i-th component, and by $(\delta z)'$ the first $n-1$ components of δz.)

It follows easily that

$$F_\lambda = R_x^n \times \Gamma^+ \times R_{(\delta\xi)_n}, \tag{1.3.3}$$

where Γ^+ is the one considered above. Finally, denote by $F_\lambda^\perp$ the symplectic polar of F_λ, i.e.

$$F_\lambda^\perp = \{(\delta y, \delta\eta) \in T_\lambda\Sigma; \ \omega((\delta y, \delta\eta),(\delta x, \delta\xi)) > 0, \forall(\delta x, \delta\xi) \in F_\lambda\}.$$

Here ω is the canonical two form and in fact $\omega((\delta y, \delta\eta),(\delta x, \delta\xi)) = \langle \delta y, \delta\xi \rangle - \langle \delta x, \delta\eta \rangle$. In view of (1.3.3) it follows therefore from $(\delta y, \delta\eta) \in F_\lambda^\perp$ that $(\delta y)_n = 0$, $\delta\eta = 0$ and that $(\delta y)' \in \Gamma^{+0}$.

5. Let us next study what theorem 1.1.6 gives when the principal part p_m of p has the form

$$p_m(\xi) = \xi_1^{s(1)}\xi_2^{s(2)} \cdots \xi_d^{s(d)}. \tag{1.3.4}$$

After an easy computation one sees that the assumptions of theorem 1.1.6 are satisfied if B is a convex closed subset in $\hat{R}^d = \{x \in R^n; x_j = 0, j > d\}$ with the property that any coordinate axis in $\hat{R}^d$ intersects B. This is essentially an analytic version of a C^∞ result from Lascar [1], in special coordinates. (In the analytic category it was proved for $d = 2$ by Tose [1].)

Working with condition I)' instead of I), we can give an invariant formulation of this result. This can be done in essentially the same way in which R.Lascar has given an invariant formulation in the C^∞-category and we describe it here for the convenience of the reader. The first assumption is then that W is a union of d homogeneous real-analytic varieties W_j, $j = 1, ..., d$, of codimension one, and for which also $W_i \cap W_j$ is regular involutive for each i, j. We assume in addition that these intersections are of codimension 2 if $i \neq j$. Further we assume that natural numbers $s(j)$ are given for $j = 1, ..., r$ so that in some conic neighborhood Γ of (x^0, ξ^0) we have

$$c'|\xi|^{m-s}(d_{W_1}(x,\xi))^{s(1)}(d_{W_2}(x,\xi))^{s(2)} \cdots (d_{W_d}(x,\xi))^{s(d)} \leq |p_m(x,\xi)| \leq$$

$$c|\xi|^{m-s}(d_{W_1}(x,\xi))^{s(1)}(d_{W_2}(x,\xi))^{s(2)} \cdots (d_{W_d}(x,\xi))^{s(d)}.$$

Here $s = \sum s(j)$ and the d_{W_i} are homogeneous distance functions from (x, ξ) to W_i. Under these assumptions it is proved in Grigis-Lascar [1] that p_m can be reduced via conjugation with a suitable analytic elliptic F.I.O. and after multiplication with a convenient elliptic analytic pseudo-differential operator to an operator which has principal symbol as in (1.3.4). One can therefore apply the result mentioned for such operators above and in fact reformulate the conclusion from there in an invariant way. We omit further details. (Cf. Lascar [1].)

6. An example which slightly generalizes the preceding one is when we are given natural numbers k_i such that

$$1 \le k_1 < k_2 < \cdots < k_j < n$$

and constant coefficient partial differential operators p_i which act only in the variables $x_{k_{i-1}+1}, ..., x_{k_i}$ and are elliptic in these variables. If then

$$p_m(\xi) = p_1(\xi) \cdots p_j(\xi),$$

then the conclusion of theorem 1.1.6 will hold if we take for B any simplex supported by some points $z^1 \neq 0, ..., z^j \neq 0$ so that $z^i_s = 0$ unless $k_{i-1} < s \le k_i$. (Actually the proof will also work in the case in which the p_i have analytic coefficients.)

7. We finally come back to an observation made just before remark 1.1.4. Let us assume that $p(D)$ is a homogeneous constant coefficient linear partial differential operator of even order m on R^{n+1}. The variables in R^{n+1} shall be denoted (x, t), those in the dual space (ξ, τ), $\xi \in R^n$, $\tau \in R$. We assume for simplicity that the operator p is strictly hyperbolic with respect to t and in fact, that for each fixed $\xi \neq 0$ there are $m/2$ strictly positive and $m/2$ strictly negative roots of $p(\xi, \tau) = 0$. (Also cf. here section 6.5.) As stated in section 1.1 it follows from the general theory of constant coefficient hyperbolic operators (under conditions less stringent than the above ones) that $p(\xi^0, \tau^0) = 0$ for $(\xi^0, \tau^0) \neq 0$ implies that grad $p(\xi^0, \tau^0)$ lies in $\Gamma^0 \cup -\Gamma^0$, where Γ is the hyperbolicity cone with respect to the direction $(0, \ldots, 0, 1)$ and Γ^0 is the polar of Γ. We want to give a direct elementary proof of this statement. We shall assume that $\tau^0 = 1$. (Actually, I think that the argument below is folklore, but I was unable to trace it down.) Let us now assume by contradiction that we can find $\xi^0 \neq 0$ with $p(\xi^0, 1) = 0$ so that grad $p(\xi^0, 1)$ is not in $\Gamma^0 \cup -\Gamma^0$. Also denote by T the tangent plane to $V = \{(\xi, \tau); p(\xi, \tau) = 0\}$ at $(\xi^0, 1)$. (It exists since p is strictly hyperbolic.) The assumption on ξ^0 now shows that T must intersect the boundary of Γ at a non-zero point. Let (ξ^1, τ^1) be such a point of intersection. It is no loss of generality to assume that $\tau^1 = 1$. By arguing in the space generated by $(\xi^0, 1), (\xi^1, \tau^1)$ and $(0, \ldots, 0, 1)$ we may assume that $n = 3$. It follows from our assumptions on the roots of p that the set

$$S = \{\xi; p(\xi, 1) = 0\}$$

consists of $m/2$ smooth simple closed algebraic curves γ_i which do not intersect mutually. (We can introduce the curves γ_i with the aid of a parametrization by the angle variable

$\omega \in [0, 2\pi)$. If ω is fixed, $\gamma_i(\omega)$ is defined to be one of the positive roots σ of $p(\sigma\omega, 1) = 0$. Note that these roots are just the reciprocals of the roots τ of the equation $p(\omega, \tau) = 0$. There are $m/2$ distinct such roots τ, so we can find $m/2$ distinct roots of type σ. The correspondence $\omega \to \sigma$ is locally smooth and two locally defined smooth pieces of curves never intersect.) For each i, $R^2 \setminus \gamma_i$ is divided into two connected components, one of which is bounded. We may assume that the labelling of the curves is made in such a way that γ_i is contained in the connected component of $R^2 \setminus \gamma_{i+1}$, $i = 1, \ldots, m/2 - 1$. We also observe that any line which intersects the bounded component of $R^2 \setminus \gamma_i$ must also intersect the curve γ_i in at least two geometrically distinct points. Let us now return to the situation above, assuming that $\xi^0 \in \gamma_{i^0}$. It is no loss in generality to assume that $i^0 > 1$. In fact, otherwise, ξ^0 lies just on the boundary of Γ, so it is clear from the fact that Γ is convex and the geometric interpretation of the polar that grad $p(\xi^0, 1)$ lies in the polar or the antipolar of Γ. If we consider the tangent line L at γ_{i^0} at the point ξ^0, then the assumption gives that L must intersect γ_1. In particular L enters the bounded component of $R^2 \setminus \gamma_i$ for any $i > 1$, so it has at least two points of intersection with any γ_i. With γ_{i^0} however it has at least 3 points of intersection when multiplicities are counted. In fact, geometrically, there are two distinct points of intersection, but the point ξ^0 is a double point since L was tangent to γ_{i^0}. This gives at least $m + 1$ points of intersection between L and S, which is a contradiction since S is given by an m-th order equation which does not contain any linear factors.

Chapter 2

Higher order wave front sets

In this chapter we introduce the higher order wave front sets on which our arguments are based. Generally speaking these wave front sets are modelled on the original definition of Hörmander of the first analytic wave front set, but when the order of localization is greater than two, then we shall consider two variants of such wave front sets which are not completely equivalent one to the other. The main result of the chapter is a micro-Holmgren type propagation theorem for the k-1 analytic wave front set, when the k-th wave front set is in good position: cf. theorem 2.1.12 below. This is closely related to the importance which (microlocal) partial analyticity has in results on propagation of the first wave front set. Indeed one of the main justifications for the introduction of theories of higher microlocalization is that partial analyticity cannot be appropriately characterized in terms of the first analytic wave front set, whereas it can be so characterized in terms of higher order ones. (This will be discussed in detail in chapter IX.) The argument in the proof of theorem 2.1.12 is by duality and the main technical point is to split entire functions which satisfy certain inequalities into sums of entire functions which satisfy related inequalities. We also show (partly in an appendix to this chapter,) in the case of second microlocalization, which is the relation between our definitions and the definitions of Sjöstrand-Lebeau based on the FBI-transform. The relation to definitions of Sato-type on the other hand, will be discussed in chapter IX. Finally, we should mention here that the way how pseudodifferential operators, respectively F.I.O. (for second microlocalization) act on our wave front sets is analyzed in the chapters III and V respectively.

2.1   Higher order wave front sets

1. In this section we introduce and discuss higher order wave front sets in special coordinates. In doing so, we shall start from Hörmander's first definition of the standard analytic wave front set (i.e. the one from Hörmander [2]), but the setting is of course also modelled on the definitions of Sjöstrand and Lebeau for higher order wave front sets in which the FBI-transform is used. A convenient feature of our definitions is that we are very close to the situation from general (so called) G_φ-microlocalization in Liess-Rodino [1,2]. In particular we shall be able to deduce a number of properties on the calculus of the higher order wave front set from related results in Liess-Rodino, loc.cit.

2. Let us at first consider a sequence of linear subspaces M_j, $j = 0, 1, ..., k$ in R^n such that $M_0 = R^n$, $M_j \subset M_{j-1}$, $M_j \neq M_{j-1}$, $M_k = \{0\}$. Denote by $\Pi_j : R^n \to M_j$ the orthogonal projection on M_j and by $\dot{M}_j = M_j \ominus M_{j+1}$ the orthogonal complement of M_{j+1} in M_j. It follows in particular that

$$R^n = \oplus_{j \geq 0} \dot{M}_j,$$

so we can write that $\xi = \sum_{j \geq 0}(\Pi_j - \Pi_{j+1})\xi$. It is then also often convenient to look at the group of variables of type $(\Pi_j - \Pi_{j+1})\xi$ separately. This suggests that one could start from the very beginning from a decomposition $R^n = \oplus X_j$ into mutually orthogonal subspaces, etc., rather than from a decreasing sequence of subspaces. One of the reasons why we prefer the present approach is that it is closer to the situation which appears in theorem 1.1.6. (However, none of the two ways of describing the situation has an invariant meaning.) Further, we consider U open in R^n, $x^0 \in U$, $\xi^j \in \dot{M}_j$, $j = 0, ..., k-1$ and $u \in D'(U)$. A function $f : (0, \infty) \to (0, \infty)$ will be called sublinear if $\forall \varepsilon$, $\exists c$ such that

$$f(t) \leq \varepsilon t + c.$$

Definition 2.1.1. *We shall say that* $(x^0, \xi^0, \xi^1, ..., \xi^{k-1}) \notin WF_A^k u$ *if we can find*
- *conic neighborhoods* $G^j \subset M_j$ *of* ξ^j,
- *constants* c, ε,
- *sublinear functions* $f_j : (0, \infty) \to (0, \infty)$, $j = 1, ..., k-1$,
and
- *a sequence of distributions* $u_i \in E'(U)$, $i = 1, 2, ...$, *such that:*

$$\text{the set } \{u_i\}_{i=1}^\infty \text{ is bounded in } E'(U),$$

$$u = u_i, \text{for } |x - x^0| < \varepsilon, \text{ and all } i$$

$$|\hat{u}_i(\xi)| \le c(ci/|\Pi_{k-1}\xi|)^i \text{ if } i = 1, 2, \dots \text{ and } \Pi_j\,\xi \in G^j, j = 0, \dots, k-1 \qquad (2.1.1)$$

$$\text{and } |\Pi_j\xi| \ge f_j(|\Pi_{j-1}\xi|), \; j = 1, \dots, k-1.$$

The last condition is considered void if k=1. With this convention, definition 2.1.1 is just Hörmander's first definition of the standard wave front set in this case. To shorten terminology later on, we now also introduce

Definition 2.1.2. *Consider* $(\xi^0, \xi^1, \dots, \xi^{k-1})$ *as before. We shall then say that* $G \subset R^n$ *is a multineighborhood of* $(\xi^0, \xi^1, \dots, \xi^{k-1})$ *if we can find open cones* $G^j \subset M_j$ *which contain* ξ^j *so that*

$$\{\xi : \Pi_j\xi \in G^j, j = 0, \dots, k-1\} \subset G.$$

Further, if G', G *are two multineighborhoods of* $(\xi^0, \xi^1, \dots, \xi^{k-1})$, *then we shall write that* $G \subset\subset G'$, *if we can find open cones* $G^j \subset\subset G'^j \subset M_j$ *which contain* ξ^j *such that*

$$G \subset \{\xi; \Pi_j\xi \in G^j, \forall j\} \subset \{\xi; \Pi_j\xi \in G'^j, \forall j\} \subset G'.$$

The main condition in definition 2.1.1 is then that the inequalities which we consider live on parts of multineighborhoods of $(\xi^0, \xi^1, \dots, \xi^{k-1})$ in the phase space.

Remark 2.1.3. *Except for the case k=1, the definition of* WF_A^k *(and of related notions introduced later on) thus depends on the fact that we have previously fixed a sequence of linear subspaces* $M_j \subset R^n$ *as above. Here we note that the* M_j *are not explicitly visible in the notation and must therefore be determined from the context. On some rare occasions we shall write*

$$(x^0, \xi^0, \xi^1, \dots, \xi^{k-1}) \notin WF_A^k u$$

(or similar) for some mutually orthogonal vectors without having introduced previously the relevant M_j. *In these situations it is understood tacitely that*

$$M_j \text{ is the orthogonal complement of span } (\xi^0, \xi^1, \dots, \xi^{j-1}).$$

It is also interesting to note that while the terminology refers to an explicit choice of vectors $(\xi^0, \xi^1, \dots, \xi^{k-1})$, *it is only the directions* $\xi^0/|\xi^0|$, $\xi^1/|\xi^1|$, $\dots$, $\xi^{k-1}/|\xi^{k-1}|$ *that matter. Alternatively we could then as well refer to the set* $\{\sum_j t_j\xi^j; t_j > 0\}$ *rather than to* $(\xi^0, \xi^1, \dots, \xi^{k-1})$.

3. Some explanations are here perhaps in order.

a) As in standard microlocalization, the inequalities from (2.1.1) are meant to mimick exponential decay in the variables from M_{k-1} for the Fourier transform of some localization of u. Let us in fact observe that

$$\min_i \ (c_i/|\Pi_{k-1}\xi|)^i \ \leq \ c' \exp[-c''|\Pi_{k-1}\xi|] \tag{2.1.2}$$

if $c'' < (ec)^{-1}$. (For fixed ξ we estimate the "min" by the value of $(c_i/|\Pi_{k-1}\xi|)^i$ for i the integer part of $|\Pi_{k-1}\xi|/(ec)$.) The reason why one has to work with a sequence of distributions and the inequalities (2.1.1) rather than with one single localization v of u which satisfies all the inequalities from (2.1.1) simultaneously, is of course that no $v \in E'(R^n)$ can exist for which we have

$$|\hat{v}(\xi)| \leq c' \exp -[c''|\Pi_{k-1}\xi|] \tag{2.1.3}$$

on the set on which the (2.1.1) are valid. This can be seen using the Phragmén-Lindelöf principle, but the heuristic reason is that v would have to be partially analytic, at a k-microlocal level, in the variables parallel to M_{k-1}. (Also cf. theorem 2.1.12 later on.)

b) It is trivial that $(x^0, \zeta^0, ..., \xi^{k-2}) \notin WF_A^{k-1}u$ implies $(x^0, \xi^0, ..., \xi^{k-2}, \eta) \notin WF_A^k u$, $\forall \eta \in M_{k-1}$. However, it is not true for $k \geq 2$ that it follows from

$$(x^0, \xi^0, \xi^1, ..., \xi^{k-2}, \eta) \notin WF_A^k u, \forall \eta \in M_{k-1}$$

that

$$(x^0, \xi^0, \xi^1, ..., \xi^{k-2}) \notin WF_A^{k-1}u.$$

(For $k = 2$ this is related to the fact that the notion which is microlocalized by two-microlocalization is not the first wave front set, but rather microlocal partial analyticity in the variables parallel to M_1. We shall discuss the relation between microlocal regularity and partial analyticity in detail in chapter IX.)

c) Conditions of type "$|\Pi_j\xi| \geq f_j(|\Pi_{j-1}\xi|)$" have been considered in higher microlocalization (using a different formulation) for the first time by Sjöstrand [2]. We discuss the relation between the conditions used in Sjöstrand loc.cit. and our conditions later on in this section. The main point with conditions of this type is that usually there is no way to localize arguments up to the axes which pass through the ξ^j. (Symbols of pseudodifferential operators will exhibit singular behaviour there, etc.) Some neighborhood of these axes has therefore to be spared out from consideration and the "width" of these neighborhoods is controlled by the "f_j". Actually it is not always appropriate to let the f_j vary in the class of all sublinear functions. One will then obtain variants of definition 1.2.1 by choosing the f_j from specific subclasses of sublinear functions and indeed also in these notes we will frequently replace the conditons $|\Pi_j\xi| \geq f_j(|\Pi_{j-1}\xi|)$ by the conditions

$$|\Pi_j\xi| \geq c_j(1 + |\Pi_{j-1}\xi|^\delta) \text{ for some } c_j > 0, \delta < 1. \qquad (2.1.4)$$

Another choice of conditions, which is natural when one works, e.g., with hyperbolic equations (cf. Lebeau [1] or Laubin-Esser [1]), is

$$|\Pi_j\xi| \geq c_j \ln(1 + |\xi|) \text{ for some } c_j > 0.$$

We shall discuss the reason why this variety of possibilities appears in chapter IX. It should be observed at this point that by taking different subclasses of sublinear functions one will obtain non-equivalent notions of wave front sets. Although it may matter sometimes with which of these variants of wave front sets one works, one should observe that these differences are at a rather technical level. On the other hand we shall consider later on in this section another variant of wave front sets which differs from the one considered above in a deeper way. (Cf. definition 2.1.9.)

d) Returning to the case of some general sublinear functions f_j we should mention the following: if Γ^{j-1} is some conic neighborhood of ξ^{j-1}, then we can find c_j so that

$$|\Pi_{j-1}\xi| \geq c_j, \ \Pi_{j-1}\xi \in G^{j-1} \setminus \Gamma^{j-1} \text{ implies } |\Pi_j\xi| \geq f_j(|\Pi_{j-1}\xi|).$$

For any fixed Γ^{j-1}, only a bounded part of $G^{j-1} \setminus \Gamma^{j-1}$ is therefore eliminated from consideration by our conditions.

In the sequel it is also convenient to write some given general sublinear function in the form

$$f(t) = \frac{t}{\rho(t)} , \text{ if } t > 0.$$

Sublinearity then means that $\lim_{t\to\infty} \rho(t) = \infty$. It is moreover convenient to reduce ourselves to the case when the ρ from before is monotonically increasing. That this is no loss of generality follows from the following elementary

Lemma 2.1.4. *Let ρ be given with $\lim_{t\to\infty} \rho(t) = \infty$. Then we can find a monotonically increasing ρ' such that $\rho'(t) \leq \rho(t)$, but still $\lim_{t\to\infty} \rho'(t) = \infty$. In particular, if $f(t) = t/\rho(t)$ and $f'(t) = t/\rho'(t)$ then*

$$\{(t,\tau); t > 0, \ \tau \geq f'(t)\} \subset \{(t,\tau); t > 0, \ \tau \geq f(t)\}.$$

Proof of lemma 2.1.4. Let $c_j \to \infty$ be a strictly increasing sequence so that $\rho(t) \geq j$ if $t \geq c_j$. We may then simply set $\rho'(t) = j$ if $c_j \leq t < c_{j+1}$.

Thus, if we denote for a moment by $\mathcal{X}$ the set of all sublinear funtions and by $\mathcal{Y}$ the subset of all sublinear functions of form $t/\rho(t)$ for some monotonically increasing ρ, then it follows that the filter of sets

$$\{(t,\tau); \tau \geq g(t)\}_{g\in\mathcal{Y}}$$

is equivalent with the filter of sets

$$\{(t,\tau); \tau \geq f(t)\}_{f\in\mathcal{X}}.$$

e) Let us also mention, for comparision, the definition of the second analytic wave front set as considered (in the involutive case) by Lebeau [1], and which is modelled on a definition from Sjöstrand [2]. Thus in this case k= 2 and, to simplify the situation notationally, we shall assume that $M_1 = \{\xi; \xi_{d+1} = \cdots = \xi_n = 0\}$ for some d and denote by $x' = (x_1, ..., x_d)$, $x'' = (x_{d+1}, ..., x_n)$. Similarily we set $\xi' = (\xi_1, ..., \xi_d)$, $\xi'' = (\xi_{d+1}, ..., \xi_n)$. To distinguish Lebeau's definition from ours notationally, we shall call it "second analytic spectrum".

Definition 2.1.5. *Consider $\xi^0 \in R^n \ominus M_1$, $\xi^1 \in M_1$, $x^0 \in U$, $u \in E'(U)$. We shall write $(x^0, \xi^0, \xi^1) \notin ss_A^2 u$ if we can find c, c', λ^0, μ^0 and a decreasing function $F : (0, \mu^0) \to R_+$ such that*

$$|u_x(e^{-\lambda\mu(y'-x')^2/2 - \lambda(y''-x'')^2/2})| \leq c e^{\lambda\mu|Im\, y'|^2/2 + \lambda|Im\, y''|^2/2 - c'\lambda\mu} \quad ,$$

for all $y \in C^n$ with $|y' - x^{0\prime} + i\xi^{1\prime}| < c'$, $|y'' - x^{0\prime\prime} + i\xi^{0\prime\prime}| < c'$, provided that $0 < \mu < \mu^0$, $\lambda > \lambda^0$ and $\lambda \geq F(\mu)$.

(The notation u_x means that u acts in the variables x.)

Remark 2.1.6. *The expression*

$$u_x(e^{-\lambda\mu(y'-x')^2/2 - \lambda(y''-x'')^2/2})$$

is called the "FBI-transform of the second kind", or FBI^2-transform, of u. Other variants of this transform will be introduced in section A.3 and chapter IX.

It is not difficult to prove that

Proposition 2.1.7. *The following two conditions are equivalent:*

a) $(x^0, \xi^0, \xi^1) \notin WF_A^2 u$.

b) $(x^0, \xi^0, \xi^1) \notin ss_A^2 u$.

This is in fact similar to the situation in the proof of the equivalence of various definitions of the standard analytic wave front set. For completeness we shall give a proof in section A.3 .

f) Let us explain here briefly how conditions of form $\lambda \geq F(\mu)$ are related to the conditions $|\Pi_1 \xi| \geq f_1(|\xi|)$ from definition 2.1.1 for the case $k = 2$. We observe then at first that in this definition we have two parameters, one λ, which corresponds to $|\xi|$, which is large, and another one, μ, which corresponds to $|\Pi_1 \xi|/|\Pi_0 \xi|$ and which is small. If f_1 is of form $f_1(t) = t/\rho(t)$ for some function ρ, then the condition $|\Pi_1\xi| \geq f_1(|\xi|)$ can be written as

$$\mu \geq 1/\rho(\lambda) \tag{2.1.5}$$

and we already know from lemma 2.1.4 that it is no loss in generality to assume that ρ is increasing. That this is essentially of the same type with $\lambda \geq F(\mu)$ follows now from

Lemma 2.1.8. *Let $\rho : (0, \infty) \to (0, \infty)$ be increasing and $F : (0, \infty) \to (0, \infty)$ decreasing. Denote by*

$$A = \{(\mu, \lambda); \lambda > 0, \mu \geq 1/\rho(\lambda)\}$$

and by

$$B = \{(\mu, \lambda); \mu > 0, \lambda \geq F(\mu)\}.$$

If one of the two functions is given, we can find the other one so that $\bar{A} = \bar{B}$.

Proof of lemma 2.1.8. The situation is almost symmetric. We may e.g. assume that ρ is given and want to find F. Let us then fix μ and denote $L = \{\lambda; \rho(\lambda) \geq (1/\mu)\}$. Since ρ is increasing, L is of form $[a, \infty)$ or (a, ∞), for some $a \geq 0$. Let us set $F(\mu) = a$. It is clear then that F is decreasing. We claim that we will always have $A \subset B$. Indeed, when $\rho(\lambda) \geq 1/\mu$, then we have $\lambda > a$ or at least $\lambda \geq a$, and in both cases $\lambda \geq F(\mu) = a$.

To prove that $B \subset \bar{A}$, assume first that

$$\lim_{\tau \to t_+} \rho(\tau) = \rho(t), \forall t > 0. \tag{2.1.6}$$

We claim that with this additional assumption actually $B \subset A$. Indeed, if a is related to μ as before, then we now have that $\rho(a) = 1/\mu$, so that $\lambda \geq F(\mu) = a$ gives $\rho(\lambda) \geq 1/\mu$.

In the general case, i.e. when we do not know wether (2.1.6) holds, we introduce the auxiliary function

$$\rho'(t) = \lim_{\tau \to t_+} \rho(\tau).$$

If $\tilde{A} = \{(\mu, \lambda); \mu \geq (1/\rho'(\lambda))\}$, then we will have $B \subset \tilde{A}$ and it is also clear that $\tilde{A} \subset \bar{A}$. This concludes the proof of the lemma.

4. Definition 2.1.5 corresponds to second microlocalization in the involutive case. Lebeau has also introduced a concept of second microlocalization in the "isotropic case". We shall need a refinement of definition 2.1.1 which is in some sense related to Lebeau's definition for the isotropic case. The main thing is to shrink the domain of validity of the inequalities from definition 2.1.1 still further. Since we do not microlocalize completely to isotropic subspaces we shall call the resulting wave front set "semi-isotropic", although we have not analyzed the relation to Lebeau's definition in detail. For simplicity we shall not consider the case of the most general sublinear functons f_j at all and immediately refer to conditions of type $|\Pi_j \xi| \geq c_j (1 + |\Pi_{j-1}\xi|^\delta)$.

Definition 2.1.9. *Let U, x^0, M_j and ξ^j be as in definition 2.1.1. We shall say that $(x^0, \xi^0, \xi^1, ..., \xi^{k-1})$ is not in the semi-isotropic analytic k-wave front set of u and write*

$$(x^0, \xi^0, \xi^1, ..., \xi^{k-1}) \notin WF^k_{A,s} u,$$

if we can find open conic neighborhoods $G^j \subset M_j$ of $\xi^j, \varepsilon, c, c_j, c'_j, 0 < \delta < 1, \beta > 0$ and a bounded sequence $\{u_i\} \subset E'(U)$ such that

$$u = u_i, \text{for } |x - x^0| < \varepsilon, \text{ and all } i$$

$$|\hat{u}_i(\xi)| \leq c(ci/|\Pi_{k-1}\,\xi|)^i \text{ if } i = 1, 2, \ldots, \Pi_j \xi \in G^j, j = 0, ..., k-1,$$

$$|\Pi_j\,\xi| \geq c_j(1 + |\Pi_{j-1}\,\xi|^\delta), j = 1, ..., k-1,$$

$$\text{and } |\Pi_{j+1}\,\xi| \geq c'_j |\Pi_j\,\xi|^{\beta+1}/|\Pi_{j-1}\,\xi|^\beta, j = 1, ..., k-2.$$

Remark 2.1.10. *a) One can of course write the conditions*

$$|\Pi_{j+1}\,\xi| \geq c'_j |\Pi_j\,\xi|^{\beta+1}/|\Pi_{j-1}\,\xi|^\beta \,, \; j = 1, ..., k-2 \tag{2.1.7}$$

in homogeneous form

$$c'_j |\Pi_j\,\xi|^\beta/|\Pi_{j-1}\,\xi|^\beta \leq |\Pi_{j+1}\xi|/|\Pi_j\,\xi| \,, \; j = 1, ..., k-2.$$

Since the quantities $|\Pi_{j+1}\xi|/|\Pi_j\,\xi|$ correspond to Sjöstrand's small parameters, this makes comparision with the theory of Sjöstrand-Lebeau easier. We stick here to the form (2.1.7) since it is more natural in propagation phenomena.

b) When $k = 3$ the conditions (2.1.7) reduce to

$$|\Pi_2\,\xi| \geq c'_1 |\Pi_1\,\xi|^{\beta+1}/|\Pi_0\,\xi|^\beta.$$

This has a tendency to localize estimates to the set $(I - \Pi_2)\xi = 0$.

c) for $k \leq 2$ *the conditions 2.1.7 are considered void. In particular, semi-isotropic microlocalization differs from the microlocalization considered above for the first time at the level of tri-microlocalization whereas in Lebeaus's paper the difference already appears for second microlocalization. Note that in our terminology the order of microlocalization is decided by the length of the chain of vectors* $(\xi^0, \xi^1, ..., \xi^{k-1})$ *and of the structure of the associated multineighborhoods, and is not visible at the level of inequalities. Actually, one could slightly generalize definition 2.1.9 to obtain a situation in which we have also semi-isotropic two-microlocalization. We do so here, but for simplicity we shall consider only the case* $k = 2$. *Consider then* $U, x^0, M_0, M_1, M_2, M_3 = \{0\}$ *and* ξ^0, ξ^1 *as in definition 2.1.9. We shall write that*

$$(x^0, \xi^0, \xi^1) \notin \tilde{W}F^2_{A,s}u,$$

if we can find open conic neighborhoods $G^j \subset M_j$ *of* ξ^j, $j = 0, 1$, $\varepsilon, c, c_j, c'_j, 0 < \delta < 1, \beta > 0$, *and a bounded sequence* $\{u_i\}_i \subset E'(U)$ *such that*

$$u = u_i \text{for } |x - x^0| < \varepsilon, \text{ and all } i$$

$$|\hat{u}_i(\xi)| \leq c(ci/|\Pi_2 \xi|)^i \text{ if } \Pi_j \xi \in G^j, j = 0, 1,$$

$$|\Pi_j \xi| \geq c_j(1 + |\Pi_{j-1} \xi|^\delta), j = 1, 2 \text{ and } |\Pi_2 \xi| \geq c'_2|\Pi_1 \xi|^{\beta+1}/|\Pi_0 \xi|^\beta.$$

It will become clear later on that

$(x^0, \xi^0, \xi^1) \notin \tilde{W}F^2_{A,s} u$ *for fixed* x^0, ξ^0, ξ^1 *is equivalent with* $(x^0, \xi^0, \xi^1, \xi^2) \notin WF^3_{A,s}u$.

$$\forall \xi^2 \in \dot{M}_2.$$

(Note that now we have a chain of three nontrivial subspaces, but still speak of two-microlocalization since localization involves the two vectors ξ^0, ξ^1.)

d) The strength of the conditions

$$|\Pi_{j+1} \xi| \geq c'_j |\Pi_j \xi|^{\beta+1}/|\Pi_{j-1} \xi|^\beta, j = 1, ..., k - 2$$

increases when β *decreases. In fact if* $\beta' = \beta + \nu$ *for some positive* ν, *then*

$$\begin{aligned}|\Pi_j \xi|^{\beta+1}/|\Pi_{j-1} \xi|^\beta &= [|\Pi_j \xi|^{\beta+\nu+1}/|\Pi_{j-1} \xi|^{\beta+\nu}] [|\Pi_j \xi|/|\Pi_{j-1} \xi|]^{-\nu} \\ &\geq |\Pi_j\xi|^{\beta+\nu+1}/|\Pi_{j-1}\xi|^{\beta+\nu},\end{aligned}$$

since $|\Pi_j\xi| \leq |\Pi_{j-1}\xi|$. *(All this is closely related to some appropriate order relation in the set of "polyindices" which we shall introduce later on in definition 3.2.5.) To use* $WF^k_{A,s}$ *in full strength, one will then tend to choose* $\beta > 0$ *small.*

5. We have now introduced $WF_A^k, WF_{A,s}^k$ and want to mention, for the moment without proofs, some of their main properties.

Proposition 2.1.11. *Consider $u \in D'(U)$ and let h be real analytic on U. Suppose that $(x^0, \xi^0, \xi^1, ..., \xi^{k-1}) \notin WF_A^k u$ (respectively that $(x^0, \xi^0, \xi^1, ..., \xi^{k-1}) \notin WF_{A,s}^k u$). Then it follows that $(x^0, \xi^0, \xi^1, ..., \xi^{k-1}) \notin WF_A^k(hu)$ (respectively that $(x^0, \xi^0, \xi^1, ..., \xi^{k-1}) \notin WF_{A,s}^k(hu)$).*

Proposition 2.1.11 is a consequence of results on general G_φ-microlocalization in Liess-Rodino [1]: the relation between general G_φ-microlocalization and higher microlocalization will be explained in section 2.2 later on. (A more general result, which also follows from results in Liess-Rodino, loc.cit., is that suitable classes of pseudodifferential operators operate well on k-wave front sets. We shall study such results in section 3.4.)

6. As mentioned above, the relation between WF_A^{k-1} and WF_A^k is not very strong for $k \geq 2$. The reason why $WF_A^{k,s}$ is useful at all in the present context, comes from the following result:

Theorem 2.1.12. *Let $U \subset R^n$ be open and consider $x^1, x^2 \in U$ with $[x^1, x^2] \subset U$, $x^1 - x^2 \in M_{k-1}$. Here and later on we denote by $[x^1, x^2]$ the segment with endpoints x^1 and x^2. Consider $\xi^0, \xi^1, ..., \xi^{k-2}$ and assume that for any $\eta \in M_{k-1}$ and any $x \in [x^1, x^2]$ it follows that*

$$(x, \xi^0, \xi^1, ..., \xi^{k-2}, \eta) \notin WF_{A,s}^k u. \qquad (2.1.8)$$

Moreover, assume that $(x^1, \xi^0, \xi^1, ..., \xi^{k-2}) \notin WF_{A,s}^{k-1} u$. Then it follows that

$$(x, \xi^0, \xi^1, ..., \xi^{k-2}) \notin WF_{A,s}^{k-1} u, \quad whatever \ x \in [x^1, x^2] \ is.$$

It also follows from this that $(x, \xi^0, \xi^1, ..., \xi^{k-2}) \notin WF_{A,s}^{k-1} u$, for all x in the connected component of $x^1 + M_{k-1} \cap U$ which contains x^1. (A similar result is true if we replace everywhere $WF_{A,s}^k$ by WF_A^k, at least when we work with wave front sets associated with the conditions $|\Pi_i \xi| \geq c_i(1 + |\Pi_{i-1}\xi|^\delta)$. Note that for $WF_{A,s}^k$ we have, for simplicity, not considered anything else.)

Note that for $k = 1$ this is just a variant of the micro-Holmgren theorem of Hörmander-Kashiwara. The case of a general k has been considered in the Lagrangian set-up in Sjöstrand [2]. The involutive case is considered, for $k = 2$ in Lebeau [1], who also proved the result (still in the case of two-microlocalization) in the isotropic situation. While the proofs of Sjöstrand and Lebeau are based on the FBI-transform, the argument which we shall use in this paper is based on an idea from Liess [2]. We shall give it in section 2.7.

Iteration of the preceding result gives the following theorem which is useful in conical refraction and which we shall prove in section 2.8.

Theorem 2.1.13. *Let X be open and convex in R^n with $0 \in X$ and consider $u \in D'(X)$. Also assume that for some fixed σ a chain of mutually orthogonal vectors $\xi^0, \xi^1, ..., \xi^{\sigma-1}$ and a closed convex subset $B \subset X$ is given with the following properties I, II, III, IV:*

I. $0 \notin B$.

II. $(x, \xi^0, \xi^1, ..., \xi^{\sigma-1}) \notin WF_{A,s}^\sigma u$, $\forall x \in B$.

III. Whatever $\eta^\sigma, ..., \eta^{n-1} \in R^n$ are given so that $\xi^0, \xi^1, ..., \xi^{\sigma-1}, \eta^\sigma, ..., \eta^{n-1}$ are mutually orthogonal and whatever x is in X, it follows that $(x, \xi^0, \xi^1, ..., \xi^{\sigma-1}, \eta^\sigma, ..., \eta^{n-1}) \notin WF_{A,s}^n u$.

IV. If $(\xi^0, \xi^1, ..., \xi^{\sigma-1}, \eta^\sigma, ..., \eta^{l-1})$ is a chain of mutually orthogonal vectors which extends $(\xi^0, \xi^1, ..., \xi^{\sigma-1})$, then we have either that

$$(x, \xi^0, \xi^1, ..., \xi^{\sigma-1}, \eta^\sigma, ..., \eta^{l-1}) \notin WF_{A,s}^l u, \forall x \in X,$$

or else there is $\tilde{x} \in B$ so that

$$\langle \tilde{x}, \xi^i \rangle = 0, \forall i, \langle \tilde{x}, \eta^j \rangle = 0, \forall j.$$

The conclusion is then that

$$(0, \xi^0, \xi^1, ..., \xi^{\sigma-1}) \notin WF_{A,s}^\sigma u.$$

Comment 2.1.14. *The usefulness of higher microlocalization in these notes is based primarily on the theorems 2.1.12 and 2.1.13, which have no full counterpart in the C^∞-category. (What is true e.g. is that the standard C^∞-wave front set will propagate if the tempered version of the second analytic wave front set is void. This has been observed by Lebeau [2].) In fact in view of this type of results, propagation of microlocal analytic singularities for solutions of some linear partial differential equation can often be proved in a two-step procedure. The first step is to prove that all solutions of the equation have some appropriate microlocal regularity and then apply the above (or related) theorems which says that some propagation result holds for any distribution which has the respective regularity. However, microlocalization to sets as small as possible in the cotangent space is an advantage in itself and has been pursued intensively in the C^∞-theory. Actually, all the evolution from the classical symbols of Kohn-Nirenberg [1] over the $S_{\rho,\delta}^m$ symbols of Hörmander to the $S_{\phi,\psi}^{m,m'}$ symbols of Beals-Fefferman [1], Beals [1] and the Weyl calculus* *of Hörmander [7] is part of this program. We should also mention here the work of Fefferman and Fefferman-Phong on the uncertainty principle. (Cf. Fefferman [1] and references cited there.)*

7. We conclude the section with a result which we have found useful in the study of F.I.O.'s in two-microlocalization:

Proposition 2.1.15. *Consider G an open bineighborhood of (ξ^0, ξ^1) and let $f \in S'(R^n)$ $(S'(R^n)$ is the space of tempered distributions on R^n) be so that $\hat{f}$ admits for suitable constants $c > 0, \delta < 1$, an holomorphic extension to the set*

$$D = \{\zeta \in C^n; Re\,\zeta \in G\,,\, 1 + |Im\,\zeta| + |Re\,\zeta|^\delta \leq c|\Pi_1\,Re\,\zeta|\,\},$$

and so that for some $c' \geq 0, b \in R$, and some convex compact $K \subset R^n$ we have that

$$|\hat{f}(\zeta)| \leq c'(1 + |\zeta|)^b\,exp(H_K(Im\,\zeta)),\ \forall \zeta \in D.$$

Then it follows that

$$(x^0, \xi^0, \xi^1) \notin WF_A^2 f, \text{for } x^0 \notin K.$$

(Here H_K is the support function of K, i.e., $H_K(\xi) = \sup_{x \in K}\langle x\,,\,\xi\rangle$.)

A similar result is valid also for higher order wave front sets. We have sticked here to the case of second microlocalization for notational reasons and since it is only this case which we need in the paper. The proof of proposition 2.1.15 will be given in section 2.9.

2.2 G_φ-**microlocalization**

1. Before we can prove the theorems stated in the introduction we must develop a microlocal calculus associated with the notions of analytic k-wave front sets introduced in the preceding section. Part of this calculus is a consequence of results from Liess-Rodino [1,2]. To explain why this is so, it is necessary to recall here the main definitions concerning G_φ-microlocalization.

Definition 2.2.1. *Let $\varphi : R^n \to R$ and consider $A \subset R^n$. We call φ a weight function on A if we can find $c > 0, c' > 0$ so that*

$$\varphi(\xi) \geq c'\,|\xi|^c\,,\ \forall \xi \in A. \tag{2.2.1}$$

We shall say that φ is Lipschitzian if there is $L \geq 0$ so that

$$|\varphi(\xi) - \varphi(\xi')| \leq L|\xi - \xi'|\,,\ \forall \xi, \forall \xi' \in R^n. \tag{2.2.2}$$

Finally, when we say that φ is a Lipschitzian weight function on A, then we mean that φ is Lipschitzian on all of R^n in the preceding sense, and that it is weight function on A.

Remark 2.2.2. *Comparision with Liess-Rodino [1] shows that we have slightly changed the set-up, in that we admit here the case when (2.2.1) holds for $\xi \in A$ only, whereas there we asked for it to be valid on all of R^n. While this fits indeed our present notational needs better, we should mention that it does not really correspond to an extension of the frame in which we work: cf. e.g. remark 2.2.5 and 2.2.7 below.*

Definition 2.2.3. *Let φ and ψ be two functions on R^n which have positive values. We shall say that φ is equivalent with ψ on A and write $\varphi \sim_A \psi$ if we can find two constants c, c' such that $c\varphi(\xi) \le \psi(\xi) \le c'\varphi(\xi)$ for all $\xi \in A$. (If we write $\varphi \sim \psi$, then A should be clear from the context.)*

Definition 2.2.4. *Let φ be a Lipschitzian weight function on A and consider $B \subset A$. When $\varepsilon > 0$ is given we denote by*

$$B_{\varepsilon\varphi} = \{\xi; dist(\xi, B) \le \varepsilon\varphi(\xi)\}.$$

A set D is called a φ-neighborhood of B if we can find $\varepsilon > 0$ so that $B_{\varepsilon\varphi} \subset D$.

Remark 2.2.5. *(Cf. Liess-Rodino [1]). a) Assume $\varphi \sim_A \psi$. Then there is a complete $\varphi-$ neighborhood $A' = A_{\varepsilon\varphi}$ of A so that $\varphi \sim_{A'} \psi$.*

b) If φ is a Lipschitzian weight function on A and if $c > 0, c' > 0$ are suitable, then $\varphi'(\xi) = max(\varphi(\xi)\, ,\, c'|\xi|^c)$ is equivalent with φ on A.

(Both parts of the remark are immediate.)

Definition 2.2.6. *a) Let φ be a Lipschitzian weight function on R^n and consider $U \subset R^n$ open, $x^0 \in U$, $u \in D'(U)$. We shall say that u is of class G_φ at x^0 if we can find c, a neighborhood $X \subset U$ of x^0, and a bounded sequence of distributions $u_j \in E'(X)$ such that*

i) $u = u_j$ on X and for all j,

ii) $|\hat{u}_j(\xi)| \le c(cj/\varphi(\xi))^j$, $j = 1, 2, \ldots, \forall \xi \in R^n$.

b) Let φ be a Lipschitzian weight function on A. For u, x^0, U, as before, we shall write that

$$(x^0 \times A) \cap WF_\varphi u = \emptyset,$$

if we can find c, c', a neighborhood X of x^0 and a bounded sequence of distributions in
$E'(X)$ such that

i) $u = u_j$ on X and for all j,

and

ii)' $|\hat{u}_j(\xi)| \leq c(cj/\varphi(\xi))^j, j = 1, 2, ..., if \xi \in A_{c'\varphi}$.

Remark 2.2.7. *a) Definition 2.2.6 has been introduced in Liess-Rodino [1]. (An equiv-*
alent definition had been considered earlier in Liess [1].) The "G" comes here from
"Gevrey". It alludes to the fact that in Liess-Rodino, loc.cit., G_φ-classes were considered
primarily as a generalization of anisotropic Gevrey classes. Indeed, when $\varphi \sim (1 + |\xi|^\delta)$
for some $\delta \leq 1$ then u is in G_φ at x^0, precisely if it is of Gevrey class δ (for $\delta = 1$ this
is the class of real analytic functions), and, more generally, when $\varphi \sim (1 + \sum_j |\xi_j|^{\delta_j})$ for
some $0 < \delta_j \leq 1$, then u is of class G_φ at x^0 if and only if we can find ϵ and c so that

$$|(\partial/\partial x)^\alpha u(x)| \leq c^{|\alpha|+1}(\alpha_1!)^{(1/\delta_1)} \cdots (\alpha_n!)^{(1/\delta_n)} \text{ for } x \text{ near } x^0.$$

Other examples of G_φ classes are given in Liess-Rodino [1]. As it will turn out later on,
G_φ-classes are sufficiently flexible to provide a convenient frame for higher microlocal-
ization as well (even in the semi-isotropic case.)

b) What is important in definition 2.2.6.b) is not so much φ itself than its equivalence
class in the set of all Lipschitzian weight functions on A. (Cf. remark 2.2.5.a).) Note
that in any such equivalence class there is a Lipschitzian weight function defined on all of
R^n. (Cf. remark 2.2.5.b).) This makes it possible to apply the results from Liess-Rodino
[1,2], even though in those papers it is always assumed that the φ' are weight functions
on all of R^n.

An important property of G_φ-regularity is that G_φ- and G_ψ-regularity together give
$G_{max(\varphi,\psi)}$-regularity:

Proposition 2.2.8. *Let $u \in D'(U)$, $x^0 \in U$, $A \subset R^n$, and two Lipschitzian weight*
functions φ and ψ on A be given. Assume that

$$(x^0 \times A) \cap WF_\varphi u = \emptyset \text{ and } (x^0 \times A) \cap WF_\psi u = \emptyset.$$

Then it follows that

$$(x^0 \times A) \cap WF_{max(\varphi,\psi)} u = \emptyset.$$

This is almost trivial, in that it would follow directly from the definitions, if we would know that we can test that $(x^0 \times A) \cap WF_\varphi u = \emptyset$ and $(x^0 \times A) \cap WF_\psi u = \emptyset$ with the same sequence of distributions u_j. That this is possible has however been observed in Liess-Rodino [1], in that we have the following result (similar to related statements for the standard wave front set):

Proposition 2.2.9. *Assume that*

$$(x^0 \times A) \cap WF_\varphi u = \emptyset$$

and let χ_j be a sequence of C^∞ functions such that

$$|D^\alpha \chi_j(x)| \leq c^{|\alpha|+1} j^{|\alpha|} \ if \ |\alpha| \leq j,$$

$$supp \ \chi_j \subset \{x; |x - x^0| \leq c'\},$$

for some constants c and c'. If c' is sufficiently small it follows that we can find $c'' > 0, d > 0$, for which

$$|\mathcal{F}(\chi_j u)(\xi)| \leq c''(c'' j/\varphi(\xi))^j, j = 1, 2, ..., \ if \ \xi \in A_{d\varphi}.$$

Let us now also recall the following result from Liess-Rodino [1]:

Proposition 2.2.10. *Consider $\varphi : R^n \to R_+$ such that $\varphi(\xi) < c(1 + \varphi(\eta))$ whenever $|\xi - \eta| < c'(1 + \varphi(\xi))$. Then we can find a Lipschitzian function φ' on R^n and a constant d so that $\varphi'(\xi) < d(\varphi(\xi) + 1)$ respectively $\varphi(\xi) < d\varphi'(\xi)$.*

Weight functions which have the property from the assumption in this proposition are called of "slow variation" and are well-known in the calculus of pseudodifferential operators in the C^∞-category: cf. e.g. Beals [1]. In fact one of the main motivations for developing a calculus of pseudodifferential and F.I.O.'s in general G_φ-classes (in Liess-Rodino [1,2]) has been precisely the attempt to find an analogue for the results from Beals [1] in the analytic category. Note however that φ does not depend on x. Also cf. here Rodino [1,2].

2. Let us now return to the situation from section 2.1 where we were given a strictly decreasing finite sequence of linear subspaces M_j with $R^n = M_0$, $M_k = \{0\}$. Assume moreover that $x^0 \in U \subset R^n$, that $\xi^j \in M_j \ominus M_{j+1}$ and that $u \in D'(U)$.

Remark 2.2.11. *There are equivalent:*

i) $(x^0, \xi^0, \xi^1, ..., \xi^{k-1}) \notin WF_{A,s}^k u$,

ii) there are constants $c_j, c'_j, \beta > 0, 0 < \delta < 1$, and open cones G^j in M_j so that with the notation

$$A = \{\xi; \Pi_j\,\xi \in G^j, |\Pi_{j-1}\,\xi| \geq c_j(1 + |\Pi_{j-1}\,\xi|^\delta), |\Pi_{j+1}\,\xi| \geq c'_j|\Pi_j\,\xi|^{\beta+1}/|\Pi_{j-1}\,\xi|^\beta\},$$

it follows that

$$(x^0 \times A) \cap WF_{|\Pi_{k-1}\,\xi|}u = \phi.$$

A similar remark is valid for WF_A^k.

As a consequence of this remark we can now, e.g., obtain from Liess- Rodino [1] a complete calculus of pseudodifferential operators which live on multineighborhoods of $(x^0, \xi^0, \xi^1, ..., \xi^{k-1})$ and act well on higher wave front sets. (For details see section 3.4.) Moreover, also for the case of F.I.O.'s we can simplify the situation significantly if we start from the results in Liess-Rodino [2], even though not all results which we need follow from that paper. Details will be elaborated in chapter V.

Another consequence is that a result parallel to proposition 2.2.9 is valid: if we want to check if $(x^0, \xi^0, \xi^1, ..., \xi^{k-1}) \notin WF_{A,\delta}^k u$, it will suffice to see if the inequalities from the definition of $WF_{A,\delta}^k$ hold for $u_j = \chi_j u$, where the χ_j are functions as in proposition 2.2.9 which have sufficiently small support and are identically one in a neighborhood of x^0.

3. We conclude the section with a result which shows how one can test G_φ-regularity by duality:

Proposition 2.2.12. *(Cf. Liess-Rodino [1].) Let $u \in D'(U)$, $x^0 \in U$, $A \subset R^n$ and a Lipschitzian weight function φ on A be given. Then there are equivalent:*

a) $(x^0 \times A) \cap WF_\varphi u = \emptyset$,

b) there are positive constants c, c', d and $b \in R$ so that

$$|u(f)| < c,$$

for any $f \in C_o^\infty(R^n)$ so that

$$|\hat{f}(\zeta)| \leq e^{d\varphi(-Re\,\varphi) + \langle x^0, Im\,\zeta\rangle + \varepsilon|Im\,\zeta| + b\,ln(1 + |\zeta|))}, \quad if\ \zeta \in C^n, Re\,\zeta \in A_{c'\varphi},$$

respectively

$$|\hat{f}(\zeta)| \leq e^{\langle x^0, Im\,\zeta\rangle + \varepsilon|Im\,\zeta| + b\,ln(1 + |\zeta|))}, \quad if\ Re\,\zeta \notin A_{c'\varphi}.$$

(The inversion of signs in the argument of φ comes from Parseval's formula:

$$u(v) = (2\pi)^{-n} \int \hat{u}(\xi)\bar{\hat{v}}(-\xi)d\xi,$$

if u and v are, e.g., in the Schwartz space $S(R^n)$.)

2.3   Splitting and approximation of entire functions

1. A number of results concerning $WF_{A,s}^k$ can be obtained starting from remark 2.2.11 and arguing by duality. The main point in the argument is then often to split some given entire function which satisfies certain inequalities into a sum of functions which satisfy related inequalities. We review here a number of results from Liess [2] which are common to all such arguments.

Lemma 2.3.1. *Let* $\varphi_i : C^n \to R$, $i = 1, 2, 3, 4$ *and* $\psi : C^n \to R$ *be functions with the following properties:*

a) $\varphi_1(\zeta) \leq \varphi_3(\zeta)$, $\varphi_2(\zeta) \leq \varphi_4(\zeta)$, $\forall \zeta \in C^n$,

b) ψ *is plurisubharmonic,*

c) $min\,(\varphi_1(\zeta), \varphi_2(\zeta)) \leq \psi(\zeta) \leq\ min\,(\varphi_3(\zeta), \varphi_4(\zeta))$,

d) there is L *such that* $|\varphi_i(\zeta^1) - \varphi_i(\zeta^2)| \leq L|\zeta^1 - \zeta^2|, i = 1, 2, 3, 4.$

Then there are constants c *and* μ *such that every* $h \in A(C^n)$ *which satisfies*

$$|h(\zeta)| \leq \exp\,[\max\,(\varphi_1(\zeta), \varphi_2(\zeta))],$$

can be decomposed in the form $h = h_1 + h_2, h_i \in A(C^n)$, *where*

$$|h_1(\zeta)| \leq c\ exp\,(\varphi_3(\zeta) + \mu \ln(1 + |\zeta|)),$$

$$|h_2(\zeta)| \leq c\ exp\,(\varphi_4(\zeta) + \mu \ln(1 + |\zeta|)).$$

(Here $A(C^n)$ is the space of entire functions on C^n.)

Lemma 2.3.1 is taken from Liess [2]. Since the proof is very short we reproduce it here for the convenience of the reader. Let us in fact denote by

$$U_1 = \{\zeta \in C^n; \varphi_2(\zeta) - \varphi_1(\zeta) \leq 1\}, U_2 = \{\zeta \in C^n; \varphi_2(\zeta) - \varphi_1(\zeta) \geq -1\}.$$

In view of d) the distance from the complement of U_1 to the complement of U_2 is greater than $1/L$. It is then easy to see that we can find $e_i \in C^\infty(C^n)$, i=1,2, so that $e_1 + e_2 \equiv 1$, supp $e_i \subset U_i$, supp $\bar{\partial} e_i \subset U_1 \cap U_2$, $e_i \geq 0$, and such that $|\bar{\partial} e_i| \leq c_1$. Then we have $\bar{\partial}(e_1 h) = -\bar{\partial}(e_2 h)$ and $|\bar{\partial}(e_1 h)| \leq c_2 \exp \psi(\zeta)$. Since ψ is plurisubharmonic, we can find $f \in C^\infty(C^n)$ such that $\bar{\partial} f = \bar{\partial}(e_1 h)$ and such that

$$|f(\zeta)| \leq c_3 \exp\,(\psi'(\zeta) + \mu \ln\,(1 + |\zeta|)),$$

where

$$\psi'(\zeta) = \sup_{|\theta| \leq 1} \psi(\zeta + \theta).$$

This is a consequence of the results on the $\bar{\partial}$-system of Hörmander [6]. (Also cf. here lemma 2.9.6, or Liess [2]). The conclusion of the lemma is then satisfied for $h_1 = e_1 h - f$, $h_2 = e_2 h + f$.

Remark 2.3.2. *Although the argument in the proof of lemma 2.3.1 is very simple, conditions of type c), (which shall play an important role in what follows) are in a deeper way related to decompositions of entire functions of the type above. For a discussion of this, cf. Liess [4].*

2. When applying lemma 2.3.1 the following two results are often useful.

Lemma 2.3.3. *Let* $\varphi : R^n \to R_+$ *satisfy* $|\varphi(\xi^1) - \varphi(\xi^2)| \leq L|\xi^1 - \xi^2|$, $\forall \xi^1, \xi^2 \in R^n$. *Then there is a plurisubharmonic function* $q : C^n \to R$ *and constants* c, c' *so that*

$$\varphi(Re\,\zeta) \leq q(\zeta) + c|Im\,\zeta| \quad and \quad q(\zeta) \leq 2\varphi(Re\,\zeta) + c|Im\,\zeta| + c'.$$

Lemma 2.3.3 is proved in Liess [1, proposition 2.1]. The proof is once more rather short, so we present it also. The argument is based on the following easy

Lemma 2.3.4. *We denote the variables in* $C^n \times R_+$ *by* (z, t). *There is a function* h *on* $C^n \times R_+$ *which is plurisubharmonic in* ζ *for each fixed* t *and which satisfies for some constants* c_i *the inequalities :*

$$h(\zeta, t) \leq c_1|Im\,\zeta| + c_2, \tag{2.3.1}$$

$$h(i\xi, t) \geq 0, if\, \xi \in R^n, \tag{2.3.2}$$

$$h(\zeta, t) \leq -|Re\,\zeta| + c_1|Im\,\zeta| + c_2 \quad if\ t/2 \leq |Re\,\zeta| \leq t, \tag{2.3.3}$$

$$h(\zeta, t) \leq -t + c_1|Im\,\zeta| + c_2 \ for\ |\zeta| \geq t. \tag{2.3.4}$$

Proof of lemma 2.3.4. (Cf. Liess [1].) Similar results are of current use in works on quasianalytic classes. We consider $\psi \in C_0^\infty(R^n)$ with $\int \psi(x)dx = 1$, $\psi \geq 0$, and try to find α, β such that the inequalities above are satisfied for $h(\zeta, t) = \alpha \ln |\hat{\psi}(\beta\zeta/t)|$. Then (2.3.1) is a consequence of the Paley-Wiener theorem and (2.3.2) follows from $\int \psi(x)dx = 1$ and $\psi \geq 0$. The remaining properties are also clear for suitable α and β.

Remark 2.3.5 *Functions of type "$h(\zeta, t)$" are one of the essential ingredients in the theory. They are often meant to do, in the frame of positive valued plurisubharmonic functions of at most linear growth, the job of partitions of unity. (Applications of this type occur in section 2.5.)*

Proof of lemma 2.3.3. (Cf. Liess [1].) We may assume that $L = 1$. Consider $\xi^0 \in R^n$ and define q_{ξ^0} by

$$q_{\xi^0}(\zeta) = \varphi(\xi^0) + 2h(\zeta - \xi^0, \varphi(\xi^0)),$$

where h is the function from lemma 2.3.4. We then have $q_{\xi^0} \leq 2(\varphi(\operatorname{Re}\zeta) + c_1|\operatorname{Im}\zeta| + c_2)$. Indeed, for $|\zeta - \xi^0| \leq \varphi(\xi^0)/2$ we have $|\varphi(\operatorname{Re}\zeta) - \varphi(\xi^0)| \leq \varphi(\xi^0)/2$ and for $|\zeta - \xi^0| \geq \varphi(\xi^0)/2$ we use (2.3.3) and (2.3.4). A function as desired is then the upper regularization of $\sup_{\xi^0 \in R^n} q_{\xi^0}$.

We may call a function $\psi : U \to R$, $U \subset C^n$, for which we can find $c > 0, c' > 0, c'' > 0$, and a plurisubharmonic function $f : U \to R$ so that

$$\psi(\zeta) \leq f(\zeta) + c|Im\,\zeta|, f(\zeta) \leq c'\psi(\zeta) + c|Im\,\zeta| + c'',$$

"essentially plurisubharmonic": as far as most needs of the present notes are concerned, it is as good as a truly plurisubharmonic function. Lipschitzian functions on R^n with positive values are then essentially plurisubharmonic when regarded as functions on C^n. The same is however only exceptionally true for Lipschitzian functions which may also have negative values. An example when it remains true is the following result:

Lemma 2.3.6. *Let $b^1 \in R$ be given and fix $\varepsilon > 0$. Then there are b^2, c and a plurisubharmonic function $\varrho : C^n \to R$ so that*

$$b^2 \ln(1 + |\zeta|) \leq \varrho(\zeta) \leq b^1 ln(1 + |\zeta|) + \varepsilon |\,Im\,\zeta| + c.$$

Lemma 2.3.6 has been proved in Liess [4]. (A related result will be considered in section 7.5.)

Remark 2.3.7. *The lemma is of course only interesting when $b^1 < 0$. In fact, for $b \geq 0$ we may use the preceding result or just observe that $b\ln(1 + |\zeta|) \sim b\ln(1 + |\zeta|^2)$ and that $b\,ln(1 + |\zeta|^2)$ is itself already plurisubharmonic.*

3. Repeated application of lemma 2.3.1 now gives, if we also take into account lemma 2.3.3 and 2.3.6:

Proposition 2.3.8. *Let $g^1, ..., g^k$ and g be Lipschitzian functions on R^n, all positive, $g(\xi) \leq max\; g^i(\xi), \forall \xi \in R^n$. Also let $b' \in R$, $d' > 0$, $\varepsilon' > 0$ and a convex compact $K \subset R^n$ be given. Then we can find $d > 0$, $b \in R$ and $c > 0$ with the following property: for any $h \in A(C^n)$ which satisfies*

$$|h(\zeta)| \leq \; exp\;(dg(Re\zeta) + H_K(Im\zeta) + b\ln(1 + |\zeta|)), \tag{2.3.5}$$

we can find $h^1, ..., h^k \in A(C^n)$ so that $h = \sum_i h^i$ and so that

$$|h^i(\zeta)| \leq c \ exp \ (d'g^i(Re\zeta) + H_K(Im\zeta) + \epsilon'|Im\zeta| + b'\ln(1 + |\zeta|)). \qquad (2.3.6)$$

4. A result which is closely related to proposition 2.3.8 is

Proposition 2.3.9. *Let $g^1, ..., g^s$ and r be Lipschitzian functions on R^n, all positive, and assume that $r(\xi) \leq min \ g^i(\xi), \ \forall \xi \in R^n$. Let also $x^1, x^2 \in R^n$ be fixed and consider $d'' > 0$, $t^i \in [x^1, x^2]$, $i = 1, ..., s$, so that $\bigcup\{t \in [x^1, x^2], |t - t^i| \leq d''\}$ contains $[x^1, x^2]$. Also consider $d' > 0, b'$.*
Then there are d, b, c so that any $h \in A(C^n)$ which satisfies

$$|h(\zeta)| \leq exp \ (dr(Re\zeta) + d|Im\zeta| + max \ (\langle x^1, Im\zeta\rangle, \langle x^2, Im\zeta\rangle) + b\ln(1 + |\zeta|)),$$

can be written in the form $h = \sum h^i$, $h^i \in A(C^n)$, where

$$|h^i(\zeta)| \leq c \ exp \ (d'g^i(Re\zeta) + d''|Im\zeta| + \langle t^i, Im\zeta\rangle + b'ln(1 + |\zeta|)).$$

Moreover, if h is of form $h = \hat{f}$ for some f in $E'(R^n)$, then the h^i can also be taken of form $h^i = \hat{f^i}$ for some $f^i \in E'(R^n)$.

(Actually we would loose nothing in generality if we assume that $g^i = r, \forall i$. To allow for $g^i \neq r$ is sometimes convenient in applications.)

To understand the geometric meaning of the inequalities involving the h and the h^i, it may be useful to recall that $\max(\langle x^1, Im \ \zeta\rangle, \langle x^2, Im \zeta\rangle)$ is the supporting function of the segement $[x^1, x^2]$, whereas $d''|Im \ \zeta| + \langle t^i, Im \ \zeta\rangle$ is the supporting function of the sphere $\{x \in R^n; |x - t^i| \leq d''\}$. Proposition 2.3.9 is a consequence of proposition 1.3.4 from Liess [2]. (Here, as opposed to Liess, loc.cit., we allow also the case $b' < 0$. That this does not affect the argument from Liess [2] is clear if we take into account lemma 2.3.6.)

5. Results of the type from proposition 2.3.8 are useful to estimate $u(v)$, $v \in C_0^\infty(R^n)$, when $h = \hat{v}$ satisfies (2.3.5), in that we can formally write that

$$u(v) = \sum u(F^{-1}h^i), \text{with } h^i \text{ satisfying (2.3.6).} \qquad (2.3.7)$$

If then we know, for some reason or another, that $u(w)$ can be estimated well for any $w \in E'(R^n)$ for which $\hat{w}$ satisfies (2.3.6) for some i, then we may hope to arrive at an estimate for $u(v)$ using (2.3.7). The following result on approximation is then useful:

Proposition 2.3.10. *Let* $g^i : C^n \to R_+, i = 1, ..., s$ *be such that* $g^i(\zeta) \le |\zeta|$ *and let* $v \in C_o^\infty(R^n)$, *supp* $\hat{v} \subset \{x; |x - x^0| < d_1\}$, *be given. Assume further that*

$$\hat{v} = \sum_{i=1}^s h^i,$$

for some entire functions h^i *which satisfy*

$$|h^i(\zeta)| \le exp\ (d_1 g^i(\zeta) + d_1 |Im\zeta| + \langle x^0, Im\zeta \rangle + b_1 \ln(1 + |\zeta|)).$$

If b is fixed and b_1 *is suitable, we can then find* $h_j^i \in A(C^n)$ *so that*

a) $\mathcal{F}^{-1} h_j^i \in E'(R^n)$,

b) $|(h^i - h_j^i)(\zeta)| \le (c/j)\ exp\ (d_1 g^i(\zeta) + 5d_1 |Im\zeta| + \langle x^0, Im\zeta \rangle + b \ln(1 + |\zeta|))$,

c) $|(\hat{v} - \sum_i h_j^i)(\zeta)| \le (c/j)\ exp\ (5d_1 |Im\zeta| + \langle x^0, Im\zeta \rangle + b \ln(1 + |\zeta|))$.

Proof. The argument is very close to the proof of proposition 3.2 in Liess [1]. Also cf. section 2.2 in Liess [2]. We sketch the argument here to make these notes easier to follow. Starting point is to observe that if we fix $b_1 \in R$, then we can find $\tilde{c}$ so that

$$|\hat{v}(\zeta)| \le \tilde{c} \exp\ (d_1 |Im\zeta| + \langle x^0, Im\zeta \rangle + b_1 ln(1 + |\zeta|)).$$

Next we choose a sequence of C^∞ functions f_j so that for some sequence of numbers η_j (which will tend to infinity)

$$|f_j(\zeta)| \le \eta_j(1 + |\zeta|) \exp\ (-d_1 |\zeta| + 3d_1 |Im\zeta|),$$

$$|\bar{\partial} f_j(\zeta)| \le (1/j)(1 + |\zeta|) \exp\ (-d_1 |\zeta| + 3d_1 |Im\zeta|),$$

$$|(1 - f_j(\zeta))| \le (1/j)(1 + |\zeta|).$$

(Here $\bar{\partial}$ is the Cauchy-Riemann operator.)

The existence of such a sequence of functions is proved in Liess [1]. A related implicit construction of the same type appears already in Hörmander [4]. Let us then denote $f_j h^i$ by H_j^i. We then have that

$$|(\hat{v} - \sum_i H_j^i)(\zeta)| = |(1 - f_j)\hat{v}(\zeta)| \le \tilde{c}(1/j)(1 + |\zeta|)\exp\ (d_1 |Im\zeta| + \langle x^0, Im\zeta \rangle +$$
$$b_1 ln(1 + |\zeta|)).$$

As for the defects from analyticity of the H_j^i we have

$$|\bar{\partial} H_j^i(\zeta)| \le (1/j)(1 + |\zeta|)\exp\ (4d_1 |Im\zeta| + \langle x^0, Im\zeta \rangle + b_1 ln(1 + |\zeta|)).$$

We can then find F_j^i so that $\bar{\partial} F_j^i = \bar{\partial} H_j^i$ and so that

$$|F_j^i(\zeta)| \leq c'(1/j)(1 + |\zeta|) \exp\left(5d_1|Im\zeta| + \langle x^0, Im\zeta\rangle + b\ln(1 + |\zeta|)\right).$$

Here we apply standard results about the solvability of the $\bar{\partial}$- operator due to Hörmander. (Cf. e.g. Liess [2], combined with lemma 1.4.3.) Denote now by $h_j^i = H_j^i - F_j^i$. The following inequalities are then consequences of the inequalities above:

$$
\begin{aligned}
|h_j^i(\zeta)| \leq \ & \eta_j(1 + |\zeta|) \exp\left(4d_1|Im\zeta| + \langle x^0, Im\zeta\rangle + b_1\ln(1 + |\zeta|)\right) + \\
& c'(1/j)(1 + |\zeta|) \exp\left(5d_1|Im\zeta| + \langle x^0, Im\zeta\rangle + b\ln(1 + |\zeta|)\right),
\end{aligned}
$$

$$
\begin{aligned}
|(h_j^i - h^i)(\zeta)| \leq \ & |(h^i(1 - f_j)| + |F_j^i(\zeta)| \leq (1/j)\exp\left(d_1 g^i(\zeta) + d_1|Im\zeta| + \right. \\
& \langle x^0, Im\zeta\rangle + (b_1 + 1)\ln(1 + |\zeta|)) + \\
& c'(1/j)\exp\left(5d_1|Im\zeta| + \langle x^0, Im\zeta\rangle + b\ln(1 + |\zeta|)\right),
\end{aligned}
$$

$$
\begin{aligned}
|(\hat{v} - \sum_{i=1}^s h_j^i)(\zeta)| \leq \ & |\sum_i F_j^i(\zeta)| + |(\hat{v} - \sum_i H_j^i)(\zeta)| \leq c''(1/j)\exp\left(5d_1|Im\zeta| + \right. \\
& \langle x^0, Im\zeta\rangle + b\ln(1 + |\zeta|)).
\end{aligned}
$$

This concludes the proof.

6. We conclude the section with the following lemma which can be proved using the argument from the proof of lemma 2.3.1

Lemma 2.3.11. *(Cf. Liess [1].) Let* $g : R^n \to R_+$ *and for every* $i \in Z^n$ *a function* $g_i : R^n \to R_+$ *be given so that*

a) $g_i(\xi) \leq g(\xi), \forall \xi$,

b) $|g_i(\xi^1) - g_i(\xi^2)| \leq c|\xi^1 - \xi^2|$, $|g(\xi^1) - g(\xi^2)| \leq c|\xi^1 - \xi^2|$, $\forall \xi^1, \xi^2 \in R^n$,

c) $\sup_i g_i(\xi) \geq g(\xi)/2$, $\forall \xi$.

If $d' > 0, \varepsilon' > 0, b' \in R$ *are given, we can then find* $c' > 0, d > 0, \varepsilon > 0, b \in R$ *so that any* $h \in A(C^n)$ *which satisfies*

$$|h(\zeta)| \leq \ exp\left[dg(Re\zeta) + \varepsilon|Im\zeta| + b\ln(1 + |\zeta|)\right],$$

can be written in the form

$$h(\zeta) = \sum_{i \in Z^n} h_i(\zeta),$$

for some $h_i \in A(C^n)$ *which satisfy*

$$|h_i(\zeta)| \leq c'(1 + |i|)^{-n-1} \ exp\left[d'g_i(Re\zeta) + \varepsilon'|Im\zeta| + b'\ln(1 + |\zeta|)\right].$$

Remark 2.3.12. *a) The factor* $(1+|i|)^{-n-1}$ *is a convergence factor:* $\sum(1+|i|)^{-n-1} < \infty$. *It is produced at the expence of a term* "$(n+1)ln(1+|\zeta|)$" *in the exponent.*

b) The decompositions from this lemma play (here, but also already in Liess [1]) a role comparable to that of dyadic decompositions in harmonic analysis.

c) In typical applications, the g_i will be related to g by the formula $g_i(\xi) = (g(i) - c|i - \xi|)_+$ for some c.

2.4 Higher microlocalization and duality

1. Higher order wave front sets of the type considered in these notes can be characterized in a convenient way by duality. The following result is a variant of proposition 2.2.12.

Proposition 2.4.1. *Consider $x^0 \in U$, M_j, $j = 0, ..., k-1$ and $\xi^j \in \dot{M}_j$ as in the definition of $WF_{A,s}^k$. Also consider $u \in D'(U)$. Then there are equivalent :*

i) $(x^0, \xi^0, \xi^1, ..., \xi^{k-1}) \notin WF_{A,s}^k u$,

ii) there is a multineighborhood G of $(\xi^0, \xi^1, ..., \xi^{k-1})$, $b \in R$, $c > 0$, $c' > 0$, $d > 0$, $\beta > 0$, $\delta < 1$ and a Lipschitzian weight function φ such that $\varphi \sim |\Pi_{k-1}|$ on

$$A = \{\xi \in G; |\Pi_{j+1}(\xi)| \geq c|\Pi_j\xi|^{\beta+1}/|\Pi_{j-1}\xi|^\beta, j = 1, ..., k-2,$$
$$|\Pi_i\xi| \geq c(1 + |\Pi_{i-1}\xi|^\delta), i = 1, ..., k-1\},$$

so that $|u(v)| \leq c'$ for any $v \in C_0^\infty(R^n)$ which satisfies

$$|\hat{v}(\zeta)| \leq \exp[d\varphi(-Re\,\zeta) + d|Im\,\zeta| + \langle x^0, Im\,\rangle + b\,ln(1 + |\zeta|)].$$

A similar result is valid for WF_A^k.

2. To shorten terminology it is now convenient to introduce two notations. Let in fact $M_0, M_1, ..., M_{k-1}, \xi^0, \xi^1, ..., \xi^{k-1}$, be as in the definition of $WF_{A,s}^k$ and consider numbers $\beta > 0$, $0 \leq \delta < 1$. We shall then denote by

$$G(\xi^0, \xi^1, ..., \xi^{k-1}, \delta, \beta),$$

respectively by

$$\mathcal{R}(\xi^0, \xi^1, ..., \xi^{k-2}, \delta, \beta),$$

the set of all functions $g : R^n \to R_+$, respectively $r : R^n \to R_+$, such that

α) $|g(\xi)| \leq |\Pi_{k-1}(\xi)|, \forall \xi, |r(\xi)| \leq |\Pi_{k-1}(\xi)|, \forall \xi,$

β) there are conic neighborhoods G^j in the M_j of the ξ^j and constants c_j such that

$$|g(\xi)| = |\Pi_{k-1}(\xi)|, \text{ if } \Pi_j\xi \in G^j, j = 0, ..., k-1, \text{ and } |\Pi_i\xi| \geq c_i(1 + |\Pi_{i-1}\xi|^\delta),$$
$$i = 0, ..., k-1, |\Pi_{j+1}\xi| \geq c_j|\Pi_j\xi|^{\beta+1}/|\Pi_{j-1}\xi|^\beta, j = 1, ..., k-2,$$

respectively

$$|r(\xi)| = |\Pi_{k-1}(\xi)|, \text{ if } \Pi_j\xi \in G^j, j = 0, ..., k-2, \text{ and } |\Pi_i\xi| \geq c_i(1 + |\Pi_{i-1}\xi|^\delta),$$
$$i = 0, ..., k-1, |\Pi_{j+1}\xi| \geq c_j|\Pi_j\xi|^{\beta+1}/|\Pi_{j-1}\xi|^\beta, j = 1, ..., k-2.$$

Note that the functions from G and $\mathcal{R}$ have very similar defining properties, in that both are compared with $|\Pi_{k-1}|$. However, the functions from G refer to $(\xi^0, \xi^1, ..., \xi^{k-1})$, whereas those from $\mathcal{R}$ refer to $(\xi^0, \xi^1, ..., \xi^{k-2})$. (The set where $r \sim |\Pi_{k-1}|$ is in particular in some sense "strictly" larger than that where $g \sim |\Pi_{k-1}|$.)

3. The main advantage of considering G is that now the definition of $WF_{A,s}^k$ becomes very short. Thus for example, proposition 2.4.1 can be formulated as :

Proposition 2.4.2. $(x^0, \xi^0, \xi^1, ..., \xi^{k-1}) \notin WF_{A,s}^k$ *is equivalent with the fact that we can find* c, d, β, δ *and* $g \in G(\xi^0, \xi^1, ..., \xi^{k-1}, \delta, \beta)$ *so that* $|u(v)| \leq c$ *for any* $v \in C_0^\infty(U)$ *which satisfies*

$$|\hat{v}(\zeta)| \leq exp\left(dg(-Re\zeta) + d|Im\zeta| + \langle x^0, Im\zeta \rangle + b\ln(1 + |\zeta|)\right).$$

On the other hand, the reason why the space of majorant functions $\mathcal{R}$ is interesting, is that we have the following

Proposition 2.4.3. *Assume that*

$$(x^0, \xi^0, \xi^1, ..., \xi^{k-2}, \eta) \notin WF_{A,s}^k u, \text{ whatever } \eta \in M_{k-1} \text{ is } .$$

Then we can find $\delta < 1, \beta > 0$, $r \in \mathcal{R}(\xi^0, \xi^1, ..., \xi^{k-2}, \delta, \beta)$ *and* $c > 0, d > 0, b$, *so that* $|u(v)| \leq c$ *for any* $v \in C_0^\infty(U)$ *which satisfies*

$$|v(u)| \leq exp\left(dr(-Re\zeta) + d|Im\zeta| + \langle x^0, Im\zeta \rangle + b\ln(1 + |\zeta|)\right). \quad (2.4.1)$$

Recall that $r(-Re\zeta) = |\Pi_{k-1}(Re\zeta)|$ on a large portion of some multineighborhood of $(\xi^0, \xi^1, ..., \xi^{k-2})$. An inequality of type (2.4.1) then roughly speaking corresponds to partial analyticity in the variables parallel to M_{k-1}. Actually it is this kind of regularity which is microlocalized in $k-$ microlocalization. We shall discuss this in detail in chapter IX.

The proposition 2.4.3 is a consequence of results from section 2.3. In fact, using the compacity of $|\eta| = 1$ and the assumptions, we can find $s, \eta^1, ..., \eta^s \in M_{k-1}, |\eta^i| = 1$, $d^i > 0, b^i, c^i, \delta^i < 1, \beta^i > 0$ and $g^i \in G(\xi^0, \xi^1, ..., \xi^{k-2}, \eta^i, \delta^i, \beta^i)$ so that

$$|u(v)| \leq c \text{ if } |\hat{v}(\zeta)| \leq \exp (d^i g^i(-Re\zeta) + d^i |Im\zeta| + \langle x^0, Im\zeta\rangle + b^i ln(1 + |\zeta|)) \quad (2.4.2)$$

and so that there is $r \in \mathcal{R}(\xi^0, \xi^1, ..., \xi^{k-2}, \delta, \beta)$ with

$$r(\xi) \leq \max_i g^i(\xi), \forall \xi.$$

(Here we may take $\delta = \max \delta^i, \beta = \min \beta^i$.)

Let us denote $b' = \min_i b^i, d' = \min_i d^i$. If $d > 0, b, c$ are suitable, it follows from proposition 2.3.9 that any $v \in C_0^\infty$ which satisfies

$$|\hat{v}(\zeta)| \leq \exp (dr(-Re\zeta) + d|Im\zeta| + \langle x^0, Im\zeta\rangle + b \ln(1 + |\zeta|))$$

can be written as

$$\hat{v} = \sum h^i,$$

with $h^i \in A(C^n)$ satisfying

$$|h^i(\zeta)| \leq \exp (d'g^i(-Re\zeta) + (d'/2)|Im\zeta| + \langle x^0, Im\zeta\rangle + b^i ln(1 + |\zeta|)).$$

Note that this gives formally

$$u(v) = \sum u(\mathcal{F}^{-1}h^i),$$

where $\mathcal{F}f = \hat{f}$ is the Fourier-Borel transform for analytic functionals and $\mathcal{F}^{-1}$ is its inverse. Still formally, we have that $|u(\mathcal{F}^{-1}h^i)| \leq c^i$ in view of (2.4.1), (2.4.2). Actually, the formal computation from before can be given a sense using the approximation result from section 2.3 :

$$u(v) = \lim_j \sum_i u(\mathcal{F}^{-1}h^i_j),$$

for suitable entire analytic functions h^i_j which approximate h_i when $j \to \infty$. The conclusion follows.

2.5   Plurisubharmonic functions related to propagation phenomena

1.   The fact that regularity for solutions of certain partial differential equations (or systems) is propagated is often related to some partial regularity which all solutions

to that equation (or system) have. There are several ways to use partial regularity in propagation. A rather strong method, developed by Sjöstrand [2] is to describe both, the partial regularity and some initial microlocal regularity of a solution, in terms of inequalities for some F.B.I.-transform of the solution. Using the maximum principle one will often be able to improve the two inequalities so obtained, and arrive in this way at an inequality for the F.B.I.-transform which corresponds to additional regularity: the regularity will have propagated. Here we shall prefer another approach, which is by duality and which is based on ideas from Liess [2,3]. One of the reasons for preferring the present approach is that it fits well with definition 1.2.4 so that in particular we shall be able to use results from Liess-Rodino [1] to obtain that much partial regularity which we need. The main step in the present approach is to show that plurisubharmonic functions with certain properties do exist.

2. We start by recalling the following result:

Lemma 2.5.1. *For any $d' > 0$ we can find $d > 0$, c, and a plurisubharmonic function $\psi : C^n \to R$ such that*

$$min(2d|Re\eta|, Im\,\tau_+) \leq \psi(\zeta) \leq min(d'|Re\eta|, Im\,\tau_+) + d'|Im\,\zeta| + d'|Re\tau| + c. \quad (2.5.1)$$

Here we have denoted the variables from C^n by $\zeta = (\eta, \tau)$, where $\eta = (\zeta_1, ...\zeta_{n-1})$, $\tau = \zeta_n$. Lemma 2.5.1 is proved in Liess [2]. The partial regularity mentioned before will be needed to deal with the term $d'|Re\,\tau|$, whereas the term $Im\,\tau_+$ will lead to propagation. Of course, the conclusion remains also true if in the extreme terms of (2.5.1) we replace $|Re\,\eta|$ by $|Re\,\zeta|$. (What is missing is essentially $|Re\,\tau|$. We can avail ourselves of the fact that this term appears in the right most term in (2.5.1).) Since the lemma is central in all propagation results in this paper, we shall give a proof of it in appendix A.2. The fact that in lemma 2.5.1 we work with $2d$ rather than with d is of course for later convenience. From lemma 2.5.1 we can deduce

Lemma 2.5.2. *For any $0 \leq d' < 1$ we can find d, c, such that for any $\zeta^0 \in R^n$ and any a there is a plurisubharmonic function $\varphi : \{C^n; |Re\zeta - \zeta^0| < d'a\} \to R$ such that*

$$min(da, Im\,\tau_+) \leq \quad \varphi(\zeta) \quad \leq min(d'a, Im\,\tau_+) + d'|Im\,\zeta| + d'|Re\,\tau| + c',$$
$$\text{for } |Re\,\zeta - \zeta^0| < d'a. \quad (2.5.2)$$

Proof. We choose ψ with (2.5.1) for $|Re\eta|$ replaced by $|Re\zeta|$ and fix some $\tilde{\zeta} \in R^n$ with $|\tilde{\zeta}| = 2a$. Next we set $\varphi(\zeta) = \psi(\zeta - \zeta^0 + \tilde{\zeta})$. When $|Re\,\zeta - \zeta^0| < d'a$ we have in view of $d' < 1$ that

$$a \leq 2a - d'a \leq |Re\zeta - \zeta^0 + \tilde{\zeta}| \leq d'a + 2a \leq 3a,$$

so the desired conclusion follows from

$$min(d|Re\,\zeta - \zeta^0 + \tilde{\zeta}|, Im\,\tau_+) \leq \varphi(\zeta) \leq min(d'|Re\,\zeta - \zeta^0 - \tilde{\zeta}|, Im\,\tau_+)$$
$$+d'|Im\,\zeta| + d'|Re\,\tau| + c'.$$

For completeness we mention the following three results, which can be obtained from lemma 2.5.1 in the same way in which we shall argue later on to deduce proposition 2.5.10 from proposition 2.5.9.

Proposition 2.5.3. *For any* $0 \leq d' < 1$ *there are* $d > 0, c'$ *such that for any* $\zeta^0 \in R^n$ *and any* $a > 0$ *we can find a plurisubharmonic function* $\varphi_1 : C^n \to R$ *which satisfies*

$$min(da, Im\,\tau_+) \leq \varphi_1(\zeta), \text{ if } |Re\,\zeta - \zeta^0| < \frac{d'}{2}a,$$

$$\varphi_1(\zeta) \leq min(d'a, Im\,\tau_+) + d'|Re\,\tau| + d'|Im\,\zeta| + c', \text{ if } |Re\,\zeta - \zeta^0| < d'a,$$

$$\varphi_1(\zeta) \leq d'|Im\,\zeta| + d'|Re\,\tau| + c', \text{ if } |Re\,\zeta - \zeta^0| > d'a.$$

Corollary 2.5.4. *Let* $g : R^n \to R_+$ *be Lipschitzian and consider* $d' > 0$. *Then there are* $d > 0, c > 0$ *and a plurisubharmonic function* χ *such that*

$$min(dg(Re\,\zeta), Im\,\zeta_+) \leq \chi(\zeta) \leq min(d'g(Re\,\zeta), Im\,\tau_+) + d'|Re\,\tau| + d'|Im\,\zeta| + c.$$

The interpretation of the corollary is that "partial analyticity" propagates any type of G_φ-regularity. (This should be compared with the results in section III.4 in Lebeau [1].)

Corollary 2.5.5. *Let* g *be as before and assume* $\psi(\zeta) \geq |Re\tau|$ *on* $\{$ supp $g\}_{c|\tau|}$. *For every* $d' > 0$ *we can then find* d, c *and a plurisubharmonic function* φ *such that*

$$min(dg(Re\,\zeta), Im\,\tau_+) \leq \varphi(\zeta) \leq min(d'g(Re\,\zeta), Im\,\tau_+) + d'\psi(\zeta) + d'|Im\,\zeta| + c.$$

($|\tau|$ stands in the notation $\{$ supp $\}_{c|\tau|}$ for the function $\zeta \to |\tau|$.)

Often one can weaken the condition that $\psi \geq |Re\tau|$ on all of $\{$supp $g\}_{c|\tau|}$ in corollary 2.5.5. We prove here a result of this type for the case in which we are directly interested, namely when g comes from a semi-isotropic wave front set. This is based on the following result :

Proposition 2.5.6. *Let* $1 \leq s < n$ *be given and denote by* $\theta = (\zeta_1, ..., \zeta_s)$. *Also consider* $\beta > 0$. *Then there are positive constants* c_i *and a plurisubharmonic function* χ *defined on the set* $U = \{\zeta\,;\,|Im\,\zeta| < c_1|Re\zeta|\}$ *such that*

$$-c_2|Re\theta|^{\beta+1}/|Re\,\zeta|^\beta \leq \chi(\zeta) \leq -|Re\theta|^{\beta+1}/|Re\,\zeta|^\beta + c_3|Im\,\zeta| + c_4 \text{ if } \zeta \in U, \quad (2.5.3)$$

Note that (2.5.3) just gives

$$-c_5|Re\zeta| \leq \chi(\zeta) \leq -c_6|Re\zeta| + c_3|Im\,\zeta| + c_4 \text{ if } \zeta \in U \text{ and } |Re\,\theta| > c_7|Re\,\zeta|, \quad (2.5.4)$$

for some constants c_5, c_6, if c_7 has been previously fixed. The term $|Re\,\theta|^{\beta+1}/|Re\,\zeta|^\beta$ is however much finer tuned near $Re\,\theta = 0$ than is $|Re\,\zeta|$.

Remark 2.5.7. *The preceding proposition follows for the case $s = n - 1$ immediately from a result proved in Liess [3]. Indeed, in that paper it is proved that there is a plurisubharmonic function λ on U such that with the notation $\zeta' = (\zeta_1, ..., \zeta_{n-1})$,*

$$-c_2|Re\zeta'|^{\beta+1}/|Re\zeta|^\beta \leq \lambda(\zeta) \leq -|Re\zeta'|^{\beta+1}/|Re\zeta|^\beta + c_3|Im\,\zeta| + c_4 \text{ if } \zeta \in U \text{ and } Re\,\zeta_n \geq 0,$$

$$-c_6|Re\,\zeta| \leq \lambda(\zeta) \leq -|Re\,\zeta| + c_3|Im\,\zeta| + c_4 \text{ if } \zeta \in U \text{ and } Re\,\zeta_n < 0.$$

The function $\chi = \max(\lambda(\zeta), \lambda(-\zeta))$ satisfies then the conclusions from the proposition for the case $s = n - 1$. One can then obtain the case of a general s from the case when $s = n - 1$ in the following way: let us choose some $n - 1$ dimensional linear subspace W of R^{n-1}, denote by Π_W the orthogonal projection from C^n onto the complexification $\tilde{W}$ of W in C^n and by χ_W a plurisubharmonic function on C^n associated with W as in the preceding proposition. Note that the existence of such a function is a consequence of Liess [3], since now we are in the situation that W has co-dimension one. We thus have

$$-c_2|\Pi_W(Re\zeta)|^{\beta+1}/|Re\zeta|^\beta \leq \chi_W(\zeta) \quad \leq \quad -|\Pi_W(Re\zeta)|^{\beta+1}/|Re\zeta|^\beta + c_3|Im\,\zeta| + c_4$$

$$if \ \zeta \in U. \qquad (2.5.5)$$

Now we define χ to be the upper regularization of $\sup \chi_W$ where the sup is for all $(n-1)-$dimensional subspaces W of R^n which contain R^s. If ζ is fixed, we can then find W of codimension one with $C^s \subset \tilde{W}$ so that $|\Pi_W(Re\,\zeta)| = |Re\,\theta|$ and in general we have $|\Pi_W(Re\,\zeta| \geq |Re\,\theta|$. We thus obtain e.g. that

$$\chi_W(\zeta) \leq \max_{W, C^s \subset W} -|\Pi_W(Re\zeta)|^{\beta+1}/|Re\zeta|^\beta + c_4|Im\,\zeta| + c_5$$

$$= -|Re\theta|^{\beta+1}/|Re\,\zeta|^\beta + c_4|Im\,\zeta| + c_5.$$

and a similar argument gives the first inequality in (2.5.3).

In the next proposition we assume that a finite sequence of subspaces M_j and of projections Π_j is given as in the definition of the k-wave front set.

Proposition 2.5.8. *There are constants C_i, C', such that for any C and any j we can find a plurisubharmonic function $f : \{\zeta \in C^n; |\Pi_j(Im\,\zeta)| \leq C_1|\Pi_j(Re\,\zeta)|\} \to R$ such that*

$$f(\zeta) \leq C'|Im\,\zeta| + C', \qquad\qquad (2.5.6)$$

$$\begin{aligned} f(\zeta) \quad\leq\quad &-|\Pi_{j+1}(Re\zeta)|^{\beta+1}/|\Pi_j(Re\zeta)|^\beta + C'|Im\,\zeta| + C', \\ &\text{if } C \leq |\Pi_{j+1}(Re\zeta)|/|\Pi_j(Re\zeta)|, \qquad (2.5.7) \end{aligned}$$

$$f(\zeta) \geq 0 \quad \text{if } |\Pi_{j+1}(Re\zeta)| \leq C|\Pi_j(Re\zeta)|/2. \qquad (2.5.8)$$

The proposition is very close to proposition 3.2 in Liess [3]. It can be reduced to proposition 2.5.6 in the same way in which proposition 3.3 from Liess [3] was reduced to proposition 3.2 in that paper. Let us in fact denote by S some orthogonal matrix acting on M_j and by χ_S a plurisubharmonic function on $\{\zeta \in C^n; |\Pi_j(Im\zeta)| \leq c|\Pi_j(Re\zeta)|\}$ for which

$$-c_2|\Pi_{j+1}(S(Re\zeta))|^{\beta+1}/|\Pi_j(Re\zeta)|^\beta \leq \chi_S(\zeta) \leq -|\Pi_{j+1}(S(Re\zeta))|^{\beta+1}/|\Pi_j(Re\zeta)|^\beta +$$

$$c_3|Im\,\zeta| + c_4 \text{ if } |\Pi_j(Im\,\zeta)| \leq c|\Pi_j(Re\zeta)|.$$

We then define f to be the upper regularization of

$$\sup_{|I-S|<\varepsilon} \chi_S(\zeta),$$

where ε is suitable and I is the identity matrix. We omit further details. (Cf. anyway the analoguous situation in Liess [3]. Note also that $f(\zeta) \geq \chi_I(\zeta) \geq 0$ if $\Pi_{j+1}(Re\,\zeta) = 0$.)

Proposition 2.5.9. *Consider $x^0 \in M_{k-1}, |x^0| = 1$, and denote by $\tau = \langle x^0, \zeta \rangle$. Also let $\beta > 0,\ 0 \leq \delta < 1,\ c > 0$ and $d' > 0$ be given.*
Then there are $c_i,\ i = 1, 2, 3, 4, 5,\ d$ and a plurisubharmonic function ψ defined on

$$|Im\,\zeta| \leq c_1|\Pi_{k-2}(Re\zeta)|$$

such that

$$min(d|\Pi_{k-2}(Re\,\zeta)|, Im\,\tau_+/2) \leq \psi(\zeta) \qquad\qquad (2.5.9)$$

if simultaneously

$$|\Pi_{k-1}(Re\,\zeta)| \geq c_2|\Pi_{k-2}(Re\,\zeta)|^{1+\beta}/|\Pi_{k-3}(Re\,\zeta)|^\beta,$$

$$|\Pi_{k-1}(Re\,\zeta)| \geq c_2(1 + |\Pi_{k-2}(Re\,\zeta)|^\delta),$$

$$|\Pi_{k-2}(Re\zeta)| < c_3|\Pi_{k-3}(Re\zeta)|,$$

respectively

$$\psi(\zeta) \leq min(d'|\Pi_{k-2}(Re\,\zeta)|, Im\,\tau_+) + d'|Re\tau| + c_4 d'|Im\,\zeta| + c_5, \qquad (2.5.10)$$

if simultaneously

$$|\Pi_{k-1}(Re\,\zeta)| \geq c|\Pi_{k-2}(Re\,\zeta)|^{1+\beta}/|\Pi_{k-3}(Re\,\zeta)|^{\beta},$$

$$|\Pi_{k-1}(Re\,\zeta)| \geq c(1 + |\Pi_{k-2}(Re\,\zeta)|^{\delta}),$$

and, finally,

$$\psi(\zeta) \leq min(d'|\Pi_{k-2}(Re\,\zeta)|, Im\,\tau_+) + c_4 d'|Im\,\zeta| + c_5, \qquad (2.5.11)$$

if

$$|\Pi_{k-1}(Re\,\zeta)| \leq c|\Pi_{k-2}(Re\,\zeta)|^{1+\beta}/|\Pi_{k-3}(Re\,\zeta)|^{\beta},$$

or at least

$$|\Pi_{k-1}(Re\,\zeta)| \leq c(1 + |\Pi_{k-2}(Re\,\zeta)|^{\delta}).$$

Moreover, the constants c_1, c_4 do not depend here on d'.

Proof of proposition 2.5.9. It follows from lemma 2.5.1 that we can find d, c_6 and a plurisubharmonic function ψ_1 for which

$$min(2d|\Pi_{k-2}(Re\,\zeta)|, Im\,\tau_+) \;\leq\; \psi_1(\zeta) \leq (min(d'|\Pi_{k-2}(Re\,\zeta)|, Im\,\tau_+) + d'|Im\,\zeta| + \\ d'|Re\,\tau| + c_6.$$

The function ψ_1 is our first step in the construction of ψ, and in fact, at this moment, the only problem is with (2.5.11). Next we observe that, as a consequence of proposition 2.5.6, we can find c_1, c_7, c_8, c_9, and a plurisubharmonic function χ on $|Im\,\zeta| < c_1|\Pi_{k-2}(Re\,\zeta)|$ such that

$$-c_7|\Pi_{k-2}(Re\,\zeta)|^{1+\beta}/|\Pi_{k-3}(Re\,\zeta)|^{\beta} \;\leq\; \chi(\zeta) \leq -|\Pi_{k-2}(Re\,\zeta)|^{1+\beta}/|\Pi_{k-3}(Re\,\zeta)|^{\beta} + \\ c_8|Im\,\zeta| + c_9.$$

Further, we consider a plurisubharmonic function κ on C^n such that

$$-c_{10}|\Pi_{k-2}(Re\,\zeta)|^{\delta} \leq \kappa(\zeta) \leq -|\Pi_{k-2}(Re\,\zeta)|^{\delta} + c_{11}|Im\,\zeta| + c_{12}.$$

The existence of such functions κ is proved in Liess-Rosu [1]. (It is essentially a consequence of the fact that one can find Gevrey-type functions with compact support.) Our second guess for a function ψ as desired is then

$$\psi_2 = \psi_1 + cd'\chi + cd'\kappa.$$

It is here immediate that ψ_2 satisfies (2.5.10), (2.5.11), but now we have problems with (2.5.9), in that instead of (2.5.9) we only have

$$\psi_2(\zeta) \geq min(2d|\Pi_{k-2}(Re\,\zeta)|, Im\,\tau_+) - cc_7d'|\Pi_{k-2}(Re\,\zeta)|^{1+\beta}/|\Pi_{k-3}(Re\,\zeta)|^{\beta}$$
$$-cc_{10}d'|\Pi_{k-2}(Re\,\zeta)|^{\delta}. \qquad (2.5.12)$$

Fortunately, this is more or less what we need when $2d|\Pi_{k-2}(Re\,\zeta)| = Im\,\tau_+$. In fact, when $|\Pi_{k-2}(Re\,\zeta)| \leq c_3|\Pi_{k-3}(Re\,\zeta)|$ and c_3 is chosen suitably, depending on the values of d, d', c, c_7, then we may assume that

$$(d/2)|\Pi_{k-2}(Re\,\zeta)| \geq cc_7d'|\Pi_{k-2}(Re\,\zeta)|^{1+\beta}/|\Pi_{k-3}(Re\,\zeta)|^{\beta}.$$

Similarily, we may assume that c_{13} is chosen so large that $|\Pi_{k-2}(Re\,\zeta)| \geq c_{13}$ implies

$$(d/2)|\Pi_{k-2}(Re\,\zeta)| \geq cc_{10}d'|\Pi_{k-2}(Re\,\zeta)|^{\delta}.$$

To use this, we apply proposition 2.5.8 and obtain a plurisubharmonic function f on $|Im\,\zeta| < c_1|\Pi_{k-2}(Re\,\zeta)|$ such that

$$f(\zeta) \geq 0 \text{ if } |\Pi_{k-2}(Re\,\zeta)| \leq (c_3/2)|\Pi_{k-3}(Re\,\zeta)|,$$

$$f(\zeta) \leq -|\Pi_{k-2}(Re\,\zeta)|^{1+\beta}/|\Pi_{k-3}(Re\,\zeta)|^{\beta} + c_{14}|Im\,\zeta| + c_{15}$$

if $|\Pi_{k-2}(Re\,\zeta)| \geq c_3|\Pi_{k-3}(Re\,\zeta)|$, and

$$f(\zeta) \leq c_{14}|Im\,\zeta| + c_{15}.$$

It is then clear from above that

$$\psi_2(\zeta) + c_{16}d'|Im\zeta| + c_{17} \geq d|\Pi_{k-2}(Re\,\zeta)| + d'cc_7f(\zeta),$$

if $2d|\Pi_{k-2}(Re\,\zeta)| = Im\,\tau_+$ and c_{16}, c_{17} are large. It follows that the function ψ defined for $|Im\,\zeta| < c_1|\Pi_{k-2}(Re\,\zeta)|$ by

$$\psi(\zeta) = max[\psi_2(\zeta) + c_{16}d'|Im\zeta| + c_{17}, d|\Pi_{k-2}(Re\,\zeta)| + cc_7d'f(\zeta)]$$

if $2d|\Pi_{k-2}(Re\,\zeta)| \leq Im\,\tau_+$, and by

$$\psi(\zeta) = max[\psi_2 + c_{16}d'|Im\,\zeta| + c_{17}, Im\,\tau_+ + cc_7d'f(\zeta)],$$

for the remaining ζ is plurisubharmonic. It is easy to see that it satisfies the desired properties.

Proposition 2.5.10. *Let* $\xi^j \in \dot{M}_j$, $j = 0, ..., k - 2$, $\beta > 0$, $0 \leq \delta < 1$, C_1 *and a multineighborhood* G' *of* $(\xi^0, \xi^1, ..., \xi^{k-2})$ *be given. Denote by*

$$
\begin{aligned}
A \quad = \quad &\{\xi \in G'; |\Pi_{j+1}(\xi)| \geq C_1 |\Pi_j(\xi)|^{1+\beta} / |\Pi_{j-1}(\xi)|^\beta, j = 1, ..., k - 2, \\
&|\Pi_i(\xi)| \geq C_1 (1 + |\Pi_{i-1}(\xi)|^\delta), i = 1, ..., k - 1\}.
\end{aligned}
$$

If $d', x^0 \in M_{k-1}, |x^0| = 1$ *are fixed, we can find* $d > 0, C_i, i = 2, 3, 4, 5$, *a multineighborhood* G *of* $(\xi^0, \xi^1, ..., \xi^{k-2})$ *and a plurisubharmonic function* φ *defined on*

$$
B = \{\zeta \in C^n; |Im\,\zeta| < C_2 |\Pi_{k-2}(Re\zeta)|\}
$$

such that

$$
min(d|\Pi_{k-2}(Re\,\zeta)|, \langle x^0, Im\,\zeta \rangle_+ / 2) \leq \varphi(\zeta), \text{ if } \zeta \in B, Re\,\zeta \in D, \tag{2.5.13}
$$

$$
\begin{aligned}
D \quad = \quad &\{\zeta \in G; |\Pi_{j+1}(\xi)| \geq C_3 |\Pi_j(\xi)|^{1+\beta} / |\Pi_{j-1}(\xi)|^\beta, j = 1, ..., k - 2, \\
&|\Pi_i(\xi)| \geq C_3 (1 + |\Pi_{i-1}(\xi)|^\delta), i = 1, ..., k - 1\},
\end{aligned}
$$

$$
\varphi(\zeta) \leq min(d'|\Pi_{k-2}(Re\,\zeta)|, \langle x^0, Im\,\zeta \rangle_+) + d'|\Pi_{k-1}(Re\,\zeta)| + C_4 d'|Im\,\zeta| + C_5, \tag{2.5.14}
$$

if $\zeta \in B, Re\,\zeta \in A$, *and*

$$
\varphi(\zeta) \leq d'C_4 |Im\,\zeta| + C_5, \text{ if } \zeta \in B, Re\,\zeta \notin A. \tag{2.5.15}
$$

Remark 2.5.11 *In the case* $k = 2$ *the conditions* $|\Pi_{j+1}(\xi)| \geq C_i |\Pi_j(\xi)|^{1+\beta} / |\Pi_{j-1}(\xi)|^\beta$ *in the definition of* A *and* D *are to be considered void. The proof of proposition 2.5.10 is then considerably shorter and in particular no reference to the results in Liess [3] is needed. This is in so far interesting that this case corresponds to the standard micro-Holmgren theorem.*

Proof of proposition 2.5.10. We start from a plurisubharmonic function ψ which is as in proposition 2.5.9. This gives us as a value for $C_2, C_2 = c_1$, where c_1 is from that proposition. What remains to be done is essentially to microlocalize ψ to the set A. (The property which makes this possible is that $\psi(\zeta) \leq 2d'|\Pi_{k-2}(Re\,\zeta)| + c_4 d'|Im\,\zeta| + c_5$.) To do so, we choose a multineighborhood G of $(\xi^0, \xi^1, ..., \xi^{k-2})$ with $G \subset\subset G'$ and denote by

$$
\begin{aligned}
E \quad = \quad &\{\xi \in G; |\Pi_{j+1}(\xi)| \geq C_1 |\Pi_j(\xi)|^{1+\beta} / |\Pi_{j-1}(\xi)|^\beta, j = 1, ..., k - 3, \\
&|\Pi_i(\xi)| \geq C_1 (1 + |\Pi_{i-1}(\xi)|^\delta), i = 1, ..., k - 2\},
\end{aligned}
$$

$$F = \{\xi \in G'; |\Pi_{j+1}(\xi)| \geq 2C_1|\Pi_j(\xi)|^{1+\beta}/|\Pi_{j-1}(\xi)|^{\beta}, j = 1, ..., k-3,$$
$$|\Pi_i(\xi)| \geq 2C_1(1 + |\Pi_{i-1}(\xi)|^{\delta}), i = 1, ..., k-2\}.$$

Note that the main difference between the sets A, D on one hand and E, F on the other, is that the defining conditions refer to different sets of indices. However, the conditions in the definition of A, D which do not appear in E, F are precisely the ones which come in in proposition 2.5.9.

Also observe that $F_{c|\Pi_{k-2}(\xi)|} \subset E$ if $c > 0$ is small. We shall now construct for each point $\theta \in F$ a plurisubharmonic function ψ_θ starting from ψ :

$$\psi_\theta(\zeta) = \psi(\zeta) + d'h(\zeta - \theta, c|\theta|)$$

where h is a function of the variables $(\zeta, t) \in C^n \times R_+$ which is plurisubharmonic in ζ for each fixed t and satisfies the inequalities :

$$h(\zeta, t) \leq C_6|Im\,\zeta| + C_7,$$

$$h(i\xi, t) \geq 0, \text{if } \xi \in R^n,$$

$$h(\zeta, t) \leq -|Re\,\zeta| + C_6|Im\,\zeta| + C_7 \text{ if } t/2 \leq |Re\,\zeta| \leq t.$$

Cf. here lemma 2.3.4. It follows from the properties of ψ and h that

$$\psi_\theta \leq C_8 d'|Im\,\zeta| + C_9, \text{if } (c/2)|\Pi_{k-2}(\theta)| \leq |Re\,\zeta - \theta| \leq c|\Pi_{k-2}(\theta)|.$$

If we define φ_θ on $\{\zeta : |Im\,\zeta| < C_2|\Pi_{k-2}(Re\,\zeta)|\}$ by

$$\varphi_\theta(\zeta) = \begin{cases} max[\psi_\theta(\zeta)\,,\,C_8|Im\,\zeta| + C_9], \text{ if } |Re\,\zeta - \theta| < c|\Pi_{k-2}(\theta)|, \\ \\ C_8|Im\,\zeta| + C_9, \text{for the remaining } \zeta, \end{cases}$$

then φ_θ is plurisubharmonic.

Finally, we define φ to be the upper regularization of $H(\zeta) = sup_{\theta \in F}\varphi_\theta(\zeta)$. It follows from the construction that

$$\varphi(\zeta) \leq C_8|Im\,\zeta| + C_9 \text{ if } Re\,\zeta \notin E,$$

which is (2.5.15) for such points. To prove (2.5.15) for the point $Re\,\zeta \in E \setminus A$, observe that for these points we have

$$|\Pi_{k-1}(Re\,\zeta)| \leq C_1|\Pi_{k-2}(Re\,\zeta)|^{1+\beta}/|\Pi_{k-3}(Re\,\zeta)|^{\beta},$$

or

$$|\Pi_{k-1}(Re\,\zeta)| \leq C_1(1 + |\Pi_{k-2}(Re\,\zeta)|^{\delta}).$$

The fact that (2.5.15) is valid is then a consequence of the inequalities for ψ. No problems appear here with (2.5.14) and (2.5.13) is valid since φ is larger than ψ for $Re\,\zeta \in F$.

Remark 2.5.12. *Proposition 2.5.10 remains valid if we replace everywhere B by C^n. (By this we mean that φ is now defined on C^n instead of on B, etc.) Let indeed, φ be a plurisubharmonic function as in proposition 2.5.10. It follows in particular from (2.5.14) and (2.5.15) that for suitable c, c',*

$$\varphi(\zeta) \leq cd'|Im\,\zeta| + c', if \; \zeta \in B \; and \; |Im\,\zeta| \geq (C_2/2)|\Pi_{k-2}(Re\,\zeta)|.$$

The function φ' defined on C^n by $\varphi'(\zeta) = max(\varphi(\zeta), cd'|Im\,\zeta| + c')$ if $\zeta \in B$ and by $\varphi'(\zeta) = cd'|Im\,\zeta| + c'$ for $\zeta \notin B$ is then plurisubharmonic and satisfies the desired properties.

We also mention here the following result, which will be used in a similar way in which proposition 2.5.8 is used:

Proposition 2.5.13. *Let $c > 0$ and $\beta > 0$ be given. Then we can find c_1, c_2 and a plurisubharmonicfunction $\varphi : \{\zeta \in C^n; \; |Im\,\zeta| < |\Pi_1(Re\,\zeta)|\} \to R$ such that*

$$\varphi(\zeta) \geq 0 \; if \; |\Pi_2(Re\,\zeta)| > c|\Pi_1(Re\,\zeta)|^{1+\beta}/|Re\,\zeta|^\beta$$

$$\varphi(\zeta) \leq c_1|Im\,\zeta| + c_2, \; if \; |\Pi_2(Re\,\zeta)| > (c/2)|\Pi_1(Re\,\zeta)|^{1+\beta}/|Re\,\zeta|^\beta$$

$$\varphi(\zeta) \leq -|\Pi_1(Re\,\zeta)|^{1+\beta}/|Re\,\zeta|^\beta + c_1|Im\zeta| + c_2, \; if \; |\Pi_2(Re\,\zeta)| \leq (c/2)|\Pi_1(Re\,\zeta)|^{1+\beta}$$
$$/|Re\,\zeta|^\beta.$$

The proof of this result is similar to that of proposition 2.5.8 and we omit the details.

2.6   An improvement of proposition 2.5.10

1. For technical reasons it is convenient to eliminate the factor $1/2$ in $\langle x^0, Im\,\zeta\rangle_+/2$ from inequality (2.5.13). We shall do so with an adaptation of an argument used in Liess [3]. To simplify the terminology we shall consider the following assertion $A(\sigma)$, in which σ is a fixed number with $0 < \sigma \leq 1$.

$A(\sigma)$: for every

$$g' \in G(\xi^0, \xi^1, ..., \xi^{k-2}, \delta, \beta), r' \in \mathcal{R}(\xi^0, \xi^1, ..., \xi^{k-2}, \delta, \beta), \; and \; d' > 0,$$

we can find

$$g \in G(\xi^0, \xi^1, ..., \xi^{k-2}, \delta, \beta), r \in \mathcal{R}(\xi^0, \xi^1, ..., \xi^{k-2}, \delta, \beta), d > 0, c > 0$$

and a plurisubharmonic function $\varphi : C^n \to R_+$ such that

$$\min(dg(-\operatorname{Re}\zeta), \sigma \langle x^0, \operatorname{Im}\zeta \rangle_+) + dr(-\operatorname{Re}\zeta) \le \varphi(\zeta) \le$$

$$\min(d'g'(-\operatorname{Re}\zeta), \langle x^0, \operatorname{Im}\zeta \rangle_+) + d'r'(-\operatorname{Re}\zeta) + d'|\operatorname{Im}\zeta| + c. \quad (2.6.1)$$

To simplify notations it shall be understood later on in this section that quantities: of type $g, g', \tilde{g}$ are always in $G(\xi^0, \xi^1, ..., \xi^{k-2}, \delta, \beta)$, of type $r, r', \tilde{r}$ in $\mathcal{R}(\xi^0, \xi^1, ..., \xi^{k-2}, \delta, \beta)$, of type $\varphi, \varphi, \tilde{\varphi}$ are plurisubharmonic functions, and of type $d, d', \tilde{d}$ are (strictly) positive numbers.

2. Our main goal in this section is to show that $A(1)$ is true. As is immediate, this is a consequence of the following three lemmas.

Lemma 2.6.1. $A(1/2)$ *is true.*

Lemma 2.6.2. *If $A(\sigma)$ holds for σ arbitrarily close to σ^0, then also $A(\sigma^0)$ is true.*

Lemma 2.6.3. *Fix $\sigma' < 1$ and denote*

$$\nu = (1 - \frac{9}{16}\sigma' + \frac{9}{16}).$$

If $A(\sigma)$ is valid and $\nu\sigma \le 1$, then also $A(\nu\sigma)$ is true.

3. **Proof of lemma 2.6.1.** The lemma is a consequence of proposition 2.5.10, respectively of the remark following it. Indeed, given d', g', r', it follows from that proposition that we can find d, g, r, c and a plurisubharmonic function $\psi : C^n \to R$ so that

$$\min(dg(-\operatorname{Re}\zeta), \langle x^0, \operatorname{Im}\zeta \rangle_+/2) + dr(-\operatorname{Re}\zeta) \le \varphi(\zeta) \le$$

$$\min(d'g'(-\operatorname{Re}\zeta), \langle x^0, \operatorname{Im}\zeta \rangle_+) + d'r'(-\operatorname{Re}\zeta)/2 + d'|\operatorname{Im}\zeta|/2 + c.$$

It is no loss of generality to assume that $r'(-\operatorname{Re}\zeta)$ is Lipschitzian, so we can find a plurisubharmonic function ρ so that

$$(d'/4)r'(-\operatorname{Re}\zeta) \le \rho(\zeta) \le (d'/2)r'(-\operatorname{Re}\zeta) + (d'/2)|\operatorname{Im}\zeta| + c_1.$$

The function $\varphi = \psi + \rho$ is then plurisubharmonic and satisfies (2.6.1) for $\sigma = 1/2$ (with $r = r'$), provided we shrink d until $d \le d'/4$.

Remark 2.6.4. *It is clear from the proof of lemma 2.6.1 that the term $dr(-\operatorname{Re}\zeta)$ from the left hand side of (2.6.1) has no deep meaning. It is however useful to make the iterative argument $A(1/2) \to A(1)$ smoother.*

4. **Proof of lemma 2.6.2.** We consider σ^0 as in the statement and want to verify the conclusion of the lemma for some d', g', r'. We shall then denote $\sigma = \sigma^0 - d'/2$, assuming as we may that $\sigma > 0$. Applying the assertion for σ we can find d, g, r and a plurisubharmonic function ψ so that

$$\min(dg(-\operatorname{Re}\zeta), \sigma\langle x^0, \operatorname{Im}\zeta\rangle_+) + dr(-\operatorname{Re}\zeta) \leq \psi(\zeta) \leq$$
$$\min(d'(-\operatorname{Re}\zeta), \langle x^0, \operatorname{Im}\zeta\rangle_+) + d'r'(-\operatorname{Re}\zeta) + d'|\operatorname{Im}\zeta|/2 + c.$$

The plurisubharmonic function $\varphi = \psi + (d'/2)\langle x^0, \operatorname{Im}\zeta\rangle_+$ satisfies then (2.6.1) for σ^0.

5. **Proof of lemma 2.6.3.** Let us assume that $A(\sigma)$ holds and fix a positive number λ so that

$$\lambda(1 - \frac{9}{16}\sigma') \leq \frac{1}{4}.$$

We also denote

$$\mu = (1 - \frac{9}{16}\sigma'),$$

and consider d', g', r'. Applying $A(\sigma)$ twice, we can find $\tilde{d}, \tilde{g}, \tilde{r}, \varphi', \tilde{c}$ and $d, g, r, \tilde{\varphi}, c$ so that

$$\min(\tilde{d}\tilde{g}(-\operatorname{Re}\zeta), \sigma\langle x^0, \operatorname{Im}\zeta\rangle_+) + \tilde{d}\tilde{r}(-\operatorname{Re}\zeta) \leq \varphi'(\zeta) \leq \quad (2.6.2)$$
$$\min(d'g'(-\operatorname{Re}\zeta), \langle x^0, \operatorname{Im}\zeta\rangle_+) + d'r'(-\operatorname{Re}\zeta)/2 + d'|\operatorname{Im}\zeta|/2 + c',$$

and

$$\min((d/\mu)g(-\operatorname{Re}\zeta), \langle x^0, \operatorname{Im}\zeta\rangle_+/2) + (d/\mu)r(-\operatorname{Re}\zeta) \leq \quad (2.6.3)$$
$$\tilde{\varphi}(\zeta) \leq \min(\lambda\tilde{d}\tilde{g}(-\operatorname{Re}\zeta), \langle x^0, \operatorname{Im}\zeta\rangle_+) + \tilde{d}\tilde{r}(-\operatorname{Re}\zeta)/2 + \tilde{d}|\operatorname{Im}\zeta|/2 + \tilde{c}.$$

It is of course no loss in generality to assume that

$$d \leq \tilde{d} \leq d'/4,\ g(\xi) \leq \tilde{g}(\xi) \leq g'(\xi),\ r(\xi) \leq \tilde{r}(\xi) \leq r'(\xi),\ \forall\xi, \quad (2.6.4)$$

and that $\tilde{r}$ is Lipschitzian. We can then in particular find a plurisubharmonic function $\tilde{\rho}$ on C^n so that

$$\tilde{r}(-\operatorname{Re}\zeta) \leq \tilde{\rho}(\zeta) \leq 2\tilde{r}(-\operatorname{Re}\zeta) + c_1|\operatorname{Im}\zeta| + c_2.$$

Finally, we shall assume that

$$c_1\tilde{d} \leq d'/2. \quad (2.6.5)$$

We now introduce the function $\varphi : C^n \to R$ by

$$\varphi(\zeta) = \varphi'(\zeta) + \tilde{d}\tilde{\rho}(\zeta) + \tilde{d}|\operatorname{Im}\zeta| + c_3 \text{ if } \tilde{d}\tilde{g}(-\operatorname{Re}\zeta) \leq (3\sigma/4)\langle x^0, \operatorname{Im}\zeta\rangle_+,$$

$$\varphi(\zeta) = \max(\varphi'(\zeta) + \tilde{d}\tilde{\rho}(\zeta) + \tilde{d}|\mathrm{Im}\,\zeta| + c_3, \mu\tilde{\varphi}(\zeta) + (9/16)\sigma\langle x^0, \mathrm{Im}\,\zeta\rangle_+)$$
$$\text{if } \tilde{d}\tilde{g}(-\mathrm{Re}\,\zeta) > (3\sigma/4)\langle x^0, \mathrm{Im}\,\zeta\rangle_+.$$

We claim that if c_3 is large enough then

I. φ is plurisubharmonic.

II. $\min(dg(-\mathrm{Re}\,\zeta), \nu\sigma\langle x^0, \mathrm{Im}\,\zeta\rangle_+) + dr(-\mathrm{Re}\,\zeta) \leq \varphi(\zeta).$

III. $\varphi(\zeta) \leq \min(d'g'(-\mathrm{Re}\,\zeta), \langle x^0, \mathrm{Im}\,\zeta\rangle_+) + d'r'(-\mathrm{Re}\,\zeta) + d'|\mathrm{Im}\,\zeta| + c_4.$

Clearly, this will conclude the proof of lemma 2.6.3.

6. Proof of I. It suffices to show that for c_3 large enough

$$\mu\tilde{\varphi}(\zeta) + \frac{9}{16}\sigma\langle x^0, \mathrm{Im}\,\zeta\rangle_+ \leq \varphi'(\zeta) + \tilde{d}\tilde{\rho}(\zeta) + \tilde{d}|\mathrm{Im}\,\zeta| + c_3 \qquad (2.6.6)$$

if

$$\tilde{d}\tilde{g}(-\mathrm{Re}\,\zeta) = \frac{3\sigma}{4}\langle x^0, \mathrm{Im}\,\zeta\rangle_+. \qquad (2.6.7)$$

To see that this is so, we note that by evaluating the minimums from (2.6.2) and (2.6.3) it follows for ζ satisfying (2.6.7) that

$$\tilde{d}\tilde{g}(-\mathrm{Re}\,\zeta) + \tilde{d}\tilde{r}(-\mathrm{Re}\,\zeta) \leq \varphi'(\zeta), \qquad (2.6.8)$$

$$\tilde{\varphi}(\zeta) \leq \lambda\tilde{d}\tilde{g}(-\mathrm{Re}\,\zeta) + \tilde{d}\tilde{r}(-\mathrm{Re}\,\zeta) + \tilde{d}|\mathrm{Im}\,\zeta| + \tilde{c}. \qquad (2.6.9)$$

Here (2.6.8) will give, still under the assumption (2.6.7),

$$\frac{9}{16}\sigma\langle x^0, \mathrm{Im}\,\zeta\rangle_+ = \frac{3}{4}\tilde{d}\tilde{g}(-\mathrm{Re}\,\zeta) \leq \frac{3}{4}\varphi'(\zeta)$$

and (2.6.9) leads to

$$\mu\tilde{\varphi}(\zeta) \leq \mu\lambda\tilde{d}\tilde{g}(-\mathrm{Re}\,\zeta) + \mu\tilde{d}\tilde{r}(-\mathrm{Re}\,\zeta) + \mu\tilde{d}|\mathrm{Im}\,\zeta| + \tilde{c} \leq \frac{1}{4}\varphi'(\zeta) + \tilde{d}\tilde{\rho}(\zeta) + \tilde{d}|\mathrm{Im}\,\zeta| + \tilde{c},$$

if we also use the condition on λ. (2.6.6) is a consequence.

7. Proof of II. We consider the two cases

$$\tilde{d}\tilde{g}(-\mathrm{Re}\,\zeta) \leq \frac{3\sigma}{4}\langle x^0, \mathrm{Im}\,\zeta\rangle_+ \text{ and } \tilde{d}\tilde{g}(-\mathrm{Re}\,\zeta) > \frac{3\sigma}{4}\langle x^0, \mathrm{Im}\,\zeta\rangle_+.$$

In the first case we have that

$$\begin{aligned}
\varphi(\zeta) &\geq \min(\tilde{d}\tilde{g}(-\mathrm{Re}\,\zeta), \sigma\langle x^0, \mathrm{Im}\,\zeta\rangle_+) + \tilde{d}\tilde{r}(-\mathrm{Re}\,\zeta) \\
&= dg(-\mathrm{Re}\,\zeta) + dr(-\mathrm{Re}\,\zeta) \\
&= \min(dg(-\mathrm{Re}\,\zeta), \nu\sigma\langle x^0, \mathrm{Im}\,\zeta\rangle_+) + dr(-\mathrm{Re}\,\zeta)
\end{aligned}$$

from $\varphi \geq \varphi'$ alone, and in the second case we have

$$\varphi(\zeta) \;\geq\; \mu\min((d/\mu)g(-\mathrm{Re}\,\zeta),\sigma\langle x^0,\,\mathrm{Im}\,\zeta\rangle_+) + dr(-\mathrm{Re}\,\zeta) + \frac{9}{16}\sigma\langle x^0,\,\mathrm{Im}\,\zeta\rangle_+$$

$$\geq\; \min(dg(-\mathrm{Re}\,\zeta),\nu\sigma\langle x^0,\,\mathrm{Im}\,\zeta\rangle_+) + dr(-\mathrm{Re}\,\zeta).$$

8. Proof of III. We consider the same two cases as in the proof of II. When $\tilde{d}\tilde{g}(-\mathrm{Re}\,\zeta) \leq (3\sigma/4)\langle x^0,\,\mathrm{Im}\,\zeta\rangle_+$ we have

$$\varphi'(\zeta) + \tilde{d}\tilde{\rho}(\zeta) + \tilde{d}|\mathrm{Im}\,\zeta| \leq \min(d'g'(-\mathrm{Re}\,\zeta),\langle x^0,\,\mathrm{Im}\,\zeta\rangle_+) + \frac{d'}{2}r'(-\mathrm{Re}\,\zeta) + \frac{d'}{2}|\mathrm{Im}\,\zeta| + c'$$

$$+ 2\tilde{d}\tilde{r}(-\mathrm{Re}\,\zeta) + \tilde{d}c_1|\mathrm{Im}\,\zeta| + \tilde{d}c_2$$

$$\leq \min(d'g'(-\mathrm{Re}\,\zeta),\langle x^0,\,\mathrm{Im}\,\zeta\rangle_+) + d'r'(-\mathrm{Re}\,\zeta) + d'|\mathrm{Im}\,\zeta| + c_5.$$

(Here we also use (2.6.4) and (2.6.5).)

This is the inequality needed for III in the first case, and the same argument also works in the second case when $\varphi = \varphi' + \tilde{d}\tilde{\rho} + \tilde{d}|\mathrm{Im}\,\zeta| + c_3$. It remains to estimate $\mu\tilde{\varphi} + (9/16)\sigma\langle x^0,\,\mathrm{Im}\,\zeta\rangle_+$ for ζ in the second case. What we obtain is

$$\mu\tilde{\varphi}(\zeta) + \frac{9}{16}\sigma\langle x^0,\,\mathrm{Im}\,\zeta\rangle_+ \leq \mu\lambda\tilde{d}\tilde{g}(-\mathrm{Re}\,\zeta) + \mu\tilde{d}\tilde{r}(-\mathrm{Re}\,\zeta) + \mu\tilde{d}|\mathrm{Im}\,\zeta| + c_6 + \frac{3}{4}\tilde{d}\tilde{g}(-\mathrm{Re}\,\zeta)$$

$$\leq d'g'(-\mathrm{Re}\,\zeta) + d'r'(-\mathrm{Re}\,\zeta) + d'|\mathrm{Im}\,\zeta| + c_6$$

in view of (2.6.3) and the condition on λ. This is "half" of III. (When we say "half", we refer to the "minimum".) The other half is

$$\mu\tilde{\varphi}(\zeta) + \frac{9}{16}\sigma\langle x^0,\,\mathrm{Im}\,\zeta\rangle_+ \leq \mu\langle x^0,\,\mathrm{Im}\,\zeta\rangle_+ + d'r'(-\mathrm{Re}\,\zeta) + d'|\mathrm{Im}\,\zeta| + c' + \frac{9}{16}\sigma\langle x^0,\,\mathrm{Im}\,\zeta\rangle_+$$

$$\leq \langle x^0,\,\mathrm{Im}\,\zeta\rangle_+ + d'r'(-\mathrm{Re}\,\zeta) + d'|\mathrm{Im}\,\zeta| + c_7,$$

once more in view of (2.6.3).

2.7   Proof of theorem 2.1.12

1.   The proof of theorem 2.1.12. is by duality and is based on arguments from the preceding three sections. As mentioned in section 2.1, it seems reasonable to expect that one can also prove it with FBI-techniques, in the spirit of the results of Sjöstrand [2], Lebeau [1], but to do so would require to prove first the equivalence of our present definitions with related definitions based on FBI-transforms. (We shall consider in this paper only the FBI-transforms which correspond to the case of second microlocalization. In this case the problem of "equivalence" will be studied in section A.3.)

Let us then assume that u, M_j, ξ^j, x^1, x^2 are as in the statement of theorem 2.1.12. It is no loss in generality to assume that $x^1 = 0$ and that $|x^2| = 1$. Next we observe that it follows from the assumption, if we also use the propositions 1.4.6 and 1.5.3 that we can find $r \in \mathcal{R}(\xi^0, \xi^1, ..., \xi^{k-2}, \delta^1, \beta^1)$, d_1, b_1, c_1 so that $|u(v)| \leq c_1$ for any $v \in E'(R^n)$ which satisfies

$$|\hat{v}(\zeta)| \leq \exp\left(d_1 r_1(-Re\,\zeta) + \langle x^2, Im\zeta \rangle_+ + d_1|Im\zeta| + b_1 ln(1 + |\zeta|)\right). \tag{2.7.1}$$

Furthermore, the assumption that $(x^1, \xi^0, \xi^1, ..., \xi^{k-2}) \notin WF_{A,s}^{k-1} u$ shows that we can find $g_2 \in G(\xi^0, \xi^1, ..., \xi^{k-2}, \delta_2, \beta_2)$, d_2, b_2, c_2 so that $|u(w)| \leq c_2$ for any $w \in E'(R^n)$ so that

$$|\hat{w}(\zeta)| \leq \exp\left(d_2 g_2(-Re\,\zeta) + d_2|Im\zeta| + b_2\, ln(1 + |\zeta|)\right).$$

For technical reasons it is here convenient to assume, as we may, that $r_1 \leq g_2$. Shrinking d_1, b_1, if necessary, we may assume that $|u(w)| \leq c_2$ if

$$|\hat{w}(\zeta)| \leq \exp\left(d_1 g_2(-Re\,\zeta) + d_1 r_1(-Re\,\zeta) + d_1|Im\zeta| + b_1 ln(1 + |\zeta|)\right). \tag{2.7.2}$$

Finally we observe that what we want to show is that if $g \in G(\xi^0, \xi^1, ..., \xi^{k-2}, \delta, \beta)$, d, b, c, are suitable, then $|u(f)| \leq c$ for any $f \in E'(R^n)$ so that

$$|\hat{f}(\zeta)| \leq \exp\left(dg(-Re\,\zeta) + \langle x^2, Im\zeta \rangle + d|Im\zeta| + bln(1 + |\zeta|)\right). \tag{2.7.3}$$

(Actually, we may assume here that $\delta_2 = \delta_1 = \delta$, $\beta_2 = \beta_1 = \beta$.)

2. The main step in the argument is the following:

Proposition 2.7.1. *Let* $r_1 \in \mathcal{R}(\xi^0, \xi^1, ..., \xi^{k-2}, \delta_1, \beta_1)$, $g_2 \in G(\xi^0, \xi^1, ..., \xi^{k-2}, \delta_2, \beta_2)$ *and* d_3, b_3 *be given. Then we can find* $c, d, b, g \in G(\xi^0, \xi^1, ..., \xi^{k-2}, \delta, \beta)$ *so that if* $f \in E'(R^n)$ *satisfies (2.7.3), then we can find* $h_1, h_2 \in A(C^n)$ *with* $f = h_1 + h_2$ *and so that*

$$|h_1(\zeta)| \leq c\, exp\left(\langle x^2, Im\zeta \rangle_+ + d_3 r_1(-Re\,\zeta) + d_3|Im\zeta| + b_3 \ln(1 + |\zeta|)\right), \tag{2.7.4}$$

$$|h_2(\zeta)| \leq c\, exp\left(d_3 g_2(-Re\,\zeta) + d_3 r_1(-Re\,\zeta) + d_3|Im\zeta| + b_3 \ln(1 + |\zeta|)\right). \tag{2.7.5}$$

3. The proof of proposition 2.7.1 is based on lemma 2.3.1. Since that lemma refers to the maximum of two weight functions, and here we have sums, it is convenient to transform at first the form of (2.7.3). This is done in the following :

Lemma 2.7.2. *Let* d', g *and* $x^2, |x^2| = 1$, *be given. Then we can find* d *so that*

$$dg(-Re\,\zeta) + \langle x^2, Im\zeta \rangle + d|Im\zeta| \leq max(d'g(-Re\,\zeta), \langle x^2, Im\zeta \rangle) + d'|Im\zeta|. \tag{2.7.6}$$

Proof of lemma 2.7.2. The conditions for d are

$$d \leq d'/2 \text{ and } 2d < (d' - d)d'. \tag{2.7.7}$$

Let us in fact at first observe that (2.7.6) is immediate if

$$dg(-Re\,\zeta) \leq (d' - d)|Im\zeta| \text{ or } (d'/2)g(-Re\,\zeta) \geq \langle x^2, Im\zeta\rangle. \tag{2.7.8}$$

It remains then to observe that in (2.7.8) we have already exhausted all possibilities if (2.7.7) is valid for d. Indeed, if ζ does not fulfil (2.7.8), then we must have

$$dg(-Re\,\zeta) > (d' - d)|Im\zeta| \text{ and } (d'/2)g(-Re\,\zeta) < \langle x^2, Im\zeta\rangle$$

simultaneously, so we also have

$$(d' - d)|Im\zeta| < (2d/d')\langle x^2, Im\zeta\rangle \leq (2d/d')|Im\zeta|. \tag{2.7.9}$$

From the second condition in (2.7.7) it follows that there are no ζ which satisfy (2.7.9), so the proof is complete.

4. We have now proved lemma 2.7.2 and return to proposition 2.7.1. In view of the preceding lemma, we may now as well replace (2.7.3) by

$$|\hat{f}(\zeta)| \leq \exp\left(max(d'g(-Re\,\zeta), \langle x^2, Im\zeta\rangle) + d'|Im\zeta| + b\ln(1 + |\zeta|)\right). \tag{2.7.10}$$

Here we recall that we have proved in section 2.6 that if

$$r_1 \in \mathcal{R}(\xi^0, \xi^1, ..., \xi^{k-2}, \delta_1, \beta_1),$$

d_3, c_1 and

$$g_2 \in G(\xi^0, \xi^1, ..., \xi^{k-2}, \delta_1, \beta_1)$$

are given, then we can find d', a plurisubharmonic function $\psi : C^n \to R$ and $g \in G(\xi^0, \xi^1, ..., \xi^{k-2}, \delta, \beta)$ so that

$$min(d'g(-Re\,\zeta), \langle x^2, Im\zeta\rangle_+) \leq \psi(\zeta) \leq min(d_3g_2(-Re\,\zeta), \langle x^2, Im\zeta\rangle_+) +$$
$$d_3r_1(-Re\,\zeta) + (d_3/3)|Im\zeta| + c_1.$$

Moreover, in lemma 2.3.6 we have also seen that we can find a plurisubharmonic function ψ' so that

$$b\,ln(1 + |\zeta|) \leq \psi'(\zeta) \leq (d_3/3)|Im\zeta| + b_4\ln(1 + |\zeta|) + c'.$$

This shows that

$$min(d'g(-Re\,\zeta), \langle x^2, Im\zeta\rangle_+) + d'|Im\zeta| + b\,ln(1 + |\zeta|) \leq (\psi + \psi')(\zeta) \leq$$
$$min[d_3g_2(-Re\,\zeta), \langle x^2, Im\zeta\rangle_+] + d_3r_1(-Re\,\zeta) + d_3|Im\,\zeta| + b_4ln(1 + |\zeta|) + c'.$$

Here we can choose b_4 as we please. (Influencing by this the choice of b). For a suitable choice of b_4, the proof of proposition 2.7.1 can now be concluded with lemma 2.3.1.

5. To conclude the proof of theorem 2.1.12, we now need a result of the type of proposition 2.3.10:

Proposition 2.7.3. *Let $d_1, b_1, r_1 \in \mathcal{R}(\xi^0, \xi^1, ..., \xi^{k-2}, \delta, \beta)$, $g_2 \in G(\xi^0, \xi^1, ..., \xi^{k-2}, \delta, \beta)$ be given. Then we can find c_1, d_3, b_3 so that if $\hat{f}$ satisfies (2.7.3) and is of form $\hat{f} = h_1 + h_2$ with $h_i \in A(C^n)$ satisfying (2.7.4) and (2.7.5), then there are sequences $h_1^j, h_2^j \in A(C^n)$ such that*

a) $\mathcal{F}^{-1}h_i^j \in E'(R^n)$,
b) $|(h_1^j - h_1)(\zeta)| \leq (c_1 c/j) \ exp \ (d_1 r_1(-Re\,\zeta) + \langle x^2, Im\,\zeta\rangle_+ + d_1|Im\,\zeta| + b_1 ln(1 + |\zeta|))$,
c) $|(h_2^j - h_2)(\zeta)| \leq (c_1 c/j) \ exp \ (d_1 g_2(-Re\,\zeta) + d_1 r_1(-Re\,\zeta) + d_1|Im\,\zeta| + b_1 ln(1 + |\zeta|))$,
d) $|(\hat{f} - h_1^j - h_2^j)(\zeta)| \leq (c_1 c/j) \ exp \ (\langle x^2, Im\,\zeta\rangle_+ + d_1|Im\,\zeta| + b_1 ln(1 + |\zeta|))$.

Proposition 2.7.3 can be proved with the arguments of the proof of proposition 2.3.10. We omit the details.

6. The proof of theorem 2.1.12 can now be concluded quite easily. In fact, if $\hat{f}$ satisfies (2.7.3), we can apply first proposition 2.7.1 and then 2.7.3. (Note that the choice of the d's, b's, must be reversed.) It follows then that

$$u(f) = \lim_j \left(u(\mathcal{F}^{-1}h_1^j) + u(\mathcal{F}^{-1}h_2^j) \right),$$

if b_1 is suitable.

On the other hand (2.7.4) together with b) shows that $|u(\mathcal{F}^{-1}h_1^j)| \leq c''$, whereas (2.7.5) together with c) shows that $u(\mathcal{F}^{-1}h_2^j)| \leq c'''$ for some constants c'', c''', which do not depend on j. We conclude that $|u(f)| \leq c'' + c'''$, as desired.

2.8 Proof of theorem 2.1.13

1. We begin the proof with the following lemma :

Lemma 2.8.1. *Let B be convex and consider $x, z \in B$. Let further y be of form tz for some $0 < t < 1$. Then there is $v \in B$ and $0 < \lambda < 1$, so that*

$$y = v - \lambda x.$$

Proof. Take $\lambda = 1 - t$ and $v = (1 - t)x + tz = \lambda x + y$.

2. **Proof of theorem 2.1.13.** We show by decreasing induction in $s \geq \sigma$ that

$$(y, \xi^0, \xi^1, ..., \xi^{\sigma-1}, \eta^\sigma, ..., \eta^{s-1}) \notin WF_{A,s}^s u, \forall y \in \bigcup_{0 \leq t \leq 1} tB, \forall \eta^\sigma, ..., \forall \eta^{s-1}. \qquad (2.8.1)$$

When $s = n$ this just follows from assumption III in the hypothesis and for $s = \sigma$ it gives the conclusion of the theorem when $y = 0$. Assume then that (2.8.1) has been checked for some $s > \sigma$ and we want to check it for s replaced by $s - 1$. Thus we fix $\eta^\sigma, ..., \eta^{s-2} \in R^n$, so that $\xi^0, \xi^1, ..., \xi^{\sigma-1}, \eta^\sigma, ..., \eta^{s-2}$ are mutually orthogonal, and want to show that

$$(y, \xi^0, \xi^1, ..., \xi^{\sigma-1}, \eta^\sigma, ..., \eta^{s-2}) \notin WF_{A,s}^{s-1} u, \forall y \in \bigcup_{0 \leq t \leq 1} tB. \qquad (2.8.2)$$

For $y \in B$ this follows in fact from assumption II. We also assume for the moment that (2.8.2) were not true. We could then apply condition IV and conclude that there is $x \in B$ which is orthogonal to the $\xi^0, \xi^1, ..., \xi^{\sigma-1}, \eta^\sigma, ..., \eta^{s-2}$. We also know that

$$(y, \xi^0, \xi^1, ..., \xi^{\sigma-1}, \eta^\sigma, ..., \eta^{s-2}, \eta^{s-1}) \notin WF_{A,s}^s u, \ \forall \eta^{s-1} \in R^n, \forall y \in \bigcup_{0 \leq t \leq 1} tB.$$

The $(s - 1)$- wave front set is then propagated along the space of variables orthogonal to $(\xi^0, \xi^1, ..., \xi^{\sigma-1}, \eta^\sigma, ..., \eta^{s-2})$: this is the content of theorem 2.1.12. It suffices therefore to observe that in view of the lemma, y lies on the line starting from some $v \in B$ and which has direction x. Thus (2.8.2) holds, contradicting our temporary assumption.

2.9   Proof of proposition 2.1.15

1. We prepare the proof with a lemma, which is formulated in such a way as to cover also k-microlocal variants of proposition 2.1.15.

Lemma 2.9.1. *Let ψ be a positive Lipschitzian function on R^n and denote for fixed $k \in Z^n$ by ψ_k the function $\psi_k(\xi) = (\psi(k) - L'|\xi - k|)_+$, where $L' > 0$. If L' is large enough, we can then find $d > 0, c, c'$, which do not depend on k and a plurisubharmonic function $\rho_k : C^n \to R$*
such that

$$-d\psi(Re\zeta) \leq \rho_k(\zeta) \leq -\psi_k(Re\zeta) + c|Im\zeta| + c'.$$

Proof of lemma 2.9.1. We choose constants c_1, c_2 so that $|\eta - \xi| \leq c_1(1 + \psi(\xi) + \psi(\eta))$ implies

$$\psi(\xi) \leq c_2(1 + \psi(\eta)) \text{ and } \psi(\eta) \leq c_2(1 + \psi(\xi)).$$

The existence of such c_1, c_2 follows from the Lipschitzianity of ψ. It follows in particular that for $L' = 3/c_1$

$$\psi(k) \leq c_2(1 + \psi(\xi)), \psi(\xi) \leq c_2(1 + \psi(k)), \text{ if } |\xi - k| \leq (3/L')\psi(k).$$

Since, on the other hand, $\psi_k(\xi) = 0$ for $|\xi - k| \geq (1/L')\psi(k)$, it is now clear that the lemma is, after a translation, a consequence of the following result for $b = (1 + \psi(k))$:

Lemma 2.9.2. *Let c_3 be given. Then there are constants c_4, c_5 such that for every $b > 0$ there is a plurisubharmonic function $\chi : C^n \to R$ which satisfies*

a) $\chi(\zeta) \leq c_4|Im\zeta| + c_5$, if $|Re\zeta| \geq c_3 b$,
b) $\chi(\zeta) \leq -b + c_4|Im\zeta| + c_5$, if $|Re\zeta| < c_3 b$,
c) $\chi(\zeta) \geq 0$, if $3c_3 b \leq |Re\zeta|$,
d) $\chi(\zeta) \geq -b$, if $|Re\zeta| \leq 3c_3 b$.

Proof of lemma 2.9.2. The last condition poses no particular problem since if we can find a plurisubharmonic function κ which satisfies the conditions a),b),c), then the function

$$\max\left(\kappa(\zeta), -b\right)$$

will be a plurisubharmonic function which satisfies a),b),c),d). In view of this remark we may disregard condition d) in what follows. The main step in the proof is then to construct for each fixed $\theta \in R^n$, $c_3 b \leq |\theta|$, a plurisubharmonic function χ_θ on C^n such that

i) $\chi_\theta(\zeta) \leq c_6|Im\zeta| + c_7$, if $Re\,\zeta| \geq c_3 b$,
ii) $\chi_\theta(\zeta) \geq 0$, if $Re\zeta = \theta$,
iii) $\chi_\theta(\zeta) \leq -b + c_6|Im\zeta| + c_7$ if $|Re\zeta| \leq c_3 b$.

In fact, once such functions have been found, a plurisubharmonic function χ which satisfies a),b),c) is the upper regularization of

$$\sup_{|\theta| \geq 3c_3 b} \chi_\theta(\zeta).$$

To find functions χ_θ with i),ii),iii), we start from the function $h : C^n \times R_+ \to R$ from lemma 2.3.4 and set

$$\chi_\theta(\zeta) = c_3^{-1} h(\zeta - \theta, c_3 b + |\theta|).$$

The properties i) and ii) are then immediate. As for iii) we first note that

$$\chi_\theta(\zeta) \leq -c_3^{-1}|Re\,\zeta - \theta| + c_6|Im\zeta| + c_7$$

if

$$(1/2)(c_3 b + |\theta|) \leq |Re\zeta| \leq (c_3 b + |\theta|). \tag{2.9.1}$$

Of course, the second inequality in (2.9.1) is satisfied trivially if $|Re\,\zeta| \leq c_3 b$ and if in addition $|\theta| \geq 3c_3 b$, then we have that

$$|Re\zeta - \theta| \geq |\theta| - |Re\zeta| \geq (1/2)(c_3 b + |\theta|).$$

The inequalities in (2.9.1) are therefore both satisfied, so we conclude that for such ζ, θ,

$$\chi_\theta(\zeta) \leq -c_3^{-1}|Re\ \zeta - \theta| + c_6|Im\zeta| + c_7 \leq -b + c_6|Im\,\zeta| + c_7.$$

2. We start the proof of proposition 2.1.15 with some general considerations. What we must show is that there are $c > 0, d > 0, b' \in R$ and $g \in G(\xi^0, \xi^1, \delta)$ so that $|v(f)| < c$ if $v \in E'(R^n)$ satisfies

$$|\hat{v}(\zeta)| \leq \exp\left(dg(-Re\zeta) + \langle x^0, Im\zeta\rangle + d|Im\zeta| + b'ln(1 + |\zeta|)\right).$$

The proof of this is based on two lemmas.

Lemma 2.9.3. *Let $d' > 0, b'' \in R, g \in G(\xi^0, \xi^1, \delta)$ be fixed and consider $L' > 0$. Then we can find $d > 0, b' \in R$ such that if v is given with*

$$|\hat{v}(\zeta)| \leq \exp\left(dg(-Re\zeta) + \langle x^0, Im\zeta\rangle + d|Im\zeta| + b'ln(1 + |\zeta|)\right),$$

then we can find for every $k \in Z^n$ some $h_k \in A(C^n)$ such that

$$\hat{v} = \sum_k h_k,$$

$$|h_k(\zeta)| \leq (1 + |k|^{n+1})^{-1}\,exp\left(d'g_k(-Re\zeta) + \langle x^0, Im\zeta\rangle + d'|Im\zeta| + b''ln(1 + |\zeta|)\right),$$

where $g_k = (g(k) - L'|\xi - k|)_+$.

This is a consequence of lemma 2.3.11.

Lemma 2.9.4. *Let f be a distribution as in proposition 2.1.15. Then there are $c, d', b'', g \in G(\xi^0, \xi^1, \delta), L' > 0$, so that $|w(f)| \leq c$ for any w for which we can find k so that*

$$|\hat{w}(\zeta)| \leq \exp\left(d'g_k(-Re\zeta) + \langle x^0, Im\zeta\rangle + d'|Im\zeta| + b''ln(1 + |\zeta|)\right). \tag{2.9.2}$$

Proof of lemma 2.9.4. We fix a bicone $G' \subset\subset G$ and write $f = f' + f''$. The equality is here in $S'(R^n)$, the support of f'' is supposed to lie in the complement of G' and f' is supposed to be continuous on R^n, to have support in G, and to have an holomorphic extension to the set

$$D' = \{\zeta; Re\zeta \in G', |Im\zeta| + |Re\zeta|^\delta \leq C'|Re\zeta'|\}.$$

Moreover, we assume that $|f'(\xi)| \leq c'(1 + |\xi|)^{b'}$ on R^n and that f' satisfies on D' an inequality similar to that satisfied by f on D. All this can of course be achieved with a partition of unity.

It is now easy to see (by writing f'' as a finite sum of derivatives of functions with polynomial growth supported in a small neighborhood of the complement of G') that if d', b'', g are suitable, then (2.9.2) implies $|w(f'')| \leq c$, for some c which depends only on d', g, b'' and f''. We may then still further shrink d', g, b'', (if this is necessary,) to estimate $w(f')$. In order to do this, we shall decompose f'.

Lemma 2.9.5. *Let f' be a continuous function on R^n, which satisfies $|f'(\xi)| \leq (1 + |\xi|)^b$ on R^n and which has an holomorphic extension to D' for which*

$$|f'(\zeta)| \leq (1 + |\zeta|)^b \ exp \ H_K(Im\zeta), \forall \zeta \in D'.$$

Also consider $d' > 0$. If $L' > 0$ (which comes in via the definition of g_k) is suitable, then we can find c, b_1, d_1 so that we can write f' for any fixed $k \in Z^n$ in the form

$$f' = h_1' + h_2'$$

where h_1' is an entire function on $A(C^n)$ which satisfies

$$|h_1'(\zeta)| \leq c \ exp \ (H_K(Im\zeta) + d'|Im\zeta| + b_1 \ln(1 + |\zeta|)),$$

and where h_2' is continuous on R^n and satisfies

$$|h_2'(\zeta)| \leq c \ exp \ (-d_1 g_k(-Re\zeta) + b_1 \ln(1 + |\zeta|)).$$

Note that h_1', h_2' depend here on k, a fact which is not made exlicit in the notation.

3. Once lemma 2.9.5 is proved, lemma 2.9.4, and therefore also proposition 2.1.15, is immediate. In fact, $v_k(\mathcal{F}^{-1}h_1') = 0$, $v_k = \mathcal{F}^{-1}h_k$, for d' small, since v_k and $\mathcal{F}^{-1}h_1'$ have disjoint supports then, (the duality $v_k(\mathcal{F}^{-1}h_1')$ makes sense if b'' is chosen suitably) and

$$|v_k(\mathcal{F}^{-1}h_2')| = |h_k(h_2')| \leq c''$$

for d', b'' small, since we can just estimate $h_k(h'_2)$ as an integral.

4. Proof of lemma 2.9.5. If $g \in G(\xi^0, \xi^1, \delta)$ is suitable, we will have that

$$g(-Re\zeta) \leq c_1(1 + |Im\zeta|),$$

if $dist(\zeta, E) \leq 1$, where E is the boundary of D' in C^n. Using a partition of unity, we can therefore decompose f' in the form

$$f' = \varphi_1 + \varphi_2, \text{ on } R^n,$$

where φ_2 is continuous on R^n and satisfies

$$|\varphi_2(\xi)| \leq (1 + |\zeta|)^b$$

and where φ_1 may be regarded as a function on C^n which satisfies

$$|\bar{\partial}\varphi_1(\zeta)| \leq \exp\left(-d_2 g(-Re\zeta) + (d'/2)|Im\zeta| + H_K(Im\zeta) + b\ln(1 + |\zeta|)\right),$$

for some $d_2 > 0$ which depends of course on d'. The desired conclusion will now follow if we can prove that for any fixed k there is function Q_k such that

$$\bar{\partial}Q_k = \bar{\partial}\varphi_1, \tag{2.9.3}$$

and for which

$$|\bar{\partial}Q_k(\zeta)| \leq c_3 \exp\left(-d_1 g_k(-Re\zeta) + d'|Im\zeta| + H_K(Im\zeta) + b_1\ln(1 + |\zeta|)\right). \tag{2.9.4}$$

Indeed, if such Q_k could be found, then we could set $h'_1 = \varphi_1 - Q_k$, $h'_2 = \varphi_2 + Q_k$. The existence of a solution of (2.9.3), (2.9.4) will now follow from standard results from the theory of the $\bar{\partial}$-operator, due to Hörmander, if we apply lemma 2.9.1, choosing L' suitably. (We shall have here $\varphi(\xi) = g(-\xi)$.) The variant of these results which we apply is the following:

Lemma 2.9.6. *Let* $\psi_i : C^n \to R, i = 1, 2, 3, 4$ *and* $\rho : C^n \to R$ *be given so that*

a) $\psi_1(\zeta) \leq \rho(\zeta) \leq \psi_2(\zeta)$,

b) $\psi_3(\zeta) \leq \psi_4(\zeta)$,

c) ρ *is plurisubharmonic,*

d) ψ_2 *and* ψ_4 *are Lipschitzian.*

Also consider $b \geq 0$. *Then there is* b_1 *and* c *with the following property:*

if w is a C^∞ function on C^n for which

$$|\bar{\partial} w(\zeta)| \leq \ exp \ (\psi_1(\zeta) + \psi_3(\zeta) + b \ln(1 + |\zeta|)),$$

then there is a C^∞ function w' on C^n such that $\bar{\partial}(w - w') = 0$ and such that

$$|w'(\zeta)| \leq c \ exp \ (\psi_2(\zeta) + \psi_4(\zeta) + b_1 \ln(1 + |\zeta|)).$$

Cf. Liess [2] for details.

Appendix

A.1 Propagation of analyticity for partially holomorphic functions.

1. In this section we consider the scalar Cauchy-Riemann operator $\bar{\partial} = (\partial/\partial x_1 + i\partial/\partial x_2)/2$, regarded as an operator on R^n. The following result is then classical:

Theorem A.1.1. *Consider X open in R^n, $x^0 \in X$ and assume $u \in D'(X)$ is a solution of $\bar{\partial} u = 0$ which is real-analytic at x^0. Then u is real-analytic at any point x in the connected component of*

$$X \cap \{x \in R^n; x_1 = x_1^0, x_2 = x_2^0\},$$

which contains x^0.

A microlocal variant of this is often referred to Kashiwara [1]. (Also cf. Kawai-Kashiwara-Kimura [1] and, for propagation of C^∞-regularity, Duistermaat- Hörmander [1]). We give here a proof of the theorem which is quite elementary, in the sense that it is very easy to see at a heuristic level why the result is true: cf. nr. 4 below. The main reason why we give the proof here in detail is however that it serves as a justification for the arguments in section A.2. For simplicity we shall assume that $u \in C^\infty$. That this implies the result for general distribution solutions follows e.g. from an argument in Rudin [1]. (Also cf. Liess [2].)

2. Before we enter the proof of theorem A.1.1, let us comment on how one can check real-analyticity for solutions of $\bar{\partial} u = 0$. Let us then assume that u solves $\bar{\partial} u = 0$ in a neighborhood of some point $t^0 \in R^n$ and assume for simplicity that $t_2^0 = 0$.

Lemma A.1.2. *Let u be a solution of $\bar{\partial} u = 0$. u is then real-analytic near t^0 if and only if we can find a holomorphic function $(z_1, z_3, ..., z_n) \to w(z_1, z_3, ..., z_n)$ defined in a neighborhood of $y^0 = (t_1^0, t_3^0, ..., t_n^0)$ such that*

$$u(x_1, 0, x_3, ..., x_n) = w(x_1, x_3, ..., x_n) \text{ near } y^0. \tag{A.1.1}$$

Moreover, if a holomorphic function with (A.1.1) exists, then the holomorphic extension v of u is given by

$$v(z_1, z_2, z_3, ..., z_n) = w(z_1 + iz_2, z_3, ..., z_n). \tag{A.1.2}$$

Proof of lemma A.1.2. Only the "if" part is a problem. Assume then that a holomorphic function w with (A.1.1) exists and define v by (A.1.2). It is clear that v is holomorphic. We claim that $u(x) = v(x)$ near t^0. Indeed, this is true for $x_2 = 0$, and must therefore also be true (by Holmgren's uniqueness theorem) in a neighborhood of t^0 since $x_2 = 0$ is noncharacteristic for $\bar{\partial}$ and since both u and $x \to v(x)$ solve $\bar{\partial} f = 0$.

3. Proof of theorem A.1.1. (First part) It suffices to show that u is real-analytic in a neighborhood of $I = \{x \in R^n; 0 < x_1 < 1, x_i = 0, i \geq 2\}$ for any u which is defined in an open set $X \supset I$, which satisfies $\bar{\partial} u = 0$ there, and which is real-analytic at $x^0 = (1, 0, ..., 0)$. To simplify the notation we shall assume $n = 3$ and denote, for fixed $\delta > 0$ and some given $d \geq 0$, by I_d the set

$$\{x \in R^n; d < x_1 < 1 + \delta, x_i = 0, i \geq 2\}.$$

We shall now use a (finite) recursion procedure to produce a finite sequence $d_j \to 0$, $d_j \geq 0$ so that u is real-analytic near I_{d_j}, if δ is small enough. The precise formulation of the induction step is :

Lemma A.1.3. *Let $\delta > 0$ be fixed and let u be a solution of $\bar{\partial} u = 0$ on the set $X = \{x; -\delta < x_1 < 1 + \delta, |x_i| < \delta\}$. Assume moreover that u admits a holomorphic extension v to some set of form*

$$Y_d = \{z \in C^n; d < Re\, z_1 < 1 + \delta, |Im\, z_1| < c_1, |z_i| \leq c_1, i \geq 2\},$$

for some c_1 and $d > 0$. Then we can find $0 \leq d' < d$, $c_2 > 0$, so that u can also be holomorphically extended to

$$Y_{d'} = \{z \in C^n; d' < Re\, z_1 < 1 + \delta, |Im\, z_1| < c_2, |z_i| < c_2, i \geq 2\},$$

Moreover, the difference $d - d'$ does not here depend on c_1 or d.

4. According to lemma A.1.2 it suffices to define the holomorphic extension of u when $z_2 = 0$. Let us then at first observe that when u is as in theorem A.1.1, then $u(z)$ has a natural meaning for all $z \in A = A' \cup A''$, where, for suitable c_3, c_4 :

$$A' = \{(x_1 + ix_2, 0, x_3); \ x_i \in R, -c_3 < x_1 < 1 + c_3, \ |x_3| < c_3\},$$

$$A'' = \{z \in C^3; z_2 = 0, \ |1 - z_1| + |z_3| < c_4\}.$$

Indeed, for $z \in A''$ this is clear, and for $z \in A'$ we may set $u(x_1 + ix_2, 0, x_3) = u(x_1, x_2, x_3)$. The main idea in the proof is then, in analogy with proofs for Hartog's extension theorem or for (Epstein's version of) the edge-of-the-wedge theorem (cf. Rudin [1]), to use Cauchy's integral formula for contours lying completely inside A, to give a meaning to $u(z)$ for points outside A. Before effectively proving lemma A.1.3, we shall at first show how one can compute the holomorphic extension of u near I in just one step, assuming that one already knows that it exists. For this purpose we introduce, for some natural number k and for $\beta \in R, s \in R$, the sets

$$T_k = \{t \in C; |t| < 1, \ -\pi/2k < \arg t < \pi/2k\},$$

and by

$$S_{\beta,s,k} = \{z \in C^3; z_1 = t \in T_k, \ z_2 = 0, \ z_3 = i\beta t^k + s\}.$$

Note that $i\beta t^k$ is never real for $t \in T_k$. If k is fixed and z is, by chance, in $S_{\beta,s,k}$, this will therefore determine β and s in an unique way. It is also clear that the union for all small (β, s) of the surfaces $S_{\beta,s,k}$ will be a neighborhood in $\{z \in C^3; z_2 = 0\}$ of I (when the latter is regarded as a subset there). The main properties of the sets $S_{\beta,s,k}$ are :

a) $S_{\beta,s,k}$ is for any fixed β, s, k an analytic surface,

b) if (β, s) is small and k is large, then the boundary of $S_{\beta,s,k}$ is contained in A.

(In fact, on the part of the boundary where $\arg z_1 = \pm\pi/2k$, z_3 will be real and we will have that $z \in A'$ if β, s, k are suitable. Moreover, on the part of the boundary where $|z_1| = 1$, we will have, once more if β, s, k are suitable, that $z \in A''$.)

Assuming now that an analytic extension v of u to $S_{\beta,s,k}$ is known to exist, we can recover it from the values of u on the boundary of $S_{\beta,s,k}$ by simply applying Cauchy's integral formula in $S_{\beta,s,k}$. (Since $S_{\beta,s,k}$ is parametrized by z_1, it is a simple matter to transfer Cauchy's formula to it.)

5. The argument above could in principle be used to prove theorem A.1.1 in the following way: at first we define an extension w of u to a complex neighborhood of I intersected

with $\{z \in C^3; z_2 = 0\}$ by defining it, for some suitably fixed k, on

$$\bigcup_{|\beta|+|s|<c_3} S_{\beta,s,k},$$

and by showing afterwards that the function so defined is analytic. It does not seem however to be very easy to prove the analyticity of w in this way and we shall therefore slightly modify the approach. In fact, rather than varying β and s, we shall consider translations in the z_1 direction of sets of type $S_{\beta,0,k}$ for some fixed β and k.

Let us, for this purpose, denote for fixed β, k, and for some given $z_1 \in C$ by

$$U_{z_1} = \{y_1 \in C; y_1 = z_1 + t, t \in C, |t| < 1 - d, -\pi/2k < \ \arg t < \pi/2k\},$$

and by

$$V_{z_1} = \{y \in C^3; y_1 \in U_{z_1}, \ y_2 = 0, \ y_3 = i\beta(y_1 - z_1)^k\}.$$

If β and k have been suitable, then V_d will be completely contained in Y_d. We also note that $\beta > 0, y \in V_{z_1}$ implies $Re\ y_3 > 0$. Conversely we claim that if $c > 0$ is fixed, then we can find $d' < d$ and c_2 (which appears in the definition of $Y_{d'}$) so that

$$(\bigcup_{|z_1-d|<c} V_{z_1}) \supset \{y \in Y_{d'} ; y_2 = 0, Im\ y_3 > 0\}. \tag{A.1.3}$$

Indeed, if $(\tau_1, 0, \tau_3) \in Y_{d'} \cap \{y; y_2 = 0, Im\ y_3 > 0\}$, is given, we can solve $\tau_3 = i\beta t^k$ for t with $-\pi/2k < \ \arg t < \pi/2k$ and then set $z_1 = y_1 - t$. When c_2 is small, τ_3 and therefore also t, is small, so we will have $|z_1 - d| < c$ if $d - d'$ was not too large.

Arguing as in nr.4, we can now use Cauchy's formula in V_{z_1} to define a function v on V_{z_1}. From (A.1.3) it follows that we can define a function $w_+(z_1, z_3)$ on

$$Y_{d'} \cap \{y; y_2 = 0, Im\ y_3 > 0\},$$

in this way, and we claim that the function so defined is analytic in (z_1, z_3). To check this, we use Hartogs' theorem on separate analyticity, using as coordinates the coordinates inside V_{z_1} and $z \to z_1$. Indeed, it is clear that w_+ is analytic when restricted to any V_{z_1} and we shall also show that it is analytic in z_1 for fixed z_3. In fact, to check the latter, it is convenient to regard all the sets V_{z_1} as translates of V_0, so we can write the Cauchy integrals as contour integrals for a fixed kernel on the contour $\Lambda = \Lambda_- \cup \Lambda_0 \cup \Lambda_+$, where:

$$\Lambda_+ = \{t \in C; |t| < 1-d, \ \arg t = \pi/2k\},$$

$$\Lambda_- = \{t \in C; |t| < 1-d, \ \arg t = -\pi/2k\},$$

$$\Lambda_0 = \{t \in C; |t| = 1-d, -\pi/2k < \ \arg t < \pi/2k\}.$$

As for the functions to which we apply the Cauchy formula, they depend parametrically on z_1 and are simply the translates in the z_1-direction of the function u: $u(\cdot + z_1, 0, z_3)$. Clearly they will depend analytically on z_1 on the pieces $\Lambda_\pm$ and Λ_0 of Λ. It is now clear that we have obtained an analytic function w_+ on $Y_{d'} \cap \{z_2 = 0, Im\, z_3 > 0\}$. In a similar way we can construct an analytic extension w_- to $Y_{d'} \cap \{z_2 = 0, Im\, z_3 < 0\}$. It is also easy to see that the boundary values of the two extensions to $Im\, y_3 = 0$ are the same. In fact, since we already know that w_+ and w_- are analytic on their respective domains of definition, we can obtain the boundary value as a limit for $\beta \to 0$ of Cauchy integrals on surfaces of type

$$\{z \in C^3; z_2 = 0, z_1 = d' + t_1, -\pi/2k < \text{ arg } t < \pi/2k, |t| < 1 - d, z_3 = i\beta t^k + x_3\},$$

and the limit will be Cauchy's integral formula on

$$\{z \in C^3; z_2 = 0, z_1 = d' + t_1, -\pi/2k < \text{ arg } t < \pi/2k, |t| < 1 - d, z_3 = x_3\},$$

which gives $u(x_1, 0, x_3)$ since u is analytic in $x_1 + ix_2$. The function w on $Y_{d'} \cap \{y_2 = 0\}$ defined by w_+, w_- for $Im z_3 \geq 0$ respectively ≤ 0, respectively by u when $Im z_3 = 0$, is then the desired analytic extension.

A.2 Proof of lemma 2.5.1

1. In section A.1 we have studied propagation of analytic regularity for solutions of $\bar{\partial}u = 0$ using contour deformations. Here we show, following more or less Liess [2], that one can use the same ideas to prove lemma 2.5.1. We start the argument with

Lemma A.2.1. *Consider $\sigma, 0 \leq \sigma < 1$, and $d' > 0$. Then we can find $d > 0, c$ and a plurisubharmonic function $\psi : C^2 \to R$ such that*

$$\psi(\zeta) \leq min(d'|Re\zeta_2|, Im\, \zeta_{1+}) + d'|Re\,\zeta_1| + d'|Im\,\zeta| + c, \text{ if } \zeta \in C^2, \qquad (A.2.1)$$

$$\psi(\zeta) \geq \sigma Im\,\zeta_{1+}, \text{ when } d\, Re\,\zeta_2 = Im\,\zeta_{1+}. \qquad (A.2.2)$$

Proof. We fix an integer k with $k \geq 2\pi/d'$ and denote by T', respectively T'', the following contours in the complex plane:

$$T' = \{t \in C; \text{ arg } t = \pi/(2k), 1 \geq |t| > 0\} \cup \{t \in C; \text{ arg } t = -\pi/(2k), 0 \leq |t| \leq 1\},$$

$$T'' = \{t \in C; |t| = 1, -\pi/(2k) < \text{ arg } t < \pi/(2k)\}.$$

(Thus $T' \cup T''$ is just the boundary of the set T_k from section A.1.)

For $f \in A(C)$ we now introduce $h_f \in A(C^2)$ by

$$h_f(\zeta) = \int_{T'} f(t) \exp[i(t-1)\zeta_1 + (d'/2)t^k\zeta_2] \, dt. \tag{A.2.3}$$

Since f is analytic, it follows that h_f also has the representation

$$h_f(\zeta) = \int_{T''} f(t) \exp[i(t-1)\zeta_1 + (d'/2)t^k\zeta_2] \, dt, \tag{A.2.4}$$

and from (A. 2.3) and (A. 2.4) we can conclude that if $|f(t)| \leq 1$ on $T' \cup T''$, then

$$\ln|h_f(\zeta)| \leq min(d'|Re\ \zeta_2|, Im\ \zeta_{1+}) + (d'/2)|Re\ \zeta_1| + (d'/2)|Im\ \zeta| + c.$$

We denote by $\mathcal{M} = \{f \in A(C)\, ; |f(t)| \leq 1 \text{ on } T' \cup T''\}$ and want to show that

$$\sup_{f \in \mathcal{M}} \ln|h_f(\zeta)| \geq \sigma\ Im\ \zeta_{1+} - (d'/2)|Re\ \zeta_1| - (d'/2)|Im\ \zeta_2| - c', \text{ if } d\ Re\ \zeta_2 = Im\ \zeta_{1+}, \tag{A.2.5}$$

provided d is sufficiently small. Once this is proved, lemma A.2.1 will follow if we define ψ to be the upper regularization of

$$\sup_{f \in \mathcal{M}} \ln|h_f(\zeta)| + (d'/2)|Im\ \zeta| + (d'/2)|Re\ \zeta_1| + c'.$$

2. To prove (A. 2.5), we first observe that we may suppose that ζ_1 is purely imaginary and that ζ_2 is real. In fact, once we have proved (A. 2.5) for such ζ, we can reduce the general case to this one by a suitable change of f.

(More precisely, we write

$$h_f(\zeta) = \exp[(d'/2)(|Re\ \zeta_1| + |Im\ \zeta_2|)] \int_{T'} f'(t) \exp[-(t-1)Im\ \zeta_1 + (d'/2)t^k Re\ \zeta_2] dt,$$

where, for fixed ζ,

$$f'(t) = f(t) \exp[i(t-1)Re\ \zeta_1 + i(d'/2)t^k Im\ \zeta_2 - (d'/2)|Re\zeta_1| - (d'/2)|Im\ \zeta_2|],$$

and have that $f' \in \mathcal{M}$ implies $f \in \mathcal{M}$.)

The proof will come to an end if we can show that

$$\ln|v(\xi)| \geq \sigma\ d\xi - c'' \text{ for } \xi \in R_+, \tag{A.2.6}$$

if

$$v(\xi) = \int_{T'} \exp[\xi(1-t)d + (d'/2)t^k)\xi] dt, \tag{A.2.7}$$

provided d is small. We have a clear case for the saddle point method. Application of this method is particularily simple here in view of the fact that $(1-t)d + (d'/2)t^k$ has precisely one simple saddle inside the sector $|\arg t| \le \pi/(2k)$, namely at $t = t^0$, with t^0 determined from $d = k(d'/2)(t^0)^{k-1}$. If we choose here d small compared with d', we can of course make t^0 as small as we please. It is now also natural to deform the contour T' to the contour defined, as a set, by

$$\{t \in T'; Re\, t > t^0\} \cup \{Re\, t = t^0; |\arg t| < \pi/(2k)\}.$$

On the part $Re\, t > t^0$, $Re\,[(1-t)d + (d'/2)t^k] < [(1-t^0)d + (d'/2)t^{0k}] - \varepsilon$ for some $\varepsilon > 0$, so we can hope that we shall remain essentially with the contribution from $Re\, t = t^0$, if this is "nontrivial". On the part $Re\, t = t^0$ we have $Re\,[(1-t)d - (d'/2)t^k] \le (1-t^0)d + (d'/2)t^{0k}$ and t^0 is the unique critical point of the imaginary part of the phase. Since the critical point is nondegenerate we can apply the method of the stationary phase (in the variable $\theta \to t^0 + i\theta$) for complex phase functions. (Actually, the situation is quite simple and the classical saddle point method applies.) What we obtain is

$$v(\xi) \sim a\xi^{1/2} \exp[(1-t^0)d\,\xi + (d'/2)t^{0k}\xi]\,(1 + O(1/\xi))$$

when $\xi \to \infty$ for some $a \ne 0$ (which can be computed). Cf. Hörmander [5] or Sjöstrand [2].

3. We can now improve lemma A.2.1 using simple tricks. The first improvement is

Remark A.2.2. *a) Lemma A.2.1 is also valid with $\sigma = 1$.*
b) We may replace (A. 2.2) by

$$\psi(\zeta) \ge Im\,\zeta_{1+}, \quad when\ d|Re\,\zeta_2| = Im\zeta_{1+}. \tag{A.2.8}$$

In fact, to obtain part a), we simply apply lemma A.2.1 for $d'/2$ and $\sigma = 1 - d'/2$ and replace the resulting function ψ by $\psi + (1-\sigma)Im\,\zeta_{1+}$. Further, to obtain part b) we start from part a) and consider two plurisubharmonic functions ψ_+, ψ_- which both satisfy (A. 2.1) and which also satisfy

$$\psi_\pm \ge Im\,\zeta_{1+}, \quad when\ \pm d\,Re\,\zeta_2 = Im\,\zeta_1.$$

The function $\psi(\zeta) = max(\psi_+(\zeta), \psi_-(\zeta))$ will then satisfy (A. 2.8).

4. Proof of lemma 2.5.1. We may assume n=2. Also fix $d' > 0$ and apply remark A.2.2 b). The function φ defined by $\varphi(\zeta) = max(\psi(\zeta), d|Re\,\zeta_2|)$ if $d|Re\,\zeta_2| \le Im\,\zeta_{1+}$

and by $\varphi(\zeta) = max(\psi(\zeta), Im\ \zeta_{1+})$ if $d|Re\ \zeta_2| > Im\ \zeta_{1+}$ is then plurisubharmonic and satisfies

$$min(d|Re\ \zeta_2|, Im\ \zeta_{1+}) \le \varphi(\zeta) \le min(d'|Re\ \zeta_2|, Im\ \zeta_{1+}) + d'|Re\ \zeta_1| + d'|Im\ \zeta| + c.$$

It is clear that this implies lemma 2.5.1.

A.3   Proof of proposition 2.1.7

1. For completeness we give a proof of proposition 2.1.7. In doing so, we shall consider another form of the FBI transform, which is closer to the standard Fourier transform. (Also cf. here chapter IX.) Let us in fact denote $Re\ y$ by t, $\lambda\mu\ Im\ y'$ by ξ', $\lambda\ Im\ y''$ by ξ'', $1/|Im\ y'|$ by α and $1/|Im\ y''|$ by β. Recall that in definition 2.1.5, $Im\ y'$ was close to ξ^1, whereas $Im\ y''$ was close to $\xi^{0''}$, which both were nonvanishing. With these notations, the estimate from definition 2.1.5 is equivalent to

$$\left| u_x(e^{[-i\langle x,\xi\rangle - \alpha|\xi'|\,|t' - x'|^2/2 - \beta|\xi''|\,|t'' - x''|^2/2]}) \right| \le c\,e^{-c'|\xi'|}.$$

As for the domains of validity, expressed in terms of t, ξ, α, β, which we obtain from definition 2.1.5, we have that ξ' is in a conic neighborhood of ξ^1, ξ'' is in a conic neighborhood of $\xi^{0''}$, $|t - x^0| < c''$, $|\xi'| < c''|\xi''|$, and $\alpha|\xi'| \ge g(\beta|\xi''|)$ for some sublinear function g.

After this discussion, it is now easy to see that proposition 2.1.7 is a consequence of the following result

Proposition A.3.1. *There are equivalent*

i) $(x^0, \xi^0, \xi^1) \notin WF_A^2 u$.

ii) For any α', α'' with $0 < \alpha' < \alpha''$ and any β', β'' with $0 < \beta' < \beta''$, we can find open cones $G^0 \ni \xi^0$, $G^1 \ni \xi^1$, constants c, c', c'', and a sublinear function g so that

$$|u_x(\exp[-i\langle x,\xi\rangle - \alpha|\xi'|\,|t' - x'|^2/2 - \beta|\xi''|\,|t'' - x''|^2/2])| \le c\exp -[c'|\xi'|] \qquad (A.3.1)$$

if $\alpha' < \alpha < \alpha''$, $\beta' < \beta < \beta''$, $|t - x^0| < c''$, $\xi \in G^0$, $\xi' \in G^1$, $|\xi'| < c''|\xi''|$, $|\xi'| \ge g(|\xi|)$.

iii) We can find α, β, open cones $\Gamma \ni \xi^0$, and $\Gamma^1 \ni \xi^1$, and a sublinear function g so that the inequality (A. 3.1) above holds for these fixed α, β and for the ξ, t specified in ii).

Remark A.3.2. a) *Results of the type that iii) implies ii) are implicit in many arguments on the equivalence of appropriate definitions of wave front sets. (An explicit formulation of a more general result in standard microlocalization is in Liess [5]. Also cf. Laubin-Esser [1], where a condition similar to ii) appears in a different context.)*

b) *In proposition A.3.1 we refer explicitly to the case when WF_A^2 is defined with respect to the complete family of sublinear functions. The proposition remains valid if we remain within the category of sublinear functions of form $f(t) = (1 + t^\delta)$, $\delta < 1$.*

2. We shall prove that i)$\Rightarrow$ ii) and that iii) $\Rightarrow$ i). Since ii) $\Rightarrow$ iii) is trivial, this will conclude the proof. To shorten notations, we shall write

$$E(x, \xi, t, \alpha, \beta) = \exp[-i\langle x, \xi \rangle - \alpha |\xi'| |x' - t'|^2/2 - \beta |\xi''| |x'' - t''|^2/2].$$

i) $\Rightarrow$ ii) The proof that this implication is valid, follows from the following two lemmas:

Lemma A.3.3. *Let v_j be a bounded sequence in $E'(R^n)$ and assume that $v_j \equiv 0$ on the set $\{x; |x - x^0| < \delta\}$. Then we can find constants c, c', b, so that*

$$|v_j(E(\cdot, \xi, t, \alpha, \beta))| \leq c(1 + |\xi|)^b \exp[-c'|\xi'|], \text{ if } |t - x^0| < \delta/2, |\xi'| \leq |\xi''|.$$

The proof of this is immediate.

Lemma A.3.4. *Let $G^0 \subset R^n$, $G^1 \subset R^d$ be open cones which contain ξ^0 and ξ^1 respectively, consider $0 < \alpha' \leq \alpha''$, $0 < \beta' \leq \beta''$ and let f be some given sublinear function. Assume further that u_j is a bounded sequence of distributions in $E'(R^n)$ so that*

$$|\hat{u}_j(\xi)| \leq c(cj/|\xi'|)^j, \text{ if } \xi \in G^0, \xi' \in G^1, |\xi'| \geq f(|\xi|). \tag{A.3.2}$$

Finally, fix cones $\Gamma^0 \subset\subset G^0$, $\Gamma^1 \subset\subset G^1$.

Then we can find constants c', c'', b and a sublinear function g so that

$$|u_j(E(\cdot, \xi, t, \alpha, \beta))| \leq c'(c'j/|\xi'|)^j + c'(1 + |\xi|)^b \exp(-c''|\xi'|)$$

if $\xi \in \Gamma^0$, $\xi' \in \Gamma^1$, $|\xi'| \geq g(|\xi|)$.

3. **Proof of lemma A.3.4.** From Parseval's identity it follows that

$$|u_j(E(\cdot, \xi, t, \alpha, \beta))| \leq (2\pi)^{-n} \int |\hat{u}_j(\eta)| |\mathcal{F}_{x \to \eta}(E(x, \xi, t, \alpha, \beta))(-\eta)| \, d\eta.$$

To compute $\mathcal{F}_{x \to \eta} E(x, \xi, t, \alpha, \beta)$, recall that the Fourier transform of

$$f(t) = e^{[-a\,t^2/2]}, t \in R,$$

is

$$\pi^{-1/2} a^{-1/2} e^{[-\tau^2/(2a)]}.$$

It follows that

$$\mathcal{F}(E(\cdot, \xi, t, \alpha, \beta))(\eta) = (2\pi)^{-n/2} |\alpha\, \xi'|^{-d/2} |\beta\xi''|^{d/2 - n/2} \exp[-|\eta' - \xi'|^2/(2\alpha|\xi'|)$$

$$-|\eta'' - \xi''|^2/(2\beta|\xi''|) - i\langle t, \eta - \xi\rangle].$$

We conclude from this that

$$|u_j(E(\cdot, \xi, t, \alpha, \beta)| \leq c_1 |\alpha\xi'|^{-d/2} |\beta\xi''|^{d/2 - n/2} \int |\hat{u}_j(\eta)| \exp[-|\eta' - \xi'|^2/(2\alpha|\xi'|)$$

$$-|\eta'' - \xi''|^2/(2\beta|\xi''|)]\, d\eta. \tag{A. 3.3}$$

Next we choose, for α', α'', etc., as in ii), $c_2 > 0$ and a sublinear function $g : (0, \infty) \to (0, \infty)$ so that, with $\alpha' \leq \alpha \leq \alpha''$, $\beta' \leq \beta \leq \beta''$,

α) $|\xi'' - \eta''| \leq c_2\beta|\xi''|$ implies $|\xi''| \sim |\eta''|$,

β) $|\xi' - \eta'| \leq c_2\alpha|\xi'|$ implies $|\xi'| \sim |\eta'|$,

γ) $|\xi'' - \eta''| \leq c_2\beta|\xi''|$, together with $|\xi' - \eta'| \leq c_2\alpha|\xi'|$ and $\xi \in \Gamma^0$ implies $\eta \in G^0$,

δ) $|\xi' - \eta'| \leq c_2\alpha|\xi'|$, $\xi' \in \Gamma^1$ implies $\eta' \in G^1$,

ε) $|\xi'| \geq g(|\xi|)$, together with $|\xi'' - \eta''| \leq c_2\beta|\xi''|$ and $|\xi' - \eta'| \leq c_2\alpha|\xi'|$ implies $|\eta'| \geq f(|\eta|)$.

In fact, the properties $\alpha), \beta), \gamma), \delta)$ are easy to satisfy and for ε) we need to know that for some previously fixed constants it follows from

$$\tau' \geq g(\tau), c_3\tau \leq t \leq c_4\tau, c_3\tau' \leq t' \leq c_4\tau',$$

that $t' \geq f(t)$.
This is clear if we set

$$g(\tau) = (1/c_3) \sup_{\theta \leq c_4\tau} f(\theta),$$

so we need only observe that the g defined by the preceding relation is sublinear.

To estimate $u_j(E(\cdot, \xi, t, \alpha, \beta))$ we shall now split the domain of integration in the integral from (A. 3.3), (for fixed α, β, $\xi \in \Gamma^0$, $\xi' \in \Gamma^1$) in the following four regions:

I. $|\xi'' - \eta''| \leq c_2\beta|\xi''|$, $|\xi' - \eta'| \leq c_2\alpha|\xi'|$,

II. $|\xi'' - \eta''| \leq c_2\beta|\xi''|$, $|\xi' - \eta'| > c_2\alpha|\xi'|$,

III. $|\xi'' - \eta''| > c_2\beta|\xi''|$, $|\xi' - \eta'| \le c_2\alpha|\xi'|$,

IV. $|\xi'' - \eta''| > c_2\beta|\xi''|$, $|\xi' - \eta'| > c_2\alpha|\xi'|$.

To study the contribution of the η from I) to the integral in (A. 3.3) we observe that for $\xi \in \Gamma^0$, $\xi' \in \Gamma^1$, $|\xi'| \ge g(|\xi|)$, we shall have $\eta \in G^0$, $\eta' \in G^1$, $|\eta'| \ge f(|\eta|)$. It follows therefore from the assumption that

$$|\hat{u}_j(\eta)| \le c_5(c_5j/|\xi'|)^j,$$

and the volume of the domain of integration can be estimated by $c_6|\xi|^n$. The contribution of the points in I) to $u_j(E(\cdot,\xi,t,\alpha,\beta))$ can therefore be estimated by

$$c_7|\xi'|^{-d/2}|\xi''|^{d/2-n/2}|\xi|^n c_5(c_5j/|\xi'|)^j.$$

Here we may assume that $g \ge 1$, so that $|\xi'|^{-d/2} \le c_8$.

In case II) the estimate (A. 3.2) is not anymore valid, but we have at least that

$$|\hat{u}_j(\eta)| \le c_9(1 + |\eta'| + |\xi''|)^{b'} \tag{A.3.4}$$

since $\{u_j\}_j$ is bounded in $E'(R^n)$ and $|\eta''| \sim |\xi''|$. Moreover, the domain of integration in η'' has volume of order $c_{10}|\xi''|^{n-d}$. As for decay, it comes from the estimates of the exponential. In fact, in this case $-|\eta' - \xi'|^2/(2\alpha|\xi'|) \le -c_{11}(|\xi'| + |\eta'|)$, so integration over the η in case II) gives a contribution of order

$$O(|\xi'|^{-d/2}|\xi''|^{d/2-n/2}(1 + |\xi|)^{b'}|\xi''|^{n-d}\exp[-c_{11}|\xi'|]).$$

The case III) can be treated in essentially the same way as the one before, taking also into account that we may assume that $|\xi'| \le |\xi''|$, and finall, in case IV) the exponential can be majorized by $\exp -c_{12}[|\xi| + |\eta|]$. Once more we obtain an estimate of the desired type after integration.

4. In the proof of iii) $\Rightarrow$i) we shall assume, to simplify notations, that $\alpha = 1$, $\beta = 1$, and write

$$E(x,\eta,t) = E(x,\eta,t,1,1).$$

It will be useful to consider the expression $u(E(\cdot,\eta,t))$ also for certain complex η and to perform a contour deformation in η. We must make then sure that the extension of E to such complex η is analytic. With the notations $\sum' = \sum_{j\le d}$, $\sum'' = \sum_{j\ge d+1}$, we shall in fact set

$$E(x,\eta,t) = \exp[-i\langle x,\eta\rangle - (\sum{}'\eta_j^2)^{1/2}|x' - t'|^2/2 - (\sum{}''\eta_j^2)^{1/2}|x'' - t''|^2/2].$$

The square root $t \to t^{1/2}$ is here assumed analytic for $t \in C \setminus R_-$, $R_- = \{\tau \in R; \tau < 0\}$ and is chosen positive for positive t.

Next we fix δ with supp $u \subset \{x; |x| \leq \delta\}$, choose a $C^\infty(R^n)$ function f for which $|x'| \geq \delta/2$ if $f(x) \neq 0$ and introduce the auxiliary function

$$w'(y) = \int_{|t| \geq \delta} \int_{\eta \in R^n} e^{i\langle y, \eta \rangle} |\eta'|^{d/2} |\eta''|^{n/2-d/2} (1+|\eta|^2)^{-n-1-m} f(t) u(E(\cdot, \eta, t)) \, dt \, d\eta, \quad \text{(A.3.5)}$$

where m is chosen larger than the order of u as a distribution. With this choice of m the integral in (A. 3.5) will be absolutely convergent. In fact

$$|u(E(\cdot, \eta, t))| \leq c(1+|\eta|)^m \exp[-\inf_{|x'|<\delta} |t' - x'|^2 |\eta'|/2 - \inf_{|x''|<\delta} |t'' - x''|^2 |\eta''|/2] \quad \text{(A.3.6)}$$

if m is larger than the order of u. Integrability in (A. 3.5) is then established by comparision with

$$\int_{\delta \leq |t| \leq M} \int_{\eta \in R^n} |\eta'|^{d/2} |\eta''|^{n/2-d/2} (1+|\eta|^2)^{-n-1} \, dt \, d\eta \, +$$

$$\int_{|t| \geq M} \int_{\eta \in R^n} |\eta'|^{d/2} |\eta''|^{n/2-d/2} (1+|\eta|^2)^{-n-1} e^{-|t'|^2|\eta'|/4 - |t''|^2|\eta''|/4} \, dt \, d\eta \, ,$$

for some large M.

The main preparation for the proof of iii) $\Rightarrow$i) is now

Lemma A.3.5. w' *is partially analytic in the variables* y' *near* 0.

(By this we mean that it satisfies Cauchy's estimates in the variables y', uniformly in y''.)

Proof of lemma A.3.5. At first we choose ε', a finite number of points $x'^1, \ldots, x'^r$, $|x'^k| = 1$ and $\varphi_l \in C_0^\infty(R^d)$, $l = 0, \ldots, r$, such that

α) supp $\varphi_0 \subset \{x'; |x'| < \delta/3\}$,
β) supp $\varphi_l \subset \{x'; |x'| > \delta/4\}$, $l = 1, \ldots, r$,
γ) $\langle y' - x', x'' \rangle \geq \delta/8$ if $x' \in$ supp φ_l, $|y'| \leq \varepsilon'$,
δ) $\sum_{l=0}^r \varphi_l(x') = 1$ for $x \in$ supp u.

We can then write $w' = \sum w_l$, where

$$w_l'(y) = u[\int_{|t| \geq \delta} \int_{\eta \in R^n} e^{i\langle y, \eta \rangle} |\eta'|^{d/2} |\eta''|^{n/2-d/2} (1+|\eta|)^{-n-1-m} f(t) \varphi_l(x') E(\cdot, \eta, t) \, dt \, d\eta].$$

$$\text{(A.3.7)}$$

Here it is clear that w_0' will be partially analytic in y' for small y, since for any γ we have

$$|D_x^\gamma [\varphi_0(x') e^{-|\eta'||x' - t'|^2/2}]| = O(\exp[-(\delta^2/36)|\eta'|]), \text{ if } t \in \text{ supp } f.$$

We would like to obtain similar estimates also for the integrals from (A. 3.7) when $l \geq 1$. This is possible if we replace integration in η' from an integration on R^d by an integration on a suitable contour from C^d. Let us in fact denote by

$$\Lambda' = \{\zeta' \in C^d; \zeta' = \eta' + ix''|\eta'|, \eta' \in R^d \}$$

and by

$$\Omega' = \{\zeta' \in C^d; \zeta' = \eta' + i\,\mu\,x''|\eta'|, \text{ where } \eta' \in R^d \text{ and } 0 \leq \mu \leq 1\}.$$

In particular, we have $|Im\,\zeta'| \leq |Re\,\zeta'|$ if $\zeta' \in \Omega'$.

When $y, |y| \leq \varepsilon'$, is real, $\zeta \in \Omega$, $x' \in \text{supp } \varphi_l$, we have

$$Re(i\langle y' - x', \zeta'\rangle - (\sum{}'\zeta_j^2)^{1/2}|x' - t'|^2/2) \leq -\delta\,|Im\,\zeta'|/8 - (|Re\,\zeta'| \\ -|Im\,\zeta'|)|x' - t'|^2)/2,$$

so

$$w_l'(y) = u[\int_{|t| \geq \delta} \int_{\zeta' \in \Lambda'} \int_{\eta'' \in R^{n-d}} \\ e^{i\langle y' - x', \zeta'\rangle + i\langle y'' - x'', \eta''\rangle - (\sum{}'\zeta_j^2)^{1/2}|x' - t'|^2/2 - |\eta''|\,|x'' - t''|^2/2} \\ |\sum{}'\zeta_j^2|^{d/2}|\eta''|^{n/2-d/2}(1 + \sum\zeta_j^2)^{-n-1-m}f(t)\varphi_l(x')\,dt\,d\zeta'\,d\eta''],$$

if we use Stokes' formula, and set $\zeta_j = \eta_j$ for $j > d$ (in the expression $(1 + \sum\zeta_j^2)^{-n-1-m}$). We parametrize here Λ' by R^d and obtain, denoting with A some suitable functional determinant (which appears when we compute $d\zeta'$ via $d\eta'$) ,

$$w_l'(y) = u[\int_{|t| \geq \delta} \int_{\eta \in R^n} e^{i\langle y - x, \theta\rangle - (\sum{}'\theta_j^2)^{1/2}|x' - t'|^2/2 - |\eta''|\,|x'' - t''|^2/2} \\ |\sum{}'\theta_j^2|^{d/2}|\eta''|^{n/2-d/2}(1 + \sum\theta_j^2)^{-n-1-m}f(t)\varphi_l(x')A(\eta)\,dt\,d\eta],$$

where $\theta = (\eta' + ix''|\eta'|, \eta'')$. It follows easily that w_l' is partially analytic in y' for $|Re\,y'| < \varepsilon'$, $|Im\,y'| < \delta/16$. In fact, for $x' \in \text{supp }\varphi_l$ and such y' we have

$$Re\,(i\langle y' - x', \eta' + ix''\,|\eta'|\rangle) \leq (\delta/16)|\eta'| - (\delta/8)\,|\eta'|.$$

so w_l' is well-defined for $|Re\,y| < \varepsilon'$, $|Im\,y'| < \delta/16$ and it is also clear that it satisfies the Cauchy-Riemann relations in y' there.

5. After these preparations we now turn to the proof of iii) $\Rightarrow$i). The assumption is that for suitable cones G^0, G^1,

$$u(E(\cdot, \eta, t)) = O(\exp[-c|\eta'|]), \text{ when } \eta' \in G^1, \eta \in G^0, |t - x^0| < \delta, |\eta'| \geq f(|\eta|),$$

for some suitable sublinear function f.

It is no loss of generality to assume that $x^0 = 0$. Next we fix $\Gamma^0 \subset\subset G^0$, $\Gamma^1 \subset\subset G^1$ and a sequence $\chi_j \in C_0^\infty(|x| < \varepsilon)$ such that

$$|D^\alpha \chi_j(x)| \le c_1(c_1 j)^{|\alpha|} \text{ for } |\alpha| \le j + m + n + 1 \qquad (A.3.8)$$

where m will be specified somewhat later. As in the proof of i) $\Rightarrow$ii) we now fix c_2 and g with the properties $\alpha)$ to $\varepsilon)$ from the proof of i) $\Rightarrow$ ii). We shall then show that if ε was suitable, then

$$|u(\chi_j e^{-i\langle x,\xi\rangle})| \le C(Cj/|\xi'|)^j(1 + |\xi|)^{m+2n} \text{ for } \xi \in \Gamma^0, \ \xi' \in \Gamma^1, \ |\xi'| \ge g(|\xi|). \qquad (A.3.9)$$

Our first remark (which is from Sjöstrand [2]) is to use

$$\int_{R^n} |\eta'|^{d/2} |\eta''|^{n/2-d/2} \exp[-|\eta'| \, |x'-t'|^2/2 - |\eta''| \, |x''-t''|^2/2] \, dt = (2\pi)^{n/2}. \qquad (A.3.10)$$

When combined with Parseval's identity, this gives

$$u(\chi_j \exp(-i\langle x,\xi\rangle)) = (2\pi)^{-n} \int\int \hat{\chi}_j(-\eta+\xi) |\eta'|^{d/2} |\eta''|^{n/2-d/2} u(E(\cdot,\eta,t)) dt \, d\eta. \qquad (A.3.11)$$

In (A. 3.11) we shall now split the integration in t into the parts $|t| < \delta$ and $|t| \ge \delta$. Furthermore we shall distinguish, when $|t| < \delta$, the following cases for the integration in η (similar to the ones already used above):

I. $|\xi'' - \eta''| \le c_2\beta|\xi''|, \ |\xi' - \eta'| \le c_2\alpha|\xi'|,$

II. $|\xi'' - \eta''| \le c_2\beta|\xi''|, \ |\xi' - \eta'| > c_2\alpha|\xi'|,$

III. $|\xi'' - \eta''| > c_2\beta|\xi''|, \ |\xi' - \eta'| \le c_2\alpha|\xi'|.$

IV. $|\xi'' - \eta''| > c_2\beta|\xi''|, \ |\xi' - \eta'| > c_2\alpha|\xi'|.$

The integral for $|t| < \delta$ can now be handled with a discussion parallel to that from the part i)$\Rightarrow$ii) of the proof . We mention the main novelties and leave details for the reader. In fact, in case I), $u(E(\cdot,\eta,t))$ admits (as a consequence of the assumptions and of the properties of c_2 and g) the estimate

$$|u(E(\cdot,\eta,t))| \le c_3 e^{-c_4|\xi'|},$$

if $\xi \in \Gamma^0$, $\xi' \in \Gamma^1$, $|\xi'| \ge g(|\xi|)$. In case II), we can only estimate $|u(E(\cdot,\eta,t))|$ by $c_5(1 + |\eta'| + |\xi''|)^m$. However, for $\hat{\chi}_j(-\eta+\xi)$ we have the estimate

$$|\hat{\chi}_j(-\eta+\xi)| \le c_6(c_6 j/|\xi-\eta|)^j \le c_7(c_7 j/(|\xi'| + |\eta'|))^j$$

and a similar estimate is valid in case III. Finally, in case IV we have $|u(E(\cdot, \eta, t))| \leq c_8(1 + |\eta|)^m$ and

$$|\hat{\chi}(-\eta + \xi)| \leq c_9(c_9 j/(|\xi| + |\eta|))^j (1 + |\xi| + |\eta|)^{-m-n-1}.$$

After the discussion of the integral (A.3.11) for $|t| < \delta$ we are now left with the estimation of

$$\int_{|t| \geq \delta} \int \hat{\chi}_j(-\eta + \xi) |\eta'|^{d/2} |\eta''|^{n/2 - d/2} u(E(\cdot, \eta, t)) dt \, d\eta.$$

In this estimation we shall essentially rely on lemma A.3.5. To prepare for the application of this lemma it is convenient to introduce a convergence factor in η. More precisely, we have

$$J(\xi) = \int_{|t| \geq \delta} \int \hat{\chi}_j(-\eta + \xi) |\eta'|^{d/2} |\eta''|^{n/2 - d/2} (1 + |\eta|^2)^{n+m+1} (1 + |\eta|^2)^{-n-m-1}$$

$$u(E(\cdot, \eta, t)) dt \, d\eta = (2\pi)^n \int w(y)(1 - \Delta_y)^{n+m+1} (\chi_j(y) e^{-i\langle y, \xi \rangle}) \, dy$$

where

$$w(y) = \int_{|t| > \delta} \int e^{i\langle y, \eta \rangle} |\eta'|^{d/2} |\eta''|^{n/2 - d/2} (1 + |\eta|^2)^{-n-m-1} u(E(\cdot, \eta, t)) dt \, d\eta$$

and where Δ_y is the Laplacian in the variables y. (m is once more larger than the order of u.) The idea is now to show that for $|y| < \varepsilon'$ and ε' small, $w = w_1 + w_2$ for two functions w_i, one partially analytic in y' and the other partially analytic in y''. Note that for both functions

$$\int w_i(y)(1 - \Delta_y)^{n+m+1} (\chi_j(y) e^{-i\langle y, \xi \rangle}) dy$$

can be estimated as desired in i).

To obtain the desired decomposition of w, we consider two continuous functions f_i, $i = 1, 2$, so that

$$f_1 + f_2 \equiv 1 \text{ on } |t| > \delta$$

$$|f_i(t)| \leq 1, \forall t,$$

$$f_1(t) \neq 0 \text{ implies } |t'| > \delta/2,$$

$$f_2(t) \neq 0 \text{ implies } |t''| > \delta/2.$$

Next we set

$$w_i(y) = \int_{|t| > \delta} \int e^{i\langle y, \eta \rangle} |\eta'|^{d/2} |\eta''|^{n/2 - d/2} (1 + |\eta|^2)^{-n-m-1} f_i(t) u(E(\cdot, \eta, t)) dt \, d\eta.$$

It follows then that w_1 is partially analytic for small y' as a consequence of lemma A.3.5. To obtain partial analyticity of w_2 in y'' we may apply once more lemma A.3.5 with the roles of y' and y'' interchanged.

Chapter 3

Pseudodifferential operators

1. In the present chapter we describe the calculus of pseudodifferential operators in k-microlocalization and prove theorem 1.1.6, together with a related result in the constant coefficient case. The calculus of pseudodifferential operators is more or less along standard lines but we will take advantage of the fact that the basic results about it are consequences of results on pseudodifferential operators in general G_φ-classes, even though an independent exposition would be much shorter than is the (general) calculus in such classes. It remains then mainly to stress a few points which are due to the additional structure which is present in k-microlocalization. In fact, the main theme of a large part of the chapter is to explain how k-microlocalization is related to what we call "poly-homogeneous structures". In particular, we shall encounter then sums of polyhomogeneous functions and the interesting thing is that we can associate a weight to their "polydegree" of polyhomogeneity, obtaining order relations in the set of these polydegrees. Actually, more choices of such order relations seem possible, according to which variant of k-wave front set one intends to use. It is remarkable that the order relation most naturally associated with $WF^k_{A,s}$ is total, so that we have a useful notion of "principal parts" in k-microlocalization for $WF^k_{A,s}$. Indeed, and we will comment on this in a moment, the order relation which we shall consider is chosen precisely with this goal in mind. Of course, polyhomogencity has direct consequences at the level of estimates and the principal part of a finite sum of polyhomogeneous functions will automatically dominate the lower order terms if it is elliptic. (Somewhat more is in fact true.) All this should be seen in contrast with the situation for WF^k_A, for which the associated order relation is not total and for which principal parts cannot be related directly to polyhomogeneity.

2. Since the order relation for degrees of polyhomogeneity is crucial in all what follows, it might be useful to explain in the case of a simple example what it is supposed to do

and how it is related to our choices of wave front sets. (A more elaborated situation is the one from section 3.5.) Assume then that p_m is a homogeneous polynomial in the variables ξ and that $\xi^0 = (0,...,0,1)$, $\xi^1 = (0,...,0,1,0)$, both in R^n. The localization polynomials $p_{m,1}$ and $p_{m,2}$ at ξ^0 and ξ^1 are then related to p_m by

$$p_m(\xi) = \xi_n^s p_{m,1}(\xi') + q_0(\xi), \quad q_0(\xi) = O(|\xi_n|^{s-1}|\xi'|^{m-s+1}),$$

$$p_{m,1}(\xi') = \xi_{n-1}^{s'} p_{m,2}(\xi'') + q_1(\xi'), \quad q_1(\xi') = O(|\xi_{n-1}|^{s'-1}|\xi''|^{m-s-s'+1}),$$

where $\xi' = (\xi_1,...,\xi_{n-1})$, $\xi'' = (\xi_1,...,\xi_{n-2})$. It follows that

$$p_m(\xi) = \xi_n^s \xi_{n-1}^{s'} p_{m,2}(\xi'') + O(|\xi_n|^s |\xi_{n-1}|^{s'-1}|\xi''|^{m-s-s'+1}) + O(|\xi_n|^{s-1}|\xi'|^{m-s+1}).$$

We have here essentially three groups of variables: ξ'', ξ_{n-1} and ξ_n. The polynomial $\xi_n^s \xi_{n-1}^{s'} p_{m,2}$ is homogeneous in each of these groups of variables, but the remainder terms are perhaps not. The q_0 and the q_1 are however polynomials, so we can write them at least as sums of monomials and these monomials are of course homogeneous in the respective groups of variables. We want then to have that the term $\xi_n^s \xi_{n-1}^{s'} p_{m,2}$ dominates all the remainder terms when p_{m-2} is elliptic in the variables ξ''. Practically we must then work, at least in the general case, in a situation in which

$$|\xi_n^s \xi_{n-1}^{s'}| \, |\xi''|^{m-s-s'} \text{ dominates } |\xi_n^{s-1}| |\xi'|^{m-s+1}, \text{ respectively } |\xi_n|^s |\xi_{n-1}|^{s'-1}|\xi''|^{m-s-s'+1},$$

and the order relation which we shall choose later on will just reflect this fact notationally. Note that the first of the above conditions says that we want $|\xi''|$ to dominate $|\xi_{n-1}|^{1+\beta}/|\xi_n|^\beta$ for $\beta = 1/(m-s-s')$. This is of course only possible if our main conditions will live on sets of type $|\xi''| \geq c|\xi_{n-1}|^{1+\beta}/|\xi_n|^\beta$. This is essentially what happened in our definition of the analytic semiisotropic wave front set in chapter II. (The second of the above conditions gives that $|\xi''|$ should be dominated by $|\xi'|$.)

3.1   Polyhomogeneous structures and polycones

1. Let M_j, $j = 0,...,k$ be a a sequence of linear subspaces of R^n as in the definition of $WF_{A,s}^k$, i.e. assume that $M_0 = R^n$, $M_j \subset M_{j-1}$, $M_j \neq M_{j-1}$, $M_k = \{0\}$ and denote, as in chapter II, by $\Pi_j : R^n \to M_j$ the orthogonal projection of R^n on M_j. For reasons which will become clear later on in this and the following section, we shall call the M_j a "polyhomogeneous" (respectively a "bihomogeneous", when we explicitly refer to the case $k = 2$) structure on R^n.

On each M_j we have a natural multiplication $(t,\xi) \to t\xi$ of elements $\xi \in M_j$ with $t \in R$. We extend this multiplication to a linear operator $\mu_j(t) : R^n \to R^n$ by setting

$$\mu_j(t)\xi = t\Pi_j(\xi) + (I - \Pi_j)(\xi), \tag{3.1.1}$$

where I is the identity operator. It follows in particular that

$$\mu_j(t)\Pi_i(\xi) = ((t-1)\Pi_j + I)\Pi_i\xi = \Pi_i((t-1)\Pi_j + I)\xi = \Pi_i(\mu_j(t)\xi), \forall i, j.$$

We also observe that μ_j leaves elements from $R^n \ominus M_j$ unchanged and that $\mu_0(t)$ is the standard multiplication on R^n. Of course, when coordinates are chosen so that $M_j = \{\xi; \xi_i = 0 \text{ for } i \geq 1 + d_j\}$, i.e. when M_j can be identified with the space of the variables $\xi_1, ..., \xi_{d_j}$, then

$$\mu_j(t)\xi = (t\xi_1, ..., t\xi_{d_j}, \xi_{1+d_j}, ..., \xi_n).$$

Note that $\mu_j(t)\mu_j(t') = \mu_j(tt')$, and that $\mu_j(1)$ is the identity, so we may regard μ_j as a multiplication operator on R^n. Moreover, it is obvious that

$$\mu_j(t)\mu_{j'}(t') = \mu_{j'}(t')\mu_j(t). \tag{3.1.2}$$

Finally, if $j > j'$, then $\mu_j(t)\mu_{j'}(t')$ comes to :

a) multiplication of the elements from M_j by tt',
b) multiplication of the elements from $M_{j'} \ominus M_j$ by t',
c) whereas the elements from $R^n \ominus M_{j'}$ are left unchanged.

In particular, $\dot{\mu}_j(t) = \mu_{j+1}(1/t)\mu_j(t)$ will act on the elements of $\dot{M}_j$ as the multiplication by t and leaves the elements from the orthogonal of $\dot{M}_j$ unchanged. As in the related situation in section 2.1, this suggests that it could sometimes be convenient to consider the multiplications $\dot{\mu}_j$ as the primary objects of study and to built up the operators μ_j from them afterwards. This is indeed the case, but we shall see in chapter IV that (at least in second microlocalization) it is the μ_j, and not the $\dot{\mu}_j$ which have the most immediate geometric meaning.

It also follows by induction that the operator

$$\mu_{k-1}(t_{k-1}/t_{k-2})\mu_{k-2}(t_{k-2}/t_{k-3})\cdots\mu_1(t_1/t_0)\mu_0(t_0), \tag{3.1.3}$$

multiplies the elements from M_{k-1} by t_{k-1} and those from $\dot{M}_j$ by t_j if $j < k-1$. (Note that actually $\dot{M}_{k-1} = M_{k-1}$.)

2. Our aim in the next section is to study functions which commute with the operators μ_j in the same way in which homogeneous functions commute with standard multiplication. Before we come to this study, we need some information on the geometry induced by the multiplications μ_j.

Lemma 3.1.1. *Let $A \subset R^n$ be fixed. Then there are equivalent:*

a) A is a cone and $\xi \in A$ implies $\mu_j(t)\xi \in A$, for all j, provided that $0 \leq t \leq 1$.

b) $\xi \in A$ implies

$$\mu_{k-1}(t_{k-1}/t_{k-2})\mu_{k-2}(t_{k-2}/t_{k-3})\cdots\mu_1(t_1/t_0)\mu_0(t_0) \in A$$

provided that $0 < t_{k-1} \leq t_{k-2} \leq \cdots \leq t_0$.

Proof. **a)** $\Rightarrow$ **b)** is by finite induction. The initial step, namely that $\xi \in A$, implies $\mu_0(t_0)\xi \in A$, is a consequence of our assumption that A is a cone. Passage from $k-1$ to k works, since the additional factor $\mu_{k-1}(t_{k-1}/t_{k-2})$ which then appears, transforms elements from A into elements from A as a consequence of a): since $0 < t_{k-1} \leq t_{k-2}$, we have $0 < t_{k-1}/t_{k-2} \leq 1$.

b) $\Rightarrow$ **a)** follows by specialization. We simply apply b) for $t_s = t$ if $s \geq j$, $t_s = 1$ for $s < j$. For $j = 0$ there is no restriction on t, so we obtain in particular that A is a cone.

Definition 3.1.2. *a) We say that A is a polycone if it satisfies one of the two equivalent conditions from lemma 2.1.1.*

b) If $A \subset R^n$ is given, we call polycone generated by A, the smallest polycone which contains A.

Remark 3.1.3. *Let $A \subset R^n$ be given and denote by Γ the polycone generated by A. Then*

$$\begin{aligned}\Gamma \;=\; &\{\mu_{k-1}(t_{k-1}/t_{k-2})\mu_{k-2}(t_{k-2}/t_{k-3})\cdots\mu_1(t_1/t_0)\mu_0(t_0)\xi \;;\\ &\quad \xi \in A, 0 < t_{k-1} \leq \cdots \leq t_0\}.\end{aligned} \tag{3.1.4}$$

Indeed, if we denote by Γ' the set defined by the right hand side of (3.1.4), then we must have $\Gamma' \subset \Gamma$, so it suffices to check that Γ' itself is a polycone. This follows from

$$\begin{aligned}\mu_{k-1}(t'_{k-1}/t'_{k-2})\cdots\mu_0(t'_0)\mu_{k-1}(t_{k-1}/t_{k-2})\cdots\mu_0(t_0) \;=\; &\mu_{k-1}(t_{k-1}t'_{k-1}/(t_{k-2}t'_{k-2})\cdots\\ &\mu_0(t_0t'_0).\end{aligned}$$

Lemma 3.1.4. *Let $A \subset R^n$ be given so that $0 < c < |(\Pi_j - \Pi_{j+1})\eta| \leq c'$ for all $\eta \in A$ and all j. Denote by Γ the polycone generated by A and assume that $\xi \in \Gamma$ and $\eta \in A$ are related by*

$$\xi = \mu_{k-1}(t_{k-1}/t_{k-2})\mu_{k-2}(t_{k-2}/t_{k-3})\cdots\mu_1(t_1/t_0)\mu_0(t_0)\eta, \tag{3.1.5}$$

for some t_j which satisfy $0 < t_{k-1} \leq \cdots \leq t_0$.

Then it follows for some constants c_1, c_2, which do not depend on ξ or η, that

$$c_1 t_j \leq |\Pi_j(\xi)| \leq c_2 t_j. \tag{3.1.6}$$

Proof. Recall that (3.1.5) means that $(\Pi_j - \Pi_{j+1})(\xi) = t_j(\Pi_j - \Pi_{j+1})(\eta)$, for all j, with the convention that $\Pi_k = 0$. This shows that

$$|(\Pi_j - \Pi_{j+1})(\xi)| = t_j|(\Pi_j - \Pi_{j+1})(\eta)|$$

so that the assumption on A implies

$$c t_j \leq |(\Pi_j - \Pi_{j+1})(\xi)| \leq c' t_j. \tag{3.1.7}$$

From (3.1.7) we get $c t_j \leq |\Pi_j(\xi)|$ since $|(\Pi_j - \Pi_{j+1})(\xi)| \leq |\Pi_j(\xi)|$, which is the first inequality from (3.1.6) for all j. On the other hand, (3.1.7) also gives the right inequality from (3.1.6) for $j = k-1$ since then $\Pi_k = 0$. We can therefore prove the second inequality from (3.1.6) by decreasing induction in j. In fact from (3.1.7) it follows that

$$|\Pi_j(\xi)| \leq |(\Pi_j - \Pi_{j+1})\xi| + |\Pi_{j+1}(\xi)| \leq c' t_j + c'_{j+1} t_{j+1}.$$

Here c'_{j+1} is a constant which depends on j. Since $t_{j+1} \leq t_j$ we can conclude the proof observing that $c' t_j + c'_{j+1} t_{j+1} \leq c'_j t_j$, where $c'_j = c' + c'_{j+1}$.

3. In the next result we relate the polyhomogeneous structure given by polycones to the multineighborhoods of form $G = \{\xi; \Pi_j \xi \in G^j\}$, G^j conic in M_j, which appear in the definition of the k-wave front sets. It will turn out in fact that these two types of sets serve the same purpose. It is of course precisely with this result in mind that we asked for the condition $t_{k-1} \leq \cdots \leq t_0$ in definition 3.1.2.

Lemma 3.1.5. *Let $\xi^j \in \dot{M}_j$ be fixed.*

a) Assume that we are given open cones $G^j \subset M_j$ which contain ξ^j and denote by

$$G = \{\xi \,;\, \Pi_j \xi \in G^j\}. \tag{3.1.8}$$

Then we can find $\sigma_j > 0, j = 0, ..., k-1$ and an open polycone Γ which contains $\theta = \sum_{j<k} \sigma_j \xi^j$ so that $\Gamma \subset G$.

b) Conversely, if $\sigma_j > 0$, $j = 0, ..., k-1$ and an open polycone Γ are given so that $\theta = \sum_{j<k} \sigma_j \xi^j \in \Gamma$, then we can find open cones $G^j \subset M_j$ which contain ξ^j so that G defined by (3.1.8) is contained in Γ.

Remark and Definition 3.1.6. *The preceding lemma reveals in particular the importance which the vectors of form $\theta = \sum \sigma_j \xi^j$, $\sigma_j > 0$, have in the present theory. For this reason we shall call any such θ a "generating element" for (or "associated with") $(\xi^0, \xi^1, ..., \xi^{k-1})$. The set of all such θ shall also be called occasionally the "polyray" associated with $(\xi^0, \xi^1, ..., \xi^{k-1})$. (It should perhaps be called "poly-halfray", since we allow only $\sigma_j > 0$. For simplicity, we shall prefer "polyray".) A set Γ will be called a polyneighborhood of $(\xi^0, \xi^1, ..., \xi^{k-1})$ precisely when we can find an open polycone $\Gamma' \subset \Gamma$ which contains a generating element θ for $(\xi^0, \xi^1, ..., \xi^{k-1})$.*

Proof of lemma 3.1.5. It is no loss in generality to assume that $|\xi^j| = 1$. To prove part a) we fix $c_j > 0$ so that

$$\begin{aligned}
\Gamma^j \;=\; & \{\xi \in M_j; |\Pi_{j+1}\xi| \le c_j|\xi - \Pi_{j+1}\xi|, \\
& |\xi^j - (\xi - \Pi_{j+1}\xi)/|\xi - \Pi_{j+1}\xi|| < c_j\} \subset G^j
\end{aligned} \tag{3.1.9}$$

and check that

i) $\Gamma = \{\xi; \Pi_j\xi \in \Gamma^j, \forall j\}$ is a polycone,

ii) for a suitable choice of $\sigma_j > 0$ it follows that $\theta = \sum_{j<k} \sigma_j \xi^j \in \Gamma$.

Proof of i). Since Γ is a cone, it suffices to show that $\xi \in \Gamma$ implies $\Pi_j(\mu_i(t)\xi) \in \Gamma^j$ for all i and j, if $0 \le t \le 1$. This will follow if we can prove that $\xi \in \Gamma^j$ implies $\mu_i(t)\xi \in \Gamma^j$ under the same conditions for i, j and t. Since $\mu_i(t)\xi = \mu_j(t)\xi$ for $i < j$ and $\xi \in M_j$ it is no loss in generality to assume that $i \ge j$. Moreover, Γ^j is a cone in M_j, so we may even assume that $i > j$.
Note then that for $0 \le t \le 1, \xi \in \Gamma^j$,

$$|\Pi_{j+1}(\mu_i(t)\xi)| \le |\Pi_{j+1}\xi| \le c_j|\xi - \Pi_{j+1}\xi| = c_j|\mu_i(t)(\xi - \Pi_{j+1}\xi)|,$$

since the μ_i leaves $\xi - \Pi_{j+1}\xi$ unchanged for $i > j$. This gives the first condition for the definition of the Γ^j. As for the second, it follows once more from the fact that μ_i leaves $\xi - \Pi_{j+1}\xi$ unchanged when $i > j$.

Proof of ii). Denote by $\eta^i = \Pi_i\theta = \sum_{j\ge i} \sigma_j \xi^j$. What we must show is that $\eta^i \in \Gamma^i$, if the $\sigma_j > 0$ are chosen suitably. This gives no restriction on σ_{k-1}, so we may choose $\sigma_{k-1} = 1$, say. We shall now find the σ_i succesively by decreasing induction, using the fact that η^i is known once the σ_j have been found for $j \ge i$ (although θ itself is not yet determined). If we have already found $\sigma_j, j > i$, for which $\eta^{i+1} \in \Gamma^{i+1}$, then we must choose σ_i with $\eta^i \in \Gamma^i$. The first condition from the definition of Γ^i gives $|\eta^{i+1}| \le c_{i+1}\sigma_i$, since $|\xi^{i+1}| = 1$. Since η^{i+1} is assumed to be known at this stage of the argument, this is a condition on σ_i. The second condition from (3.1.9) is trivially satisfied since $(\eta^i - \Pi_{i+1}\eta^i)/|\eta^i - \Pi_{i+1}\eta^i| = \xi^i$. It is then clear that we can choose the σ_i as desired, using decreasing induction.

Proof of lemma 3.1.5 b). Let Γ, θ be as in the assumption. Since

$$R^n = \oplus_{j\geq 1}(\Pi_{j-1} - \Pi_j)R^n$$

and Γ is open, we can find $c > 0$ so that

$$A = \{\xi; |(\Pi_{j-1} - \Pi_j)\xi/\sigma_j - \xi^j| < c, \forall j\} \subset \Gamma. \tag{3.1.10}$$

Consider next for $j = 0, ..., k-1$ the cones

$$\begin{aligned} G^j &= \{\xi \in \dot{M}_j; |\Pi_{j+1}(\xi)| < c'|\xi - \Pi_{j+1}(\xi)|, \tag{3.1.11}\\ &\quad ||\xi - \Pi_{j+1}(\xi)|^{-1}(\xi - \Pi_{j+1}(\xi)) - \xi^j| < c'\}. \end{aligned}$$

We claim that if c', $0 < c' \leq c$ is sufficiently small, then

$$G = \{\xi; \Pi_j(\xi) \in G^j\} \tag{3.1.12}$$

is contained in the polycone generated by A. Recall that this means that for any $\xi \in G$ we can find $0 < t_{k-1} \leq t_{k-2} \leq ... \leq t_0$ and $\eta \in A$ such that

$$\xi = \mu_{k-1}(t_{k-1}/t_{k-2})\cdots\mu_0(t_0)\eta. \tag{3.1.13}$$

Also recall from the proof of lemma 3.1.4 that (3.1.13) means that

$$t_j^{-1}(\Pi_j - \Pi_{j+1})\xi = (\Pi_j - \Pi_{j+1})\eta. \tag{3.1.14}$$

What we then want is to find $t_j, t_j \geq t_{j+1} \geq 0$, so that $\xi \in G$ implies

$$|\xi^j - \sigma_j^{-1}t_j^{-1}(\Pi_j - \Pi_{j+1})\xi| < c, \forall j. \tag{3.1.15}$$

Of course, (3.1.15) will follow from the second inequality in (3.1.11) if we assume $c' \leq c$ and that

$$t_j\sigma_j = |(\Pi_j - \Pi_{j+1})\xi|,$$

but we must also make sure that $t_{j-1} \geq t_j$. In fact, the preceding relation shows that this will be valid, if we can make sure that

$$|(\Pi_j - \Pi_{j+1})\xi|/\sigma_j \leq |(\Pi_{j-1} - \Pi_j)\xi|/\sigma_{j-1}. \tag{3.1.16}$$

This can be deduced from the first inequality in (3.1.11) if we choose c' small. In fact, for suitable c' we may assume at first that $|\Pi_j \xi|/2 \leq |(\Pi_j - \Pi_{j+1})\xi| \leq 2|\Pi_j \xi|$, $\forall j$. The first inequality from (3.1.11) then gives

$$|(\Pi_j - \Pi_{j+1})\xi|/\sigma_j \leq 2|\Pi_j \xi|/\sigma_j \leq (2c'/\sigma_j)|(\Pi_{j-1} - \Pi_j)\xi|,$$

which gives (3.1.16) if $2c'/\sigma_j \leq 1/\sigma_{j-1}$.

5. Given the importance which the operators $\mu_j(t)$ will aquire later on, we conclude the section with a simple-minded remark on families of linear operators $\mu_j(t) : R^n \to R^n$, $t \in R$, for which, in analogy with the above, $\mu(t)\mu(t') = \mu(tt')$ and $\mu(1) = id$. (Actually, it would suffice for all appliacations to consider multiplications for $t > 0$. This would reduce us to representations of R_+ in the set of matrices. Multiplications are studied form this point of view in von Grudzinski [1].)

Let us at first observe that $\mu(0)$ is a projection. If $A = ker(id - \mu(0))$, $B = ker\mu(0)$ it follows that

$$R^n = A \oplus B.$$

Moreover, if $\xi \in A$ then $\mu(t)\xi = \mu(t)\mu(0)\xi = \mu(0)\xi = \xi$, so the restriction to A of $\mu(t)$ is the identity for any t.

We may then wonder, under what additional assumptions we could conclude that

$$\mu(t)\xi = t(\mu(1) - \mu(0))\xi + \mu(0)\xi. \tag{3.1.17}$$

(Recall, that this was the structure of the multiplication operators μ_j considered in the above.) It is easy to see that (3.1.17) will hold, if we assume that $t \to \mu(t)\xi$ is affine for any fixed ξ. In fact, in this case we would have at first that $\mu(t)\xi = t\xi^1 + \xi^2$, for some $\xi^1, \xi^2 \in R^n$ (which depend on ξ), but it is then clear that $\xi^2 = \mu(0)\xi$. For $t = 1$ we must then also have $\xi = \xi^1 + \mu(0)\xi$, so that in fact, $\xi^1 = (\mu(1) - \mu(0))\xi$. On the other hand, we should note that (3.1.17) does not follow if we assume only the continuity of $t \to \mu(t)\xi$. (Indeed, when e.g. $n = 1$, then $t \to \sqrt{|t|}\xi = \mu(t)\xi$ defines a continuous map such that $\mu(1) = id$, $\mu(t)\mu(t') = \mu(tt')$ which is not affine.)

3.2   Polyhomogeneity and polyhomogeneous symbols

1. Polyhomogeneous functions are related to a polyhomogeneous structure defined by a sequence of subspaces $M_j \subset M_{j-1}$ in much the same way in which homogeneous functions are associated with the standard multiplication in R^n.

Definition 3.2.1. *Assume that $f : A \to C$, where A is a polycone (for some fixed sequence of linear subspaces M_j as in section 3.1.) Let also $m = (m_0, ..., m_{k-1})$, $m_i \in R$ be given. We say that f is positively homogeneous of "polydegree" m if*

$$f(\mu_j(t)\xi) = t^{m_j}f(\xi), \forall t, 0 \le t \le 1 \ if \ j \ge 1, \ and \ \forall t > 0 \ if \ j = 0. \tag{3.2.1}$$

(We shall sometimes, by abuse, speak about "polyhomogeneous functions" without that the domain of definition of the respective function has been specified. In such situations the function is normally defined in a polyneighborhood of some point $(\xi^0, \xi^1, ..., \xi^{k-1})$ and it is assumed that it is polyhomogeneous on an open polycone which contains some generating element for $(\xi^0, \xi^1, ..., \xi^{k-1})$.)

Obviously, when f, g are positively polyhomogeneous of some polydegrees m, m', then $f \cdot g$ is polyhomogeneous of polydegree $m + m' = (m_0 + m'_0, ..., m_{k-1} + m'_{k-1})$ and $1/f$ is polyhomogeneous of polydegree $-m = (-m_0, ..., -m_{k-1})$, if $f \ne 0$. Finally we observe that if f is positively polyhomogeneous of polydegree $(m_0, ..., m_{k-1})$, then it is positively homogeneous of degree $m_j - m_{j+1}$ in the variables from $\dot{M}_j$.

Lemma 3.2.2. *Let f be polyhomogeneous of polydegree m in a polyneighborhood G of $(\xi^0, \xi^1, ..., \xi^{k-1})$. Then we can find a polyneighborhood Γ of $(\xi^0, \xi^1, ..., \xi^{k-1})$ and a constant c so that*

$$|f(\zeta)| \le c|\Pi_{k-1}(\xi)|^{m_{k-1}}|\Pi_{k-2}(\xi)|^{m_{k-2} - m_{k-1}} \cdots |\Pi_0(\xi)|^{m_0 - m_1}, \ for \ \xi \in \Gamma. \tag{3.2.2}$$

b) Conversely, if f does not vanish on G, then we may choose $c' > 0$ and Γ so that also

$$|f(\xi)| \ge c'|\Pi_{k-1}(\xi)|^{m_{k-1}}|\Pi_{k-2}(\xi)|^{m_{k-2} - m_{k-1}} \cdots |\Pi_0(\xi)|^{m_0 - m_1}, \xi \in \Gamma.$$

c) If, moreover, f is real analytic on G, then we can find c, Γ so that

$$|\partial_\xi^\alpha f(\xi)| \le c^{|\alpha|+1}\alpha!|\Pi_{k-1}(\xi)|^{-|\alpha|}|\Pi_{k-1}(\xi)|^{m_{k-1}}|\Pi_{k-2}(\xi)|^{m_{k-2} - m_{k-1}} \cdots |\Pi_0(\xi)|^{m_0 - m_1},$$
$$for \ \xi \in \Gamma. \tag{3.2.3}$$

Proof of lemma 3.2.2. We choose Γ to be the polycone generated by some small neighborhood A of some generating element $\theta = \sum \sigma_j \xi^j$, $\sigma > 0$. Let us also fix c_1, c_2, c_3 so that

$$|f(\xi)| \le c_1, |f(\xi)| \ge c_2 > 0, |\partial^\alpha f(\xi)| \le c_3^{|\alpha|+1}\alpha!, \ for \ \xi \in A,$$

in the cases a),b) and c) respectively. Of course such c_i will exist, if we shrink A suitably around some generating element $\theta = \sum \sigma_j \xi^j$ of $(\xi^0, \xi^1, ..., \xi^{k-1})$. We can then write

$$f(\xi) = f(\mu_{k-1}(t_{k-1}/t_{k-2}) \cdots \mu_1(t_1/t_0)\mu_0(t_0)(\eta)) = t_{k-1}^{m_{k-1}}t_{k-2}^{m_{k-2} - m_{k-1}} \cdots t_0^{m_0 - m_1}f(\eta),$$

for suitable $\eta \in A$ and $0 \leq t_{k-1} \leq t_{k-2} \leq \cdots \leq t_0$. The assertions from a) and b) then follow if we recall that according to lemma 3.1.4 we may assume that $t_j \sim |\Pi_j(\xi)|$. In the case of c) it is useful to choose first coordinates so that $M_j = \{\xi; \xi_i = 0 \text{ for } i \geq d_j + 1\}$. We then write that

$$
\begin{aligned}
|f^{(\alpha)}(\xi)| &= |f^{(\alpha)}(\mu_{k-1}(t_{k-1}/t_{k-2}) \cdots \mu_1(t_1/t_0)\mu_0(t_0)(\eta))| \\
&= t_{k-1}^{\beta_{k-1}} t_{k-2}^{m_{k-2} - n_{k-1} - \beta_{k-2}} \cdots t_0^{m_0 - \beta_0} |f^{(\alpha)}(\eta)| \\
&\leq t_{k-1}^{-|\alpha|} t_{k-1}^{m_{k-1}} t_{k-2}^{m_{k-2} - m_{k-1}} \cdots t_0^{m_0 - m_1} |f^{(\alpha)}(\eta)|,
\end{aligned}
$$

where $\beta_{k-1} = |\Pi_{k-1}\alpha|$, $\beta_j = |(\Pi_j - \Pi_{j+1})\alpha|$, $j \leq k - 2$. The argument can then be concluded as before.

Remark 3.2.3. *a) Part c) of the preceding lemma shows in which way the polyhomogeneity will become useful later on. In fact, using the homogeneity of f alone, we already can obtain that*

$$
|\partial^\alpha f(\xi)| \leq c_1^{|\alpha|+1} \alpha! \, |\Pi_0 \xi|^{m_0 - |\alpha|}, \tag{3.2.4}
$$

if we stay with ξ in a cone $\tilde{\Gamma} \subset\subset G$. (Thus note that while G is a polycone, $\tilde{\Gamma}$ is only assumed conic. Of course the relation " $\subset\subset$ " refers here to the inclusion of cones.) If $|\alpha|$ is large, this is of course much better than (3.2.3) $-$ on $\tilde{\Gamma}$. The main point in (3.2.3) is however that the polycones on which (3.2.3) are valid are much larger than cones of type $\tilde{\Gamma} \subset\subset G$.

b) Estimates of type (3.2.2), (3.2.3), on the other hand, are more precise than is (3.2.4), even in the case when we can estimate our quantities up to the axes. A case when this will be useful is described in the following lemma:

Lemma 3.2.4. *Let p_m be a homogeneous analytic symbol of order m defined in a conic neighborhood Γ of $\xi^0 = (0, ..., 0, 1)$ and which does not depend on x. Assume that p_m vanishes of order s precisely at ξ^0 and denote by $p_{m,1}$ the localization of p_m at ξ^0. Let $q = p_m - \xi_n^{m-s} p_{m,1}$ and fix a conic neighborhood $\Gamma' \subset\subset \Gamma$ of ξ^0. We can then find $c > 0$ so that*

$$
|\partial^\alpha q(\xi)| \leq c^{|\alpha|+1} \alpha! |\xi|^{m-s-1} |\xi'|^{s+1-|\alpha|}, \ \ if \ \xi \in \Gamma'. \tag{3.2.5}
$$

Proof. a) Assume first that $|\alpha'| \leq s$, where $\alpha' = (\alpha_1, ..., \alpha_{n-1})$. The function $\partial^\alpha q$ vanishes then of order $s + 1 - |\alpha'|$ when $\xi' = 0$. When $|\xi| = 1$ it follows from the Cauchy inequalities that

$$
|\partial^\alpha q(\xi)| \leq c^{|\alpha|+1} \alpha! |\xi'|^{s+1-|\alpha'|} \text{ near } \xi' = 0.
$$

We conclude after homogeneization that

$$|\partial^\alpha q(\xi)| \le c^{|\alpha|+1} \alpha! |\xi|^{m-s-1-\alpha_n} |\xi'|^{s+1-|\alpha'|}$$

in a conic neighborhood Γ'' of ξ^0. This gives (3.2.5) on Γ''. On $\Gamma' \setminus \Gamma''$ we obtain (3.2.5) since q is an analytic symbol there.

b) When $|\alpha'| > s$, then

$$\partial^\alpha q = \partial^\alpha p_m,$$

so the best possible estimate is

$$|\partial^\alpha q(\xi)| \le c^{|\alpha|+1} \alpha! |\xi|^{m-|\alpha|} \text{ for } \xi \in \Gamma'.$$

It remains to observe that for such α,

$$|\xi|^{m-|\alpha|} \le |\xi|^{m-s-1} |\xi'|^{s+1-|\alpha|}.$$

2. On the set of polyindices we have two types of order relations. The first is only a partial ordering and is related to the calculus of WF_A^k. The second is more powerful in that it is total, but it can be used only for $WF_{A,s}^k$.

Definition 3.2.5. *a) Let $m = (m_0, m_1, ..., m_{k-1})$ and $m' = (m'_0, m'_1, ..., m'_{k-1})$ be given. We shall write $m \prec m'$ if $m_0 < m'_0$ or if $m_0 = m'_0$ and $m_i \ge m'_i, i > 0$, but $m \ne m'$. (The reason for the apparently strange inversion of magnitudes will become clear in remark 3.2.6.)*

b) We shall write $m < m'$ if $m_0 < m'_0$ or if we can find $s \ge 1$ so that $m_i = m'_i$ for all $i < s$, but $m'_s < m_s$.

Thus "$<$" is essentially the lexicographic ordering and when $k = 2$, then the two orderings are just the same. The main thing when considering the order relation $<$ is that one can indeed work with it. It is clear that $m \prec m'$ implies $m < m'$. It is also immediate that if $m \prec m'$, respectively $m < m'$, then $-m' \prec -m$, respectively $-m' < -m$. Moreover, if m'' is another polyindex, then $m + m'' \prec m' + m''$, respectively $m + m'' < m' + m''$. Finally, in the same conditions, $tm \prec tm'$, respectively $tm < tm'$ if $t > 0$.

Remark 3.2.6. *a) Recall that if f is a positively polyhomogeneous function of polydegree m, then $m_j - m_{j+1}$ is the degree of homogeneity of f in the variables from M_j. If $m < m'$ and $m_i = m'_i$ for $i < s$, $m_s > m'_s$, then*

$$m_i - m_{i+1} = m'_i - m'_{i+1}, \text{ for } i < s-1$$

but

$$m_{s-1} - m_s < m'_{s-1} - m'_s.$$

If now f is positively polyhomogeneous of polydegree m and f' of polydegree m', then these polydegrees cannot be distinguished with respect to the polyhomogenous structure induced by $M_0, ..., M_{s-1}$, whereas $(m_0, ..., m_s) < (m'_0, ..., m'_s)$. Absolute preference is thus given in definition 2.2.4 b) to the variables from $\dot{M}_j$ over those from M_{j+1}. In definition 2.2.4. a) this is not anymore true , although even there the variables from M_j are given a bigger weight than those from M_{j+1}.

b) To understand the meaning of definition 3.2.5 a) let us denote for $s \geq 1$ by $\nu^s = (\nu_0, \nu_1, ..., \nu_{k-1})$ the polyindices defined by $\nu_i = 0$ for $i < s$ or $i > s$ and $\nu_s = 1$. It is clear that $\nu^s < 0$. We conclude that $m \prec m'$ means when $m_0 = m'_0$ that we can find $\alpha_s \geq 0$ for which

$$m' = m + \sum \alpha_s \nu^s.$$

Also note that a typical polyhomogeneous function of polydegree ν^s is

$$|\Pi_s(\xi)|/|\Pi_{s-1}(\xi)|.$$

c) One can obtain a similar statement for the order relation "$<$" : we shall consider a more intuitive statement of this type in section 3.5. In fact a set of generators of multiindices which are < 0 is given by the ν^s above, if we add to them the multiindices m built by the prescription: $m_i = 0$ if $i < s$ or $i > s+1, m_s = \beta$, $m_{s+1} = -1$. Here $0 < s \leq k - 2$ and β runs through the positive reals. (Some redundancies are discussed, implicitly, in section 3.5.)

d) When working k-microlocally, we may shrink our polyneighborhood around $(\xi^0, \xi^1, ..., \xi^{k-1})$ as much as we please. It follows that if c_1 is fixed and if we choose a suitable polyneighborhod of $(\xi^0, \xi^1, ..., \xi^{k-1})$, then we may assume that

$$|\Pi_s(\xi)|/|\Pi_{s-1}(\xi)| \leq c_1 \ , \ |\Pi_0(\xi)| \geq 1/c_1.$$

We conclude that if $m \prec \mathbf{0} = (0, ..., 0)$ then

$$|\Pi_{k-1}(\xi)|^{m_{k-1}} |\Pi_{k-2}(\xi)|^{m_{k-2}-m_{k-1}} \cdots |\Pi_0(\xi)|^{m_0-m_1} =$$
$$\prod_{j>0} (|\Pi_j \xi|/|\Pi_{j-1}\xi|)^{m_j} |\Pi_0(\xi)|^{m_0} \leq c_1^{\sum m_j}.$$

This gives the following lemma:

Lemma 3.2.7. *Assume that $m \prec m'$ and let g, f be positively homogeneous functions of polydegree m, m' defined in a polyneighborhood G of $(\xi^0, \xi^1, ..., \xi^{k-1})$. Assume that Γ is the polycone generated by an open neighborhood A of some generating element $\theta = \sum \sigma_j \xi^j$ associated with $(\xi^0, \xi^1, ..., \xi^{k-1})$. Assume further that $|g(\xi)| \leq c$ on A and that $|f(\xi)| \geq c'$ on A. If C is fixed we can then find a polyneighborhood Γ' of $(\xi^0, \xi^1, ..., \xi^{k-1})$ so that*

$$|g(\xi)| \leq C^{\sum (m_j - m_j') + 1} |f(\xi)|, \ \ if \ \xi \in \Gamma', |\xi| \geq 1/C.$$

Moreover, C does not depend here explicitly on g and f.

The lemma thus expresses in a quantitative way that functions of smaller degree of poly-homogeneity -with respect to the order relation $\prec$ - have a smaller weight in estimates. The estimate makes it possible sometimes to consider infinite series of polyhomogeneous functions which as homogeneous functions have the same degree of homogeneity. It is this situation which appears in second microlocalization, for example when one expands an homogeneous function into a Taylor series in part of the variables. Results of this type can be found in Laurent [1]. Here we shall be interested mainly in the order relation $<$, but we shall only consider finite sums of polyhomogeneous functions explicitly : cf. section 3.5 below.

3.3 Symbol classes and pseudodifferential operators in general G_φ-microlocalization

1. The results on pseudodifferential operators in higher microlocalization which we need in this paper can be deduced from the results on pseudodifferential operators in general G_φ- classes from Liess-Rodino [1]. For the convenience of the reader, we recall in this section those definitions and results from Liess-Rodino, loc.cit., which are relevant in the present context. Actually the situation is here, to a certain extent (and for some aspects), simpler than in Liess-Rodino [1]. Consequently, we have also simplified a number of definitions and statements of results; we refer to Liess-Rodino [1] for more general statements.

Starting point is that we are given an open set X in R^n, a Lipschitzian weight function φ on R^n and an unbounded set $G \subset R^n$. To simplify the situation we shall assume at first that G is open.

Definition 3.3.1. *Consider $m \in R$, and $p : X \times G \to C$.*

a) We shall say that p is a symbol in $S_\varphi^m(X, G)$ if we can find for any $X' \subset\subset X$, some ε, c, and an extension $\tilde{p} : X' \times G_{\varepsilon\varphi} :\to C$ such that

$$|\partial_x^\alpha \partial_\xi^\beta \tilde{p}(x, \xi)| \le c^{|\alpha|+|\beta|+1} \alpha! \beta! \varphi(\xi)^{m-|\alpha|-|\beta|}, \forall \alpha, \forall \beta, \forall x \in X', \forall \xi \in G_{\varepsilon\varphi}. \qquad (3.3.1)$$

b) We shall say that p is a symbol in $\tilde{S}_\varphi^m(X, G)$ if for any $X' \subset\subset X$ we can find ε, c, c', and an extension $\tilde{p} : X' \times G_{\varepsilon\varphi} :\to C$ such that (3.3.1) holds for all α, all $x \in X'$, all $\xi \in G_{\varepsilon\varphi}$, and all β for which $c'|\beta| \le \varphi(\xi)$.

c) The principal symbol of a symbol $p \in S_\varphi^m(X, G)$ is by definition the class of p in $S_\varphi^m(X, G)/S_\varphi^{m-1}(X, G)$.

Remark 3.3.2. *a) If $\tilde{p}$ satisfies the estimates (3.3.1) near some point (x^0, ξ^0), then it is clear that $\tilde{p}$ admits an analytic extension to a set of form*

$$A = \{(z, \zeta) \in C_z^n \times C_\zeta^n; |z - x^0| < c, |\zeta - \xi^0| < c\varphi(\xi^0)\}.$$

The converse is of course also true: if p admits an analytic extension to a set of type A, and if this extension is "suitably" estimable, then we have estimates of form (3.3.1) for (x^0, ξ^0). (This follows e.g. from the Cauchy inequalities and is sometimes a very convenient way to check estimates for symbols.)

b) When $\varphi(\xi) \sim |\xi|$ and Γ is an open cone, the symbols from $S_\varphi^m(X, \Gamma)$ are just the symbols from standard first microlocalization. Conversely, of course, all statements from this and the following section are closely related to what one usually does in standard microlocalization.

Definition 3.3.3. *a) $\hat{S}F_\varphi^m(X, G)$ is the space of all formal sums $\sum_{j \ge 0} p_j$ of functions $p_j : X \times G \to C$ which have the property that for any $X' \subset\subset X$ we can find ε, c, c' and extensions $\tilde{p}_j$ of p_j to $X' \times G_{\varepsilon\varphi}$ such that*

$$|\partial_x^\alpha \partial_\xi^\beta \tilde{p}_j(x, \xi)| \le c^{|\alpha|+|\beta|+j+1} \alpha! \beta! j! \varphi(\xi)^{m-|\beta|-j}, \forall x \in X', \forall \xi \in G_{\varepsilon\varphi}, \forall \alpha,$$
$$\forall \beta \text{ with } c'(|\beta| + j) \le \varphi(\xi). \qquad (3.3.2)$$

b) $\sum p_j \in \hat{S}F_\varphi^m(X, G)$ and $\sum q_j \in \hat{S}F_\varphi^m(X, G)$ are said to be equivalent if for all $X' \subset\subset X$ we can find ε, c, c' and extensions $\tilde{p}_j, \tilde{q}_j : X' \times G_{\varepsilon\varphi} \to C$ such that

$$|\partial_x^\alpha \partial_\xi^\beta \sum_{j < s}(\tilde{p}_j - \tilde{q}_j)(x, \xi)| \le c^{|\alpha|+|\beta|+s+1} \alpha! \beta! s! \varphi(\xi)^{m-|\beta|-s}, \forall x \in X', \forall \xi \in G_{\varepsilon\varphi}, \forall \alpha,$$
$$\forall \beta \text{ with } c'(|\beta| + s) \le \varphi(\xi). \qquad (3.3.3)$$

In this situation we shall also write that $\sum p_j \sim \sum q_j$, for short.

c) If $p \in \tilde{S}_\varphi^m(X, G)$ we associate with it $\sum p_j \in \tilde{S}F_\varphi^m(X, G)$ by $p_0 = p$, $p_j = 0$, if $j > 0$. In particular it makes sense to write that $p \in \tilde{S}_\varphi^m(X, G)$ is equivalent to some $\sum p_j \in \tilde{S}F_\varphi^m(X, G)$.

Proposition 3.3.4. *(Cf. Liess-Rodino [1]) Let $\sum p_j \in \tilde{S}F_\varphi^m(X, G)$ be given. If $X' \subset\subset X$ is fixed, we can then find $p \in \tilde{S}_\varphi^m(X', G)$ such that $p \sim \sum p_j$ in $\tilde{S}F_\varphi^m(X', G)$.*

2. When $p \in \tilde{S}_\varphi^m(X, G)$ and $q \in \tilde{S}_\varphi^{m'}(X, G)$ is given, we define a formal symbol $\sum r_j$ by the (standard) formula

$$r_j = \sum_{|\alpha|=j} \partial_\xi^\alpha p(x, \xi)(-i\partial_x)^\alpha q(x, \xi)/\alpha!.$$

It is easy to check that $\sum r_j \in \tilde{S}F_\varphi^{m+m'}(X, G)$. As is common practice, we denote $\sum r_j$ by $p \circ q$ and call it the composition of p with q.

Also the next result is standard:

Proposition 3.3.5. *Consider $p \in \tilde{S}_\varphi^m(X, G)$ and assume that we can find $m', \varepsilon, c, c', c'',$ c''' such that*

$$|p(x, \xi)| \geq c\varphi(\xi)^{m'}, \text{ if } x \in X, \xi \in G_{\varepsilon\varphi}, |\xi| \geq c', \tag{3.3.4}$$

and

$$|\partial_x^\alpha \partial_\xi^\beta p(x, \xi)| \leq c''^{|\alpha|+|\beta|+1}\alpha!\beta! \, |p(x, \xi)| \, \varphi(\xi)^{-|\beta|}, \tag{3.3.5}$$

if $x \in X, \xi \in G_{\varepsilon\varphi}$, $|\xi| \geq c', c'''|\beta| \leq \varphi(\xi)$.

Also fix $X' \subset\subset X$. Then we can find $q \in \tilde{S}_\varphi^{-m'}(X', G)$ such that $q \circ p \sim 1$.

3. We now consider pseudodifferential operators associated with our symbols. When $p \in \tilde{S}_\varphi^m(X, G)$, we fix $X' \subset\subset X$ and choose an extension $\tilde{p}$ of p to $X' \times G_{\varepsilon\varphi}$ for some $\varepsilon > 0$ as in definition 3.3.3. We then consider $h \in C_0^\infty(R^n)$ such that for some $c' < c'' < \varepsilon$ and some c_1,

i) $h(\xi) = 1$ if $\xi \in G_{c'\varphi}$, $|\xi| > c_1$,

ii) $h(\xi) = 0$, for $\xi \notin G_{c''\varphi}$,

iii) $|\partial_\xi^\alpha h(\xi)| \leq c_\alpha, \forall \alpha$, for some constants c_α.

We then set for $u \in C_0^\infty(X')$:

$$p(x, D)u = \int e^{i\langle x, \xi \rangle} \tilde{p}(x, \xi)h(\xi)\hat{u}(\xi)d\xi.$$

Thus the definition of $p(x, D)u$ depends on the choice of X', of h and of the extension $\tilde{p}$ of p. However, the class modulo distributions which have no wave front set at $X' \times G$, does not depend on these choices. More generally, it is proved in Liess-Rodino [1] that if p and p' in $\tilde{S}_\varphi^m(X, G)$ are equivalent as symbols in $\tilde{S}F_\varphi^m(X, G)$ and if $p(x, D), p'(x, D)$ are obtained by the above prescriptions, then

$$(X' \times G) \bigcap WF_\varphi(p(x, D) - p'(x, D))u = \emptyset, \forall u \in C_0^\infty(X').$$

If $\sum p_j \in \tilde{S}F_\varphi^m(X, G)$ is given, it follows that we can associate an operator

$$(\sum p_j(x, D)) : C_0^\infty(X') \to C_0^\infty/\mathcal{N},$$

where $\mathcal{N} = \{f \in D'(X'); (X' \times G) \cap WF_\varphi f = \emptyset\}$ with $\sum p_j$, by the prescription

$$(\sum p_j)(x, D)u = p(x, D)u.$$

($C_0^\infty/\mathcal{N}$ is thus essentially the space of microfunctions which have C^∞ representants. We shall discuss microfunctions in more detail in chapter IX.)

Here p is chosen so that $p \sim \sum p_j$ in $\tilde{S}_\varphi^m(X', G)$. (Cf. proposition 3.3.4.)

4. Let us further consider $p \in \tilde{S}_\varphi^m(X, G)$, $q \in \tilde{S}_\varphi^{m'}(X, G)$. It is then proved in Liess-Rodino [1] that

$$(X' \times G) \cap WF_\varphi(p(x, D)(gq(x, D)u) - (p \circ q)(x, D)u) = \emptyset,$$

for any $u \in C_0^\infty(X')$ and any $g \in C_0^\infty(X)$, provided that g is identically one in a neighborhood of X'.

3.4   Pseudodifferential operators in k-microlocalization

1. As we have seen in section 2.2, k-microlocalization is essentially φ-microlocalization with $\varphi = |\Pi_{k-1}|$ on appropriate sets. If we work with $WF_{A,s}^k$ near $(\xi^0, \xi^1, ..., \xi^{k-1})$, then the relevant sets in the ξ- space on which we localize have the form :

$$E = \{\xi \in G \;\; ; \;\; |\Pi_{j+1}(\xi)| \geq c|\Pi_j(\xi)|^{1+\beta}/|\Pi_{j-1}(\xi)|^\beta, j = 1, ..., k-2,$$
$$|\Pi_i(\xi)| \geq c(1 + |\Pi_{i-1}(\xi)|^\delta), i = 1, ..., k-1 \}.$$

Somewhat different sets appear for other variants of k-wave front set: in the case of WF_A^k, we have, e.g., sets of type

$$\tilde{E} = \{\xi \in G; |\Pi_i(\xi)| \geq f_i(|\Pi_{i-1}\xi|), i = 1, \ldots, k-1 \},$$

where the f_i vary in the class of sublinear functions, etc. Here G is some multineighborhood of $(\xi^0, \xi^1, ..., \xi^{k-1})$, $\beta > 0$, $\delta < 1$ and $c \geq 0$.

We mention explicitly two results, which are consequences of remark 2.2.11, when combined with the results of section 3.3.

Proposition 3.4.1. *Let E —for some $c, \delta < 1, \beta > 0$—, be as before, consider X open in R^n and $p \in \tilde{S}^m_{|\Pi_{k-1}|}(X, E)$. Also fix $X' \subset\subset X$ and let $p(x, D)$ be a pseudodifferential operator associated with p as in section 3.3.*

a) If $x^0 \in X'$, $(x^0, \xi^0, \xi^1, ..., \xi^{k-1}) \notin WF^k_{A,s}u$, then $(x^0, \xi^0, \xi^1, ..., \xi^{k-1}) \notin WF^k_{A,s}p(x, D)u$.

b) If we can find constants c', c'' such that

$$|\partial_x^\alpha \partial_\xi^\beta p(x, \xi)| \leq c'^{|\alpha|+|\beta|+1} \alpha! \beta! \, |p(x, \xi)| \, |\Pi_{k-1}(\xi)|^{-|\beta|}$$

for $x \in X'$, $\xi \in E_{c''|\Pi_{k-1}|}$, then it follows from

$$(x^0, \xi^0, \xi^1, ..., \xi^{k-1}) \notin WF^k_{A,s}p(x, D)u \text{ that } (x^0, \xi^0, \xi^1, ..., \xi^{k-1}) \notin WF^k_{A,s}u$$

for all $x^0 \in X'$.

2. Recall from section 3.3 that the symbols considered in proposition 3.4.1 were introduced in the context of φ-microlocalization, which was designed to provide a modality to microlocalize on φ-neighborhoods. In the case of higher microlocalization, φ-neighborhoods are not always as natural as are multineighborhoods. On the other hand φ-microlocalization is somewhat sharper than microlocalization on multineighborhoods, as is clear from the following easy result, the proof of which we leave to the reader:

Proposition 3.4.2. *Let G, G' be multineighborhoods which contain $(\xi^0, \xi^1, ..., \xi^{k-1})$ and assume that $G \subset\subset G'$. Then we can find c so that $G_{c|\Pi_{k-1}|} \subset G'$. The "converse" is not true in general: if G is a given multineighborhood of $(\xi^0, \xi^1, ..., \xi^{k-1})$ then it is not true in general that $G \subset\subset G_{c|\Pi_{k-1}|}$.*

3. When Fourier integral operators are considered, symbols of type $\tilde{S}^m_\varphi(X, G)$ do not quite suffice to build up a complete calculus: at certain instances one actually has to work with symbols from $S^m_\varphi(X, G)$. It is then an interesting question to see wether or not one can find for given $p \in \tilde{S}^m_\varphi(X, G)$ some $q \in S^m_\varphi(X, G)$ so that $p \sim q$. We deal with a question of this type in the following result, but we shall restrict our attention to the case of two-microlocalization, which is the only case for which we have an explicit use later on. As usual we shall assume that $M_1 = \{\xi \in R^n; \xi_{d+1} = \cdots = \xi_n = 0\}$ and denote for given $\xi \in R^n$ by $\xi' = (\xi_1, ..., \xi_d)$.

Proposition 3.4.3. *Let* (ξ^0, ξ^1) *be given and let* G *be an open bineighborhood of* (ξ^0, ξ^1). *Denote by*

$$E = \{\xi \in G\,;\; |\xi'| > c_1(1 + |\xi|^\delta)\}.$$

Also consider $X' \subset\subset X$, *both open. Then we can find an open bineighborhood* G' *of* (ξ^0, ξ^1) *and* c_2, δ' *so that for*

$$E' = \{\xi \in G'\,;\; |\xi'| > c_2(1 + |\xi|^{\delta'})\},$$

we have the following property:

for any $p \in \tilde{S}^m_{|\Pi_1|}(X, E)$, *there is* $q \in S^m_{|\Pi_1|}(X', E')$ *so that* $p \sim q$ *in* $\check{S}F^m_{|\Pi_1|}(X', E')$.

4. Proposition 3.4.3 is a consequence of proposition 1.3.6 and of remark 1.3.7, both from Liess-Rodino [2]. The proof of those results is based on $\bar{\partial}$-arguments, and in order to make the respective arguments work, one needs to work on suitable domains of holomorphy and to have estimates which involve plurisubharmonic weight functions. The existence of the relevant domains of holomorphy and plurisubharmonic functions is required explicitly in the hypothesis of proposition 1.3.6, respectively remark 1.3.7 in Liess-Rodino, loc. cit. That these conditions are satisfied in the present situation is precisely the content of the following two propositions.

Proposition 3.4.4. *Let* ξ^0, ξ^1 *be given. Then we can find an open bineighborhood* G *of* (ξ^0, ξ^1) *and a plurisubharmonic function* $\rho : V = \{\zeta; Re\,\zeta \in G\} \to R$ *such that for some* $c > 0, c', c''$,

$$- |Re\,\zeta'| \le \rho(\zeta) \le -c|Re\,\zeta'| + c'|Im\,\zeta| + c'', \text{ for } \zeta \in V. \tag{3.4.1}$$

Proposition 3.4.5. *Let* $c_1 > 0, c_2 \ge 0, c_3 \ge 0$, *some rational number* $\delta < 1$ *and an open convex bineighborhood* G *of* (ξ^0, ξ^1) *be given so that* $|\xi'| \le c' \langle \xi^1, \xi \rangle$ *if* $\xi \in G$. *Then we can find* c_4, c_5, c_6 *and a domain of holomorphy* Ω *so that*

$$\{\zeta \in C^n \;\; ; \;\; Re\,\zeta \in G, |Im\,\zeta| < c_4|Re\,\zeta'|, |Re\,\zeta'| > c_5 + c_6|Re\,\zeta|^\delta\} \subset \Omega \tag{3.4.2}$$
$$\subset \;\{\zeta \in C^n; Re\,\zeta \in G, |Im\,\zeta| < c_1|Re\,\zeta'|, |Re\,\zeta'| > c_2 + c_3|Re\,\zeta|^\delta\}$$

5. The proof of proposition 3.4.4 is rather trivial. We may in fact assume that $\xi \in G$ implies that $|\xi'| \le c_1\langle \xi, \xi^1\rangle$. The function $-c_2\langle Re\,\zeta, \xi^1\rangle$ is then for a suitable c_2 a plurisubharmonic function with the desired properties. As for proposition 3.4.5, we first prove

Lemma 3.4.6. *Let $c > 0$ and consider G some open convex bineighborhood of (ξ^0, ξ^1) such that $\xi \in G$ implies $|\xi'| < c'\langle \xi^1, \xi \rangle$ for some c'. Then we can find $c'' > 0$ and a domain of holomorphy $U \subset C^n$ for which*

$$\{\zeta \in C^n; Re\zeta \in G, |Im\zeta| < c''|Re\zeta'|\} \subset U \subset \{Re\zeta \in G, |Im\zeta| < c|Re\zeta'|\}. \quad (3.4.3)$$

Proof. We may assume that $|\xi^1| = 1$ and fix some constant $c_1 < c/n$. Let us then introduce for $j = 1, ..., n$ the $2n$ domains of holomorphy

$$U_j^\pm = \{\zeta; Re\,\zeta \in G, \pm Im\zeta_j < c_1\,Re\,\langle \xi^1, \zeta \rangle\}.$$

We claim that

$$U = \bigcap_j (U_j^+ \cap U_j^-)$$

satisfies (3.4.3) if c'' is chosen so that $c'c'' \leq c_1$. In fact the condition $Re\zeta \in G$ is present in the definition of all three sets which enter in (3.4.3). Regarding the first inclusion in (3.4.3), we then observe that $Re\zeta \in G$ and $|Im\zeta| < c''|Re\zeta'|$ imply $\pm Im\zeta_j \leq |Im\zeta| < c''|Re\zeta'| < c'c''Re\langle \xi^1, \zeta \rangle$, so $\zeta \in U_j^\pm$, and conversely, when $\zeta \in U$, then $|Im\zeta| \leq \sum |Im\zeta_j| \leq nc_1Re\langle \xi^1, \zeta \rangle \leq c|Re\zeta'|$.

6. To continue, we shall now assume that $\delta = \mu/\sigma$, for two rational numbers μ, σ, with $\mu < \sigma$. Our next remark towards the proof of proposition 3.4.5 is then that

$$Q = \{\zeta \in C^n; Re\,(\sum_{j \leq d} \zeta_j^2)^\sigma \geq c_7 + c_8 Re\,(\sum_{j \leq n} \zeta_j^2)^\mu\}$$

is a domain of holomorphy. Here c_7, c_8 are still to be determined and coordinates and d are assumed to be such that $M_1 = \{\xi; \xi_i = 0$ if $i > d\}$. To simplify the notation we shall write $\sum'$ in the remainder of this section for $\sum_{j \leq d}$ and $\sum$ for $\sum_{j \leq n}$. We want to show that for suitable choices of the constants c_4, c_5, c_6, c_7, c_8 and c (in lemma 3.4.6) $\Omega = U \cap Q$ will satisfy the conclusion of proposition 3.4.5.

To see this, observe that we can find c_9, c_{10} so that

$$||Re\zeta'|^{2\sigma} - Re(\sum{}'\zeta_j^2)^\sigma| \leq c_9 c^2 |Re\zeta'|^{2\sigma},$$

$$||Re\zeta'|^{2\mu} - Re(\sum{}'\zeta_j^2)^\mu| \leq c_{10}\,c^2 |Re\zeta'|^2 |Re\zeta|^{2\mu-2},$$

$$||Re\,\zeta|^2 - Re\sum\zeta_j^2| \leq c^2 |Re\,\zeta'|^2,$$

if $|Im\zeta| < c|Re\zeta'|$. It follows in particular that

$$|Re(\sum{}'\zeta_j^2)^\sigma|^{1/2\sigma}/2 \leq |Re\zeta'| \leq 2|Re(\sum{}'\zeta_j^2)^\sigma|^{1/2\sigma},$$

$$|Re(\sum \zeta_j^2)^\mu|^{1/2\mu}/2 \le |Re\zeta| \le 2|Re(\sum \zeta_j^2)^\mu|^{1/2\mu},$$

$$|Re \sum \zeta_j^2|/2 \le |Re\,\zeta|^2 \le 2|Re \sum \zeta_j^2|,$$

if $|Im\zeta| < c|Re\zeta'|$ and c is small enough. We shall here assume in addition that $c < c_1$. This gives a condition on c and application of lemma 3.4.6 for this c will give us some c''. We shall then assume that c_4 is fixed with $c_4 \le c''$.

To conclude the proof of proposition 3.4.5 let us consider at first the case of the second inclusion in (3.4.2). Thus assume that $\zeta \in U \cap \Omega$. It follows that $Re\zeta \in G$, $|Im\zeta| \le c|Re\zeta'|$ and that

$$|Re\zeta'| \ge |Re(\sum{}'\zeta_j^2)^\sigma|^{1/2\sigma}/2 \ge c_7^{1/2\sigma}/4 + c_8^{1/4\sigma}|Re \sum \zeta_j^2|^{\delta/2}/2 \ge c_7^{1/2\sigma}/4 + (c_8^{1/2\sigma}/8)|Re\zeta|^\delta.$$

If $c_7^{1/2\sigma}/4 \ge c_2$, $c_8^{1/2\sigma}/8 \ge c_3$, this will give the second inclusion in (3.4.2.) As for the first inclusion there, we already know from lemma 3.4.6 that $\zeta \in U$ if ζ is in the set from the left hand side of (3.4.2). We need then only show that ζ also lies in Q for suitable choices of constants. Here we observe that $|Re\,\zeta'| > c_5 + c_6|Re\,\zeta'|^\delta$ implies $Re(\sum{}'\zeta_j^2)^\sigma \ge |Re\,\zeta'|^{2\sigma}/2^{2\sigma} \ge (c_5 + c_6|Re\,\zeta|^\delta)^{2\sigma}/2^{2\sigma} \le c_{11} + c_{12}|Re\,\zeta|^{2\mu} \ge c_{13} + c_{14}e(\sum \zeta_j^2)^\mu$ with c_{13} and c_{14} proportional to c_5, respectively c_6. Once c_7 and c_8 have been fixed to give the second inclusion in (3.4.2), we can therefore also find c_5 and c_6 to obtain the first inclusion there.

7. In the preceding we have mainly discussed the situation which referred to the case when wave front sets were associated with conditions of type $|\Pi_i\xi| \ge c(1 + |\Pi_{i-1}\xi|^\delta)$, rather than $|\Pi_i\xi| \ge f_i(|\Pi_{i-1}\xi|)$ for some general sublinear functions f_i or $|\Pi_i\xi| \ge c\ln(2 + |\Pi_{i-1}\xi|)$. We refer here of course to definition 2.1.1 and the discussions following definition 2.1.2 . As far as the case $|\Pi_i\xi| \ge c\ln(2 + |\Pi_{i-1}\xi|)$ is concerned, we should say that it is not formally included in the results from Liess -Rodino [1] and we have not checked if the analogues of the propositions from this section are all true. In the case of $|\Pi_i\xi| \ge f_i(|\Pi_{i-1}\xi|)$, for general f_i, we can however apply once more the results from Liess- Rodino [1], to obtain results on regularity and composition of pseudodifferential operators as considered above. We do not state these results explicitly. The analogue of proposition 3.4.5 merits some attention. It reads in the present situation as follows:

Proposition 3.4.7. *Let a convex cone $G \subset R^n$, c_1 and a sublinear function f be given. Then we can find c_2, a sublinear function g and a domain of holomorphy Ω so that*

$$\{\zeta \in C^n \;\; ; \;\; Re\zeta \in G, |Im\zeta| < c_2|Re\zeta'|, |Re\zeta'| > g(|Re\zeta|)\} \subset \Omega$$
$$\subset \{\zeta \in C^n; Re\zeta \in G, |Im\zeta| < c_1|Re\zeta'|, |Re\zeta'| > f(|Re\zeta|)\}.$$

Proof of proposition 3.4.7. We may assume $d = 1$, so that $\zeta' = \zeta_1$ and can apply lemma 3.4.6 as in the proof of proposition 3.4.5. It suffices then to prove the following result:

Lemma 3.4.8. *Let c_1 and a sublinear function f be given. Then there is a sublinear function g and a domain of holomorphy U so that*

$$\{\zeta \in C^n \quad ; \quad |Im\,\zeta_1| < |Re\,\zeta_1|, |Im\zeta''| < c_1|Re\zeta''|, |Re\zeta_1| > g(|Re\zeta''|)\} \subset U$$
$$\subset \{\zeta \in C^n; |Im\,\zeta_1| < |Re\,\zeta_1|, |Im\zeta''| < c_1|Re\zeta''|, |Re\zeta_1| > f(|Re\zeta''|)\}.$$

Note that we have replaced here the conditions $|Re\zeta'| > f(|Re\zeta|)$, $|Re\zeta'| > g(|Re\zeta|)$ by conditions of type $|Re\zeta'| > f(|Re\zeta''|)$, $|Re\zeta'| > g(|Re\zeta''|)$, which is possible if we also simultaneously change f and g slightly.

Proof of lemma 3.4.8. We base our proof on the remark that if W is a domain of holomorphy, and if $F, G : W \to C$ are holomorphic functions, then

$$\{\zeta \in W; |F(\zeta)| > |G(\zeta)|\}$$

is a domain of holomorphy. (Cf. e.g. Hörmander [6].)

Let us next choose an entire analytic function $J : C^{n-1} \to C$ so that

$$|J(\zeta'')| \geq c_3 e^{f(|Re\,\zeta''|)} \text{ for } |Im\,\zeta''| < |Re\,\zeta''|/\sqrt{3} \tag{3.4.4}$$

and so that $J(\zeta'')$ is of infraexponential type. By "infraexponential" we mean that

$$|J(\zeta'')| \leq e^{g'(|Re\,\zeta''|)+g'(|Im\,\zeta''|)} \tag{3.4.5}$$

for some suitable sublinear function g', which it is no loss in generality to assume increasing. The existence of such functions J has been established in another context in Kawai [1], Kaneko [1]. (For the precise form of 3.4.4 cf. Liess [4]). Entire analytic functions of infraexponential type will also play an important role in chapter IX.

Let us now also choose constants c', c'' so that

$$c'e^{|Re\,\zeta_1|} \leq |e^{\zeta_1} + e^{-\zeta_1}| \leq c''e^{|Re\,\zeta_1|} \text{ for } |Im\,\zeta_1| < |Re\,\zeta_1|$$

and denote by

$$\tilde{U} = \{\zeta \in C^n; |e^{\zeta_1} + e^{-\zeta_1}|/c'' > |J(\zeta'')|\},$$

which is a domain of holomorphy.

It is now clear that $\zeta \in \tilde{U}$ together with $|Im\, \zeta_1| < |Re\, \zeta_1|$, $|Im\, \zeta''| < |Re\, \zeta''|/\sqrt{3}$ implies

$$e^{|Re\, \zeta_1|} \geq |e^{\zeta_1} + e^{-\zeta_1}|/c'' > |J(\zeta'')| \geq e^{f(|Re\, \zeta''|)}.$$

On the other hand, if c''' is large, then $c' \exp c''' \geq c''$. We will therefore have that $|Re\, \zeta_1| > c''' + 2g'(|Re\, \zeta''|)$ together with $|Im\, \zeta_1| < |Re\, \zeta_1|$, $|Im\, \zeta''| < |Re\, \zeta''|/\sqrt{3}$ implies

$$|e^{\zeta_1} + e^{-\zeta_1}| \geq c' e^{|Re\, \zeta_1|} > c'' e^{g'(|Re\, \zeta''|) + g'(|Im\, \zeta''|)} \geq c''|J(\zeta'')|.$$

This concludes the proof of lemma 3.4.8, if we set $g(\xi) = c''' + 2g'(\xi)$.

3.5   Polyhomogeneity and symbols.

1. In this section we consider two copies of R^n, one in which the variables are denoted x or y and one in which they are denoted by ξ, η or τ. We assume that in the second copy we are given a polyhomogeneous structure defined by $M_0 \supset M_1 \supset \cdots \supset M_k$, and denote the corresponding multiplication operators by μ_j. Our first concern is to extend the $\mu_j(t)$ to operators on $R^{2n} = R^n_x \times R^n_\xi$. We shall in fact set

$$\mu_j(t)(x,\xi) = (x, \mu_j(t)\xi). \tag{3.5.1}$$

Thus in (3.5.1), $\mu_j(t)$ has an "old" and a "new" meaning. Of course, in this and most of the following, it is not important that the spaces of the x and of the ξ variables are of the same dimension. Indeed, later on we shall encounter situations when this actually is not the case. For notational reasons we shall only consider however the case when the two dimensions are equal. We can now extend the definitions from the sections 2.1, 2.2 to the case of sets in R^{2n} and to functions defined on sets $U \subset R^{2n}$.

Definition 3.5.1. *a)* $G \subset R^{2n}$ *is called a polycone if*

$$\mu_0(t_0)\mu_1(t_1/t_0) \cdots \mu_{k-1}(t_{k-1}/t_{k-2})(x,\xi) \in G$$

for any $(x, \xi) \in G$ *and* $0 < t_{k-1} \leq t_{k-2} \leq \cdots \leq t_0$.

b) If $x^0, \xi^j \in \dot{M}_j$ *and* $\sigma_j > 0$ *are given, we say that* $(x^0, \sum \sigma_j \xi^j)$ *is a generating element for* $(x^0, \xi^0, \xi^1, ..., \xi^{k-1})$.

c) Consider $G \subset R^{2n}$, $x^0 \in R^n$ *and* $\xi^j \in \dot{M}_j$. *We say that* G *is a polyneighborhood of* $(x^0, \xi^0, \xi^1, ..., \xi^{k-1})$, *if we can find a generating element* (x^0, θ) *for* $(x^0, \xi^0, \xi^1, ..., \xi^{k-1})$ *and an open polycone* $G' \subset G$ *so that* $(x^0, \theta) \in G'$.

d) Assume that $G \subset R^{2n}$ is a polycone and consider $f : G \to R$. We say that f is polyhomogeneous of polydegree $(m_0, ..., m_{k-1})$ if

$$f(\mu_j(t)(x, \xi)) = t^{m_j} f(x, \xi), \text{ for } t > 0 \text{ if } j = 0 \text{ and } 0 < t \leq 1 \text{ if } j > 0.$$

Remark 3.5.2. Thus in all the above, x plays the role of a parameter. Wether or not we use our terminology in R^{2n} or R_ξ^n must be clear from the context. A typical example of a polycone is of course $U \times \Gamma$ where U is in R_x^n and Γ is a polycone in R_ξ^n. More generally, if $G \subset R^{2n}$ is a polycone and $(x^0, \xi) \in G$ for some ξ, then $\{\eta; (x^0, \eta) \in G\}$ is a polycone in R_ξ^n. Polycones in R^{2n} are thus of form

$$\cup_x (x, \Gamma_x),$$

where x is in the projection of G onto R_x^n, and where Γ_x is polyconic in R_ξ^n. Note that (x, Γ_x) may be regarded as built up from "generating" elements.

2. Let now f be a C^∞-function defined on some polyneighborhood G of $(x^0, \xi^0, \xi^1, ..., \xi^{k-1})$. It follows from the computations in section 2.2 (cf. lemma 3.2.2) that

$$
\begin{aligned}
|\partial_x^\beta \partial_\xi^\alpha f(\xi)| &\leq c^{|\beta|+|\alpha|+1} \alpha! \beta! |\Pi_{k-1}(\xi)|^{-|\alpha|} |\Pi_{k-1}(\xi)|^{m_{k-1}} |\Pi_{k-2}(\xi)|^{m_{k-2}-m_{k-1}} \cdots \\
&\quad |\Pi_0(\xi)|^{m_0-m_1},
\end{aligned}
\tag{3.5.2}
$$

in some (possibly somewhat smaller) polyneighborhood G' of $(x^0, \xi^0, \xi^1, ..., \xi^{k-1})$. We would like to regard f as a symbol in k-microlocalization, which, as we have seen in section 3.4, is essentially G_φ-localization associated with the weight function $\varphi = |\Pi_{k-1}|$. We see in particular that estimates are improving after derivation in the appropriate way. However, as already observed, $|\Pi_{k-1}|$ is not an admissible weight function on a full polyneighborhood of $(x^0, \xi^0, \xi^1, ..., \xi^{k-1})$ and, for a closely related reason, we cannot dominate the factor

$$T = |\Pi_{k-1}(\xi)|^{m_{k-1}} |\Pi_{k-2}(\xi)|^{m_{k-2}-m_{k-1}} \cdots |\Pi_0(\xi)|^{m_0-m_1}$$

by $|\Pi_{k-1}(\xi)|^m$ (for some suitable m), on a full polyneighborhood of $(x^0, \xi^0, \xi^1, ..., \xi^{k-1})$. It is for these reasons that we shall microlocalize on smaller sets than full polyneighborhoods. Actually, T is dominated by $|\Pi_{k-1}(\xi)|^m$ for some large m, as soon as we restrict our attention to sets of type

$$\{(x, \xi); (x, \xi) \in G, (1 + |\Pi_{i-1}(\xi)|)^\delta \leq c|\Pi_i(\xi)|, i = 1, ..., k-1\},$$

for some suitable polyneighborhood G of $(x^0, \xi^0, \xi^1, ..., \xi^{k-1})$ and some suitable $c > 0$ and $\delta < 1$. Moreover, on such sets φ is an admissible weight function.

This is the first remark which makes it possible to use polyhomogeneous functions successfully in higher microlocalization. Actually, we restrict in the definition of $WF_A^{k,s}$ to still smaller sets. This is useful if rather than working with one single polyhomogeneous symbol, one has to work with sums of such symbols. We discuss this in detail in the following section.

3.6   Finite sums of polyhomogeneous functions and principal parts. Relative Poisson brackets.

1. Let $(x^0, \xi^0, \xi^1, ..., \xi^{k-1})$ be given and consider a finite set of polyindices I. For each $m \in I$ we assume that a function f^m is given in a multineighborhood of $(x^0, \xi^0, \xi^1, ..., \xi^{k-1})$ and that f^m is positively polyhomogeneous of polydegree m there. We shall call

$$\sum_{m \in I} f^m$$

a finite sum of polyhomogeneous functions for short.

Our first remark is that the terms f^m in such a sum are uniquely determined.

Lemma 3.6.1. *Let G be a polycone and let I be a finite set of (distinct) polyindices. Consider further for each $m \in I$ a function $f^m : G \to C$ which is positively polyhomogeneous of polydegree m. If $\sum_{m \in I} f^m \equiv 0$ it follows that $f^m \equiv 0$ for all m.*

Proof. This can be obtained from lemma 3.2.2. It can also be obtained inductively from the corresponding statement for positively homogeneous functions. We may in fact conclude by induction that if $\sum f^m \equiv 0$ and if $i \leq k-1$ and $\beta_0, ..., \beta_i$ are fixed, then it follows that $\sum_{m \in I'} f^m \equiv 0$, where I' is the subset in I of those multiindices m for which $m_j = \beta_j$ for $j \leq i$. The functions f^m from the sums for I' are now homogeneous of degree m_{i+1} with respect to the multiplication μ_{i+1} and therefore it follows for fixed β_{i+1} that $\sum_{m \in I''} f^m \equiv 0$, where $I'' = \{m \in I'; m_{i+1} = \beta_{i+1}\}$, etc.

Definition 3.6.2. *Let $F = \sum f^m$ be a finite sum of polyhomogeneous functions. Denote by I' the subset of those polyindices m in I for which f^m does not vanish identically and consider $\nu = max_{m \in I'} \, m$, where the maximum is for the order relation $<$. f^ν is then called the principal part of F. (Thus the principal part is associated with the order relation $<$. Recall that the order relation $\prec$ was not total, so in general the maximum of a finite set of polyindices does not make sense for $\prec$.)*

3. Definition 3.6.2 is justified by the following result:

Proposition 3.6.3. *Let m, m' be two polyindices and assume that $m < m'$. Let further f and g be positively homogeneous functions of polydegrees m and m' respectively. Assume finally that $g(x, \xi) \neq 0$ for (x, ξ) in a suitable multineighborhood of $(x^0, \xi^0, \xi^1, ..., \xi^{k-1})$. If c is given, we can then find $\beta > 0, \delta < 1, c'$ and a multineighborhood G of $(x^0, \xi^0, \xi^1, ..., \xi^{k-1})$ such that*

$$|f(x, \xi)| \leq c\, |g(x, \xi)|, \qquad (3.6.1)$$

if $(x, \xi) \in G$, $|\Pi_{j+1}(\xi)| \geq c'|\Pi_j(\xi)|^{1+\beta}/|\Pi_{j-1}(\xi)|^{\beta}$, $j = 1, ..., k - 2$, $|\Pi_i(\xi)| \geq c'(1 + |\Pi_{i-1}(\xi)|^{\delta})$, $i = 1, ..., k - 1$.

Remark 3.6.4. *The proposition gives a useful interpretation of how the conditions*

$$|\Pi_{j+1}\xi| \geq c'|\Pi_j\xi|^{1+\beta}/|\Pi_{j-1}\xi|^{\beta}$$

come in in our theory (of $WF^k_{A,s}$): under these conditions we have

$$|\Pi_{j+1}\xi|^{-1}|\Pi_j\xi|^{1+\beta}|\Pi_{j-1}\xi|^{-\beta} < 1/c'.$$

When c' is large, this means that $|\Pi_{j+1}\xi|^{-1}|\Pi_j\xi|^{1+\beta}/|\Pi_{j-1}\xi|^{\beta}$ is "negligible" when compared with the function $f \equiv 1$.

4. To simplify the notations in the proof of proposition 3.6.3, we shall denote $|\Pi_j(\xi)|$ by t_j, $m_j - m_{j+1}$ by α_j for $j \leq k - 1$ and m_{k-1} by α_{k-1}. Before we enter the proof of proposition 3.6.3 we prove:

Lemma 3.6.5. *Let $\beta' > 0, c', l < s < r, l, s, r$ natural numbers smaller than $k - 1$, be given. Then there are $\beta > 0, c$ so that $t_{j+1} \geq ct_j^{1+\beta}/t_{j-1}^{\beta}$ for $j = 0, ..., k - 2$ implies $t_r \geq c't_s^{1+\beta'}/t_l^{\beta'}$.*

Proof of lemma 3.6.5. When $s = l + 1, r = s + 1$, the conclusion follows directly from the assumption. We may then prove the lemma by finite induction in $s - l$ and $r - s$. To show how one can increase $s - l$, respectively $r - s$, changing constants and β, assume that simultaneously

$$t_{j+1} \geq ct_j^{1+\beta}t_{j-1}^{-\beta}, \text{ and } t_j \geq ct_{j-1}^{1+\beta}t_{j-2}^{-\beta}.$$

We shall show that then

$$t_{j+1} \geq c^{2+\beta}t_{j-1}^{1+\beta'}/t_{j-2}^{\beta'},$$

where $\beta' = \beta(1 + \beta)$. In fact this follows from

$$t_{j+1}^{-1}t_{j-1}^{1+\beta'}t_{j-2}^{-\beta'} = [t_j^{1+\beta}t_{j-1}^{-\beta}t_{j+1}^{-1}]\,[t_{j-1}^{1+\beta}t_{j-2}^{-\beta}t_j^{-1}]^{1+\beta}. \qquad (3.6.2)$$

Similarily we have for $\beta'' = \beta^2/(1+\beta)$ that

$$t_{j+1} \geq c' t_j^{1+\beta''}/t_{j-2}^{\beta''},$$

in view of

$$t_j^{1+\beta''} t_{j-2}^{-\beta''} t_{j+1}^{-1} = [t_j^{-1} t_{j-2}^{-\beta} t_{j-1}^{1+\beta}]^{\beta/(1+\beta)} [t_j^{1+\beta} t_{j-1}^{-\beta} t_{j+1}^{-1}]. \tag{3.6.3}$$

The lemma follows.

5. Proof of proposition 3.6.3. Preparations. We may divide both sides of (3.6.1) by g. Practically this means that it suffices to consider the case when $g \equiv 1$, $m' = (0, ..., 0)$, so that $m < 0$. Also recall from section 3.2 that (maintaining here and later on in this section the notations $\alpha_j = m_j - m_{j+1}$, respectively $\alpha_{k-1} = m_{k-1}$, introduced immediately after remark 3.6.4)

$$|f(x,\xi)| \leq c_1 t_0^{\alpha_0} \cdots t_{k-1}^{\alpha_{k-1}},$$

and observe that shrinking the multineighborhood G we may assume that $(x, \xi) \in G$ implies $t_i/t_{i-1} \leq c'$ if c' has been fixed previously. It suffices therefore to prove the following result:

Lemma 3.6.6. *Let m be a polyindex which is < 0. If $c > 0$ is given we can find $c' > 0, \beta > 0, \delta < 1$, so that*

$$I = t_0^{\alpha_0} \cdots t_{k-1}^{\alpha_{k-1}} < c, \tag{3.6.4}$$

if

$$t_i/t_{i-1} < c', i = 1, ..., k-1, \tag{3.6.5}$$

$$t_{j+1}^{-1} t_j^{1+\beta} t_{j-1}^{-\beta} < c', 1 \leq j \leq k-2, \tag{3.6.6}$$

and

$$t_i \geq c'(1 + t_{i-1}^\delta). \tag{3.6.7}$$

Moreover, condition (3.6.7) is here only needed in the case that $m_0 = \sum \alpha_i < 0$.

Remark 3.6.7. *a) In particular it is clear from this result that the main point in proposition 3.6.3 was not the polyhomogeneity of the functions f and g, but the fact that we could estimate them as in lemma 3.2.2.*

b) It is useful to note for later purpose that part of the conditions from (3.6.5), (3.6.6) are redundant. In fact if we know that

$$t_{k-1}/t_{k-2} < c' \tag{3.6.8}$$

and that (3.6.6) is valid, then we can deduce that

$$t_i/t_{i-1} < c'', \ i = 1, ..., k-2,$$

for some c'' which becomes small with c'. This can be seen inductively. Thus for example we have

$$\begin{aligned} t_{k-2}/t_{k-3} &= [t_{k-2}^\beta/t_{k-3}^\beta]^{1/\beta} = [t_{k-2}^{1+\beta}/(t_{k-3}^\beta t_{k-2})]^{1/\beta} \\ &= [t_{k-2}^{1+\beta}/(t_{k-3}^\beta t_{k-1})]^{1/\beta}[t_{k-1}/t_{k-2}]^{1/\beta} \le (c'^2)^{1/\beta}, \end{aligned} \qquad (3.6.9)$$

if (3.6.6) and (3.6.8) are valid.

Proof of lemma 3.6.6. We have to consider two cases: $I) : \sum \alpha_i = 0, II) : \sum \alpha_i < 0$.
I. The case $\sum \alpha_i = 0$. From the assumption it follows that the first nonvanishing α_i is negative, and it suffices to argue for the case when already $\alpha_0 < 0$. Denote by s the smallest index so that $\alpha_s > 0$ and by r the largest index so that $\alpha_r < 0$. If $i \ge 1$ is any other index, we can estimate $t_i^{\alpha_i}$ for $t_j/t_{j-1} < c'$ by

$$c'^{\alpha_i(i-s)}t_s^{\alpha_i} \text{ if } \alpha_i > 0, \text{ and by } c'^{-\alpha_i(r-i)}t_r^{\alpha_i}, \text{ if } \alpha_i < 0.$$

Now, while there is certainly some $s > 0$ with $\alpha_s > 0$ (since $\sum \alpha_i = 0$ and $\alpha_0 < 0$), it may actually happen that $r = 0$. In this case

$$I \le c'' \, t_0^{\alpha_0} \, t_s^{-\delta},$$

where $\delta = \sum_{i>0} \alpha_i = -\alpha_0$, so that I may be made arbitrarily small if t_s/t_0 is small enough. In the remaining (sub-)case we have in a similar way

$$I \le c'' \, t_0^{\alpha_0} t_s^{\delta_+} t_r^{-\delta_-},$$

where $\delta_+ = \sum_{i>0}(\alpha_i)_+$, $\delta_- = \sum_{i>0}(\alpha_i)_-$. Our assumption here is that $\delta_- > 0$, and from $\sum \alpha_i = 0$ it follows that $\alpha_0 + \delta_+ - \delta_- = 0$. We can therefore write that

$$t_0^{\alpha_0} t_s^{\delta_+} t_r^{-\delta_-} = (t_r^{-1} t_s^{1+\beta'} t_0^{-\beta'})^{\delta_-}$$

where $\beta' = -\alpha_0/\delta_-$. Lemma 3.6.5 now shows that $t_r^{-1}t_s^{1+\beta'}t_0^{-\beta'}$ may be made arbitrarily small, if we choose $\beta > 0$ suitably, shrink c' sufficiently and assume that (3.6.6) is valid.

II. In the case $\sum \alpha_i < 0$ it is not anymore clear that $\alpha_0 < 0$, but we can argue essentially as before to estimate I by $\tilde{c} t_s^{\gamma'} t_r^{-\gamma''}$, where $\gamma' = \sum_{i \ge 0}(\alpha_i)_+$, $\gamma'' = \sum_{i \ge 0}(\alpha_i)_-$. The assumption for this case shows that $\gamma' < \gamma''$ so that $\delta' = \gamma'/\gamma'' < 1$. It remains to rewrite $t_s^{\gamma'} t_r^{\gamma''}$ as $(t_s^{\delta'}/t_r)^{\gamma''}$ and to observe that the conditions (3.6.7) imply that $t_s^{\delta'}/t_r$ is arbitrarily small, if c', δ are suitable. (This is in analogy with lemma 3.6.5, but simpler.) We omit the details.

Remark 3.6.8. *In the case $m_0 = 0$ it follows from the proof of lemma 3.6.6 that if β and some $\mu_j \geq 0$ are chosen suitably, then we can write I in the form*

$$I = (t_{k-1}/t_{k-2})^{\mu_0} \prod_{j \geq 1} \left(t_{j+1}^{-1} t_j^{1+\beta} t_{j-1}^{-\beta} \right)^{\mu_j}. \tag{3.6.10}$$

(Here we take also into account the proof of lemma 3.6.5 and of remark 3.6.7 b): cf. e.g. (3.6.2), (3.6.3), (3.6.9)).

Moreover, if such a representation is possible for some β, it is possible for any smaller β. (To see why this is so, observe that

$$t_{j+1}^{-1} t_j^{1+\beta+\nu} t_{j-1}^{-\beta-\nu} = t_{j+1}^{-1} t_j^{1+\beta} t_{j-1}^{-\beta} (t_j/t_{j-1})^{\nu},$$

and that t_j/t_{j-1} has, once more by remark 3.6.7, a representation of the type from the right hand side of (3.6.10) for any β.

At the level of polyindices this is related to remark 3.2.6 c).

6. We conclude the section with some simple minded remarks on Poisson brackets which should justify some constructions performed later on in chapter IV. To start, we consider two polyhomogeneous functions p and q defined on some polyneighborhood of $(x^0, \xi^0, \xi^1, ..., \xi^{k-1})$. Also denote by $p(x, D)$, respectively $q(x, D)$ the k-pseudodifferential operators associated with p and q. In the corresponding G_φ-calculus, the commutator of $p(x, D)$ and $q(x, D)$ has a symbol for which the formal asymptotitc expansion starts with the Poisson bracket

$$\{p, q\}(x, \xi) = \sum_{j=1}^{n} [(\partial p/\partial x_j)(\partial q/\partial \xi_j) - (\partial p/\partial \xi_j)(\partial q/\partial x_j)] \tag{3.6.11}$$

of p and q. We also assume that coordinates are chosen so that $M_j = \{\xi; \xi_i = 0 \text{ if } i \geq 1 + d_j\}$ for some suitable d_j. Each individual term in the preceding sum is then polyhomogeneous, but their degrees of polyhomogeneity are not the same. Indeed, the degree of polyhomogeneity of a term

$$(\partial p/\partial x_j)(\partial q/\partial \xi_j) - (\partial p/\partial \xi_j)(\partial q/\partial x_j)$$

is equal with that of a term

$$(\partial p/\partial x_k)(\partial q/\partial \xi_k) - (\partial p/\partial \xi_k)(\partial q/\partial x_k)$$

precisely when we can find some i so that $d_i < j, k \leq d_{i+1}$. We may now however regard (3.6.11) as a finite sum of polyhomogeneous functions and as such it has a polyhomogeneous principal part. Indeed, it is easy to see that the polyhomogeneous principal part of (3.6.11) is

$$\{p, q\}(x, \xi) = \sum_{j=1}^{d_{k-1}} [(\partial p/\partial x_j)(\partial q/\partial \xi_j) - (\partial p/\partial \xi_j)(\partial q/\partial x_j)]. \tag{3.6.12}$$

Given the importance of commutators in the calculus, it is now intuitively clear that the expression from (3.6.12) must play a significant role in any invariant theory. Here we shall deal with invariant formulations only in the case k=2. In that case, (3.6.12) was first considered by Laurent [1] and called "relative" Poisson bracket. We shall come back to it with more details in chapter IV.

3.7 Successive localizations

1. Let P be a homogeneous polynomial in the variables ξ of degree m_0 and assume that P vanishes precisely of order s at $\xi^0 \neq 0$. We may assume that linear coordinates are chosen so that $\xi^0 = (0, ..., 0, 1)$ and consider the localization p_1 of P at ξ^0. Explicitly, p_1 is the Taylor expansion of order s of P at ξ^0 and we have the relation

$$P = \xi_n^{m_0-s} p_1 + q_0, \tag{3.7.1}$$

where p_1 is a polynomial of degree s which depends only on the variables $\xi_1, ..., \xi_{n-1}$ and where

$$q_0 = 0(|\xi_n|^{m_0-s-1} |(\xi_1, ..., \xi_{n-1})|^{1+s}). \tag{3.7.2}$$

Let us denote by $M_1 = \{\xi \in R^n; \xi_n = 0\}$ and consider on R^n the bihomogeneous structure given by (R^n, M_1). Since q_0 is a polynomial, we can write it in a unique way as a sum of bihomogeneous terms and we denote a fixed, but otherwise generic, term from this sum by q'. Let further (m_0, m_1) be the bidegree of bihomogeneity of q'. The corresponding bidegree of bihomogeneity of $\xi_n^{m_0-s} p_1$, on the other hand, is (m_0, s). From (3.7.2) it follows now that we must have $m_1 > s$, so that $(m_0, m_1) < (m_0, s)$.

In later constructions we shall pass from the bihomogeneous structure just described to more refined polyhomogeneous structures. In general, neither $\xi_n^{m_0-s} p_1$ nor q' will be polyhomogeneous in the new structure, but since they are polynomials, both can be written as finite sums of polyhomogeneous summands. Any summand which appears from $\xi_n^{m_0-s} p_1$ will remain bihomogeneous of bidegree (m_0, s) and any summand which appears from q' will remain bihomogeneous of bidegree (m_0, m_1). In this process therefore all terms which come from $\xi_n^{m_0-s} p_1$ will dominate in the order relation $<$ all terms which come from the remainder term q. It follows that the polyhomogeneous principal part of P

will come from the summands of $\xi_n^{m_0-s}p_1$ and will be $\xi_n^{m_0-s}$ times the principal part of p_1 computed in the polyhomogeneous structure on M_1 induced from our polyhomogeneous structure on R^n.

2. Let us now consider a fixed chain of vectors $(\xi^0, \xi^1, ..., \xi^{k-1})$, of natural numbers $(s(0), s(1), ..., s(k-1))$, and of polynomials $(p_0, p_1, ..., p_k)$, with the properties a),b),c) from section 1.1, i.e. such that

a) $p_0 = P$ and p_0 vanishes of multiplicity $s(0)$ at ξ^0,

b) p_{j-1} vanishes of multiplicity $s(j-1)$ at ξ^{j-1} and p_j is the localization of p_{j-1} at ξ^{j-1}, $j = 1, ..., k$,

c) the vectors $\xi^0, \xi^1, ..., \xi^{k-1}$ are linearly independent.

We may then choose linear coordinates so that $\xi^j = (0, ..., 0, 1, 0, ..., 0)$, where the 1 sits on position $n - j$. In particular, p_j does not depend on the variables $\xi_{n-j+1}, ..., \xi_n$ effectively. Also denote by $M_j = \{\xi; \xi_i = 0, i > n - j\}$. On R^n we now consider the polyhomogeneous structure induced by $M_0, M_1, ..., M_{k-1}$. Recall that p_j is the Taylor expansion of order $s(j-1)$ of p_{j-1} at ξ^{j-1}, so that in analogy with (3.7.1) we have

$$p_{j-1}(\xi) = \xi_{n-j+1}^{s(j-2)-s(j-1)}p_j(\xi) + q_{j-1}(\xi), 1 \le j \le k, \qquad (3.7.3)$$

where for convenience we have set $s(-1) = m_0$. As for the remainder term q_{j-1} it satisfies

$$q_{j-1}(\xi) = 0(|\Pi_j(\xi)|^{s(j-1)+1}|\xi_{n-j+1}|^{s(j-2)-s(j-1)-1}). \qquad (3.7.4)$$

It follows in particular that

$$\begin{aligned}
P &= q_0 + \xi_n^{s(-1)-s(0)}q_1 + \xi_{n-1}^{s(0)-s(1)}\xi_n^{s(-1)-s(0)}q_2 + \cdots \\
&\quad + \xi_{n-j+1}^{s(j-2)-s(j-1)} \cdots \xi_{n-1}^{s(0)-s(1)}\xi_n^{s(-1)-s(0)}q_j + \cdots \\
&\quad + \xi_{n-k+2}^{s(k-3)-s(k-2)} \cdots \xi_n^{s(-1)-s(0)}q_{k-1} + \xi_{n-k+1}^{s(k-2)-s(k-1)} \cdots \xi_n^{s(-1)-s(0)}p_k. \quad (3.7.5)
\end{aligned}$$

3. In the preceding relation, all q_i are homogeneous polynomials. We may therefore regard them as finite sums of polyhomogeneous terms for the polyhomogeneous structure of the $(M_0, M_1, ..., M_{k-1})$ and note that

$$Q = \xi_{n-k+1}^{s(k-2)-s(k-1)} \cdots \xi_n^{s(-1)-s(0)}p_k ,$$

already is polyhomogeneous. We claim that Q is the polyhomogeneous principal part of P. This may be seen directly, by computing the polydegree of polyhomogeneity of

Q and comparing it with the worst cases which can appear from the remainder terms. It follows however also from the discussion in nr. 1 in this section. In fact, each step in our successive localizations is of the type from nr.1, so, according to the discussion from there, at step j the polyhomogeneous principal part of P for the polyhomogeneous structure $M_0, ..., M_j$ is $\xi_{n-j+1}^{s(j-2)-s(j-1)} \cdots \xi_n^{s(-1)-s(0)} p_j$.

4. We have assumed above that P had been a polynomial. When we apply the present discussion to the situation in theorem 1.1.6, we need a similar analysis for the case when P is just a constant coefficient analytic symbol which is defined in a conic neighborhood of (x^0, ξ^0). Actually, $p_{m,1}$ will be a polynomial even in this situation, so all terms $p_{m,i}, q_i$, $i \geq 1$, can be treated as before. As for q_0 it will not be a polynomial in general, but we have seen in section 3.2 that it satisfies an estimate of form

$$|\partial_\xi^\alpha q_0(\xi)| \leq c^{|\alpha|+1} \alpha! |\xi|^{m-s-1} |\xi'|^{s+1-|\alpha|}$$

in a conic neighborhood Γ' of ξ^0. Arguing essentially as before we can then conclude that q_0 is dominated by $\xi_{n-k+1}^{s(k-2)-s(k-1)} \cdots \xi_n^{s(-1)-s(0)} p_k$ on a polyneighborhood of $(\xi^0, \xi^1, ..., \xi^{k-1})$, if we intersect this with sets of type

$$|\xi_i| \geq c_i |\xi_{i-1}|^{1+\beta} / |\xi_{i-2}|^\beta,$$

for suitable $\beta > 0$ and if p_k is elliptic as an operator on R^{n-k}. It will follow from this that the regularity results needed in the proof of theorem 1.1.6 are valid in this case too.

3.8 Regularity theorems I.

1. We explicitly mention in this section two regularity theorems which are consequences of the theory from the preceding sections and which we use in the proof of theorem 1.1.3. The first of them is standard in the theory of two-microlocalization. The assumptions which we make here are that a classical analytic pseudodifferential (or just differential) operator $p(x, D)$ is given on $X \times G$ which satisfies the conditions (1.1.1) and (1.1.2) for some analytic homogeneous involutive manifold Σ and that u is a solution of $p(x, D)u = 0$ on $X \times G$, i.e. that $WF_A p(x, D)u \cap X \times G = \emptyset$.

Theorem 3.8.1. *Consider* $\lambda^0 \in \Sigma$ *and let* $v^0 \in R^d \subset R^n$. *Assume that* (λ^0, v^0) *is micro-noncharacteristic for* p_m *(the principal part of* p*) relative to* Σ. *Then it follows that*

$$(\lambda^0, v^0) \notin WF_A^2 u.$$

Proof. We can write p^T in the form

$$p^T(x, \xi'', v) = \sum_{|\alpha|=s} \partial_\xi^\alpha p_m(x, 0, \xi'') v^\alpha / \alpha! \; .$$

Thus p^T is the first nontrivial homogeneous term in the Taylor expansion of p_m and we have the following relation

$$p(x, \xi) = p^T(x, \xi'', \xi') + R(x, \xi) + Q(x, \xi) \qquad (3.8.1)$$

where $Q(x, \xi) = p_m(x, \xi) - p^T(x, \xi'', \xi')$ is the remainder term in the Taylor expansion which gives p^T and R consists of the lower order terms of p. We have thus the estimates

$$|\partial_x^\alpha \partial_\xi^\beta R(x, \xi)| \leq c_1^{|\alpha|+|\beta|+1} \alpha! \beta! (1 + |\xi|)^{m-1-|\beta|}$$

and

$$|\partial_x^\alpha \partial_\xi^\beta Q(x, \xi'', \xi')| \leq c_2^{|\alpha|+|\beta|+1} \alpha! \beta! \sum_{j \leq \min(s+1, |\beta|)} (1 + |\xi'|)^{s+1-j}(1 + |\xi|)^{m-s-1+j-|\beta|}$$

In the bihomogeneous structure given by $M_0 = R^n$, $M_1 = \{\xi, \xi'' = 0\}$, p^T is (m,s) bihomogeneous and the assumption on (λ^0, v^0) is that $p^T(\lambda^0, v^0) \neq 0$. It is then clear that p^T dominates $R + Q$ on a bineighborhood of (λ^0, v^0), whence the theorem.

2. In our second result we assume that a polyhomogeneous structure $M_0 = R^n$, $M_j \subset M_{j-1}$, $M_j \neq M_{j-1}$, $M_k = \{0\}$, is given and consider $x^0 \in R^n$, $\xi^j \in \dot{M}_j$. Further we consider an open polycone G which contains $(\xi^0, \xi^1, ..., \xi^{k-1})$ and let $p_0, p_1, \ldots, p_s$ be real analytic functions defined on a set of form

$$E = \{(x, \xi); |x - x^0| < c, \xi \in G, |\Pi_i(\xi)| > c'(1 + |\Pi_{i-1}(\xi)|^\delta)\},$$

for some constants c,c' and $\delta < 1$. We assume moreover that a finite set of polyindices $m^0, ..., m^s$ is given so that
$m^0 > m^i$, $\forall i \geq 1$ and so that

$$|\partial_x^\alpha \partial_\xi^\beta p_i(x, \xi)| \leq c''^{|\alpha|+|\beta|+1} \alpha! \, \beta! \, |\Pi_{k-1}\xi|^{-|\beta|+m_{k-1}^i} |\Pi_{k-2}\xi|^{m_{k-2}^i - m_{k-1}^i} \cdots |\Pi_0\xi|^{m_0^i - m_1^i}, \forall i.$$

Assume finally that

$$|p_0(x, \xi)| \geq c''' |\Pi_{k-1}\xi|^{m_{k-1}^0} |\Pi_{k-2}\xi|^{m_{k-2}^0 - m_{k-1}^0} \cdots |\Pi_0\xi|^{m_0^0 - m_1^0}.$$

Theorem 3.8.2. *Denote* $p(x, \xi) = \sum_{i \leq s} p_i(x, \xi)$ *and assume that*

$$(x^0, \xi^0, \xi^1, ..., \xi^{k-1}) \notin WF_{A,s}^k p(x, D)u.$$

Then it follows that

$$(x^0, \xi^0, \xi^1, ..., \xi^{k-1}) \notin WF_{A,s}^k u.$$

The theorem is a consequence of proposition 3.3.5.

3.9 Regularity theorems II. Proof of theorem 1.1.6

1. We assume in this section that P is a constant coefficient classical analytic symbol which is defined in a conic neighborhood of ξ^0. All notations and constructions are at first as in section 3.7. We shall add a "lower order" term $R(x, D)$ to $P(D)$ and want to study the regularity of solutions of $p(x, D)u = 0$, $p(x, D) = P(D) + R(x, D)$. Our first result refers to the situation from theorem 1.1.6. In that case, R is a classical analytic pseudodifferential operator associated with a symbol $R(x, \xi)$ which is defined in a conic set of form $X \times \Gamma$, Γ an open cone which contains ξ^0, and X open in R^n with $0 \in X$. Explicitly this means that

$$|\partial_x^\alpha \partial_\xi^\beta R(x, \xi)| \le c^{|\alpha| + |\beta| + 1} \alpha! \beta! \, (1 + |\xi|)^{m_0 - 1 - |\beta|}, \text{ if } x \in X' \subset\subset X, \xi \in G. \tag{3.9.1}$$

Let us also denote by

$$
\begin{aligned}
E \; = \; & \{ \xi \in G; |\Pi_{j+1}(\xi)| \ge c \, |\Pi_j(\xi)|^{1+\beta} / |\Pi_{j-1}(\xi)|^\beta, j = 1, ..., k - 2, \\
& |\Pi_i(\xi)| \ge c(1 + |\Pi_{i-1}(\xi)|^\delta), \, i = 1, ..., k - 1 \},
\end{aligned}
$$

for some $\delta < 1, \beta > 0, c > 0$ and some multineighborhood G of $(\xi^0, \xi^1, ..., \xi^{k-1})$. Assume finally that p_k is elliptic as a polynomial on M_k, i.e. that

$$|\Pi_k(\xi)|^{s(k-1)} \le c'|p_k(\xi)| \le c''|\Pi_k(\xi)|^{s(k-1)}, \forall \xi \in M_k.$$

(Note that here $M_k = \{ \xi \in R^n; \xi_i = 0 \text{ for } i > n - k \}$. In particular we do not assume in this section that $M_k = \{0\}$.)

For the term

$$Q = \xi_{n-k+1}^{s(k-2) - s(k-1)} \cdots \xi_n^{s(-1) - s(0)} p_k,$$

which appears in our constructions in section 3.6, this implies that

$$|Q(\xi)| \sim |\xi_{n-k+1}^{s(k-2) - s(k-1)}| \cdots |\xi_n^{s(-1) - s(0)}| \, |\Pi_k(\xi)|^{s(k-1)} \text{ on } E, \tag{3.9.2}$$

if G, c, β are suitable. Recall that Q was just the polyhomogeneous principal part of P when P was a polynomial. On the other hand, we have on G that

$$|\partial_\xi^\beta P(\xi)| \le c_1^{|\beta| + 1} \beta! \, |\xi_{n-k+1}^{s(k-2) - s(k-1)}| \cdots |\xi_n^{s(-1) - s(0)}| \, |\Pi_k(\xi)|^{s(k-1) - |\beta|}. \tag{3.9.3}$$

(Here we apply lemma 3.2.2 and argue as in section 3.6.) Moreover, if $X' \subset\subset X$ is fixed, we can choose (and this is a parametric variant of section 3.6) G, δ, β, c so that $|R(x, \xi) + P(\xi)| \sim |P(\xi)|$.

We conclude that

$$
\begin{aligned}
|\partial_x^\alpha \partial_\xi^\beta (P(\xi) + R(x,\xi))| \;\leq\;& c_2^{|\alpha|+|\beta|+1} \alpha! \beta! ((1+|\xi|)^{m_0-1-|\beta|} \\
&+ |\xi_{n-k+1}^{s(k-2)-s(k-1)}| \cdots |\xi_n^{s(-1)-s(0)}|) \, |\Pi_k(\xi)|^{s(k-1)-|\beta|} \\
\leq\;& c_3 c_2^{|\alpha|+|\beta|+1} \alpha! \beta! \, |P(\xi) + R(x,\xi)| .
\end{aligned}
$$

Finally we observe that $|\xi_{n-k+1}^{s(k-2)-s(k-1)}| \cdots |\xi_n^{s(-1)-s(0)}| \leq c_4 |\Pi_k(\xi)|^b$ on E if b is large enough.

All this together shows that $P(\xi) + R(x,\xi)$ is invertible as a symbol in $\tilde{S}_{|\Pi_k|}^b(X,E)$ for some large b. (Cf. proposition 3.3.5). We conclude from this that we have

Theorem 3.9.1. *If $p(x,D)u = 0$ near (x^0,ξ^0), (i.e. if $(x^0,\xi^0) \notin WF_A\, p(x,D)u$), it follows for $\eta \in M_k$ that*

$$(x^0, \xi^0, \xi^1, ..., \xi^{k-1}, \eta) \notin WF_{A,s}^k\, u .$$

2. The assumption on R was not needed in full strength in theorem 3.9.1. It suffices to assume that R is defined on a set of type $X \times E_{c|\Pi_k|}$ and that it satisfies

$$|\partial_x^\alpha \partial_\xi^\beta R(x,\xi)| \leq c^{|\alpha|+|\beta|+1} \alpha! \beta! \, |\xi_{n-k+1}^{\alpha_{k-1}}| \cdots |\xi_n^{\alpha_0}| \, |\Pi_k(\xi)|^{\alpha_k - |\beta|},$$

there, for some polyindex $(m_0', ..., m_k')$ for which $m_k' = \alpha_k$, $m_i' - m_{i+1}' = \alpha_i, i < k$ and which is $<$ than the polydegree of polyhomogeneity of the principal part of P. Moreover, it is also clear that we have not used the assumption that P was with constant coefficients in an essential way: a certain type of ellipticity would have sufficed. (Cf. section 3.6.)

3. We can now also conclude the proof of theorem 1.1.6. In fact the theorem is a consequence of theorem 3.9.1, when combined with theorem 2.1.13. To see this, let us in fact assume that p is an operator on $X \times G$ as in theorem 1.1.6 and let $u \in D'(X)$ be a solution of $p(x,D)u = 0$ such that

$$tB \times \{\xi^0\} \cap WF_A u = \emptyset$$

for some $t > 0$ so that $tB \subset X$. It is no loss in generality to assume that $t = 1$. The main thing is now to show that the assumptions of theorem 2.1.13 are fulfilled with $\sigma = 1$. This is of course trivial for the conditions I and II in those assumptions. As for condition III, we consider some chain of form $\xi^0, \eta^1, ..., \eta^{n-1}$. Of these, some part, say $\xi^0, \eta^1, ..., \eta^{\nu-1}$, is a chain of vectors which appears in the procedure of succesive localizations as in the statement, but $\xi^0, \eta^1, ..., \eta^{n-1}$ is certainly not. We may then assume that $\xi^0, \eta^1, ..., \eta^{\nu-1}$ was maximal with the above property, i.e. we assume that $\xi^0, \eta^1, ..., \eta^{\nu-1}$ is a chain of localization vectors, but that already $p_{m,\nu}(\eta^\nu) \neq 0$ if $p_{m,\nu}$ is the localization of $p_{m,\nu-1})$ at $\eta^{\nu-1}$. It will follow therefore from theorem 3.9.1 that

$$(x, \xi^0, \eta^1, ..., \eta^\nu) \notin WF^{\nu+1}_{A,s}u,$$

whatever $x \in X$ is. This gives of course also that $(x, \xi^0, \eta^1, ..., \eta^{n-1}) \notin WF^n_{A,s}u$. It remains to check that condition IV is valid. Here we may assume that $\xi^0, \eta^1, ..., \eta^{l-1}$ is a chain of vectors which did appear in our localization procedure and that $p_{m,l-1}(\eta^{l-1}) = 0$. (Otherwise we would have as before $(x, \xi^0, \eta^1, ..., \eta^{l-1}) \notin WF^l_{A,s}u$.) We continue then the process of successive localizations (in whatever way) until we arrive at a chain $(\xi^0, \eta^1, ..., \eta^\rho)$ so that $p_{m,\rho} \neq 0$. The assumption on B is now that it contains some vector x which is orthogonal to the lineality of $p_{m,\rho}$. This vector is then also orthogonal to the vectors from the chain $\xi^0, \eta^1, ..., \eta^{l-1}$, since these vectors lie in that lineality. This gives IV. Theorem 2.1.13 is therefore applicable and the proof of theorem 1.1.6 is complete.

3.10 Improvements in the case of constant coefficient operators

1. When $p(D)$ is a constant coefficient operator, then the results from the preceding section can be strengthened. We prove a result of this type in the present section. Consequently one can also obtain an improved version of theorem 1.1.6 in this case, but since no new ideas are involved in the passage from partially regularity to propagation, we shall not make the corresponding result explicit.

Let us then assume throughout this section, that $p = P + R$ is a polynomial in the variables $(\xi_1, ..., \xi_n)$, with principal part P and lower order term R. With P we associate chains $(\xi^0, \xi^1, ..., \xi^{k-1})$, $(p_0, p_1, ..., p_k)$, $(s(0), s(1), ..., s(k-1))$ which satisfy the assumptions a),b),c) from section 3.8. As there we shall assume for simplicity that coordinates have been chosen so that $\xi^j = (0, ..., 0, 1, 0, ..., 0)$, where the 1 sits on position $n - j$, and introduce the notation M_j for $\{\xi \in R^n; \xi_i = 0 \text{ for } i > n - j\}$. Here we assume that $j \leq k+1$, so the sequence M_j is now also defined for $j = k$ and $j = k+1$. (And of course we do not have in general that $M_k = \{0\}$.) As for the condition d) from section 1.1, we shall not exclude here -and it is of course at this point that the results from this section are more general than those from section 3.8 -, that there are vectors $\xi^k \in M_k \setminus \{0\}$ for which $p_k(\xi^k) = 0$. We fix one such vector and assume that $\xi^k = (0, ..., 0, 1, 0, ..., 0)$ with 1 on position $n - k$. In particular, $\xi^k \in M_k$. However, in order to obtain any results at all for $(\xi^0, \xi^1, ..., \xi^{k-1}, \xi^k)$ we shall assume that

$\tilde{d}$) $t \to p_k(tx^0 + \xi^k) \not\equiv 0$, and that we can find c, c' and a conic neighborhood Γ of ξ^k in $\dot{M}_k$ so that $\zeta \in C^n$, $p_k(\Pi_k\zeta) = 0$ implies

$$|Re\,\langle x^0, \zeta\rangle| \le c|Im\Pi_k\zeta| + (|\Pi_{k+1}\zeta|^{1+\beta}/|\Pi_k\zeta|^\beta) \text{ if } |\Pi_{k+1}\zeta| \le c'|\Pi_k\zeta| \text{ and } \Pi_k\,Re\,\zeta \in \Gamma.$$
$$(3.10.1)$$

(Note that this is compatible with $p_k(\xi^k) = 0$.)

Remark 3.10.1. *When p_k is elliptic in the variables from M_k, then $p_k(\zeta) = 0$ will imply*

$$|\Pi_k Re\zeta| \le c\,|\Pi_k Im\zeta| \text{ for all } \zeta.$$

In (3.10.1) we have weakened this in three respects: the fact that instead of estimating $|\Pi_k Re\zeta|$, we just estimate $|\langle x^0, Re\zeta\rangle|$ corresponds to our final intention in that in the end we only aim at obtaining partial regularity in the variables parallel to x^0. Moreover, the fact that the inequality (3.10.1) is only valid in a conic neighborhood of ξ^k, will also be reflected in the conclusion. Finally, the additional term

$$|\Pi_{k+1}\zeta|^{1+\beta}/|\Pi_k\zeta|^\beta,$$

leads to a loss of regularity which is exactly the one which we can tolerate if we want to propagate the wave front set $WF_{A,s}^k$.

2. To simplify the notations we shall assume in what follows that $x^0 = (1, 0, ..., 0)$. Our main result from this section is

Proposition 3.10.2. *Let $(\xi^0, \xi^1, ..., \xi^{k-1}, \xi^k)$, $(p_0, p_1, ..., p_k)$, $(s(0), s(1), ..., s(k-1))$ be as before and consider a solution $u \in D'(X)$ of $p(D)u = 0$. We can then find c and a polycone G which contains $(\xi^0, \xi^1, ..., \xi^{k-1}, \xi^k)$, so that*

$$(x, A) \cap WF_{|\xi_1|}u = \emptyset, \forall x \in U,$$

where

$$A = \{\xi \in G; |\xi_1| \ge c|(\xi_2, ..., \xi_{n-k})|^{1+\beta}/|\Pi_{k-1}\xi|^\beta, |\Pi_{j+1}\xi| \ge c|\Pi_j\xi|^{1+\beta}/|\Pi_{j-1}\xi|^\beta,$$
$$j = 1, ..., k-1, |\Pi_i\xi| \ge c(1 + |\Pi_{i-1}\xi|)^\delta, i = 1, ..., k, |\xi_1| \ge c(1 + |\Pi_k\xi|)^\delta\}.$$

Proposition 3.10.2 is a consequence of the following general result on regularity in G_φ-spaces:

Proposition 3.10.3. *Let $p(D)$ be a constant coefficient linear partial differential operator (not necessarily with the properties from before) and denote by $V = \{\zeta \in C^n; p(-\zeta) =$*

$0\}$. *Let also* $\varphi : R^n \to R_+$ *be some given Lipschitzian weight function and consider an unbounded set* $A \subset R^n$. *Assume finally that there is a constant* $c > 0$ *so that*

$$\zeta \in V \ \text{together with} \ Re \ \zeta \in A \ \text{implies} \ \varphi(Re \ \zeta) \leq c(1 + |Im \ \zeta|).$$

It follows then that

$$(x^0 \times A) \cap WF_\varphi u = \emptyset,$$

for any solution $u \in D'(X)$ *of* $p(D)u = 0$ *which is defined in a neighborhood* X *of* x^0.

Remark 3.10.4. *Actually proposition 3.10.3 remains valid with almost no change for general systems of linear partial differential operators with constant coefficients.*

Proposition 3.10.3 is a rather standard consequence of the results from the general theory of constant coefficient partial differential operators. Since it does not appear explicitly in the published literatur, we shall briefly explain in the following section how it is proved. As for proposition 3.10.2 it is clear that it is a consequence of proposition 3.10.3, when used together with the estimate from the following proposition.

Proposition 3.10.5. *In the assumptions from proposition 3.10.2 there is a polyneighborhood* Γ *of* $(\xi^0, \xi^1, ..., \xi^k)$, *and* c, c', c'', β, δ *so that*

$$
\begin{aligned}
|Re\zeta_1| &\leq c(1 + |Im\zeta|), \ if \ \zeta \in C^n, p(\zeta) = 0, Re\zeta \in \Gamma, \\
|\Pi_{j+1}\zeta| &\geq c'|\Pi_j\zeta|^{1+\beta}/|\Pi_{j-1}\zeta|^\beta, j = 1, ..., k-1, \\
|Re\zeta_1| &\geq c'|\Pi_k\zeta|^{1+\beta}/|\Pi_{k-1}\zeta|^\beta, \ and \\
|\Pi_i\zeta| &\geq c''(1 + |\Pi_{i-1}\zeta|)^\delta, i = 1, ..., k \\
|\zeta_1| &\geq c''(1 + |\Pi_k(\zeta)|)^\delta.
\end{aligned}
\tag{3.10.2}
$$

Remark 3.10.6. *It suffices to prove (3.10.2) under the additional assumption that* $|Im \ \zeta| < c''|Re \ \zeta|$. *Indeed, on the complementary region* $|Im \ \zeta| > c''|Re \ \zeta|$ *the estimate (3.10.2) is trivial.*

3. In the proof of proposition 3.10.5 we shall use notations from section 3.6. It is also convenient to introduce the notation η for $(\xi_1, ..., \xi_{n-k})$, η' for $(\xi_2, ..., \xi_{n-k})$, and τ for ξ_1. Thus in particular, $\eta = (\tau, \eta')$. From the assumption $\tilde{d})$ it follows that we can find $s' \geq 1$ so that

$$(\partial/\partial\tau)^j p_k(\xi^k) = 0, \ \text{for} \ j < s', \ \text{but} \ (\partial/\partial\tau)^{s'} p_k(\xi^k) \neq 0.$$

(Here we use that $t \to p_k(t, 0, ..., 0, 1) \not\equiv 0$).

We can therefore write, using the Weierstrass preparation theorem, that

$$p_k(\eta) = q'(\eta)q''(\eta), \qquad (3.10.3)$$

for η in a complex neighborhood of $\eta^k = (0,...,0,1) \in R^{n-k}, q'(\eta^k) \neq 0$, and where q'' is of form

$$q''(\eta) = \tau^{s'} + a_1(\eta')\tau^{s'-1} + \cdots + a_{s'}(\eta'), \qquad (3.10.4)$$

for some analytic functions a_i which are defined in a neighborhood of $\eta'^k = (0,...,0,1) \in R^{n-k-1}$ and satisfy $a_i(\eta'^k) = 0$.

By homogeneity, and using the uniqueness of the Weierstrass decomposition, we can extend (3.10.3) to hold in a complex conic neighborhood of η^k, with q' positively homogeneous of degree $s(k-1) - s'$. It follows in particular that

$$|\eta|^{s(k-1)-s'} = O(\,|q'(\eta)|\,), \qquad (3.10.5)$$

in a complex conic neighborhood G^k of η^k.

4. To simplify the notation, we shall now introduce the expression $I(\zeta)$ by

$$\begin{aligned} I(\zeta) \;=\; & [p(\zeta) - \zeta_{n-k+1}^{s(k-2)-s(k-1)} \cdots \zeta_{n-1}^{s(0)-s(1)} \zeta_n^{s(1)-s(0)} p_k] / [\zeta_{n-k+1}^{s(k-2)-s(k-1)} \cdots \\ & \zeta_{n-1}^{s(0)-s(1)} \zeta_n^{s(1)-s(0)}], \end{aligned} \qquad (3.10.6)$$

provided that $\zeta_{n-j} \neq 0$ for $j = 0,...,k-1$. Our interest in this expression comes from the fact that

$$q''(\eta) = I(\zeta)/q'(\eta) \text{ when } p(\zeta) = 0, \qquad (3.10.7)$$

which is just another way of writing that $I = p_k$ then. We also have the following very elementary estimate

$$|I(\zeta)| = O(|\eta|^{s(k-1)}), \text{ if } p(\zeta) = 0, \qquad (3.10.8)$$

which follows from the fact that p_k is homogeneous of degree $s(k-1)$ and depends only on the variables η. Our goal is now to obtain estimates for ζ with $p(\zeta) = 0$, by using (3.10.6) and the properties of p_k. The elementary estimate (3.10.8) for I is then too weak and the first thing to do is to evaluate I against q' in a complex conic neighborhood G'^k of η^k. Here we can use (3.10.5) and conclude that

$$I^{(s(k-1)-s')/s(k-1)}/q'$$

is bounded in a conic neighborhood G'^k of η^k. We can therefore conclude that

$$q''(\eta) = O(\,|I^{s'/s(k-1)}(\zeta)|\,) \text{ on } G'^k \text{ if } p(\zeta) = 0. \qquad (3.10.9)$$

To exploit this, we can now write q'' in the form $q''(\eta) = (\tau - \tau_1(\eta')) \cdots (\tau - \tau_{s'}(\eta'))$. If (3.10.9) is to be true, then there must be at least one j so that

$$|\tau - \tau_j(\eta')| \leq c_1 |I|^{1/s(k-1)}. \tag{3.10.10}$$

Of course all this is interesting for us only when η is in a complex conic neighborhood of η^k. It is then clear from (3.10.10) that $(\tau_j(\eta'), \eta')$ will stay in any prefixed complex conic neighborhood G' of η^k if we choose G suitably.

To sum up, we have thus proved the following main intermediate result:

Proposition 3.10.7. *Let I be the expression defined in (3.10.6) and consider complex conic neighborhoods G, G' of η^k. If G is suitably small and ζ satisfies $p(\zeta) = 0, \eta = (\tau, \eta') \in G$, then we can find $\mu \in C$ with $p_k(\mu, \eta') = 0$, $(\mu, \eta') \in G'$ and so that*

$$|\tau - \mu| \leq c_1 |I^{1/s(k-1)}(\zeta)|.$$

In particular

$$|Re\,\tau| \leq |Re\,\mu| + c_1 |I^{1/s(k-1)}(\zeta)|, \tag{3.10.11}$$

$$|Im\,\mu| \leq |Im\,\tau| + c_1 |I^{1/s(k-1)}(\zeta)|. \tag{3.10.12}$$

5. It follows that we will be able to estimate $|Re\tau|$ if we dispose of suitable estimates for $|Re\,\mu|$ and of $I^{1/s(k-1)}$. The estimates for $|Re\,\mu|$ which we need are precisely those from (3.10.2). Indeed, if G' is suitable, then $(\mu, \eta') \in G'$ will imply

$$|Re\,\mu| \leq c'(|Im\,\eta| + |Im\,\mu| + |(\zeta_2, ..., \zeta_{n-k-1})|^{1+\beta}/|\eta|^\beta),$$

(if $(\mu, \eta') \in G', \eta \in G$, then $|\eta| \sim |(\mu, \eta')|$).
We can here estimate $|Im\,\mu|$ by (3.10.1) and therefore conclude that

$$|Re\,\tau| \leq c_2 |Im\,\eta| + 2c_1 |I^{1/s(k-1)}(\zeta)| + |(\zeta_2, ..., \zeta_{n-k-1})|^{1+\beta}/|\eta|^\beta. \tag{3.10.13}$$

The proof of proposition 3.10.5 now comes to an end if we can prove:

Lemma 3.10.8. *There is c, β, δ so that if*

$$|\tau| \geq c|(\zeta_2, ..., \zeta_{n-k-1})|^{1+\beta}/|\eta|^\beta, |\Pi_{j+1}\zeta| \geq c|\Pi_j\zeta|^{1+\beta}/|\Pi_{j-1}\zeta|^\beta, j = 1, ..., k-1,$$

$$|\Pi_i\zeta| \geq c(1 + |\Pi_{i-1}\zeta|)^\delta, i = 1, ..., k, |\tau| \geq c(1 + |\Pi_k(\zeta)|)^\delta,$$

then

a) $|(\zeta_2, ..., \zeta_{n-k-1})|^{1+\beta}/|\eta|^\beta \leq (1/4)|\tau|,$
b) $2c_1 |I^{1/s(k-1)}(\zeta)| \leq (1/4)|\tau|.$

Proof of lemma 3.10.8. Part a) of the lemma is obvious. Rather than proving part b), we shall prove that for any previously fixed number $\varepsilon > 0$

$$|I(\zeta)| \le \varepsilon |\tau|^{s(k-1)} \text{ for the } \zeta \text{ under consideration },\qquad (3.10.14)$$

provided c, β, δ are suitable.

The idea of the proof is now that I can be estimated by a sum of polyhomogeneous functions which all have degrees of polyhomogeneity strictly smaller (in the " $<$ " sense) than the function $\tau^{s(k-1)}$. (The degree of polyhomogeneity of the latter is $(s(k-1), ..., s(k-1))$.) The proof is then concluded with the aid of lemma 3.2.2. To estimate I we use the definition of I and relation (3.7.5). It follows that

$$|I(\zeta)| \;\le\; c_3(1 + \sum_j |\zeta_j|^{m-1})|\zeta_{n-k+1}^{s(k-1)-s(k-2)} \cdots \zeta_n^{s(0)-s(-1)}| \qquad (3.10.15)$$
$$+|\zeta_{n-k+1}^{s(k-1)-s(k-2)} \cdots \zeta_n^{s(0)-s(-1)} q_0(\zeta)|$$
$$+|\zeta_{n-k+1}^{s(k-1)-s(k-2)} \cdots \zeta_{n-1}^{s(1)-s(0)} q_1(\zeta)| + \cdots$$
$$+|\zeta_{n-k+1}^{s(k-1)-s(k-2)} \zeta_{n-k+2}^{s(k-2)-s(k-3)} q_{k-2}(\zeta)| + |\zeta_{n-k+1}^{s(k-1)-s(k-2)} q_{k-1}(\zeta)|.$$

The first term on the right hand side of the preceding inequality is already a sum of positively polyhomogeneous functions. As positively homogeneous functions, the respective degrees of homogeneity are $s(k-1)-1$. (Recall that $s(-1) = m$.) It follows that in the category of sums of polyhomogeneous functions, this term is dominated by $\tau^{s(k-1)}$ in the desired sense. As for the remaining terms we use the estimates (3.7.4) for the q_j. For the generic term in the right hand side of (3.10.15) this gives

$$|\zeta_{n-k+1}^{s(k-1)-s(k-2)} \zeta_{n-k+2}^{s(k-2)-s(k-3)} \cdots \zeta_{n-j-1}^{s(j+1)-s(j)} \zeta_{n-j}^{s(j)-s(j-1)} q_j(\zeta)|$$
$$\le c_4 |\zeta_{n-k+1}^{s(k-1)-s(k-2)} \zeta_{n-k+2}^{s(k-2)-s(k-3)} \cdots \zeta_{n-j-1}^{s(j+1)+1} \zeta_{n-j}^{-1}|.$$

(For the last term we obtain $|\zeta_{n-k+1}^{s(k-1)-s(k-2)} q_{k-1}(\zeta)| \le c_4|\zeta_{n-k}^{s(k-1)} \zeta_{n-k+1}^{-1}|$.) As a homogeneous function the expression from the last line has degree of homogeneity $s(k-1)$, and this is also the degree of homogeneity of the expression as a function on M_i, for all $i \le j$. However, as a homogeneous function on M_{j+1}, the degree of homogeneity is $s(k-1)+1$, which is stricly bigger than the corresponding degree of $\tau^{s(k-1)}$. It follows therefore that the degree of polyhomogeneity of the expressions from (3.10.15) is $<$ than that of $\tau^{s(k-1)}$ and the proof of proposition 3.10.5 is complete.

3.11   Proof of proposition 3.10.3

1. Proposition 3.10.3 is a consequence of the following result:

Proposition 3.11.1. *Let $K \subset R^n$ be a convex compact set and denote by H_K the support function of K. Also consider $c > 0, b' \in R$ and $\varepsilon' > 0$.*

Then we can find s, b, c with the following property: for any $h \in A(C^n)$ which satisfies

$$|\partial^\alpha h(\zeta)| \leq exp\left(H_K(Im\ \zeta) + b\ln(1 + |\zeta|)\right) \text{ for } |\alpha| \leq s, \zeta \in V,$$

there is $f \in A(C^n)$ such that

$$|f(\zeta)| \leq c\exp(H_K(Im\ \zeta) + \varepsilon'|Im\ \zeta| + b'\ln(1 + |\zeta|)), \text{ for } \zeta \in C^n,$$

$$h(\zeta) = f(\zeta) + p(-\zeta)g(\zeta) \text{ for some } g \in A(C^n).$$

(A somewhat sharper result will be stated later on in this section.)

2. We briefly show how one can deduce proposition 3.10.3 from proposition 3.11.1. We shall then assume that all assumptions are as in proposition 3.10.3. For simplicity we assume that $x^0 = 0$. That u is defined in a neighborhood of 0 then gives that there are c', ε', b' so that

$$|u(g)| \leq c' \text{ for any } g \in C_0^\infty \text{ such that } |\hat{g}(\zeta)| \leq c'\exp(\varepsilon'|Im\ \zeta| + b'\ln(1 + |\zeta|)). \quad (3.11.1)$$

and the condition $p(D)u = 0$ may conveniently be written as

$$u\ ({}^tp(D)g) = 0\ , \ \forall g \in C_0^\infty(|x| < \varepsilon'),$$

where ${}^tp$ denotes the formal adjoint of p.

Next we observe that what we need to show is that we can find a Lipschitzian function $\psi : R^n \to R_+$ and constants c, d, b, ε so that $|u(v)| \leq c$ for any $v \in C_0^\infty(R^n)$ such that

$$|\hat{v}(\zeta)| \leq \exp(d\psi(-Re\ \zeta) + \varepsilon|Im\ \zeta| + b\ln(1 + |\zeta|)).$$

The main thing is now that for suitable d, ε and b, v can be written in the form

$$v = v_1 + {}^t p(D)v_2, \quad (3.11.2)$$

where v_1, v_2 are distributions supported in $|x| \leq \varepsilon'/2$ with

$$|\hat{v}_1(\zeta)| \leq c_1 \exp\left((\varepsilon'/2)|Im\ \zeta| + b'\ln(1 + |\zeta|)\right).$$

This is a consequence of proposition 3.11.1 applied for $h = \hat{v}$, if we set $f = \hat{v}_1, g = \hat{v}_2$. (That $\mathcal{F}^{-1}g$ is a distribution with compact support follows from a result of B.Malgrange: if $u = {}^t p(D)w$ in the category of analytic functionals and if u is a distribution, then also w is a distribution. Moreover, the convex hull of the supports of u and of w are equal.)

From (3.11.2) it now follows formally that $u(v) = u(v_1)$ and (3.11.1) is a formal consequence. That $u(v) = u(v_1)$ is actually true (the problem is with the fact that u and v_2 are both distributions) can be seen either by convolving (3.11.2) with a sequence of $C_0^\infty(|x| < \epsilon'/2)$ functions f_j, $j = 1, 2, ...$, which approximate the Dirac distribution at 0, or by estimating $\hat{v}_2$ from $\hat{v}_2 = (\hat{v} - \hat{v}_1)/p(-\zeta)$ and choosing b' in the beginning so, that the duality $u(^t p(D) v_2)$ is actually meaningful. We can argue in a similar way to estimate $u(v_1)$ starting from (3.11.1).

3. Proposition 3.11.1 is part of what is called the "fundamental principle" for constant coefficient linear partial differential operators. (Cf. Ehrenpreis [1] and Palamodov [1].) For the convenience of the reader, we mention here for the case of one scalar equation a version of the extension form of this principle which contains proposition 3.11.1 and which covers also all results in this direction which we need later on. (Actually, there are two main forms in which this principle can appear: an "extension form" of the type of proposition 3.11.1, in which, practically, holomorphic functions and inequalities are extended from an algebraic variety to C^n, and a "representation form", in which solutions of $p(D)u = 0$ are written as superpositions of polynomial- exponential solutions to the equation. We shall encounter an instance of the representation form in chapter IX.) Let then p be a polynomial in the n variables $\zeta \in C^n$ and write

$$p = p_1^{r_1} \cdots p_k^{r_k}$$

for the decomposition of p into irreducible factors p_i. Denote

$$V = \{\zeta \in C^n; p(\zeta) = 0\}, \quad V_i = \{\zeta \in C^n; p_i = 0\},$$

such that $V = \cup V_i$. Also choose a direction $\theta \in R^n$ so that

$$(\partial/\partial\theta)p_i(\zeta)_{|V_i} \not\equiv 0, \forall i.$$

When h, h_1, h_2 are analytic functions it also follows that

$$(\partial/\partial\theta)^j h(\eta) = (\partial/\partial\theta)^j h_2 \text{ if } \eta \in V_i \text{ and } j < r_i \qquad (3.11.3)$$

and if

$$h(\zeta) = p(\zeta)h_1(\zeta) + h_2(\zeta) \text{ in a neighborhood of } \eta.$$

Conversely, if (3.11.3) holds for η in a neighborhoood of $\zeta^0 \in V$ and for given holomorphic h, h_2, then

$$h(\zeta) = p(\zeta)h_1(\zeta) + h_2(\zeta)$$

for some holomorphic h_1 in a neighborhood of ζ^0. The following extension theorem is then a variant of the fundamental principle for scalar equations.

Theorem 3.11.2. *Consider g and φ, where $g : R^n \to R_+$ is Lipschitz-continuous, whereas $\varphi : C^n \to R$ is Lipschitz-continuos and plurisubharmonic. If $\varepsilon' > 0, d' > 0, b' \in R$ are given, we can then find $c, d > 0, \varepsilon > 0, b \in R$ with the following property:*

assume that $h \in A(C^n)$ is given so that for $i = 1, ..., k$

$$|(\partial/\partial\theta)^j h(\zeta)| \le exp(dg(Re\,\zeta) + \varphi(\zeta) + \varepsilon|Im\,\zeta| + b\,ln(1 + |\zeta|))\ for\ \zeta \in V_i, j < r_i.$$

Then there is $h' \in A(C^n)$ so that

$$|h'(\zeta)| \le c\ exp(dg(Re\,\zeta) + \varphi(\zeta) + \varepsilon|Im\,\zeta| + b\,ln(1 + |\zeta|))\ for\ \zeta \in C^n$$

and so that

$$h - h' \in p(\zeta)A(C^n).$$

(The prototype of a function which will play the role of φ is $H_K(Im\,\zeta)$ where K is a convex compact in R^n.)

Remark 3.11.3. *The couples $((\partial/\partial\theta)^j, V_i)$; $i = 1, ..., k$, $j < r_i$ shall be called "a collection of Noetherian operators" associated with p. (The terminology is from Palamodov [1].) We shall refer to them in general as "$(\partial^1, V^1), ..., (\partial^s, V^s)$", where the ∂^l is of form $(\partial/\partial\theta)^j$ and $V^l \in \{V_1, ..., V_k\}$, with $l < r_i$ if $V^l = V_i$. (More general forms of "Noetherian operators" are considered in Ehrenpreis and Palamodov, loc.cit., but will not be needed here.)*

3.12   Proof of the propositions 1.1.9 and 1.1.12

1. a) $\Rightarrow$ b).   We shall assume throughout this section that $\xi^j = (0, ..., 0, 1, 0, ..., 0)$ with "1" on position $n - j$. The situation is closely related to the proof of theorem 3.9.1. We write, with (by now) obvious notations, that

$$p_{m,j-1}(\xi) = \xi_{n-j+1}^{s(j-2)-s(j-1)} p_{m,j}(\xi) + q_{j-1}(\xi), 1 \le j \le k - 1. \tag{3.12.1}$$

(This is similar to relation (3.7.3)). We obtain from this that

$$p_m(\xi) = q_0 + \xi_n^{s(-1)-s(0)}q_1 + \cdots + \xi_{n-k+3}^{s(k-4)-s(k-3)} \cdots \xi_n^{s(-1)-s(0)}q_{k-2} + \xi_{n-k+2}^{s(k-3)-s(k-2)} \cdots \xi_n^{s(-1)-s(0)}p_{m,k-1}. \tag{3.12.2}$$

The last term in the right hand side of (3.12.2) is the polyhomogenous principal part of p_m and the assumption is that it is elliptic on a polyneighborhood of $(\xi^0, \xi^1, ..., \xi^{k-2}, \sigma)$. If $\beta > 0$ is chosen small enough it will then dominate the other terms in (3.12.2) on sets of the type which appear in proposition 1.1.9 b). In particular, p_m will not vanish then. We omit further details.

2. b) $\Rightarrow$ a).   We shall argue by contradiction and prove the fact that $i) \Rightarrow ii)$ in proposition 1.1.11. Assume then that $p_{m,k-1}(\sigma) = 0$. We shall show that we can find a sequence $\{\zeta^r\}_r \subset C^n$ with the following properties:

α) The ζ^r_{n-j} are real positive numbers for $j < k-1$ and $\zeta^r_{n-k+1} = 1$,

β) $p_m(\zeta^r) = 0$,

γ) $|(\zeta^r_1, ..., \zeta^r_{n-k})| \to 0$,

δ) $\zeta^r_{n-j}/\zeta^r_{n-j+1} \to 0$, if $j \leq k-1$,

ε) $(1/\zeta^r_{n-j})(\zeta^r_{n-j+1})^{1+\beta}(\zeta^r_{n-j+2})^{-\beta} \to 0$, if $j \leq k-1$.

It is then clear in particular that

$$\Pi_i(\zeta^r)/|\Pi_i(\zeta^r)| \to \Pi_i(\xi^i).$$

3.   To find a sequence ζ^r with the properties above, we shall apply the Weierstrass preparation theorem in suitably chosen variables. Before we show how this is done, we observe that ζ is characteristic for p_m precisely if, in the notations from the first part of the proof,

$$0 = F(\zeta) = \zeta_{n-k+2}^{-s(k-3)+s(k-2)} \cdots \zeta_n^{-s(-1)+s(0)} q_0(\zeta) + \cdots + \zeta_{n-k+2}^{-s(k-3)+s(k-2)} q_{k-2}(\zeta) + p_{m,k-1}(\zeta).$$
$$(3.12.3)$$

Here we note that $p_{m,k-1}(\sigma) = 0$, but that $p_{m,k-1}(\xi_1, ..., \xi_{n-k}, 1) \not\equiv 0$. It follows that we can find $\theta \in R^{n-k}$ so that $t \to p_{m,k-1}(t\theta, 1) \not\equiv 0$. Arguing in the space generated by $(\theta, 0), \sigma, \xi^j, 0 \leq j \leq k-2$, we may assume then from the very beginning that $n = k+1$. In this situation we will have $\xi_1 \leftrightarrow \theta$, of course.

The idea is now to solve (3.12.3) in θ for suitable real positive $\zeta_3, ..., \zeta_n, \zeta_2 = 1$, and the problem is to do this in such a way as to obtain a small solution θ if the ζ_j , $j \geq 3$, are related by the conditions $\delta), \varepsilon)$.

It is here convenient to choose another set of independent variables. Let us in fact introduce μ_i by

$$
\begin{aligned}
\mu_3 &= 1/\zeta_3 = \zeta_2/\zeta_3, \\
\mu_4 &= \zeta_3^{1+\beta}/\zeta_4^{\beta} = \zeta_2^{-1}\zeta_3^{1+\beta}/\zeta_4^{\beta}, \\
\mu_5 &= \zeta_3^{-1}\zeta_4^{1+\beta}\zeta_5^{-\beta},
\end{aligned}
$$

$$\cdot$$

$$\cdot$$

$$\cdot$$

$$
\mu_n = \zeta_{n-2}^{-1}\zeta_{n-1}^{1+\beta}\zeta_n^{-\beta},
$$

where β is some positive rational number still to be specified.

We now claim that if the μ_i are all small, then the expressions

$$\zeta_{n-j}/\zeta_{n-j+1},\ j \le k-1, \tag{3.12.4}$$

$$\zeta_{n-j}^{-1}\zeta_{n-j+1}^{1+\beta}\zeta_{n-j+2}^{-\beta},\ 2 \le j \le k-1, \tag{3.12.5}$$

are small simultaneously. In fact this is obvious for (3.12.5) and also for (3.12.4) in the case $j = k-1$. That it is true for the remaining j in (3.12.4) can then be seen as in remark 3.6.7.

We can now rewrite $F(\theta, 1, \zeta_3, ..., \zeta_n)$ as a function of θ, μ:

$$F(\theta, 1, \zeta_3, ..., \zeta_n) = G(\theta, \mu),$$

and must show that $G(\theta, \mu) = 0$ admits a small solution θ for every small μ, if $\beta > 0$ has been chosen small enough.

This will follow from the following assertions:

I. $G(\theta, \mu) \not\equiv 0$,

II. G can be written as a finite sum of terms of form

$$c\theta^{\delta_1}\mu_3^{\delta_3}\cdots\mu_n^{\delta_n}, \tag{3.12.6}$$

where all δ_j are rational. Moreover if β is small enough, then they are ≥ 0 and at least one of them is strictly larger than zero .

III. $G(0,0) = 0$.

The first of these assertions is obvious and the third follows from the second. It is also clear that $G(\theta, \mu)$ is a sum of terms as in (3.12.6) for some rational δ_i. The only problem is then to prove, assuming β small enough, that these δ_i are ≥ 0, with one of them being > 0. Let us then look at first at the part of G which comes from a term

$$\zeta_3^{-s(k-3)+s(k-2)}\cdots\zeta_{n-j}^{-s(j-1)+s(j)}q_j = H_j(\zeta).$$

Since the polydegree of polyhomogeneity of H_j is strictly smaller than that of $p_{m,k-1}$ it follows that the polydegree of polyhomogeneity of $\zeta_2^{-s(k-2)} H_j$ is strictly smaller than 0. (Recall that $s(k-2)$ is the degree of $p_{m,k-1}$.) The same is then true for any term of form

$$c' \zeta_3^{-s(k-3)+s(k-2)} \cdots \zeta_{n-j}^{-s(j-1)+s(j)} \zeta_2^{-s(k-2)} \; [\theta^{\delta'} \zeta_2^{\delta''}]$$

if we write q_j as a sum of terms of form $c'\theta^{\delta'}\zeta_2^{\delta''}$. It follows now from remark 3.6.7 that if we write

$$\zeta_3^{-s(k-3)+s(k-2)} \cdots \zeta_{n-j}^{-s(j-1)+s(j)}$$

in the form $\mu_3^{\delta_3} \cdots \mu_n^{\delta_n}$, then all δ_j must be ≥ 0, if β is small enough. Moreover, since this expression must become arbitrarily small with μ, at least one of the δ_j is strictly larger than zero. This takes care of the part of G which comes from the q_j. To treat the part which comes from $p_{m,k-1}$, we need only observe that $p_{m,k-1}(\theta, 1)$ is a sum of form $\sum c_j \theta^j$, and c_0 must be zero since $p_{m,k-1}(\sigma)$ was zero.

We have now established the properties I,II,III, and can conclude the proof applying the Weierstrass preparation theorem for the function

$$\check{G}(\theta, \nu_3, ..., \nu_n) = G(\theta, \nu_3^{t_3}, ..., \nu_n^{t_n}),$$

where the t_i are natural numbers for which $\check{G}(\theta, \nu)$ is analytic in the variables (θ, ν). It follows that $\check{G} = G'G''$ for two functions which are analytic in a neighborhood of $(0,0)$ and satisfy $G''(0,0) \neq 0$,

$$G'(\theta, \nu) = \sum_{0 \leq j \leq s} a_j(\nu)\theta^j, a_j(0) = 0 \text{ for } j > 0, \; a_0 \equiv 1.$$

If μ is small we can now choose ν so that $(\nu_3^{t_3}, ..., \nu_n^{t_n}) = \mu$ and solve $G'(\theta, \nu) = 0$ with some small θ: it is this the desired θ. The proof is complete.

3. Proof of proposition 1.1.12. We assume $\xi^0 = (0, ..., 0, 1)$ and write, with the notation $\xi' = (\xi_1, ..., \xi_{n-1})$,

$$p_m(\xi) = \xi_n^s p_{m,1}(\xi') + \sum_{0 \leq j < s} \xi_n^j q_j(\xi'),$$

where the q_j are homogeneous of order $m - j$ (in ξ') and $p_{m,1}$ is homogeneous of order $m - s$. It follows in particular that

$$\operatorname{grad}_{\xi'} p_m(\xi) = \xi_n^s \operatorname{grad}_{\xi'} p_{m,1}(\xi') + \sum_{0 \leq j < s} \xi_n^j \operatorname{grad}_{\xi'} q_j(\xi'),$$

whereas

$$(\partial/\partial\xi_n)p_m(\xi) = s\xi_n^{s-1}p_{m,1}(\xi') + \sum_{1 \le j < s} j\xi_n^{j-1}q_j(\xi^{r'}).$$

The proposition will follow if we can prove the following three relations

$$\lim_{r \to \infty} \frac{\text{grad}_{\xi'}q_j(\zeta^{r'})(1/\zeta_n^r)^{s-j}}{|\text{grad}_{\xi'}p_{m,1}(\zeta^{r'})|} \to 0,$$

$$\lim_{r \to \infty} \frac{p_{m,1}(\zeta^{r'})(1/\zeta_n^r)}{|\text{grad}_{\xi'}p_{m,1}(\zeta^{r'})|} \to 0,$$

$$\lim_{r \to \infty} \frac{q_j(\zeta^{r'})(1/\zeta_n^r)^{s-j+1}}{|\text{grad}_{\xi'}p_{m,1}(\zeta^{r'})|} \to 0.$$

All these relations are now immediate consequences of the assumptions. In fact, since the expressions involved in them are homogeneous, we can renormalize to the situation that $|\zeta^{r'}| = 1$. It follows that $\zeta_n^r \to \infty$ and that

$$|\text{grad}_{\xi'}p_{m,1}(\zeta^{r'})| \to |\text{grad}_{\xi'}p_{m,1}(\sigma'/|\sigma'|)| \ne 0.$$

In particular $(1/\zeta_n^r)^{s-j} \to 0$, $(1/\zeta_n^r) \to 0$, $(1/\zeta_n^r)^{s-j+1} \to 0$, for the s and j under consideration. It remains to observe that $|\text{grad}_{\xi'}q_j(\zeta^{r'})|$, $|p_{m,1}(\zeta^{r'})|$ and $|q_j(\zeta^{r'})|$ remain bounded on $|\zeta'| = 1$.

3.13   Remarks on localization of polynomials. Proof of proposition 1.1.13

Our first step in the proof of proposition 1.1.13 is an analysis of what happens at simple zeros, but for general C^1-functions rather than for polynomials. We start with a definition of localization polynomials of functions which covers this case. Let then f be, for some natural number s, a C^s-function on an open cone G in R^n, which is positively homogeneous of some degree m and consider $\xi \in G$. We assume that f vanishes precisely of order s at ξ^0, i.e. we assume that

$$\partial^\alpha f(\xi^0) = 0, \ \forall \ |\alpha| < s, \ \sum_{|\alpha|=s} |\partial^\alpha f(\xi^0)| \ne 0.$$

The localization f^T of f at ξ^0 is then defined by

$$f^T(\xi) = \sum_{|\alpha|=s} \partial^\alpha f(\xi^0)(\xi - \xi^0)^\alpha/\alpha! .$$

Thus, by definition, $f^T \not\equiv 0$. Of course, homogeneity is not strictly speaking needed in this definition. It is convenient however to ask for it, since we can then use Euler's relation for homogeneous functions later on. As in the case of localization of polynomials, it is easy to observe that $f^T(\xi + t\xi^0) = f^T(\xi)$, $\forall t \in R$. It follows then that in fact f^T is homogeneous of degree s. Another remark which is often useful is that if f_1 and f_2 are two functions on G as before, then

$$(f_1 f_2)^T = f_1^T f_2^T. \tag{3.13.1}$$

(All localizations are at ξ^0.)

In fact, if f_1 is C^{s_1} and ξ^0 is a zero of order s_1 for f_1, respectively if f_2 is C^{s_2} with ξ^0 a zero of order s_2 for f_2, then

$$f_1(\xi) = f_1^T(\xi) + o(|\xi - \xi^0|^{s_1}), \quad f_2(\xi) = f^T(\xi) + o(|\xi - \xi^0|^{s_2}),$$

near ξ^0, so

$$(f_1 f_2)(\xi) = (f_1^T f_2^2)(\xi) + o(|\xi - \xi^0|^{s_1 + s_2}),$$

which gives the desired conclusion if we assume that $f_1 f_2$ is in $C^{s_1 + s_2}$. Similar remarks are also valid if we consider products of more than two functions.

2. The first step in the proof of proposition 1.1.13 is the following elementary

Lemma 3.13.1. *Assume that f is a C^1-function in a conic neighborhood of ξ^0 so that $f(\xi^0) = 0$, grad $f(\xi^0) \neq 0$. Also assume that $f(t\xi) = tf(\xi)$ if $t > 0$ and denote by g the linear form*

$$g(\xi) = \langle grad\ f(\xi^0), \xi \rangle.$$

If σ satisfies $\langle \sigma, \xi^0 \rangle = 0$, $g(\sigma) = 0$, we can then find a sequence $\xi^r \in R^n$ with $f(\xi^r) = 0$, $\xi^r \to \xi^0$, and

$$\Pi_1(\xi^r)/|\Pi_1(\xi^r)| \to \Pi_1(\sigma)/|\Pi_1(\sigma)|,$$

where Π_1 is the orthogonal projection on the space $X = \{\xi \in R^n; \langle \xi, \xi^0 \rangle = 0\}$.

Proof of lemma 3.13.1. We may assume $\xi^0 = (0, ..., 0, 1)$, that $|\sigma| = 1$, and that $|\text{grad}\ f(\xi^0)| = 1$. Since $\langle \text{grad}\ f(\xi^0), \xi^0 \rangle = 0$ (by Euler's relation for homogeneous functions), we may assume that coordinates are so that grad $f(\xi^0) = (0, ..., 0, 1, 0)$. The zeros of $f(\xi) = 0$ can then be written in a conic neighborhood of ξ^0 in the form

$$\xi_{n-1} = T(\xi_1, ..., \xi_{n-2}, \xi_n),$$

for some positively homogeneous C^1-function T. The fact that $\xi^0 = (0, ..., 0, 1)$ gives here

$$T(0, ..., 0, 1) = 0 \tag{3.13.2}$$

and from grad $f(\xi^0) = (0, ..., 0, 1, 0)$ we obtain

$$(\partial/\partial\xi_j)T(0, ..., 0, 1) = 0 \text{ if } j \leq n - 2 \text{ or } j = n. \qquad (3.13.3)$$

Let us now assume that $\langle\sigma, \xi^0\rangle = 0$ and that $g(\sigma) = 0$, i.e. we assume $\sigma_{n-1} = \sigma_n = 0$. We need then find a sequence ξ^r so that

i) $\xi_j^r \to 0$ if $j \leq n - 1$, $\xi_n^r = 1$,

ii) $\xi_{n-1}^r = T(\xi_1^r, ..., \xi_{n-2}^r, 1)$,

iii) $\xi_j^r/|\xi^{r\prime}| \to \sigma_j$, for $j \leq n - 2$,

iv) $\xi_{n-1}^r/|\xi^{r\prime}| \to 0$,

where $\xi' = \Pi_1(\xi)$.

A sequence with these properties is for example

$$
\begin{aligned}
\xi_j^r &= \sigma_j/r, \text{ for } j \leq n - 2, \\
\xi_{n-1}^r &= T(\sigma_1/r, ..., \sigma_{n-2}/r, 1), \\
\xi_n^r &= 1.
\end{aligned}
$$

The property ii) is then a definition and i) is in part a definition and in part immediate, if we also take into account (3.13.2). Moreover, once iv) is checked, iii) is also immediate, so it remains to check iv). To do so, it suffices to show that $r\xi_{n-1}^r \to 0$. That this is true follows now from (3.13.3). In fact using (3.13.3) and once more (3.13.2) it follows that

$$
\begin{aligned}
r\xi_{n-1}^r &= r\left[T(\xi_1^r, ..., \xi_{n-2}^r, 1) - T(0, ..., 0, 1)\right] \\
&= r\sum_{j \leq n-2}(\partial/\partial\xi_j)T(0, ..., 0, 1)\xi_j^r + r|(\xi_1^r, ..., \xi_{n-2}^r, 0)|\,o(|(\xi_1^r, ..., \xi_{n-2}^r, 0)|) \\
&= r(1/r)o(1/r),
\end{aligned}
$$

where $o(\alpha)$ is a function with $\lim_{\alpha \to 0} o(\alpha) = 0$. The property iv) now follows.

Proof of proposition 1.1.13 . As in the proof of the proposition 1.1.9, we may assume that $n = k + 1$, i.e. that n=3, since $k = 2$ here. We assume then also that coordinates are chosen so that $N = (0, 0, 1)$ and denote (ξ_1, ξ_2) by ξ'. Since ξ^0 is not parallel to N it is clear that $\xi^{0\prime} = (\xi_1^0, \xi_2^0) \neq 0$ and it is no loss in generality to assume then that $|\xi^{0\prime}| = 1$. We shall now reduce the dimension of our problem by one, using the homogeneity. To do so, we first study $p_m(\xi', \xi_3)$ on $|\xi'| = 1$. Then $p_m(\xi', \xi_3)$ is a real-analytic function which is a polynomial in ξ_3 and has the property that $p_m(\xi', \xi_3) = 0$ has only real roots if $|\xi'| = 1$, $\xi' \in R^2$. Let us denote by $\xi_3^1(\xi'), ..., \xi_3^m(\xi')$ the roots of $\xi_3 \to p_m(\xi', \xi_3) = 0$, counted with multiplicities. According to a result of Bronstein [1], the labeling of the $\xi_3^i(\xi')$ can be done (the fact that the roots are real for real arguments is of course essential in this) in such a way as to obtain differentiable functions in a neighborhood of $\xi^{0\prime}$. It follows by homogeneity that we have

$$p_m(\xi', \xi_3) = \prod_{i=1}^m (\xi_3 - |\xi'| \xi_3^i(\xi'/|\xi'|)), \text{ if } \xi' \in R^2.$$

In this relation it may of course happen that $\xi_3^i(\xi') \equiv \xi_3^j(\xi')$ for some $i \neq j$.

Now, while Bronstein's theorem says that the ξ_3^i may be chosen differentiable, they will also admit a Puiseux expansion and therefore have continuous derivatives. We denote $|\xi'| \xi_3^i(\xi'/|\xi'|)$ once more by $\xi_3^i(\xi')$ and have then succeeded to find C^1- functions ξ_3^i, defined in a conic neighborhood of $\xi^{0\prime}$, so that

$$p_m(\xi) = \prod_i (\xi_3 - \xi_3^i(\xi')).$$

Let us now assume here that $s \geq 1$ is choosen so that for a suitable labeling of the functions ξ_3^i we have

$$\xi_3^i(\xi^{0\prime}) = \xi_3^0, \text{ for } i \leq s, \text{ and that } \xi_3(\xi^{0\prime}) \neq \xi_3^0 \text{ for } i > s.$$

We next denote by p', respectively p'', the functions

$$p'(\xi) = \prod_{i \leq s} (\xi_3 - \xi_3^i(\xi')), \, p''(\xi) = \prod_{i > s} (\xi_3 - \xi_3^i(\xi')).$$

Both functions are then homogeneous C^∞−functions in some conic neighborhood of ξ^0 and $p''(\xi^0) \neq 0$. As for the localization polynomial $p_{m,1}$ of p_m at ξ^0, we obtain

$$p_{m,1}(\xi) = p''(\xi^0) \prod_{i \leq s} (\xi_3 - \langle \, \mathrm{grad}_{\xi'} \xi_3^i(\xi^{0\prime}), \xi' \rangle) \, .$$

In fact, as is clear from the discussion from the beginning of this section, $p_{m,1}$ is the product of the localizations of the factors $(\xi_3 - \xi_3^i(\xi'))$, so we need only observe that the localization of $(\xi_3 - \xi_3^i(\xi'))$ is

$$\xi_3 - \langle \, \mathrm{grad}_{\xi'} \xi_3^i(\xi^{0\prime}), \xi' \rangle \, ,$$

when $i \leq s$, whereas it is $p''(\xi^0)$ for the product of the remaining factors. If now σ satisfies $p_{m,1}(\sigma) = 0$ and is orthogonal to ξ^0, then we will have that

$$\sigma_3 - \langle \, \mathrm{grad}_{\xi'} \xi_3^i(\xi^{0\prime}), \sigma' \rangle = 0 \, ,$$

for some i. We may then apply the lemma from the beginning of the section for the factor $(\xi_3 - \xi_3^i(\xi'))$, thus concluding the proof.

Chapter 4

Bi-symplectic geometry and multihomogeneous maps

In the present chapter we study some of the geometric aspects which are underlying the calculus of F.I.O.s in second microlocalization. Our true goal, to which we come in chapter V, is to develop a local calculus of F.I.O.s which is similar to the local calculus of standard F.I.O.s in the C^∞-category, although some important differences appear. The most important of these differences is that the calculus of classical F.I.O. is associated with the homogeneous structure of conic sets in T^*X, whereas in second microlocalization of the type considered here, it is more natural to work in the normal bundle $N\Sigma$ of the given homogeneous involutive submanifold Σ in T^*X, with respect to which we two-microlocalize, and to consider the natural bihomogeneous structure on $N\Sigma$. (For definitions see section 4.4 and also 1.1.) As we have seen in chapter III, this leeds to a new concept of principal parts, and, in relation to this, to a modified definition of Poisson brackets. Actually, these brackets can be derived from a natural fundamental two-form in much the same way in which the standard Poisson bracket can be derived from the canonical two-form in T^*X. Moreover, it will turn out that in the present context, we dispose of two interesting two-forms, and for this reason we shall say, following Laurent, that $N\Sigma$ admits a "bi-symplectic" structure. Of course, coordinate transformations should then leave invariant these forms, etc. The central problem in the second part of the chapter is to relate the symplectic structure of T^*X with the bi-symplectic structure of $N\Sigma$. The reason why this problem appears, comes of course from the theory of F.I.O. Suppose in fact, e.g., that a standard analytic F.I.O. is given in a conic neighborhood of some point in the cotangent space. We may then consider this operator as an operator in first microlocalization, but we shall also want to be able to regard him as an operator in second microlocalization. In the latter case, the underlying geometry is now that of

$N\Sigma$. The operator on the other hand, once given explicitly, is just there and will not care much for the interpretation which we intend to give him, so the problem which we mentioned before appears. The reason why we can reconcile these two points of view is that we may regard $N\Sigma$ microlocally as the "bihomogeneous approximation" of T^*X.

4.1   Polyhomogeneous changes of coordinates in R^n.

1. To prepare for the study of bicanonic maps associated with some F.I.O. we shall study in this section polyhomogeneous changes of coordinates in R^n_ξ. Although the present considerations are given for general k, we should say that later on we shall actually assume that $k = 2$. Moreover, the discussion here is of a somewhat preliminary nature in that actually we are interested in polyhomogeneous changes of coordinates in $R^n_x \times R^n_\xi$, to which we shall turn in the next section.

Let us then assume that on R^n_ξ we are given a polyhomogeneous structure $M_0 \supset M_1 \supset \cdots \supset M_k = \{0\}$, $M_i \neq M_{i+1}$, with corresponding multiplications μ_j. Let us also assume that $\Gamma \subset R^n$ is a polycone and consider a map $\chi : \Gamma \to R^n$.

Definition 4.1.1. *We say that χ is polyhomogeneous if*

$$\chi(\mu_j(t)\xi) = \mu_j(t)\chi(\xi), \forall t \in [0,1] \ if \ j \geq 1, \ and \ \forall t \geq 0 \ if \ j = 0. \tag{4.1.1}$$

(For simplicity, we shall not consider maps which relate two different polyhomogenous structures.)

Clearly the image of a polycone $\Gamma' \subset \Gamma$ under χ is also a polycone and if $\chi : \Gamma \to \Gamma''$ is polyhomogeneous and $f : \Gamma'' \to C$ is a polyhomogeneous function of some polydegree, then $f \circ \chi$ is also polyhomogeneous of the same polydegree. Another property of polyhomogeneous maps is that their graph is left invariant under the action of the canonic multiplications on R^{2n}. A trivial example of a polyhomogeneous map is the identity. To discuss a less trivial situation, assume that $M_j = \{\xi \in R^n; \xi_i = 0 \ for \ i \geq 1 + d_j\}$ for some natural numbers $d_0 = n > d_1 > \cdots > d_k = 0$. Also denote for $1 \leq i \leq n$ by $j(i)$ the numbers defined by the condition

$$d_{j(i)+1} < i \leq d_{j(i)},$$

i.e.,

$$j(i) = \sup_{j,d_j \geq i} j.$$

Thus for example, $j(1) = k-1$, $j(n) = 0$. Consider now a homogeneous map $\chi : \Gamma \to R^n$, $\chi = (\chi_i)_{i=1}^n$, so that the components χ_i depend only on the variables from $M_{j(i)}$. χ satisfies then a good deal of the conditions related to polyhomogeneity. Indeed, the assumption can be written as

$$\Pi_j \chi(\xi) = \Pi_j \chi(\Pi_j \xi),$$

so we obtain

$$\Pi_j \chi(\mu_j(t)\xi) = \Pi_j \chi(\Pi_j \mu_j(t)\xi) = \Pi_j \chi(t\Pi_j \xi) = t\Pi_j \chi(\xi),$$

since χ is homogeneous. In order to have (4.1.1) it suffices to require then in addition that

$$(I - \Pi_j)\chi(\mu_j(t)\xi) = (I - \Pi_j)\chi(\xi).$$

2. Let us now fix $\xi^i \in \dot{M}_i$ and let Γ be a polyneighborhood of $(\xi^0, \xi^1, ..., \xi^{k-1})$. Consider two generating elements θ and θ' of $(\xi^0, \xi^1, ..., \xi^{k-1})$ and denote

$$\tau = \chi(\theta) \ , \quad \tau' = \chi(\theta').$$

Let further η^j, respectively η'^j, denote the $\dot{M}_j$ components of τ, respectively τ'. We can then find $\lambda_j > 0$ so that $\eta^j = \lambda_j \eta'^j$. Indeed, since θ and θ' are both generating elements for the same $(\xi^0, \xi^1, ..., \xi^{k-1})$, it follows from section 3.1 that we can find $t_0, ... t_{k-1}$, $t'_0, ... t'_{k-1}$, all strictly larger than 0, such that

$$\mu_0(t_0) \cdots \mu_{k-1}(t_{k-1})\theta = \mu_0(t'_0) \cdots \mu_{k-1}(t'_{k-1})\theta' = \theta''$$

and we may also assume that $\theta'' \in G$. It follows then immediately that

$$\eta''^j = t_0 \cdots t_j \eta^j = t'_0 \cdots t'_j \eta'^j.$$

We conclude in particular, that if $\eta^j \neq 0$ for all j, then $\eta'^j \neq 0$ for all j. Moreover, both, $\chi(\theta)$ and $\chi(\theta')$ are generating elements for $(\eta^0, \eta^1, ..., \eta^{k-1})$. By abuse, we shall then also sometimes write that

$$(\eta^0, \eta^1, ..., \eta^{k-1}) = \chi(\xi^0, \xi^1, ..., \xi^{k-1}). \tag{4.1.2}$$

In this way, χ aquires, here and in related situations, two meanings: one as a map $\chi : \Gamma \to R^n$, and one between generating elements. Also note that, much in the same way in which homogneous maps transform half-rays into half-rays, polyhomogeneous maps thus transform polyrays into polyrays. (Cf. remark 3.1.6.)

3. To simplify the situation, we shall assume henceforth in this section that linear coordinates have been chosen in R^n so that $M_j = \{\xi; \xi_i = 0 \text{ for } i \geq 1 + d_j\}$ for some suitable d_j. It is then clear that (4.1.1) comes to

$$\chi_s(\mu_j(t)\xi) = t\chi_s(\xi) \text{ for } s \leq d_j, \tag{4.1.3}$$

$$\chi_s(\mu_j(t)\xi) = \chi_s(\xi) \text{ for } s \geq 1 + d_j. \tag{4.1.4}$$

This means of course that χ_s is $(1, 1, ..., 1, 0, ..., 0)$ polyhomogeneous, where the last "1" sits on position $1 + j(s)$.

4. Assume that θ is a generating element for $(\xi^0, \xi^1, ..., \xi^{k-1})$ and denote $\chi(\theta)$ by τ. Further, denote by η^j the $\dot{M}_j$ component of τ. We assume that $\eta^j \neq 0$ for all j. Then $\chi(\Gamma)$ is a polycone which contains τ. It is not obvious that $\chi(\Gamma)$ contains a neighborhood of τ. If we assume that χ is C^1, this will be the case for example if $\det \chi'(\theta) \neq 0$, where χ' is the Jacobian (matrix) of χ. Note that the condition $\eta^j \neq 0$ is satisfied for some j for example if

$$\det \left(\frac{\partial \chi_i}{\partial \xi_s} \right)_{d_{j+1} < i, s \leq d_j} \neq 0.$$

This follows indeed from Euler's relation for homogeneous polynomials.

5. Conditions on $\det \chi'$ are important later on. It is therefore useful to understand the polyhomogeneous structure of χ' and $\det \chi'$. Let us then observe that, as a function of the variables M_μ, $\partial \chi_i / \partial \xi_r$ is homogeneous of order

$$1 \text{ if } i \leq d_\mu, r > d_\mu,$$

$$-1 \text{ if } i > d_\mu, r \leq d_\mu,$$

$$0 \text{ in all other situations.}$$

Actually we have thus 4 regions:

$$\begin{pmatrix} i \leq d_\mu, & r \leq d_\mu \\ i \leq d_\mu, & r > d_\mu \\ i > d_\mu, & r \leq d_\mu \\ i \geq d_\mu, & r > d_\mu \end{pmatrix}$$

As for $\det \chi'$ we have the following result which we shall admit for the moment, but mention that we shall prove a more general proposition in the next section.

Proposition 4.1.2. *$\det \chi'$ is polyhomogeneous of polydegree $(0, ..., 0)$.*

6. We are interested in establishing conditions under which a bihomogeneous map is injective. Here we shall study for simplicity only a very simple situation of this kind, when χ_j is known to depend only on part of the variables from R^n.

Proposition 4.1.3. *Let Γ be a polyneighborhood of $(\xi^0, \xi^1, ..., \xi^{k-1})$ and assume that χ_i depends only on the variables from $M_{j(i)}$ or from the orthogonal complement of $M_{j(i)+1}$. Assume further that*

$$det \left(\frac{\partial \chi_i}{\partial \xi_s}(\theta) \right)_{d_{j+1} < i, s \leq d_j} \neq 0, \tag{4.1.5}$$

for some generating element θ of $(\xi^0, \xi^1, ..., \xi^{k-1})$.

Then we can find a polyneighborhood Γ' of $(\xi^0, \xi^1, ..., \xi^{k-1})$ so that χ is bijective on Γ'. Moreover, there are constants c, c' so that

$$|\chi_i(\xi)| \leq c' \ |\Pi_j(\xi)|, \ if \ i \leq d_j, \ \xi \in \Gamma'. \tag{4.1.6}$$

$$|\Pi_j(\xi)| \leq c \sum_{d_{j+1} < i \leq d_j} |\chi_i(\xi)|, \ if \ \xi \in \Gamma'. \tag{4.1.7}$$

Proof of proposition 4.1.3. (4.1.6) is just a consequence of the results of section 3.2 and to prove (4.1.7) we use Euler's relation for homogeneous polynomials to write that

$$\chi_i(\xi) = \sum_{s=d_{j+1}+1}^{d_j} \frac{\partial \chi_i(\xi)}{\partial \xi_s} \xi_s \ , \ d_{j+1} + 1 \leq i \leq d_j.$$

(Here we observe that $\chi_i(\mu_j(t)\xi) = t\chi_i(\xi)$ for $i \leq d_j$, whereas $\chi_i(\mu_{j+1}(t)\xi) = \chi_i(\xi)$ if $i \geq d_{j+1} + 1$. It follows that χ_i is positively homogeneous of degree 1 in the variables from $\dot{M}_j$.) (4.1.7) will now follow if we show that if we shrink Γ' suitably, then

$$\left| det \left(\frac{\partial \chi_i}{\partial \xi_s}(\xi) \right)_{d_{j+1} < i, s \leq d_j} \right| \geq c_1 > 0, \ for \ \xi \in \Gamma'.$$

Indeed, this is true for $\xi = \theta$, so it is also true in a polyneighborhood of $(\xi^0, \xi^1, ..., \xi^{k-1})$, since $det \ (\partial \chi_i / \partial \xi_s)_{d_{j+1} < i, s \leq d_j}$, is homogeneous of degree 0 in M_j.

It remains to check the injectivity. Here we observe that if $\chi(\xi) = \chi(\xi^*)$ for some ξ and ξ^* which lie both in a small polyneighborhood Γ' of $(\xi^0, \xi^1, ..., \xi^{k-1})$, then it follows that

$$|\Pi_j \xi| \leq c_1 |\Pi_j \xi^*| \leq c_2 |\Pi_j \xi|, \forall \xi, \xi^* \in \Gamma'. \tag{4.1.8}$$

This is in fact a consequence of the part of the proposition which we have already proved. We may also assume that Γ' is convex, so we shall have that

$$\sum_j \frac{\partial \chi_i}{\partial \xi_j}(\theta')(\xi_j - \xi_j^*) = 0, \tag{4.1.9}$$

for some points θ^i which lie on the segment $[\xi, \xi^*]$. We must therefore show that

$$\left| \det \left(\frac{\partial \chi_i}{\partial \xi_j}(\theta^i) \right)_{i,j=1,\dots,n} \right| \neq 0, \tag{4.1.10}$$

if (4.1.8) is valid, if ξ and ξ^* lie in a small polyneighborhood of $(\xi^0, \xi^1, \dots, \xi^{k-1})$ and if the θ^i lie on the segment $[\xi, \xi^*]$. Here we shall now use that χ_i depends only on the variables from $M_{j(i)}$, or from the orthogonal complement of $M_{j(i)+1}$, so that (4.1.10) comes to

$$\left| \det \left(\frac{\partial \chi_i}{\partial \xi_s}(\theta^i) \right)_{d_{j+1} < i,s \leq d_j} \right| \neq 0, \forall j.$$

It remains then to apply the assumption and the continuity.

Definition 4.1.4. χ *is called nondegenerate if* $\forall \Gamma' \subset\subset \Gamma$ *we can find* c, c' *such that*

$$|\Pi_j(\xi)| \leq c \sum_{d_{j+1} \leq i \leq d_j} |\chi_i(\xi)| \leq c' |\Pi_j(\xi)| \ \textit{if } \xi \in \Gamma'.$$

4.2   Polyhomogeneous changes of coordinates in R^{2n}

1. As in section 2.5 we consider here two copies of R^n, one in which the variables are denoted x or y and one in which they are denoted by ξ, η or τ. We assume that in the second copy we are given a polyhomogeneous structure defined by $M_0 \supset M_1 \supset \cdots \supset M_k = \{0\}$, $M_i \neq M_{i+1}$, and denote the corresponding multiplication operators by μ_j. Moreover, we set once more

$$\mu_j(t)(x, \xi) = (x, \mu_j(t)\xi). \tag{4.2.1}$$

2. We can now extend the definition of polyhomogeneous maps.

Definition 4.2.1. *Let* G *be a polycone in* R^{2n} *and consider* $\chi : G \to R^{2n}$. *We say that* χ *is a polyhomogeneous map if*

$$\begin{aligned}
\chi(\mu_j(t)(x, \xi)) &= \mu_j(t)\chi(x, \xi), \ \forall j, \ \forall t, 0 < t \leq 1 \ \textit{if } j \geq 1, \\
&\quad \forall t > 0 \ \textit{if } j = 0.
\end{aligned} \tag{4.2.2}$$

The map is called $\xi-nondegenerate$ *if* $\xi \to \chi_\eta(x, \xi)$ *is nondegenerate for every fixed* x.

All remarks following the definition of polyhomogeneous maps from section 4.1 can now be transferred, after appropriate modification, to the present situation. We mention the most important ones. Let us then choose coordinates in R_ξ^n so that $M_j = \{\xi; \xi_i = 0$ for $i \geq 1 + d_j\}$. It is then clear that (4.2.2) comes to

$$\chi_s(\mu_j(t)(x,\xi)) = \chi_s(x,\xi), \text{ if } s \leq n, \qquad (4.2.3)$$

$$\chi_{n+s}(\mu_j(t)(x,\xi)) = t\chi_{n+s}(x,\xi) \text{ for } 1 \leq s \leq d_j, \qquad (4.2.4)$$

$$\chi_{n+s}(\mu_j(x,\xi)) = \chi_{n+s}(x,\xi) \text{ for } s \geq 1 + d_j. \qquad (4.2.5)$$

In particular, χ_s is a $(0,...,0)$ polyhomogeneous function for $s \leq n$, whereas χ_{n+s} is $(1,1,...,1,0,...,0)$ polyhomogeneous, the last "1" sitting on position $1 + j(s)$, $j(s) = \sup_{d_j \geq s} j$.

3. From (4.2.3),(4.2.4),(4.2.5) it is also clear which is the homogeneity of $\partial\chi_i/\partial x_j$, respectively $\partial\chi_i/\partial\xi_r$, in the variables from M_ν. In fact, the homogeneity of the function $\partial\chi_i/\partial x_j$ is the same with that of χ_i, and for $\partial\chi_i/\partial\xi_r$ it is

$$-1 \text{ if } i \leq n, \, r \leq d_\nu,$$

$$0 \text{ if } i \leq n, \, r > d_\nu,$$

$$1 \text{ if } n < i \leq n + d_\nu, r > d_\nu,$$

$$-1 \text{ if } i > n + d_\nu, \, r \leq d_\nu,$$

$$0, \text{ in the remaining situations.}$$

As for det χ', we have

Proposition 4.2.2. *det χ' is polyhomogeneous of polydegree $(0,...,0)$.*

Proof. We fix $\nu \geq 0$ and show that det χ' is homogeneous of degree 0 in the variables from M_ν. Let us in fact observe at first that the entries a_{ij} of χ' ,

$$a_{ij} = \frac{\partial\chi_i}{\partial x_j}, \, i \leq 2n, j = 1,...,n,$$

$$a_{i,n+l} = \frac{\partial\chi_i}{\partial\xi_l}, \, i \leq 2n, l = 1,...,n,$$

are homogeneous of degree 0 except (possibly) for the case when either $n < i \leq n + d_\nu$ or $n < j \leq n + d_\nu$. Moreover, away from the region $n < i \leq n + d_\nu, n < j \leq n + d_\nu$, the effect of multiplying the variables from M_ν by t produces a factor t for the elements on the lines $n < i \leq n + d_\nu$ and a factor t^{-1} for $n < j \leq n + d_\nu$. Inside the region $n < i \leq n + d_\nu, n < j \leq n + d_\nu$ the entries remain unchanged, which is the same as saying that we multiply them by $t^{-1}t$. The way in which χ' is changed when ξ is replaced

by $\mu_\nu(t)\xi$ is therefore the same as if we would first multiply the d_ν lines $n < i \le n + d_\nu$
by t and then the d_ν columns $n < j \le n + d_\nu$ by t^{-1}. The complessive effect of this is of
course to leave det χ' unchanged.

4. For simplicity we shall assume henceforth in this paragraph that $k = 2$, that $M_1 = \{\xi; \xi_i = 0 \text{ for } i > d\}$ and denote by $R^n = R^d_{\xi'} \times R^{n-d}_{\xi''}$. Coordinate changes appear often in
a situation when part of the new coordinate map is prescribed and part of it has still to
be found. A typical example is when $W \subset G$ is some bihomogeneous smooth variety with
defining equations $f_j = 0, j = 1, ..., q$, and when we want to choose coordinates in which
the f_j are components of the coordinate map. To be more precise, let us assume that
W is a C^1, bihomogeneous, variety in a bineighborhood of (x^0, ξ^0, ξ^1), $\xi^{0'} = 0$, and that
$(x^0, \theta) \in W$, where θ is a generating element for (ξ^0, ξ^1). We assume that at (x^0, θ) the
equations of W are $f_i(x, \xi', \xi'') = 0$, $i = 1, ..., q$, where the f_i are $(1,1)$- bihomogeneous,
that $q \le d$ and that the differentials $d_{\xi'} f_i$ are linearly independent at (x, θ). We want
then to find g_i, $i \le n$, bihomogeneous of bidegree $(0,0)$, f_i, $i = q+1, ..., d$, bihomogeneous
of bidegree $(1,1)$ and f_i, $i = d + 1, ..., n$, bihomogeneous of bidegree $(1,0)$ so that

$$(x, \xi) \to (g, f)$$

is a coordinate change. If we denote the new variables by (y, η), then W will now just
have the equations $\eta_i = 0$, $i \le q$. If no other conditions are imposed on the map
$(x, \xi) \to (g, f)$, then suitable g_i, $i \le n$ and f_i, $i > q$ are easy to find. We may in fact
simply set $g_i = x_i$, $i \le n$, $f_j = \xi_j$, $j > d$ and choose f_i, $q < i \le d$, as linear functions in
$\xi_1, ..., \xi_d$ so that

$$\det \left(\frac{\partial f_i}{\partial \xi_j}\right)(x^0, \theta)_{i,j=1,...,d} \ne 0.$$

As an application of this discussion, let us assume that W and f_i, $i \le q$, are related as in
the above and consider some function ψ defined in a bineighborhood of (x^0, ξ^0, ξ^1) which
is bihomogeneous of bidegree (m,s) and which satisfies $\psi(x, \xi) = 0$ when $(x, \xi) \in W$. We
can then write

$$\psi(x, \xi) = \sum_{j=1}^{q} a_j(x, \xi) f_j(x, \xi), \tag{4.2.6}$$

for (x, ξ) in some bineighborhood of (x^0, ξ^0, ξ^1) and with $a_j(x, \xi)$ bihomogeneous of bidegree $(m - 1, s - 1)$. Indeed, to arrive at (4.2.6), we may first change coordinates in a
bihomogeneous way to arrive at a situation in which $f_j(x, \xi) = \xi_j, j \le d$, in which case
$\sum_{j=1}^{q} a_j(x, \xi)\xi_j$ simply corresponds to the remainder term in Taylor expansion of order
0 in the variables ξ'. The fact that the a_j are bihomogeneous of order $(m - 1, s - 1)$
then follows from the explicit expression such remainder terms have. In particular, if ψ

is (1,1) bihomogeneous, it will follow from (4.2.6) that

$$|\psi(x,\xi)| = O(\sum_{j=1}^{q} |f_j|), \text{ on a biconic neighborhood of } (x^0, \xi^0, \xi^1),$$

and another remark on the same vein is that if $\psi_1, ..., \psi_q$ is another set of (1,1)-bihomogeneous equations which define W and for which the differentials $d_{\xi'}\psi_j$ are linearly independent, then

$$|\psi| \sim |f|, \text{ on a bineighborhood of } (x^0, \xi^0, \xi^1).$$

Finally, let us assume in the situation from before, that P is (m,s)-bihomogeneous on a bineighborhood of some point $(x^0, \xi^0, \sigma^0) \in W$, $W = \{(x,\xi); f_j(x,\xi) = 0, j \leq q\}$, and f_j with the properties above. Assume moreover that P vanishes of some order s' on W. Arguing as before, we can then find for any $\beta = (\beta_1, ..., \beta_q)$ with $|\beta| = s'$ some $(m - s', s - s')$ -bihomogeneous functions $a_\beta(x, \xi', \xi'')$ so that

$$P(x, \xi', \xi'') = \sum_{|\beta|=s'} a_\beta(x, \xi', \xi'') f(x, \xi', \xi'')^\beta.$$

4.3 Bihomogeneous approximations of homogeneous maps

1. Assume that in R^n there is given a bihomogeneous structure M_0, M_1, μ_0, μ_1, and that coordinates are chosen so that $M_1 = \{\xi; \xi'' = 0\}$. Consider further G biconic in R^{2n} and let $f : G \to C$ be given. We assume that f is positively homogeneous of order m. Extending slightly the range of some definition from section 3.6 we shall say that f admits a bihomogeneous principal part of bidegree (m, s) if we can find a bihomogeneous function $f' : G \to C$ and $\beta > 0$ so that

a) f' is bihomogeneous of bidegree (m, s),
b) $f - f' = 0(|\xi'|^{s+\beta}|\xi|^{m-s-\beta})$.

In this situation, f' is also called the principal part of f. Of course, a) and b) together imply that

$$f(x, \xi) = O(|\xi'|^s |\xi|^{m-s}). \tag{4.3.1}$$

Note that b) is fulfilled for example when $f - f'$ is $(m, s + \beta)$-bihomogeneous. What we want to say in the above is thus roughly speaking that f coincides with f', up to bihomogeneous terms of lower order. When we are also interested in estimates for derivatives (in the case that $f \in C^1$), we shall use a variant of the above terminology: we say that f' is the C^1 principal part of f if, in addition to a) and b), we also have

c) $(\partial/\partial\xi')(f - f') = O(|\xi'|^{s-1+\beta}|\xi|^{m-s-\beta})$,

d) $(\partial/\partial\xi'')(f - f') = O(|\xi'|^{s+\beta}|\xi|^{m-s-1-\beta})$.

Remark 4.3.1. *a) As in the case of the standard calculus of pseudodifferential operators it may happen that $f' \equiv 0$: this will be the case for example, if instead of (4.3.1) we have a better estimate of bihomogeneous type for f.*

b) At this moment, once more, it would be tempting to consider expansions of f into infinite sums of bihomogeneous terms, whenever this is possible. We are however not interested in doing so, since anyway all we do later on is to develop a calculus of bihomogeneous approximations which is precise up to principal parts only.

c) One may consider related notions also in the case of polyhomogeneous structures for $k > 2$. How to do this is clear from section 3.5, but since we have no use for this, we shall not give details.

2. We now introduce

Definition 4.3.2. *Consider the homogeneous map $\chi^* : G \to R^{2n}$. We say that χ^* admits a bihomogeneous approximation if there is a bihomogeneous map $\chi : G \to R^{2n}$ and $\beta > 0$ such that*

$$\chi^*_{(y)}(x,\xi) - \chi_{(y)}(x,\xi) = O(|\xi'|^{\beta}/|\xi|^{\beta}), \tag{4.3.2}$$

$$\chi^*_{(\eta')}(x,\xi) - \chi_{(\eta')}(x,\xi) = O(|\xi'|^{1+\beta}/|\xi|^{\beta}), \tag{4.3.3}$$

$$\chi^*_{(\eta'')}(x,\xi) - \chi_{(\eta'')}(x,\xi) = O(|\xi'|^{\beta}|\xi|^{1-\beta}). \tag{4.3.4}$$

(Here and later on we denote by $F_{(t)}$ the t-components of the map F.)

The idea underlying the relations (4.3.2),(4.3.3),(4.3.4) is very simple: we want the components of χ to be the bihomogeneous principal parts of the corresponding components of χ^*.

Remark 4.3.3. *a) Bihomogeneous approximations, if they exist, are unique. Indeed, if e.g. χ_y^1 and χ_y^2 are both $(0,0)$-bihomogeneous and satisfy*

$$\chi_y^*(x,\xi) - \chi_y^i(x,\xi) = 0(|\xi'|^{\beta}/|\xi|^{\beta}) \text{ for } i = 1,2,$$

then $\chi_y^1(x,\xi) - \chi_y^2(x,\xi) = O(|\xi'|^\beta/|\xi|^\beta)$ *and is* $(0,0)$-*bihomogeneous. It follows from lemma 3.2.2 that* $\chi_y^1 - \chi_y^2 \equiv 0$.

One of the reasons why we are interested in bihomogeneous approximations of some given homogeneous map is of course that in this way we relate the initial map to the bihomogeneous structure. Here we recall for example that bihomogeneous maps transform polyrays into polyrays and that they map a sufficiently small bineighborhood of some given $(\tilde{x}, \tilde{\xi}, \tilde{\xi}')$ *into any previously fixed bineighborhood of some well-defined* $(\tilde{y}, \tilde{\eta}, \tilde{\eta}')$. *(Cf. here 4.1.2.) . A homogeneous map which admits a bihomogeneous approximation will then "almost" have this property.*

3. In definition 4.3.2 we have not asked that χ be nondegenerate, but we shall ask for it in all applications. As an immediate consequence of (4.3.2),(4.3.3),(4.3.4) it then follows that:

Lemma 4.3.4. *Let* G *be a bineighborhood of* (x^0, ξ^0, ξ^1) *and assume that* χ^* *admits a nondegenerate bihomogeneous approximation* χ *on* G. *If we shrink* G *it will follow that*

$$|\chi_{(\eta')}(x,\xi)| \sim |\tilde{\chi}_{(\eta')}(x,\xi)| \sim |\chi_{(\eta')}^*(x,\xi)|, \forall(x,\xi) \in G, \tag{4.3.5}$$

$$|\chi_{(\eta'')}(x,\xi)| \sim |\tilde{\chi}_{(\eta'')}(x,\xi)| \sim |\chi_{(\eta'')}^*(x,\xi)|, \forall(x,\xi) \in G, \tag{4.3.6}$$

for any $\tilde{\chi}(x,\xi)$ *which lies on the segment* $[\chi_{(\eta)}(x,\xi), \chi_{(\eta)}^*(x,\xi)]$.

Proof. If we shrink G it will follow from (4.3.3) that there is $c > 0$ for which

$$|\chi_{(\eta')}(x,\xi) - \chi_{(\eta')}^*(x,\xi)| \le c|\xi'|^{1+\beta}|\xi|^{-\beta}, \text{ if } (x,\xi) \in G. \tag{4.3.7}$$

On the other hand, since χ is nondegenerate, we may assume that

$$|\xi'| \le c_1|\chi_{(\eta')}(x,\xi)| \le c_2|\xi'|, \text{ if } (x,\xi) \in G. \tag{4.3.8}$$

It follows from (4.3.8) that we can fix c_3 so that

$$cc_1^{1+\beta}|\chi_{(\eta')}(x,\xi)|^\beta|\xi|^{-\beta} \le 1/2, \text{ if } |\xi'|/|\xi''| \le c_3. \tag{4.3.9}$$

Returning to (4.3.7) and using once more (4.3.8), we can conclude that

$$|\chi_{(\eta')}(x,\xi) - \chi_{(\eta')}^*(x,\xi)| \le (1/2)|\chi_{(\eta')}(x,\xi)|, \text{ if } (x,\xi) \in G,$$

and (4.3.5) is an immediate consequence. (4.3.6) is proved with a similar argument.

Proposition 4.3.5. *Assume G is a bineighborhood of (x^0, ξ^0, ξ^1) in $R_x^n \times R_\xi^n$. Let further $\chi^* : G \to R^{2n}$ be a homogeneous map which admits a nondegenerate bihomogeneous approximation $\chi : G \to R^{2n}$. Denote by η^0, η^1 the M_0 and M_1 component of $\chi(x^0, \theta)$, where θ is a generating element for ξ^0, ξ^1 and by $y^0 = \chi_{(y)}(x^0, \theta)$. Consider a bineighborhood U of (y^0, η^0, η^1). Then we can find a bineighborhood G' of (x^0, ξ^0, ξ^1) so that $\chi^*(G') \subset U$.*

Proof. Let us fix a bineighborhood $U' \subset\subset U$ of (y^0, η^0, η^1). Since χ is continuous we can then find a bineighborhood G'' of (x^0, ξ^0, ξ^1) so that

$$\chi(G'') \subset U'.$$

We claim that shrinking G'' still further it follows that $\chi^*(G'') \subset U$.

Let us in fact assume that c is chosen so that $(y, \eta) \in U'$, $|y - y^*| < c$, together with $|\eta' - \eta^{*'}| < c|\eta'|$, and $|\eta'' - \eta^{*''}| < c|\eta''|$ implies $(y^*, \eta^*) \in U$. Such a c exists since $U' \subset\subset U$. We can then conclude the argument by shrinking G'' until $(x, \xi) \in G''$ implies that $|\xi'|/|\xi|$ is sufficiently small. Indeed, in that case we may assume that

$$|\chi_{(y)}(x, \xi) - \chi^*_{(y)}(x, \xi)| = O(|\xi'|^\beta / |\xi|^\beta) < c,$$

$$|\chi_{(\eta')}(x, \xi) - \chi^*_{(\eta')}(x, \xi)| = O(|\xi'|^{1+\beta} / |\xi|^\beta) < c_1 |\chi_{(\eta')}(x, \xi)| (|\xi'|/|\xi|)^\beta \le c|\chi_{(\eta')}(x, \xi)|,$$

$$|\chi_{(\eta'')}(x, \xi) - \chi^*_{(\eta'')}(x, \xi)| = O(|\xi'|^\beta / |\xi|^\beta) |\xi| < c|\chi_{(\eta'')}(x, \xi)|,$$

if we also use that χ is nondegenerate. Since $\chi(x, \xi) \in U'$, it will follow therefore that $\chi^*(x, \xi) \in U$.

Proposition 4.3.6. *Let all the assumptions be as in proposition 4.3.5 and assume in addition that f is a function which is defined in a bineighborhood U of (y^0, η^0, η^1). Assume that f is homogeneous of order m and that it has a bihomogeneous principal part f' of bidegree (m, s) which is a C^1 function. Then $f \circ \chi^*$, regarded as a function defined in a bineighborhood of (x^0, ξ^0, ξ^1) (we assume tacitely that the situation is as in proposition 4.3.5), admits $f' \circ \chi$ as a bihomogeneous principal part.*

Proof. We have $f(\chi^*(x, \xi)) = f'(\chi^*(x, \xi)) + (f - f')(\chi^*(x, \xi))$. Here

$$(f - f')(\chi^*(x, \xi)) = O(|\chi^*_{(\eta')}(x, \xi)|^{s+\beta} |\chi^*_{(\eta'')}(x, \xi)|^{m-s-\beta}) =$$

$$= O(|\chi_{(\eta')}|^{s+\beta} |\chi_{(\eta'')}(x, \xi)|^{m-s-\beta}) = O(|\xi'|^{s+\beta} |\xi|^{m-s-\beta}),$$

according to lemma 4.3.4 and the fact that χ is nondegenerate. As for $f'(\chi^*(x, \xi))$, we use Taylor expansion to write

$$f'(\chi^*(x, \xi)) = f'(\chi(x, \xi)) + \text{grad}_y f'(\tilde{\chi}(x, \xi))(\chi^*_{(y)} - \chi_{(y)})(x, \xi) +$$

$$\text{grad}_{\eta'} f'(\tilde{\chi}(x,\xi))(\chi^*_{(\eta')} - \chi_{(\eta')})(x,\xi) + \text{grad}_{\eta''} f'(\tilde{\chi}(x,\xi))(\chi^*_{(\eta'')} - \chi_{(\eta'')})(x,\xi),$$

where $\tilde{\chi}(x,\xi)$ lies on the segement $[\chi(x,\xi), \chi^*(x,\xi)]$. For the factors $\chi^*_{(y)} - \chi_{(y)}, \chi^*_{(\eta')} - \chi_{(\eta')}, \chi^*_{(\eta'')} - \chi_{(\eta'')}$, we have the estimates (4.3.2),(4.3.3),(4.3.4). On the other hand we have, arguing as before,

$$\text{grad}_y f'(\tilde{\chi}(x,\xi)) = O(|\check{\chi}_{(\eta')}(x,\xi)|^s |\check{\chi}_{(\eta'')}(x,\xi)|^{m-s}) = O(|\xi'|^s |\xi|^{m-s}),$$

etc. We omit further details.

4.4   The bihomogeneous structure of $N\Sigma$.

1. In this section we begin the study of the geometric structures which are related to the Fourier integral operators which we shall use in second microlocalization. Our starting object is a regular involutive homogeneous subvariety Σ in the cotangent bundle of some n-dimensional manifold. Practically we shall always assume that the manifold is an open set X in R^n, so we assume, with standard notations, $\Sigma \subset T^*X$. It is natural then, to study canonical transformations χ defined in some conic neighborhood Γ of $(x^0, \xi^0) \in \Sigma$ which leave Σ invariant, in the sense that $\chi(\Gamma \cap \Sigma) \subset \Sigma$. An important mathematical object associated with Σ in this context is the normal bundle $N\Sigma \to \Sigma$, which has for $\lambda \in \Sigma$ the fibers $N_\lambda \Sigma = T_\lambda T^*X / T_\lambda \Sigma$.

To justify our interest in $N\Sigma$, let us observe that any χ with the properties above induces two tangent maps

$$\chi' : T\Gamma \to TT^*X \text{ and } \chi' : T(\Sigma \cap \Gamma) \to T\Sigma,$$

so that, by factorization, we also obtain a map

$$\chi' : N(\Sigma \cap \Gamma) \to N\Sigma.$$

Later on it is of course too restrictive to consider canonical transformations which are defined in full conic neighborhoods of (x^0, ξ^0), since second microlocalization is itself in fact associated with much smaller sets, namely bineighborhoods of (x^0, ξ^0, σ^0) for some σ^0. It seems therefore appropriate to introduce a bihomogeneous structure on $N\Sigma$ and we shall do so in the present section. In addition it will turn out that, in analogy with the canonical two-form on T^*X, we have two natural two-forms on $N\Sigma$ which are relevant in second microlocalization. The study of the associated "bisymplectic" structure is left for the following section.

2. When studying $N\Sigma$ it is often useful to choose canonical coordinates so that $\Sigma = \{(x,\xi); \xi' = 0\}$, $\xi' = (\xi_1, ..., \xi_d)$, in a conic neighborhood G of (x^0, ξ^0), for some d. We may then take (the classes of) $\partial/\partial\xi_1, ..., \partial/\partial\xi_d$, as a basis in $N_{(x,\xi)}\Sigma$ for $(x,\xi) \in G$. In fact, $\sigma_j = d\xi_j(v)$ is well-defined for $v \in N_{(x,\xi)}\Sigma$ for $j \leq d$ since $d\xi_j$ vanishes on $T_{(x,\xi)}\Sigma$ for such j and we have $v = \sum_{j=1}^{d} \sigma_j \partial/\partial\xi_j$. We shall denote by (x, ξ'', σ) the coordinates introduced in this way. To introduce a bihomogeneous structure on $N\Sigma$, let us now observe that on $N\Sigma$ we have two natural multiplications. The first, called here μ^0, is just the tangent map of the standard multiplication in T^*X. In the special coordinates considered above, we thus have that $\mu^0(t)(x, \xi'', \sigma) = (x, t\xi'', t\sigma)$ for $t \geq 0$. The second multiplication, which we call μ^1, is the multiplication in the fibers of $N\Sigma \to \Sigma$; in coordinates it is given by $\mu^1(t)(x, \xi'', \sigma) = (x, \xi'', t\sigma)$. The situation is thus very similar with the one encountered in section 3.1, but of course here the definitions are defined in a geometrically invariant way. It is now natural to endow $N\Sigma$ with the preceding two multiplications simultaneously. After doing so we shall say, in analogy with section 3.1, that we have defined a bihomogenous structure on $N\Sigma$. Biconic sets, bineighborhoods, bihomogeneous functions and maps, are now defined in complete analogy to the corresponding definitions in section 3.1 and we omit details. The fact that the bihomogeneous structure on $N\Sigma$ has a geometric meaning implies in particular that if we choose special coordinates as before for Σ and if we consider a homogeneous canonical transformation χ in $T^*X \cap G$ which leaves Σ invariant, then the map induced on $N\Sigma$ by χ' is bihomogeneous (in the sense of section 4.2) in the chosen coordinates.

3. Although the geometric objects which we consider here and in section 4.5 are defined in $N\Sigma$, the F.I.O. which we shall consider later on are often defined in terms of phases and amplitudes which are directly related to conic sets in T^*X. It is therefore necessary to translate statements from T^*X to statements in $N\Sigma$ and viceversa. In fact, if coordinates are fixed in T^*X so that $\Sigma = \{\xi' = 0\}$ in a conic neighborhood G of some point $(x^0, \xi^0) \in \Sigma$, then we can identify $(x, \xi) \in G$ with the corresponding point $(x, \xi'', \xi') \in N\Sigma$ in the coordinates from $N\Sigma$ associated with those from G:

$$(x, \xi', \xi'') \in G \longleftrightarrow (x, \xi'', \xi') \in N\Sigma. \tag{4.4.1}$$

We have thus obtained an identification of G with a subset of $N\Sigma$, but this identification depends on the particular choice of coordinates which we have used to write Σ in the form $\{\xi' = 0\}$. Our main aim in this section is to show that this identification is good enough to allow for a calculus of principal parts in the bihomogeneous structure of $N\Sigma$.

Remark 4.4.1. *The reader may wonder why we have not defined Σ as $\xi'' = 0$, in which case the coordinates could have been written (x, ξ', ξ'') in both situations. Actually it* *is precisely the inversion $(\xi', \xi'') \to (\xi'', \xi')$ which seemed appealing, since it gives the possibility to specify where we are placing our argument. We admit however that we have not systematically taken advantage of this possibility, and we shall not distinguish too strongly between (ξ', ξ'') and (ξ'', ξ') in what follows.*

To understand the dependence of (4.4.1) on the choice of coordinates, let us assume that coordinates (x, ξ) are fixed so that $\Sigma \cap G \subset \{\xi' = 0\}$. We consider a positively homogeneous change of coordinates $(x, \xi) \to (y, \eta) = \chi(x, \xi)$, for (x, ξ) in a still smaller neighborhood of (x^0, ξ^0). We want that in the new coordinates Σ be given by $\eta' = 0$, so we must have

$$\chi_{n+j}(x, 0, \xi'') = 0, \ \forall (x, \xi''), \ j = 1, ..., d. \tag{4.4.2}$$

Since

$$\frac{\partial \chi_{n+j}}{\partial x_i}(x, 0, \xi'') \equiv 0, \frac{\partial \chi_{n+j}}{\partial \xi_k}(x, 0, \xi'') \equiv 0, \ \text{for } j \le d, i \le n, k \ge d+1,$$

χ can only be a change of coordinates if

$$\det \left(\frac{\partial}{\partial \xi_j} \chi_{n+i}(x, 0, \xi'') \right)_{i,j=1}^d \ne 0.$$

If now $P = (x, \xi)$ is given, then the point in $N\Sigma$ associated with P, using the variables (x, ξ) for identification, is $Q = (x, \xi'', \xi')$. The actual vector in $N\Sigma$ associated with Q is of course $v = \sum_{j=1}^d \xi_j \partial / \partial \xi_j$, regarded as a vector in $N_{(x, 0, \xi'')}\Sigma$. In the new coordinate system (y, η), P now has the coordinates $(y, \eta) = \chi(x, \xi)$ which, in the identification $(y, \eta', \eta'') \leftrightarrow (y, \eta'', \eta')$ leads to the vector $w = \sum_{j=1}^d \chi_{n+j}(x, \xi) \partial / \partial \eta_j$ at $(\chi_{(y)}(x, \xi), 0, \chi_{(\eta'')}(x, \xi))$. On the other hand, transformed by the tangent map, $((x, 0, \xi''), v)$ is mapped to

$$((\chi_{(y)}(x, 0, \xi''), \chi_{(\eta'')}(x, 0, \xi''), \Sigma \sigma_j \partial / \partial \eta_j)$$

where

$$\sigma_j = \sum_{s=1}^d \frac{\partial \chi_{n+j}}{\partial \xi_s}(x, 0, \xi'') \xi_s ,$$

which also shows immediately that the tangent map is nondegenerate. We study next the distance between $(\chi_{(y)}(x, \xi), \chi_{(\eta'')}(x, \xi), \chi_{(\eta')}(x, \xi))$ and $(\chi_{(y)}(x, 0, \xi''), \chi_{(\eta'')}(x, 0, \xi''), \sum_{s=1}^d (\partial / \partial \xi_s) \chi_{n+j}(x, 0, \xi'') \xi_s, j \le d)$. Here we use the following relations

$$\chi_{(y)}(x, \xi) - \chi_{(y)}(x, 0, \xi'') = O(|\xi'|/|\xi''|), \tag{4.4.3}$$

$$\chi_{(\eta'')}(x, \xi) - \chi_{(\eta'')}(x, 0, \xi'') = O(|\xi'|), \tag{4.4.4}$$

$$\chi_{n+j}(x, \xi) - \sum_{s=1}^d \frac{\partial \chi_{n+j}}{\partial \xi_s}(x, 0, \xi'') \xi_s = O(|\xi'|^2/|\xi''|), j \le d, \tag{4.4.5}$$

when ξ is in a conic neighborhood of $(0, \xi'')$. In fact all these relations follow from Taylor expansion, if we observe that $\chi_{(y)}$ is positively homogeneous of order 0, $\chi_{(\eta'')}$ positively homogeneous of order 1 and $\chi_{n+j}(x, 0, \xi'') = 0$ for $j \leq d$. At this moment we note that the map

$$(x, \xi) \to (\chi_{(y)}(x, 0, \xi''), \; (\sum_{s=1}^{d} (\partial \chi_{n+j}/\partial \xi_s)(x, 0, \xi'') \xi_s)_{j \leq d}, \; \chi_{(\eta'')}(x, 0, \xi'') \;)$$

is bihomogeneous. What the relations (4.4.3),(4.4.4),(4.4.5) then say is that this map is the bihomogeneous approximation of the map $(x, \xi) \to (\chi_{(y)}(x, \xi), \chi_{(\eta')}(x, \xi)\chi_{(\eta'')}(x, \xi),)$. Another way of saying this is the following: if T^*X is identified with $N\Sigma$ by the above procedure in two different ways, then we obtain a map

$$\kappa : N\Sigma \cap G \to N\Sigma,$$

and this map is a bihomogeneous approximation of the identity. (Bi-homogeneity is used here in two senses: in R^{2n} and in $N\Sigma$.)

4. To see that the identification above is useful, let us consider a funtion $f : G \to C$ which vanishes on $G \cap \Sigma$ and is positively homogeneous of order one. The differential df of f defined for points $\lambda \in \Sigma$ gives then a function on $T_\Sigma T^*X$ which vanishes on $T\Sigma$. After factorization we obtain in this way a well-defined function $\tilde{f}$ on $N\Sigma$. Obviously, $\tilde{f}$ is just the Taylor expansion of order one at points $\lambda \in \Sigma$, and in the special coordinates used above we have

$$\tilde{f}(x, \xi'', \xi') = \sum_{j=1}^{d} \frac{\partial f}{\partial \xi_j}(x, 0, \xi'') \xi_j, \tag{4.4.6}$$

$$f(x, \xi) = \tilde{f}(x, \xi'', \xi') + O(|\xi'|^2/|\xi|). \tag{4.4.7}$$

More generally, if $p : G \to C$ is a positively homogeneous function of order m which vanishes of order s on Σ, then Taylor expansion of order s for points $\lambda \in \Sigma$ will give an invariantly defined function $p^T : T_{\Sigma \cap G} T^*X \to C$, which has the property that

$$p^T(\lambda)(v + w) = p^T(\lambda)(v), \text{ if } w \in T_\lambda \Sigma.$$

(Recall that when $\lambda \in \Sigma$ and $v \in T_\lambda G$ we may in fact just set $p^T(\lambda)(v) = V^s p(\lambda)$, where V is any C^∞ vectorfield on G such that $V(\lambda) = v$. Moreover, when $w \in T\Sigma$ - which we identify with a subspace in TG using $\Sigma \subset G \cap \Sigma$ -, then $p^T(\lambda)(v + w) = p^T(\lambda)(v)$, since we can conclude from $\partial^\alpha p(\lambda) = 0$ for $|\alpha| < s$ that $V^1 V^2 \cdots V^s p(\lambda) = 0$ if the V^j are smooth vector fields, of which at least one is tangent to Σ.) Once more we obtain by factorization a natural map $p^T : N(\Sigma \cap G) \to C$. In the special coordinates above we shall have, in analogy with (4.4.6), (4.4.7),

$$p^T(x, \xi'', \xi') = \sum_{|\alpha|=s} (\partial_{\xi'}^\alpha p)(x, 0, \xi'')(\xi')^\alpha / \alpha!, \qquad (4.4.8)$$

and

$$p(x, \xi) = p^T(x, \xi'', \xi') + O(|\xi'|^{s+1} / |\xi|^{m-s-1}). \qquad (4.4.9)$$

It is worthwhile to insist on the fact that while the identification $T^*X \leftrightarrow N\Sigma$ depends on the special coordinates used to write Σ as $\{\xi' = 0\}$, the functions $\tilde{f}$ and p^T from (4.4.6), respectively (4.4.8), are well-defined geometric objects and (4.4.7), respectively (4.4.9), are valid in any identification. Let us also observe that $\tilde{f}$ is (1,1)- bihomogeneous and that p^T is (m, s)- bihomogenous. The relations (4.4.7) and (4.4.9) then express the fact that $\tilde{f}$ and p^T are the bihomogeneous principal parts of f and p and the geometric meaning of these principal parts relates well to the fact that the tangent map of χ is the bihomogeneous approximation of χ.

4.5   The canonical foliation and the relative tangent space of $N\Sigma$.

1. The restriction of the fundamental two-form $\omega = \sum_j dx_j \wedge d\xi_j$ in $(\bigwedge^2 T^*)T^*X$ to Σ defines a two-form on Σ which we denote by ω''. If canonical coordinates are chosen so that $\Sigma = \{\xi' = 0\}$, then of course we have

$$\omega'' = \sum_{j \geq d+1} dx_j \wedge d\xi_j \; .$$

Let us also note that we can lift ω'' to a two-form on $N\Sigma$ by pulling it back via $N\Sigma \to \Sigma$, which is useful if we have to work on subsets of $N\Sigma$ which do not intersect the zero section of $N\Sigma$. Anyway, at the moment, ω'' is defined on Σ and if χ is a canonical map which preserves Σ, then we will of course have that the cotangent map χ_* preserves ω''.

It is a classical fact now that ω'' defines a natural foliation on Σ. Since a number of important objects are associated with this foliation, we briefly recall how it is obtained, although the construction is wellknown. Let us then consider at first

$$ker \; \omega'' = \{\lambda \in T\Sigma \; ; \; \omega''(\lambda, \mu) = 0, \forall \mu \in T\Sigma\}.$$

Since Σ is involutive, $ker \; \omega''$ coincides with

$$(T\Sigma)^\perp = \{\lambda \in TT^*X; \omega(\lambda, \mu) = 0, \forall \mu \in T\Sigma\}.$$

Note that $ker\ \omega''$ is invariantly defined -in the sense that it is preserved under canonical maps which leave Σ invariant-, and it is in fact generated by the Hamiltonian fields H_{r_i} of the defining functions r_i of Σ: if μ is a tangent vector to Σ and if we regard μ and H_{r_i} also as vectors in TT^*X, then we have $\omega''(H_{r_i}, \mu) = dr_i(\mu)$, so it remains to observe that $dr_i(\mu) = 0$, $i \leq d$, are precisely the defining equations of $T\Sigma$ in TT^*X. Here the H_{r_i} satisfy the Frobenius integrability conditions and span in each point (x, ξ) a linear space of dimension d. It follows that in Σ we have a foliation associated with $ker\ \omega''$ for which the tangent space of the foliation is precisely $ker\ \omega''$. Note that all this construction is canonical, so in particular the leaves of the foliation are invariantly defined. In the special coordinates in which $\Sigma = \{\xi' = 0\}$, the leaves of the foliation are of course of form

$$L = \{(x, \xi) \in T^*X \, ; \, x'' = const, \xi'' = const, \xi' = 0\}.$$

(The only variable parameter is thus x'.)

2. It is common practice at this step to identify $N\Sigma$ with the cotangent space of the foliation constructed just before. (In the present context this identification has perhaps first been considered by Boutet de Monvel [1]. Also cf. Bony [1]). One of the formal reasons for performing this identification is that in this way $N\Sigma$ aquires the structure of a "cotangent" object and is therefore similar to our initial T^*X (where standard microlocalization lives). Let us in fact consider for each leaf $L \subset \Sigma$ the cotangent bundle T^*L. We thus obtain a natural bundle $T_F^*\Sigma$ over Σ which, as a set, is just the union of the cotangent spaces of the L's. Here "F" stands for "foliation". We can now identify $T_F^*\Sigma$ with $N\Sigma$ using the canonical two-form ω. Let us in fact define for each fixed $(x, \xi) \in \Sigma$ a map $t \to \tau$, $t \in N_{(x,\xi)}\Sigma$, $\tau \in (T_F^*)_{(x,\xi)}\Sigma$, by

$$\tau(t') = \omega(t'', t'), \forall t' \in (T_F)_{(x,\xi)}\Sigma, \tag{4.5.1}$$

where t'' is a representant of t in $T_{(x,\xi)}T^*X$ and t' itself is regarded as an element in $T_{(x,\xi)}T^*X$ (using the inclusions $(T_F)_{(x,\xi)}\Sigma \subset T_{(x,\xi)}\Sigma \subset T_{(x,\xi)}T^*X$). Note that (4.5.1) is well-defined: if t^1 is another representant of t, then $t'' - t^1 \in T_{(x,\xi)}\Sigma$, so $\omega(t'' - t^1, t') = 0$ since $(T_F)_{(x,\xi)}\Sigma = (T_{(x,\xi)}\Sigma)^\perp = ker\ \omega''$. Moreover, the map $t \to \tau$ is injective: if $\omega(t'', t') = 0$, $\forall t' \in (T_F)_{(x,\xi)}\Sigma$, then $t'' \in (T_{(x,\xi)}\Sigma)^{\perp\perp} = T_{(x,\xi)}\Sigma$, since ω is nondegenerate. This gives the desired identification of $N\Sigma$ with $T_F^*\Sigma$. Note that when coordinates in T^*X are such that $\Sigma = \{\xi' = 0\}$, then we may choose coordinates for $N\Sigma$ to be (x', x'', ξ'', σ) and for $T_F^*\Sigma$, (x', μ', x'', ξ''). The identification is then just $\mu' = \sigma$.

3. The construction of the foliation of Σ associated with ω'' can now be repeated almost word for word, to define a canonical foliation on $N\Sigma$. This shall be associated with the

pull-back to $N\Sigma$ of the canonical two-form ω'' to $N\Sigma$, which we shall also denote by ω''. Note that the tangent space at $(x, \xi'', \sigma) \in N\Sigma$ to the leaves of this foliation is now given by

$$T_{rel}N\Sigma = \{t \in T_{(x,\xi'',\sigma)}N\Sigma;\ \omega''(t,t') = 0, \forall t' \in T_{(x,\xi'',\sigma)}N\Sigma\},$$

and in the special coordinates in which $\Sigma = \{\xi' = 0\}$, these leaves have the form $\{(x, \xi'', \sigma); x'' = const, \xi'' = const\}$. Following Laurent [1], $T_{rel}N\Sigma$ is called the "relative tangent space". It is also convenient to introduce for general subvarieties M of $N\Sigma$ the notation

$$T_{rel}M = T_{rel}N\Sigma \cap TM.$$

Practically, the tangent vectors in $T_{rel}M$ can be computed by considering the tangent vectors for all curves in M which are completely contained in leaves of the canonical foliation of $N\Sigma$. Conversely, if $TM \subset T_{rel}M$, then M itself is completely contained in some leaf. Later on we shall be interested in a class of changes of coordinates χ - called "bicanonical" - which, among other things, do not mix up the foliation of $N\Sigma$. A convenient way to achieve this is to assume that

$$\chi_*\omega'' = \omega'', \tag{4.5.2}$$

and, in fact, we shall include (4.5.2) as one of our requirements for bicanonicity.

4.6 The bisymplectic structure of $N\Sigma$.

1. We have seen in section 4.5 that on $N\Sigma$ a natural two-form ω'' is given, which in special coordinates has the form

$$\omega'' = \sum_{j \geq d+1} dx_j \wedge d\xi_j.$$

The part $\sum_{j \leq d} dx_j \wedge d\xi_j$ of the canonical two-form ω of T^*X, which is missing in ω'', will also play an important role in what follows. We shall denote it ω_r and shall call it, following Laurent [1], "relative". In fact, while ω'' is a form on $(\bigwedge^2 T^*)(N\Sigma)$, it has been noted by Laurent that the new form ω_r has an invariant meaning only on $T_{rel}N\Sigma$ and this must always be kept in mind. We shall introduce it using special coordinates for Σ and study its invariance properties afterwards. Let us then at first fix coordinates (x, ξ) in T^*X so that $\Sigma = \{\xi' = 0\}$ in a conic neighborhood of some $(x^0, \xi^0) \in \Sigma$ and let (x, ξ'', σ) be the corresponding coordinates in $N\Sigma$. We may then introduce, in these fixed coordinates, the form

$$\omega_r = \sum_{j \leq d} dx_j \wedge d\sigma_j, \tag{4.6.1}$$

so that we have defined a two-form in $(\bigwedge^2 T^*)N\Sigma$. Since the definition depends on our special choice of coordinates, we have to show that $\omega_r(t,t')$ has an invariant meaning if t and t' are in $T_{rel}N\Sigma$. Let us then consider a canonical transformation χ with $\chi(\Sigma) \subset \Sigma$ and assume that in a conic neighborhood of (x,ξ), Σ is given by $\xi_i = 0, i \leq d$. We denote $\chi(x,\xi)$ by (y,η), i.e. we shall set $y_j = \chi_j, j \leq n$, $\eta_j = \chi_{n+j}, j \leq n$. The fact that $\chi(\Sigma) \subset \Sigma$ then comes to

$$\chi_{n+j}(x,0,\xi'') = 0, \text{ for } j \leq d, \tag{4.6.2}$$

and the fact that χ preserves the leaves of Σ gives

$$\frac{\partial}{\partial x_i}\chi_j(x,0,\xi'') = \frac{\partial}{\partial x_i}\chi_{n+j}(x,0,\xi'') = 0, \forall i \leq d, \forall j \geq d+1. \tag{4.6.3}$$

Moreover, we also write down explicitly, in the Lagrangian formalism, that χ is canonical, and obtain (cf. Caratheodory [1]):

$$\sum_{i \leq n}\frac{\partial \chi_i}{\partial x_j}\frac{\partial \chi_{n+i}}{\partial x_k} - \frac{\partial \chi_i}{\partial x_k}\frac{\partial \chi_{n+i}}{\partial x_j} = 0, \forall k, \forall j, \tag{4.6.4}$$

$$\sum_{i \leq n}\frac{\partial \chi_i}{\partial x_j}\frac{\partial \chi_{n+i}}{\partial \xi_k} - \frac{\partial \chi_i}{\partial \xi_k}\frac{\partial \chi_{n+i}}{\partial x_j} - \delta_{jk} = 0, \forall k, \forall j, \tag{4.6.5}$$

$$\sum_{i \leq n}\frac{\partial \chi_i}{\partial \xi_j}\frac{\partial \chi_{n+i}}{\partial \xi_k} - \frac{\partial \chi_i}{\partial \xi_k}\frac{\partial \chi_{n+i}}{\partial \xi_j} = 0, \forall k, \forall j. \tag{4.6.6}$$

(δ_{jk} is the Kronecker symbol.) Let us now denote by (x,ξ'',σ) the coordinates in $N\Sigma$ and by (y,η'',τ) the coordinates after the transformation χ. We then have

$$y_j = \chi_j(x,0,\xi''), \quad j \leq n,$$

$$\eta_j = \chi_{n+j}(x,0,\xi''), \quad j \geq d+1,$$

$$\tau_j = \sum_{k \leq d}\frac{\partial}{\partial \xi_k}\chi_{n+j}(x,0,\xi'')\sigma_k, \quad j \leq d.$$

The question is now to study $\sum_{j \leq d} dy_j \wedge d\tau_j$ in the coodinates (x,ξ'',σ), disregarding terms which contain dx_j, or $d\xi_j$, with j larger than $d+1$. (Such terms vanish on vectors from $T_{rel}N\Sigma$.) The remaining part of $\sum_{j \leq d} dy_j \wedge d\tau_j$ then gives

$$\sum_{i \leq d}\{(\sum_{r \leq d}\frac{\partial \chi_i}{\partial x_r}(x,0,\xi'')dx_r) \wedge [\sum_{k \leq d}\sum_{l \leq d}\frac{\partial}{\partial \xi_k}\frac{\partial \chi_{n+i}}{\partial x_l}(x,0,\xi'')\sigma_k dx_l +$$

$$\sum_{k \leq d}\frac{\partial \chi_{n+i}}{\partial \xi_k}(x,0,\xi'')d\sigma_k]\}. \tag{4.6.7}$$

To simplify this, we use the relations (4.6.2),(4.6.3),(4.6.4),(4.6.5),(4.6.6). Note for example that for $r \leq d, k \leq d$

$$\sum_{i \leq d} \frac{\partial \chi_i}{\partial x_r}(x,0,\xi'')\frac{\partial \chi_{n+i}}{\partial \xi_k}(x,0,\xi'') = \sum_{i \leq n}[\frac{\partial \chi_i}{\partial x_r}(x,0,\xi'')\frac{\partial \chi_{n+i}}{\partial \xi_k}(x,0,\xi'') -$$
$$\frac{\partial \chi_i}{\partial \xi_k}(x,0,\xi'')\frac{\partial \chi_{n+i}}{\partial x_r}(x,0,\xi'')] = \delta_{rk}. \tag{4.6.8}$$

In fact, all the terms which we have added to the left hand side of (4.6.8) to make of it the second term in (4.6.8) vanish either by (4.6.2) or by (4.6.3). The last equation in (4.6.8) is of course a consequence of (4.6.5). Similarily, we can check that for $r \leq d, l \leq d$,

$$\sum_{i \leq d} \left[\frac{\partial \chi_i}{\partial x_r}(x,0,\xi'')\frac{\partial}{\partial \xi_k}\frac{\partial \chi_{n+i}}{\partial x_l}(x,0,\xi'') - \frac{\partial \chi_i}{\partial x_l}(x,0,\xi'')\frac{\partial}{\partial \xi_k}\frac{\partial \chi_{n+i}}{\partial x_r}\chi_{n+i}(x,0,\xi'') \right] = 0. \tag{4.6.9}$$

Indeed, we may start from

$$\sum_{i \leq n} \left[\frac{\partial \chi_i}{\partial x_r}(x,\xi)\frac{\partial \chi_{n+i}}{\partial x_l}(x,\xi) - \frac{\partial \chi_i}{\partial x_l}(x,\xi)\frac{\partial \chi_{n+i}}{\partial x_r}(x,\xi) \right] = 0. \tag{4.6.10}$$

When $i \geq d+1$, both $(\partial \chi_i/\partial x_r)(x,\xi)$ and $(\partial \chi_{n+i}/\partial x_l)(x,\xi)$ vanish for $(x,\xi) \in \Sigma$ in view of (4.6.3). The equality from (4.6.9) then follows if we derivate (4.6.10) and restrict to Σ, using also (4.6.2). Putting all this together, it is now clear that the part of $\sum_{j \leq d} dy_j \wedge d\tau_j$ which interests us, coincides with $\sum_{j \leq d} dx_j \wedge d\sigma_j$ in the coordinates (x,ξ'',σ). This proves that it is invariantly defined, as desired.

2. Next we want to study how ω_r is transformed under diffeomorphisms (of a special form). Let us then assume that $U \subset N\Sigma$ is open and let $\chi : U \to N\Sigma$ be a C^∞-map which has the property that if L is a canonical leaf of $N\Sigma$ with respect to ω'', then $\chi(L \cap U)$ is contained again in such a canonical leaf. In particular we then have that the tangent map $\chi' : TU \to TN\Sigma$ maps $(T_{rel})_{(x,\xi)}N\Sigma$ into $(T_{rel})_{\chi(x,\xi)}N\Sigma$ for $(x,\xi) \in U$.

Definition 4.6.1. *The pull-back* $\chi_*(\omega_r)$ *of* ω_r *is defined by the (natural) formula*

$$\chi_*(\omega_r)(t^1,t^2) = \omega_r(\chi'(t^1),\chi'(t^2)),$$

where $t^1, t^2 \in (T_{rel})_{(x,\xi)}N\Sigma$.

Recall that the condition that $\chi(L \cap U)$ is contained in a canonical leaf of $N\Sigma$ for every canonical leaf L of $N\Sigma$ is fulfilled if $\chi_*\omega'' = \omega''$. This gives sense to the following definition:

Definition 4.6.2. *(Laurent)* $\chi : U \to N\Sigma$ *is called bicanonical if* $\chi_*\omega'' = \omega''$ *and* $\chi_*\omega_r = \omega_r$.

3. A natural example of a bicanonical map is obtained in the present context in the following way: let $V \subset T^*X$ be open, let $\chi : V \to T^*$ be canonical and assume that $\chi(\Sigma \cap V) \subset \Sigma$. We have seen that the tangent map induces a map $\rho : N(\Sigma \cap V) \to N\Sigma$.

Proposition 4.6.3. *ρ is bicanonical.*

The proof of this result is straightforward. In fact it is obvious that $\rho_* \omega'' = \omega''$. As for the verification that $\rho_* \omega_r = \omega_r$, it is similar to the computations which showed that ω_r is well-defined. We omit the details.

The bicanonical maps which we shall actually use in this paper will be compositions of such tangent bicanonical maps with bicanonical maps which have the form

$$(x, \xi) \to \begin{pmatrix} \chi_{(x')}(x, \xi) \\ x'' \\ \chi_{(\xi')}(x, \xi) \\ \xi'' \end{pmatrix},$$

$(x', \xi') \to (\chi_{(x')}(x', x'', \xi', \xi''), \chi_{(\xi')}(x', x'', \xi', \xi''))$ being canonical in (x', ξ') for every fixed (x'', ξ'').

4. Associated with our two-forms we can now introduce corresponding Hamiltonian fields and Poisson brackets. We discuss them here for the case ω_r in detail, since they shall be important later on. Let us then consider $U \subset N\Sigma$ open and consider some C^1-function $f : U \to R$. The differential df is then a section of T^*U but, by restriction, it also defines a functional $\bar{d}f$ on $T_{rel}U$. We can now use ω_r to associate with $\bar{d}f$ a vector field H_f^r in $T_{rel}U$, in analogy with the definition of standard Hamiltonian vector fields. In fact the "relative" Hamiltonian vector field of f at the point $(x, \xi) \in N\Sigma$ is defined to be the vector H_f^r in $T_{rel}U$ such that

$$\omega_r(H_f^r, t) = (\bar{d}f)(t), \forall t \in T_{rel}U.$$

Note that if coordinates are chosen so that $\Sigma = \{\xi' = 0\}$, and if the coordinates in $N\Sigma$ are (x, ξ'', σ), then

$$\bar{d}f = \sum{}' \frac{\partial f}{\partial x_j} dx_j + \sum{}' \frac{\partial f}{\partial \sigma_j} d\sigma_j,$$

where, here and later on in this section, $\sum' = \sum_{j \leq d}$. If now we write $H_f^r = \sum' \alpha_j \partial/\partial x_j + \sum' \beta_j \partial/\partial \sigma_j$, then it is immediate that we must have $\alpha_j = \partial f/\partial \sigma_j$, $\beta_j = -\partial f/\partial x_j$. Relative Poisson brackets are introduced correspondingly by the relation:

$$\{f, g\}_r = H_f^r g$$

so that we also have

$$\{f,g\}_r = dg(H_f^r) = \bar{d}g(H_f^r) = \omega_r(H_g^r, H_f^r).$$

In the special coordinates from before we have of course

$$\{f,g\}_r = \sum{}' \big[\frac{\partial f}{\partial \sigma_j} \frac{\partial g}{\partial x_j} - \frac{\partial f}{\partial x_j} \frac{\partial g}{\partial \sigma_j} \big].$$

In particular it is then clear that we have Iacobi's identity

$$\{g, \{h, f\}_r\}_r + \{h, \{f, g\}_r\}_r + \{f, \{g, h\}_r\}_r \equiv 0,$$

which implies the relation

$$[H_f^r, H_g^r] = H_{\{f,g\}_r}^r,$$

for the commutator $[H_f^r, H_g^r] = H_f^r H_g^r - H_g^r H_f^r$ of H_f^r and H_g^r. Note that this implies in particular that H_f^r and H_g^r commute if $\{f,g\}_r = 0$ or if $\{f,g\}_r \equiv 1$. The latter are of course the relations which one requires for coordinates in the Darboux theorem. We also observe that if f is $(1,1)$–bihomogeneous, then $H_f^r g$ is (m,s)–bihomogeneous, if so is g, i.e. H_f^r operates well in relation with bihomogeneity. Finally, it is standard to check that $\chi_* \omega_r = \omega_r$ is equivalent with the fact that relative Poisson brackets are preserved under χ: cf. Laurent [1]. We omit the details.

5. Associated with our relative Hamiltonian fields is the notion of relative bicharacteristics. In fact, if $f : U \to R$, U open in $N\Sigma$, is given, we have seen that it defines a relative Hamiltonian field in $(T_{rel})N\Sigma$. The associated integral curves are called relative bicharacteristics. Of course, such bicharacteristics are contained in leaves. It is easy to see that they are defined invariantly up to bicanonical diffeomorphisms. More precisely, we have the following result:

Proposition 4.6.4. *Let $f : U \to R$, $U \subset N\Sigma$ open, and consider $(x^0, \xi^0) \in U$. Let furthermore $\chi : V \to U$ be a bicanonical map and let γ be the relative bicharacteristic of $f \circ \chi$ which passes through $\chi^{-1}(x^0, \xi^0)$. Then $\chi \circ \gamma$ is the relative bicharacteristic of f which passes through (x^0, ξ^0).*

To prove this result, we need only observe that $\chi' H_{f \circ \chi}^r = H_f^r$. To prove the latter, we observe that for any $t \in T_{rel}U$ we have $(df)(\chi' t) = d(f \circ \chi)(t) = \omega_r(H_{f \circ \chi}^r, t) = \chi_* \omega_r(H_{f \circ \chi}^r, t) = \omega_r(\chi' H_{f \circ \chi}^r, \chi' t)$. Therefore $(df)(t') = \omega_r(\chi' H_{f \circ \chi}^r, t')$ for any vector t' of form $\chi' t$, $t \in T_{rel}U$. The fact that $\chi' H_{f \circ \chi}^r = H_f^r$ then follows if we observe that any vector $t' \in (T_{rel})_P U$ is of form $\chi' t$, $t \in (T_{rel})_Q U$, if the base points P, Q are related by $P = \chi(Q)$.

4.7   The relative bicharacteristic foliation of Λ

1. In this section we consider a real analytic bihomogeneous variety $\Lambda \subset N\Sigma$ which lives in a bineighborhood Γ of some point $(\lambda^0, \sigma^0) \in N\Sigma$ and fix analytic homogeneous canonical coordinates so that $\Sigma = \{(x, \xi) \in T^*X; \ \xi' = 0\}$. We assume that we are given real analytic functions $f_j : \Gamma \to R$, $j = 1, ..., d'$, with the following properties:

a) $\Lambda = \{(\lambda, \sigma) \in N\Sigma \cap \Gamma; f_i(\lambda, \sigma) = 0, \ i = 1, ..., d'\}$,

b) the f_i are (1,1)-bihomogeneous,

c) $\{f_i, f_j\}_r = 0, \forall i, \forall j$,

d) the differentials $\bar{d}f_i$, $i \leq d'$, and the 1-form $\Sigma_{s \leq d'} \sigma_s \, dx_s$ are linearly independent,

e) rank $(\partial f_i / \partial \sigma_j)_{i \leq d', j \leq d} = d'$.

In the sequel we shall often write σ for ξ' and (x, ξ'') instead of λ.

Definition 4.7.1. *If the f_i can be found with the properties a),b),c),d),e), then Λ is called a nonradial regularly involutive subvariety of $N\Sigma$. (We shall often omit to specify the "nonradiality" later on, since we shall never consider radial situations. Moreover, instead of "regularly" we shall often say that Λ is in regular position.)*

2. Some comments on the conditions c),d),e) seem here in order, and we start with a comment on condition c). Note then at first that it follows from a) and c) that

$$T_{rel,\lambda}\Lambda = \{t \in T_{rel,\lambda}N\Sigma; \bar{d}f_j(t) = 0, \ j = 1, ..., d'\}.$$

Next we observe that $\bar{d}f_j(H_{f_k}^r) = \{f_j, f_k\}_r = 0$, so that

$$\text{span } (H_{f_1}, ..., H_{f_{d'}}^r) \subset T_{rel}\Lambda. \tag{4.7.1}$$

The following lemma is wellknown in standard symplectic geometry

Lemma 4.7.2. *Denote by*

$$A = \{t \in T_{rel}N\Sigma; \omega_r(t, t') = 0, \forall t' \in T_{rel}\Lambda\} = (T_{rel}\Lambda)^{\perp_r}, \tag{4.7.2}$$

where the last equality is a notation. Then $A = \text{span } (H_{f_1}^r, ..., H_{f_{d'}}^r)$.

Proof. From dimensionality considerations it is clear that it suffices to show that span $(H_{f_1}^r, ..., H_{f_{d'}}^r) \subset A$. To prove this, observe that if $t \in T_{rel}\Lambda$, then $0 = \bar{d}f_j(t) = \omega_r(H_{f_j}^r, t)$.

Combining (4.7.1) and (4.7.2) we also have that

$$(T_{rel}\Lambda)^{\perp_r} \subset T_{rel}\Lambda,$$

which is the standard way to write an involutivity condition. Here we shall work, for simplicity, with the seemingly stronger condition c) from definition 4.7.1.

Our next concern is to discuss condition d), which is the "non-radiality" condition. The importance of such conditions in homogeneous symplectic geometry is also well-known. Assume in fact that f is (1,1)-bihomogeneous, that g is (0,0)-bihomogeneous and that we want to have $\{f,g\}_r \equiv 1$. Assume moreover that $\bar{d}f$ were proportional to $\Sigma'\xi_s dx_s$. It would follow that

$$\{f,g\}_r = \bar{d}f(H_g^r) = \alpha \sum {}'\xi_s dx_s(H_g^r) = \alpha \sum {}'\xi_s(\partial/\partial\xi_s)g$$

for some proportionality factor α. Here $\sum'\xi_s(\partial/\partial\xi_s)g \equiv 0$ by Euler's relation for homogeneous functions, so we will never be able to get $\{f,g\}_r \equiv 1$.

Another standard way to express condition d) is that the Hamiltonian vector fields $H_{f_i}^r$, $i \leq d'$, and the relative radial vector field $\rho = \sum_{i\leq d}\xi_i\partial\xi_i$ are linearly independent. Note that the relative radial field satisfies $\rho f(x,\xi'',\xi') = (d/dt)f(x,\xi'',t\xi')_{|t=1}$, where $(x,\xi'',\xi') \to (x,\xi'',t\xi')$ is the multiplication in the fibers. This gives ρ an invariant meaning. It is then also clear that condition d) has an invariant meaning.

Our last comment in this section refers to condition e). It is for this condition that we say that Λ is in regular position. We have not seriously tried to eliminate it, so we do not know which is its true importance. (Note anyway that the corresponding condition is often superfluous in standard first order microlocalization.) We shall show here at least that condition e) is invariantly defined. Assume then that for some special choice of homogeneous canonical coordinates for which $\Sigma = \{\xi' = 0\}$ we have succeeded in writing Λ in the form a) for some f_i which satisfy e) and let us fix some other set of homogeneous canonical coordinates so that $\Sigma' = \{\eta' = 0\}$. We want to show that in these coordinates we can write Λ in the form

$$\Lambda = \{(y,\eta'',\eta');\ g_i(y,\eta'',\eta') = 0, i \leq d'\}$$

where rank $(\partial g_i/\partial\eta_j') = d'$. The relation between the coordinates (x,ξ'',ξ') and (y,η'',η') is now that

$$(x,\xi'',\xi') = \chi'(y,\eta'',\eta')$$

where $\chi' : N(\Sigma \cap \Gamma) \to N\Sigma$ is the tangent map of some homogeneous canonical map $\chi : \Gamma \to T^*X$ which leaves Σ invariant. What we need to prove then is the following elementary result:

Lemma 4.7.3. *Let* $\Sigma = \{(y,\eta) \in T^*X; \eta' = 0\}$ *and fix* $(y^0,\eta^0) \in \Sigma$. *Consider further some conic neighborhood* Γ *of* (y^0,η^0) *and let* $\chi : \Gamma \to T^*X$ *be a homogeneous canonical* *transformation so that* $\chi(\Sigma \cap \Gamma) \subset \Sigma$. *Denote by* χ' *the induced tangent map* χ' : $N(\Sigma \cap \Gamma) \to N\Sigma$. *If functions* $f_i : A \subset N\Sigma \to R$ *are given so that* $\mathrm{rank}\,(\partial f_i/\partial \xi_j)_{i \le d', j \le d} = d'$, *then it follows that the rank of the matrix* $(\partial g_i/\partial \eta_j)_{i \le d', j \le d}$ *is* d', *if* $g_i(y, \eta'', \eta') = f_i(\chi'(y, \eta'', \eta'))$.

Proof. Explicitly χ' is given, as we have seen, by

$$x_j = \chi_j(y, 0, \eta''), \quad j \le n,$$

$$\xi_j = \chi_{n+j}(y, 0, \eta''), \quad j \ge d+1,$$

$$\xi_j = \sum_{k \le d} \frac{\partial}{\partial \eta_k} \chi_{n+j}(y, 0, \eta'') \eta_k, \quad j \le d.$$

It follows that we have

$$\frac{\partial}{\partial \eta_j} f_i(\chi'(x, \eta'', \eta')) = \sum_{s \le d} \frac{\partial f_i}{\partial \xi_s}(\chi'(x, \eta'', \eta')) \frac{\partial \chi_{n+s}}{\partial \eta_j}(x, 0, \eta''),$$

so the desired conclusion is a consequence of the fact that (as already observed) $\det\,(\partial \chi_{n+s}/\partial \eta_j)_{s \le d, j \le d} \ne 0$.

3. A foliation on Λ is now introduced in the same way in which we have introduced the canonical foliation on Σ. What we want to exhibit are smooth subvarieties $L \subset \Lambda$ so that

$$TL = (T_{rel}\Lambda)^{\perp r}. \tag{4.7.3}$$

(TL is here the complete tangent space to L.) We have already observed in section 4.5 that (4.7.3) implies that L is contained in some leaf F of the canonical foliation of $N\Sigma$. It is clear that this foliation has an invariant meaning. The same is then true for the projections of the leaves of the foliation onto Σ, since the projection $N\Sigma \to \Sigma$ is a geometrically well-defined map. To compute the foliation in Λ we may then in particular choose bicanonical homogeneous coordinates as we please. If coordinates are given so that $\Sigma = \{(x, \xi); \xi' = 0\}$, $\Lambda = \{\xi_1 = 0, ..., \xi_{d'} = 0\}$, then the leaves L have locally the form $x_i = const$ for $i > d'$, $\xi'' = const$, $\xi' = const$.

4.8   Phase functions and bihomogeneous approximations of canonical maps

1. In section 4.4 we have seen that the tangent map of a canonical map χ with $\chi(\Sigma) \subset \Sigma$ is a bihomogeneous approximation of χ. In the present section we study another class

of homogeneous canonical transformations for which the bihomogeneous approximation can also be computed from the data at hand. We consider $R^{2n} = R_x^n \times R_\eta^n$ and assume that $R_\eta^n = R_{\eta'}^d \times R_{\eta''}^{n-d}$. Setting $M_1 = \{\eta; \eta'' = 0\}$ we obtain a bihomogeneous structure on R_η^n and therefore also on R^{2n}. Consider $\eta^0 \in R^n, \eta^1 \in \dot{M}_1$ and let Γ be an open bineighborhood of (x^0, η^0, η^1) in R^{2n}. Further we consider $S : \Gamma \to R$, a real-analytic function which is positively bihomogeneous of bidegree $(1,1)$ and assume that there is a constant c_1 so that

$$| \det S_{x'\eta'}(x,\eta)| \geq c_1 \text{ on } \Gamma. \tag{4.8.1}$$

Recall that the fact that S is bihomogeneous of bidegree $(1,1)$ also gives estimates "from the above" for S and derivatives of S. After shrinking Γ, if necessary, we may in fact assume that

$$|S_x(x,\eta)| \leq c_2|\eta'|, \text{ on } \Gamma, \tag{4.8.2}$$

$$|S_{x\eta'}(x,\eta)| \leq c_2 \text{ on } \Gamma, \tag{4.8.3}$$

$$|S_{x\eta''}(x,\eta)| \leq c_3|\eta'|/|\eta''|, \text{ on } \Gamma. \tag{4.8.4}$$

We denote by $S^*(x,y,\eta)$ the function

$$S^*(x,y,\eta) = S(x,\eta) + \langle x'', \eta''\rangle - \langle y, \eta\rangle = S(x,\eta) - \langle y', \eta'\rangle + \langle x'' - y'', \eta''\rangle,$$

which is a nondegenerate phase function in the sense of Hömander if $|\eta'|/|\eta''|$ is small. (Cf. (4.8.1),(4.8.3),(4.8.4).) With S^* we can associate a canonical transformation which we shall denote T^*. In fact, T^* is the map which corresponds to the canonical relation

$$\left(x, S_{x'}(x,\eta), \eta'' + S_{x''}(x,\eta), S_{\eta'}(x,\eta), x'' + S_{\eta''}(x,\eta), \eta\right). \tag{4.8.5}$$

We denote by (y,η) the initial and by (x,ξ) the final variables of the transformation, so that (4.8.5) comes to

$$\begin{aligned} \xi' &= S_{x'}(x,\eta), \\ \xi'' &= \eta'' + S_{x''}(x,\eta), \end{aligned} \tag{4.8.6}$$

$$\begin{aligned} y' &= S_{\eta'}(x,\eta), \\ y'' &= x'' + S_{\eta''}(x,\eta). \end{aligned} \tag{4.8.7}$$

Note that the canonical relation is not bihomogeneous, in that $S_{\eta''}$ is $(0,1)$- bihomogeneous rather than $(0,0)$-bihomogenous and $S_{x''}$ is $(1,1)$ rather than $(0,1)$-bihomogeneous. It is the main goal of this section to show that, still, we can define T^* on a biconic set and that we can find a natural bihomogeneous approximation T for T^* on that set. All this is of course based on the fact that the map $(x, \eta) \to (y, \xi)$ defined by the relations (4.8.6), (4.8.7) has a very simple bihomogeneous approximation, namely

$$(x, \eta) \to (S_{\eta'}(x, \eta), x'', S_{x'}(x, \eta), \eta'').$$

As a consequence, the construction of T is remarkably simple: it will leave x'' and ξ'' unchanged and will be for fixed $x'' = y'', \xi'' = \eta''$ the canonical transformation $(x', \xi') \leftrightarrow (y', \eta')$ associated with the phase function $S(x', x'', \eta', \eta'') - \langle y', \eta' \rangle$. In particular, thus, T is bicanonical.

Remark 4.8.1. *a) It is interesting to note that the canonical relation associated with a phase function of form $S(x, \eta) - \langle y, \eta \rangle$ with S (1,1)-bihomogeneous cannot be useful in the present theory so that the modification of the phase to*

$$S(x, \eta) + \langle x'', \eta'' \rangle - \langle y, \eta \rangle$$

is not only for opportunistic reasons. In fact the relations (4.8.6)and (4.8.7) read for the phase function $S(x, \eta) - \langle y, \eta \rangle$ as

$$\xi' = S_{x'}(x, \eta), \xi'' = S_{x''}(x, \eta), y' = S_{\eta'}(x, \eta), y'' = S_{\eta''}(x, \eta),$$

and from bihomogeneity it would follow that

$$|\xi''| \leq c|\eta'| \leq c'|\xi'|$$

respectively that

$$|y''| \leq c|\eta'|/|\eta''|.$$

Since we may shrink $|\eta'|/|\eta''|$ as much as we please in our theory, this would leed to $y'' \sim 0$ whatever x is. Moreover, we also want to be able to shrink $|\xi'|/|\xi''|$ to arbitrarily small values, which is in opposition to $|\xi''| \leq c'|\xi'|$.

b) We should also mention that the estimates for derivatives of the phases which we need are slightly weaker than those for derivatives of (1,1)- bihomogeneous functions and are needed only from second derivatives on. This is the reason why the term $\langle x'', \eta'' \rangle$ is not creating problems (in any theory of F.I.O.).

2. Let us at first recall the construction of a canonical map associated with a phase function. In fact, what one does is simply to solve (4.8.7) for $x = x(y, \eta)$ and insert the result in (4.8.6). (This will be carried out afterwards in detail.) It is however a priori not obvious that the map which one obtains in this way can be defined on a full

bineighborhood of (y^0, η^0, η^1), so we have to embark on a more systematic study. To do so, we denote by F^* and G^* the following maps defined on Γ:

$$F^*(x, \eta) = (S_{x'}(x, \eta), \eta'' + S_{x''}(x, \eta)) = (\xi', \xi''),$$

$$G^*(x, \eta) = (S_{\eta'}(x, \eta), x'' + S_{\eta''}(x, \eta)) = (y', y'').$$

Our first remark is now that the Jacobian JF^* of F^* in η, respectively the Jacobian JG^* of G^* in x, are

$$\begin{pmatrix} S_{x'\eta'}, & S_{x'\eta''} \\ S_{x''\eta'}, & Id + S_{x''\eta''} \end{pmatrix},$$

respectively

$$\begin{pmatrix} S_{\eta'x'}, & S_{\eta'x''} \\ S_{\eta''x'}, & Id + S_{\eta''x''} \end{pmatrix}.$$

It is easily checked that (4.8.1) implies that these Jacobians are nondegenerate if $|\eta'|/|\eta''|$ is small.

Let us then fix $(\tilde{x}, \tilde{\eta}) \in \Gamma$ with $|\tilde{\eta}'|/|\tilde{\eta}''|$ small and denote by $\tilde{y}^* = G^*(\tilde{x}, \tilde{\eta})$ and by $\tilde{\xi}^* = F^*(\tilde{x}, \tilde{\eta})$. From the implicit function theorem it is now clear that we can find a map $H^*(y, \eta)$ defined in a neighborhood of $(\tilde{y}^*, \tilde{\eta})$ such that (4.8.7) is equivalent with $x = H^*(y, \eta)$ if (x, y, η) is in a neighborhood of $(\tilde{x}, \tilde{y}^*, \tilde{\eta})$. H^* is already the x–component of T^*.

The ξ–component of T^* is now obtained inserting this into (4.8.6). Thus

$$T^*_{(\xi')}(y, \eta) = S_{x'}(H^*(y, \eta), \eta),$$

$$T^*_{(\xi'')}(y, \eta) = \eta'' + S_{x''}(H^*(y, \eta), \eta).$$

Here we have denoted by $T^*_{(\xi')}$, $T^*_{(\xi'')}$ the ξ'-, respectively ξ''-components of T^*. (Similar notations have been already used before and shall also be used later on.) Unfortunately, the maps H^*, T^* are not yet what we wanted to obtain, in that they are defined on a neighborhood of $(\tilde{y}, \tilde{\eta})$ rather than on a bineighborhood of (y^0, η^0, η^1). Moreover, it is not clear how the choice of $(\tilde{y}, \tilde{\eta})$ affects the definition of H^*, T^*. The same questions are much easier to answer for the quasihomogeneous approximation T of T^* and we turn our attention to the study of T for a moment.

3. The construction of the bihomogeneous approximation T of T^* is similar to that of T^*. In fact, T will be the map associated with the relation

$$(x, S_{x'}(x, \eta), \eta'', S_{\eta'}(x, \eta), x'', y'', \eta) = (x, \xi', \xi'', y', y'', \eta),$$

which, for fixed x'', η'' is canonical in the variables (x', ξ', y', η') . To define T let us introduce

$$F(x, \eta) = (S_{x'}(x, \eta), \eta'') = (\xi', \xi''),$$

$$G(x, \eta) = (S_{\eta'}(x, \eta), x'') = (y', y'').$$

We can then solve $y = G(x, \eta)$ for x -keeping η fixed- and obtain $x = H(y, \eta) = T_{(x)}(y, \eta)$. Introducing this into $\xi = F(x, \eta)$, we obtain $\xi = T_{(\xi)}(y, \eta) = F(H(y, \eta), \eta)$ for the ξ - component of T.

Once more the existence of H follows locally from the implicit function theorem. If we start from (x^0, θ^0), where θ^0 is a generating element for (η^0, η^1), then T will be defined initially in a neighborhood of (y^0, θ^0), where $y^0 = G(x^0, \theta^0)$. However, since the components of G are bihomogeneous of bidegree $(0, 0)$, we may extend the components of H to bihomogeneous functions of bidegree $(0, 0)$ such that $y = G(x, \eta)$ is equivalent with $x = H(y, \eta)$ for (x, y) in a neighborhood of (x^0, y^0) and η in a bineighborhood of (η^0, η^1). It is then clear that T itself extends to a bihomogeneous map on a bineighborhood of (y^0, η^0, η^1). Finally let us introduce I^*, I, by the conditions $I^*(x, \xi) = \eta^*$, $I(x, \xi) = \eta$, where η^* and η solve $\xi = F^*(y, \eta^*)$, respectively $\xi = F(y, \eta)$. (That this is welldefined if η^* and η are fixed in a small bineighborhood of (η^0, η^1) follows e.g. from lemma 4.8.2 below.)

4. In our later arguments, we need a number of properties of the $F^*, F, G^*, G, H^*, H, I^*, I$.

Lemma 4.8.2. *If we shrink Γ around (x^0, η^0, η^1), then F^* and F are injective for fixed x whereas G^* and G are injective for fixed η.*

Remark 4.8.3. *It follows in particular that the maps H^* and T^* are well-defined on the set $\{(y, \eta); y = G^*(x, \eta), (x, \eta) \in \Gamma\}$.*

Proof of lemma 4.8.2. It is no loss in generality to assume that Γ is of form $\Gamma = U \times \Gamma_\eta$, $U \subset R_x^n$, $\Gamma_\eta \subset R_\eta^n$. We shall prove the assertion for F^*, the argument being similar or simpler for the others. Observe then that $F^*(x, \eta) = F^*(x, \tilde{\eta})$ comes to $S_{x'}(x, \eta) = S_{x'}(x, \tilde{\eta})$, $\eta'' + S_{x''}(x, \eta) = \tilde{\eta}'' + S_{x''}(x, \tilde{\eta})$. We can deduce from this that for $j = 1, ..., n$ we can find θ^j in the segement $[\eta, \tilde{\eta}]$ such that

$$0 = S_{x_j \eta}(x, \theta^j)(\eta - \tilde{\eta}) \text{ for } j = 1, ..., d,$$

$$0 = S_{x_j \eta}(x, \theta^j)(\eta - \tilde{\eta}) - (\eta_j - \tilde{\eta}_j) \text{ for } j = d + 1, ..., n.$$

Here we assume of course that Γ is convex in the fibers. The problem is then to show that if Γ is shrinked still further, then $\eta, \tilde{\eta} \in \Gamma$, $\theta^j \in [\eta, \tilde{\eta}]$ implies $\det D \neq 0$ where

$$
D = \begin{pmatrix}
S_{x_1,\eta}(x, \theta^1) \\
\cdot \\
\cdot \\
\cdot \\
S_{x_d,\eta}(x, \theta^d) \\
S_{x_{d+1},\eta}(x, \theta^{d+1}) - \delta_{d+1,j} \\
\cdot \\
\cdot \\
\cdot \\
S_{x_n,\eta}(x, \theta^n) - \delta_{n,j}
\end{pmatrix},
$$

and where δ_{ij} is the Kronecker delta. To see this we observe that we may assume that

$$
\left| \det \begin{pmatrix} S_{x'\eta'}(x,\eta) & 0 \\ 0 & Id \end{pmatrix} \right| \geq c_1 > 0 \text{ for } (x,\eta) \in \Gamma.
$$

To conclude the argument, we shall now show that $\det D$ is a small perturbation of $\det S_{x'\eta'}(x, \theta^1)$. To simplify the notation, we shall denote by $o(1)$ quantities which can be made arbitrarily small by shrinking Γ and by $O(1)$ quantities which are bounded on Γ. Bihomogeneity considerations then show that

a) $S_{x_i,\eta'}(x, \theta^1) - S_{x_i,\eta'}(x, \theta^i) = o(1)$, for $i = 1, ..., d$.

Indeed, S_{x_i,η_j} is for $j \leq d$ bihomogeneous of bidegree $(0,0)$, so we can normalize to a neighborhood of (x^0, θ), where θ is some primitive element for η^0, η^1. Moreover, we can make this neighborhood small if we shrink Γ. It suffices therefore to use the continuity of the S_{x_i,η_j}. (Note that by assumption the θ^i stay in Γ.)

b) $S_{x_i,\eta''}(x, \theta^i) = o(1)$ for all i. Here we use 4.8.4

c) $S_{x_i,\eta'}(x, \theta^i) = O(1)$ for all i. Here we use 4.8.3

It follows that

$$
D = \begin{pmatrix} S_{x'\eta'}(x, \theta^1) + o(1), & o(1) \\ O(1), & Id + o(1) \end{pmatrix}.
$$

It is then immediate that

$$
\det D = \det S_{x'\eta'}(x, \theta^1) + o(1).
$$

Lemma 4.8.4. *There are constants c, c' so that*

$$|H^*(y, \eta) - H(y, \eta)| \le c$$

implies

$$|H^*(y, \eta) - H(y, \eta)| \le c' |\eta'| / |\eta''|.$$

Proof. Denote $x^* = H^*(y, \eta)$, $x = H(y, \eta)$. By the definition of H^*, H, this means if c is small that

$$y = G^*(x^*, \eta) = G(x, \eta).$$

It follows from this that

$$G(x^*, \eta) - G(x, \eta) = G(x^*, \eta) - G^*(x^*, \eta) = O(|\eta'| / |\eta''|).$$

The last equality is here a consequence of $G(x^*, \eta) - G^*(x^*, \eta) = (0, -S_{\eta''}(x^*, \eta))$.

It remains then to observe that $G(x^*, \eta) - G(x, \eta) = O(|\eta'| / |\eta''|)$ implies $x^* - x = O(|\eta'| / |\eta''|)$ if x^* is close to x. In fact, $G(x^*, \eta) - G(x, \eta) = (S_{\eta'}(x^*, \eta) - S_{\eta'}(x, \eta), x^{*''} - x'') = (S_{\eta_1, x}(z^1, \eta)(x^* - x), ..., S_{\eta_d, x}(z^d, \eta)(x^* - x), x^{*''} - x'')$ for some $z^j \in [x, x^*]$. We conclude from this at first that $x^{*''} - x'' = O(|\eta'| / |\eta''|)$, and that

$$(S_{\eta_1, x'}(z^1, \eta)(x^{*'} - x'), ..., S_{\eta_d, x'}(z^d, \eta)(x^{*'} - x')) = O(|\eta'| / |\eta''|).$$

Next we observe that, since $S_{\eta_i, x'}$ is $(0, 0)$-bihomogeneous for $i \le d$, $S_{\eta_i, x'}(z^i, \eta) = S_{\eta_i, x'}(z^1, \eta) + O(|x^{*'} - x'|)$. The desired conclusion then follows if we also use (4.8.1).

Remark 4.8.5. *With the above notations we have*

$$|\eta'| \sim |\xi'|, \ |\eta''| \sim |\xi''|$$

in a suitable bineighborhood of (x^0, η^0, η^1).

In fact we have $|\xi'| = |S_{x'}(x, \eta)| = O(|\eta'|)$ and $|\xi''| = |\eta''| + O(|\eta'|)$. When $|\eta'| = o(|\eta''|)$ the last equality gives $|\xi''| \sim |\eta''|$. The fact that $|\eta'| = O(|\xi'|)$ follows from (4.8.1).

Next we prove

Lemma 4.8.6.

$$I^*_{(\eta')}(x, \xi) - I_{(\eta')}(x, \xi) = O(|\xi'|^2 / |\xi''|),$$

$$I^*_{(\eta'')}(x, \xi) - I_{(\eta'')}(x, \xi) = O(|\xi'|).$$

Proof. Let $\eta = I(x, \xi)$, $\eta^* = I^*(x, \xi)$. By the definition of I, I^*, this means $\xi = F(x, \eta)$ $= F^*(x, \eta^*)$. We conclude that

$$F(x, \eta^*) - F(x, \eta) = F(x, \eta^*) - F^*(x, \eta^*).$$

It follows from this that

$$S_{x'}(x, \eta^*) - S_{x'}(x, \eta) = 0 \qquad (4.8.8)$$

and that

$$\eta^{*''} - \eta'' = -S_{x''}(x, \eta^*) = O(|\eta^{*'}|).$$

As a consequence of (4.8.8) we obtain, using (4.8.1) and Euler's relation for homogeneous functions, that

$$|\eta^{*'}| \leq c_1 |\eta'| \leq c_2 |\eta^{*'}|. \qquad (4.8.9)$$

It follows in particular that $\eta^{*''} - \eta'' = O(|\xi'|)$ as desired. Moreover, we can shrink Γ to make $|\eta'|/|\eta''|$ as small as we please. From $\eta^{*''} - \eta'' = O(|\eta^{*'}|) = O(|\eta'|)$ it will therefore follow that

$$|\eta^{*''}| \leq c_3 |\eta''| \leq c_4 |\eta^{*''}|. \qquad (4.8.10)$$

If τ is a point on the segment $[\eta, \eta^*]$ we will now have

$$|\tau'|/|\tau''| \leq C \min(|\eta'|/|\eta''|, |\eta^{*'}|/|\eta^{*''}|)$$

for a suitable constant C. In fact, $|\tau'|/|\tau''| \leq (\max(|\eta'|/|\eta^{*'}|)/(\min(|\eta''|, |\eta^{*''}|)))$, etc. After these preparations we write for $i = 1, ..., d$ that

$$0 = S_{x_i}(x, \eta^*) - S_{x_i}(x, \eta) = S_{x_i, \eta'}(x, \tau^i)(\eta^{*'} - \eta') + S_{x_i, \eta''}(x, \tau^i)(\eta^{*''} - \eta'')$$

$$= S_{x_i, \eta'}(x, \tau^1)(\eta^{*'} - \eta') + o(1)(\eta^{*'} - \eta') + O((|\tau^{i'}|/|\tau^{i''}|)(\eta^{*''} - \eta'')),$$

where τ^i lies in $[\eta, \eta^*]$. (The notation $o(1)$ has been introduced in the proof of lemma 4.8.2.) Inserting here that $|\tau^{i''}|/|\tau^{i'''}| = O(|\eta^{*'}|/|\eta^{*''}|)$ we obtain from (4.8.1) that

$$\eta^{*'} - \eta' = O(|\eta^{*'}|/|\eta^{*''}|)(\eta^{*''} - \eta'') = O(|\xi'|^2/|\xi''|).$$

4. We now want to estimate the ranges of F^* and G^*. Let us then fix a generating element, θ^0 for (η^0, η^1) and denote $S_{x'}(x^0, \theta^0)$ by ξ^1. Note that it follows from the bihomogeneity of $S_{x'}$ that $\xi^1/|\xi^1|$ does not depend on θ^0. The map $\eta \to F(x^0, \eta)$ being bihomogeneous, it follows that $L = F(x^0, \Gamma)$ is biconic if Γ is a bicone. In view of the continuity of F, we can for any bineighborhood Λ' of (η^0, ξ^1) find Γ and $\varepsilon > 0$ so that

$$F(x, \Gamma) \subset \Lambda' \text{ if } |x - x^0| < \varepsilon.$$

If now $\Lambda' \subset\subset \Lambda$ it will follow that

$$F^*(x, \Gamma) \subset \Lambda$$

if $|x - x^0|$ is small enough and Γ is shrinked still further. Indeed, we know for $\xi = F(x, \eta)$ that $|\xi'| \sim |\eta'|$, $|\xi''| \sim |\eta''|$ and we have if $\xi^* = F^*(x, \eta)$ that $\xi' = \xi^{*\prime}$ and $|\xi'' - \xi^{*\prime\prime}| = O(|\eta'|) = o(|\xi''|)$. In a similar way we can also prove that if Y is a fixed neighborhood of $y^0 = (S_{x'}(x^0, \theta^0), x^{0\prime\prime})$, then we can find a neighborhood X of x^0 and a bineighborhood Γ of (η^0, η^1) so that

$$G^*(x, \eta) \in Y \text{ for } x \in X \text{ and } \eta \in \Gamma.$$

Our next concern is now to study statements in the opposite direction and here we shall use a quantitative version of the inverse mapping theorem:

Lemma 4.8.7. *Let* $L : U = \{t \in R^l; |t| < c_1\} \to R^l$ *be a* C^2 *map such that* $L(0) = 0$. *Assume that* $|\det L'(0)| \geq c_2$ *and that*

$$|(\partial^2 / \partial t_i \partial t_j) L(t)| \leq c_3, \forall i, \forall j, \text{ if } |t| < c_1.$$

Then there is c_4, *which depends only on* c_1, c_2, c_3, *such that*

a) $U' = \{t \in R^l; |t| < c_4\} \subset L(\{t \in R^l; |t| \leq c_2\})$,
b) there is a unique map $L^{-1} : U' \to U$ *such that* $L \circ L^{-1}$ *is the identity on* U'.

The lemma is a standard consequence of the proofs of the inverse mapping theorem. Cf. anyway lemma 3.4.1 in Liess-Rodino [2]. There is also a complex analytic version of this result (and it is this version which we shall actually apply) when L is holomorphic on $\{t \in C^l; |t| < c_1\} \to C^l$, which we do not state explicitly.

5. Our next result is a consequence of lemma 4.8.7, after a rescaling.

Proposition 4.8.8. *Let* $\tilde{\eta} \in \Gamma$ *be given with* $|\tilde{\eta}''| = 1$. *If* $|\tilde{\eta}'|$ *is small enough, we can then for every* $c_1 > 0$ *find* $c_2 > 0$ *so that*

$$\begin{aligned} F^*(x^0, \{\eta; |\eta' - \tilde{\eta}'| \ &< \ c_1 |\tilde{\eta}'|, |\eta'' - \tilde{\eta}''| < c_1\}) \ \supset \ \{\xi; |\xi' - F^*_{(\xi')}(x^0, \tilde{\eta})| \\ &\leq c_2 |F^*_{(\xi')}(x^0, \tilde{\eta})|, |\xi'' - \tilde{\eta}''| < c_2\}. \end{aligned} \tag{4.8.11}$$

Moreover, c_2 *depends here on* c_1 *and* Γ, *but not explicitly on* $\tilde{\eta}$.

Remark 4.8.9. *The problem is here mainly with the second derivatives of* F^* *in* η *when* η' *is small. To be able to reduce ourselves explicitly to lemma 4.8.7 (and not only to its proof) we shall rescale to a neighborhood of some point* $\tilde{\tau}$ *with* $|\tilde{\tau}'| = c > 0$.

Proof of proposition 4.8.8. We fix $c > 0$ so that $\eta \in \Gamma$ implies $(c\eta'/|\eta'|, \eta'') \in \Gamma$. Next, we replace our initial map F^* defined in a neighborhood of $(x^0, \tilde{\eta})$, by

$$\Phi(\tau) = \begin{pmatrix} S_{x'}(x^0, (\tau', \tau'')) \\ \tau'' + S_{x''}(x^0, (|\tilde{\eta}'|\tau'/c, \tau'')) \end{pmatrix},$$

which we consider defined in a neighborhood of $(\tilde{\tau}', \tilde{\tau}'')$, where $\tilde{\tau}' = c\tilde{\eta}'/|\tilde{\eta}'|, \tilde{\tau}'' = \tilde{\eta}''$. (The rescaling is thus $(\tau', \tau'') \rightarrow (|\tilde{\eta}'|\tau'/c, \tau'')$. Note that

$$S_{x'}(x^0, \eta', \eta'') = (|\tilde{\eta}'|/c) S_{x'}(x^0, \tau', \eta'')$$

then.) If we can show that Φ maps

$$\{\tau; |\tau' - \tilde{\tau}'| < c_1, |\tau'' - \tilde{\tau}''| < c_1\} \text{ onto } \{\theta; |\theta' - \Phi_{(\tau')}(\tau^0)| < c_2, |\theta'' - \Phi_{(\tau'')}(\tilde{\tau})| < c_2\},$$

then $F^*(x^0, \cdot)$ will map $\{\eta; |\eta' - \tilde{\eta}'| < c_1|\tilde{\eta}'|/c, |\eta'' - \tilde{\eta}''| < c_1\}$ onto

$$\{\xi; |\xi' - F_{(\xi')}(x^0, \tilde{\eta})| < c_2|\tilde{\eta}'|/c, \quad |\xi'' - F_{(\xi'')}(x^0, \tilde{\eta})| < c_2\},$$

which is, after a renotation, the desired result, since $|F_{(\xi')}(x^0, \tilde{\eta})| < c_3|\tilde{\eta}'|$.

The aforementioned assertion about the range of Φ is now however a consequence of lemma 4.8.7. Indeed,

$$\Phi'(\tilde{\tau}) = \begin{pmatrix} S_{x',\tau'}(x^0, \tilde{\tau}), & S_{x',\tau''}(x^0, \tilde{\tau}) \\ \frac{|\tilde{\eta}'|}{c} S_{x'',\eta'}(x^0, \frac{|\tilde{\eta}'|}{c}\tilde{\tau}', \tilde{\tau}''), & Id + S_{x'',\eta''}(x^0, \frac{|\tilde{\eta}'|}{c}\tilde{\tau}', \tilde{\tau}'') \end{pmatrix},$$

which is easily seen to satisfy $|\det \Phi'(\tilde{\tau})| \geq c'$, if c had been chosen small enough and $|\tilde{\eta}'|/|\tilde{\eta}''|$ is suitably small. (We use here the bihomogeneity of the various entries of Φ'.) It remains to estimate Φ'' on $|\tau' - \tilde{\tau}'| < c_1, |\tau'' - \tilde{\tau}''| < c_1$. In fact, if we choose $c_1 < c/2$, it follows that $|\tau' - \tau^{0'}| \leq c_1$ implies $|\tau'| \geq c/2$. This shows that $(\partial^2/\partial \tau^2) S_{x'}(x^0, \tau)$ remains bounded for the τ under consideration. In a similar way

$$(\partial^2/\partial \tau'^2) S_{x''}(x^0, \frac{|\tilde{\eta}'|}{c}\tau', \tau'') = \frac{|\tilde{\eta}'|^2}{c^2} S_{x'',\eta'\eta'}(x^0, \frac{|\tilde{\eta}'|}{c}\tau', \tau'')$$

is bounded for such τ and the same also holds by a similar argument for

$$(\partial^2/\partial \tau'' \partial \tau) S_{x''}(x^0, |\tilde{\eta}'|\tau'/c, \tau'').$$

The proof is thus complete.

In proposition 4.8.8 we have assumed $|\tilde{\eta}''| = 1$. We can remove that restriction using the homogeneity of F^*.

Proposition 4.8.10. *If $|\tilde{\eta}'|/|\tilde{\eta}''|$ is small we can find for every c_1 some c_2 so that*

$$F^*(x^0, |\eta' - \tilde{\eta}'| < c_1|\tilde{\eta}'|, |\eta'' - \tilde{\eta}''| < c_1|\tilde{\eta}''|\}) \supset \{\xi; |\xi' - F^*_{(\xi')}(x^0, \tilde{\eta})| <$$
$$c_2|F^*_{(\xi')}(x^0, \tilde{\eta})|, |\xi'' - F^*_{\xi''}(x^0, \tilde{\eta})| < c_2|\tilde{\eta}''|\}.$$

Indeed, we can simply apply proposition 4.8.8 for $\tilde{\eta}^0 = \tilde{\eta}/|\tilde{\eta}''|$, and transfor n back by $\eta \to |\tilde{\eta}''|\eta$.

Remark 4.8.11. *In all the above x^0 may also be regarded as a parameter. If $\cdot$ is small enough, the same argument therefore shows that if $|x - x^0| < c_1$, then*

$$F^*(x, \{|\eta' - \tilde{\eta}'| < c_1|\tilde{\eta}'|, |\eta'' - \tilde{\eta}''| < c_1|\tilde{\eta}''|\}) \supset \{\xi; |\xi' - F^*_{(\xi')}(x^0, \tilde{\eta})|$$
$$< c_2|F^*_{(\xi')}(x^0, \tilde{\eta})|, |\xi'' - F^*_{\xi''}(x^0, \tilde{\eta})| < c_2|\tilde{\eta}''|\}.$$

Proposition 4.8.12. *If c_3 is small enough, then $F^*(x, \Gamma)$ contains, for x satisfying $|x - x^0| < c_3$, a fixed bineighborhood of $(\tilde{\eta}, \xi^1)$. Similarily, if Γ is small enough, then $G(|x - x^0| < c_3, \eta)$ contains for $\eta \in \Gamma$ a fixed neighborhood of y^0.*

Proof. Consider a bineighborhood $\Gamma' \subset\subset \Gamma$ of (η^0, η^1). We can find c so that $|\theta' - \eta'| < c|\eta'|$, and $|\theta'' - \eta''| < c|\eta''|$ together with $\eta \in \Gamma'$ implies $\theta \in \Gamma$. Next we observe that $F(x, \Gamma')$ contains an open bineighborhood Γ'' of (η^0, ξ^1) if $|x - x^0| < c_3$. This is immediate. We want to show that

$$F^*(x, \Gamma) \supset \Gamma'' \cap \{\xi; |\xi'|/|\xi''| < c_3\},$$

if c_3 is suitable. Let us in fact choose (x, ξ) with $|x - x^0| < c_3, \xi \in \Gamma''$. By the way in which Γ'' has been introduced we can then find $\eta \in \Gamma'$ with $\xi = F(x, \eta)$. Also denote $\xi^* = F^*(x, \eta)$. It follows from the definition of F^* and F that $\xi' = \xi^{*\prime}$ and $|\xi^{*\prime\prime} - \xi''| < c_4|\eta'|$. Here we recall that $|\eta'| < c_5|\xi'|$ since $\eta = I(x, \xi)$, so we must also have $|\xi^{*\prime\prime} - \xi''| \le c_6|\xi''|$ for any previously fixed c_6 if $|\xi'|/|\xi''|$ is sufficiently small. At this moment we apply the preceding remark. If c_6 is suitable we can conclude from there that there is θ with $\xi = F^*(x, \theta), |\eta' - \theta'| < c|\eta'|, |\eta'' - \theta''| < c|\eta''|$. By the above this implies $\theta \in \Gamma$, as desired. This proves the assertion from the proposition for F^*. The assertion for G^* can be proved with similar arguments.

It is now clear from all the above that T^*, T can both be defined on a bineighborhood of (y^0, η^0, η^1). We complete this result with the following

Proposition 4.8.13. *We have the following relations*

$$|T^*_{(x)}(y, \eta) - T_{(x)}(y, \eta)| = O(|\eta'|/|\eta''|),$$
$$|T^*_{(\xi')}(y, \eta) - T_{(\xi')}(y, \eta)| = O(|\eta'|^2/|\eta''|),$$
$$|T^*_{(\xi'')}(y, \eta) - T_{(\xi'')}(y, \eta)| = O(|\eta'|).$$

In other words, T is the bihomogeneous approximation of T^*.

Proof. The relation for the x—components is expressed in lemma 4.8.4 and the relation for the ξ'—components follows also from that lemma. In fact from that lemma we obtain

$$
\begin{aligned}
|T^*_{(\xi')}(y,\eta) - T_{(\xi')}(y,\eta)| &= |S_{x'}(H^*(y,\eta),\eta) - S_{x'}(H(y,\eta),\eta)| \\
&\leq \sup |S_{x'x}(z,\eta)||H^*(y,\eta) - H(y,\eta)| \leq C|\eta'|^2/|\eta''|,
\end{aligned}
$$

where the supremum is for $z \in [H^*(y,\eta), H(y,\eta)]$.

Similarily

$$
|T^*_{(\xi'')}(y,\eta) - T_{(\xi'')}(y,\eta)| = |S_{x''}(H^*(y,\eta),\eta)| \leq C|\eta'|.
$$

4.9   Construction of the phase

1. In this section we assume that Λ is a real analytic regular involutive bihomogeneous subvariety in $N\Sigma$ near some point $(x^0, \xi^0, \sigma^0) \in N\Sigma$, with x^0, $\xi^0 \neq 0$, $\sigma^0 \neq 0$. We recall from section 4.7 that this means (also cf. the introduction) that we can find a bineighborhood Γ of (x^0, ξ^0, σ^0) and functions $f_1, ..., f_q : \Gamma \rightarrow R$, $q \leq d$, with the following properties:

a) $\Lambda = \{(x, \xi'', \xi') \in N\Sigma \cap \Gamma; f_j(x, \xi'', \xi') = 0, \forall j\}$,

b) the f_j are bihomogeneous of bidegree $(1,1)$,

c) $\{f_i, f_j\}_r = 0, \forall i, j$,

d) the f_j are real analytic,

e) the $\bar{d}f_j$ and $\sum' \xi_s dx_s$ are linearly independent,

f) rank $(\partial f_i/\partial \xi_j)_{i \leq q, j \leq d} = q$.

Here we have assumed of course already that real-analytic canonical coordinates have been chosen so that $\Sigma = \{(x, \xi); \xi_i = 0, i \leq d\}$ and used (x, ξ'', ξ') as coordinates in $N\Sigma$. In particular, ξ' corresponds to σ. In the sequel we will have however also to use the identification $N\Sigma \leftrightarrow T^*X$, $(x, \xi'', \xi') \leftrightarrow (x, \xi', \xi'')$ since canonical transformations are more naturally written in the coordinates (x, ξ', ξ''). As a consequence we shall then also write, here and later on in similar situations, $f(x, \xi', \xi'')$ rather than $f(x, \xi'', \xi')$, etc.

What we want to obtain is a bihomogeneous phase function S of bidegree $(1,1)$ which is defined in a bineighborhood Γ' of (x^0, ξ^0, σ^0) and which is so that the canonical transformation χ^* associated with $S(x, \eta) + \langle x'', \eta'' \rangle$ has the property

$$(f_i \circ \chi^*)(y, \eta) = \eta_i + O(|\eta'|^2/|\eta|), i \leq q. \tag{4.9.1}$$

2. Since bihomogeneity considerations are important in what follows it is useful to dispose of a convenient way to test homogeneity. Our main concern is here of course the homogeneity in the variables ξ', since the variables ξ'' will essentially be parameters. We recall then that a C^1-function is positively homogeneous of degree μ in ξ' if and only if it satisfies Euler's relation

$$\rho f = \mu f,$$

where ρ is the relative radial field $\Sigma' \xi_s \partial \xi_s$ considered in section 4.7. It is then useful to note that if f is homogeneous of degree μ in ξ', then we have

$$[H_f^r, \rho] = (1 - \mu) H_f^r,$$

for the commutator $[H_f^r, \rho] = H_f^r \rho - \rho H_f^r$ of H_f^r and ρ. This is indeed clear in our special coordinates. (For an "invariant" calculation in standard symplectic geometry, see Hörmander [5,vol. III, chapt. XXII].) In particular it follows if $\mu = 1$ that

$$[H_f^r, \rho] = 0. \tag{4.9.2}$$

3. For fixed x'', ξ'' the functions f_i, f_j are in involution in the variables x', ξ'. The main idea is then (in analogy with arguments from the preceding section) to consider for fixed x'', ξ'' a canonical transformation $(y', \eta') \rightarrow (x', \xi')$ under which f_i is transformed to η_i. For each fixed $x'' = y'', \xi'' = \eta''$, we obtain a phase function $S^{x'',\eta''}$ and we denote $S(x, \eta) = S^{x'',\eta''}(x', \eta')$. It remains then to show that S has the desired properties. The construction of $S^{x'',\eta''}$ is of course classical, but we review it here in some detail, to see that it satisfies indeed the properties of a phase required in section 4.8.

4. The first step in the argument is to apply Darboux's theorem in its homogeneous form and with parameters. We may in fact find a bineighborhood Γ' of (x^0, ξ^0, σ^0) and $f_j, q + 1 \leq j \leq d, g_j, j \leq d$, defined on Γ' such that

α) the g_j are bihomogeneous of bidegree $(0,0)$,

β) the f_j are bihomogeneous of bidegree $(1,1)$,

γ) $\{f_j, f_k\}_r = \{g_j, g_k\}_r = \{f_j, g_k\}_r - \delta_{jk} = 0, \forall j, \forall k$,

δ) the f_j and g_k are real-analytic,

ε) $\det (\partial f_i / \partial \xi_j)_{i,j=1}^d (x^0, \sigma^0, \xi^{0\prime\prime}) \neq 0$.

Note that the existence of such f_j, g_k can be proved with the classical proofs of the homogeneous version of Darboux's theorem. (Cf. e.g. Duistermaat-Hörmander [1, prop. 6.1.3.] or Hörmander [5,vol.III, theorem 21.1.9.]) We briefly recall the main steps in the construction. The procedure is in fact to introduce the f_j, g_k, successively, one after the other. If we want to construct f_{q+1}, say, then what we must do, is find some real-analytic function f_{q+1} which is

- homogeneous of degree 0 in ξ'',
- homogeneous of degree 1 in ξ',
- satisfies

$$H^r_{f_j} f_{q+1} = 0, \text{ for } j \leq q \tag{4.9.3}$$

and has the property that

$$\text{rank} \left(\frac{\partial f_i}{\partial \xi_j} \right)_{i \leq q+1, j \leq d} = q + 1. \tag{4.9.4}$$

We also recall that the second property can be conveniently written as

$$\rho f_{q+1} = f_{q+1}. \tag{4.9.5}$$

What we want here initially is to find f_{q+1} on a bineighborhood of (x^0, ξ^0, σ^0) but if we use the bihomogeneity it will actually suffice to find f_{q+1} on a neighborhood of some point $(x^0, t\sigma^0, \xi^{0''})$, $t > 0$. The first of the above properties does not pose any particular problem, since x'', ξ'' may be regarded as parameters. As for the conditions (4.9.3), (4.9.5), we regard them as an overdetermined system of first order partial differential equations for the unknown function f_{q+1}. It is important to observe that

$$[H_{f_i}, H_{f_j}] = 0 \text{ if } i, j \leq q, \text{ respectively that } [H_{f_k}, \rho] = 0, \text{ if } k \leq q$$

as a consequence of the conditions b) and c).(Cf. (4.9.2).) The system (4.9.3), (4.9.5) is thus integrable and the Frobenius theorem can be applied. It follows that we can solve the system with arbitrarily given "initial conditions" on some previously fixed subvariety which is transversal to the fields $H^r_{f_j}$, $j \leq q$, and ρ. In particular we can ask for $H_{f_{q+1}} = \varepsilon$, where ε is chosen so that

$$\text{rank} \left(\begin{array}{c} \frac{\partial f_i}{\partial \xi_j}(x^0, \sigma^0, \xi^{0''}), j \leq d, \\ \varepsilon_{(\xi')} \end{array} \right) = q + 1.$$

(Here $\varepsilon_{(\xi')}$ is the ξ'-component of ε.)
How this is done is clear e.g. from the proof of theorem 21.1.9 in Hörmander [5, vol.III]. We omit further details.

The map

$$(x,\xi) \to \begin{pmatrix} f_j(x,\xi',\xi''), j \leq d \\ x'' \\ g_j(x,\xi',\xi''), j \leq d \\ \xi'' \end{pmatrix} = \kappa(x,\xi)$$

is then locally a bihomogeneous and bicanonical diffeomorphism from a bineighborhood of (x^0,ξ^0,σ^0) to a bineighborhood of (y^0,η^0,η^1),

$$y^0 = (f_1(x^0,\sigma^0,\xi^{0''}),...,f_d(x^0,\sigma^0,\xi''^0),x''^0),$$

$$\eta^1 = g_1(x^0,\sigma^0,\xi''^0),...,g_d(x^0,\sigma^0,\xi^{0''}), \ \eta^0 = (0,\xi^{0''}).$$

(To see that the map is locally a diffeomorphism, we use γ). Recall in fact that γ) implies

$$\det \begin{pmatrix} \frac{\partial g}{\partial x'} & \frac{\partial g}{\partial \xi'} \\ \frac{\partial f}{\partial x'} & \frac{\partial f}{\partial \xi'} \end{pmatrix} \neq 0.$$

A convenient way to check this is to assume that for some real numbers φ_k, ψ_k we have that

$$v = \sum \varphi_i H^r_{f_j}(x^0,\sigma^0,\xi^{0''}) + \sum \psi_k H^r_{g_k}(x^0,\sigma^0,\xi^{0''}) = 0.$$

It follows that $\varphi_i = \omega_r(v, H^r_{g_i}) = 0, \ \psi_k = \omega_r(v, H^r_{f_k}) = 0.$) We denote the inverse map by χ. It clearly has the property that $f_j \circ \chi = \eta_j, j \leq q$.

The relation γ) codifies the behaviour of Poisson brackets. It is well-known that similar relations must also hold for Lagrange brackets. This gives a set of additional relations for expressions involving the f and g. (Cf. Caratheodory [1].) Let us in fact denote by F and by G the y' and η' components of κ. We then have the following matrix relations:

$$^tF_{x'}G_{x'} - \, ^tG_{x'}F_{x'} = 0, \tag{4.9.6}$$

$$^tF_{\xi'}G_{\xi'} - \, ^tG_{\xi'}F_{\xi'} = 0, \tag{4.9.7}$$

$$^tF_{x'}G_{\xi'} - \, ^tG_{x'}F_{\xi'} = I, \tag{4.9.8}$$

where ^tA denotes the adjoint of A, $F_{x'}$ the Jacobian of F in the variables x', etc. Next we want to find a generating function $S^{x'',\eta''}(x',\eta')$ from which χ comes. Also here the argument is classical. We solve at first

$$\eta' = F(x,\xi',\xi'')$$

for ξ', i.e., we consider $\xi' = H(x,\eta',\xi'')$ such that

$$\eta' = F(x,\eta'',H(x,\eta)), \ \forall x, \ \forall \eta. \tag{4.9.9}$$

Of course, for this to be possible, we need to know that $\det F_{\xi'} \neq 0$, which is property ε). It is clear that we may choose H bihomogeneous of bidegree $(1,1)$. We then need to find S so that

$$H(x,\eta) = \nabla_{x'} S(x,\eta), \quad T = G(x,\eta'',H(x,\eta)) = \nabla_{\eta'} S(x,\eta). \qquad (4.9.10)$$

Of course, for (4.9.10) to have a solution, we must show that

$$\begin{pmatrix} \nabla_{x'} H, & \nabla_{x'} T \\ \nabla_{\eta'} H, & \nabla_{\eta'} T \end{pmatrix} \qquad (4.9.11)$$

is symmetric. This can be proved starting from (4.9.9). In fact, it is clear that (4.9.9) implies

$$0 = F_{x'} + F_{\xi'} H_{x'}, I = F_{\xi'} H_{\eta'}. \qquad (4.9.12)$$

It follows from (4.9.12) that $H_{x'} = -F_{\xi'}^{-1} F_{x'}$ and that $H_{\eta'} = F_{\xi'}^{-1}$. Incidentally it is now also clear that if S exists, then

$$\det \nabla_{\eta'} \nabla_{x'} S(x^0,\eta^0) = \det H_{\eta'}(x^0,\eta^0) = \det F_{\xi'}^{-1}(x^0,\xi''^0,\xi'^0) \neq 0$$

in view of ε). This is the condition for the nondegeneracy of S from section 4.8. Let us now return to the study of the symmetry from (4.9.11). The symmetry of $\nabla_{x'} H$ comes to

$$F_{\xi'}^{-1} F_{x'} = {}^t F_{x'} {}^t F_{\xi'}^{-1}.$$

This is equivalent to $F_{x'} {}^t F_{\xi'} = F_{\xi'} {}^t F_{x'}$, which is another way of writing that $\{f_j, f_k\}_r = 0$. Furthermore, $T_{\eta'} = G_{\xi'} H_{\eta'} = G_{\xi'} F_{\xi'}^{-1}$, so symmetry of $T_{\eta'}$ comes to

$$G_{\xi'} F_{\xi'}^{-1} = {}^t F_{\xi'}^{-1} {}^t G_{\xi'}, \text{ or } {}^t F_{\xi'} G_{\xi'} = {}^t G_{\xi'} F_{\xi'},$$

which is precisely (4.9.7). Finally we want to check that ${}^t \nabla_\eta H = \nabla_{x'} T$, which comes to ${}^t F_{\xi'}^{-1} = G_{x'} + G_{\xi'} H_{x'}$. We obtain that ${}^t F_{\xi'}^{-1}$ must be equal to $G_{x'} - G_{\xi'} F_{\xi'}^{-1} F_{x'}$, or that

$$I = {}^t F_{\xi'} G_{x'} - {}^t F_{\xi'} G_{\xi'} F_{\xi'}^{-1} F_{x'}.$$

Using (4.9.7) we can rewrite the last expression as

$${}^t F_{\xi'} G_{x'} - {}^t G_{\xi'} F_{\xi'} F_{\xi'}^{-1} F_{r'} = {}^t F_{\xi'} G_{x'} - {}^t G_{\xi'} F_{x'},$$

so we get indeed I as a consequence of (4.9.8).

We have thus proved that the matrix from (4.9.11) is symmetric. It follows that there is a solution S of (4.9.10). It is also clear that we may choose the solution to be bihomogeneous.

5. We have now proved the existence of S and want to show that $f_j \circ \chi^* = \eta_j + O(|\eta'|^2/|\eta''|)$, if χ^* is the canonical transformation associated with S. Let us in fact denote once more by χ the bihomogeneous approximation of χ^*. Thus $f_j \circ \chi = \eta_j$ and we have that

$$\chi^*_{(x)} - \chi_{(x)} = O(|\eta'|/|\eta''|),$$
$$\chi^*_{(\xi')} - \chi_{(\xi')} = O(|\eta'|^2/|\eta''|),$$
$$\chi^*_{(\xi'')} - \chi_{(\xi'')} = O(|\eta'|).$$

The desired estimate for $f_j \circ \chi^* - \eta_j$ now follows from Taylor expansion. (The situation is similar to the one from section 4.8).

Definition 4.9.1. *Let $\Lambda \subset N\Sigma$ be a real analytic bihomogeneous variety. We say that Λ is regular involutive if we can write it locally in the form $\{(x, \xi'', \xi'); f_j(x, \xi'', \xi') = 0\}$, where the f_j satisfy the conditions b), c), d), e) from the beginning of this section.*

4.10   Reduction of p_m to model form, inequalities and a regularity theorem

1. Our final goal in studying the theory of bicanonical transformations is to simplify the structure of the principal part of some pseudodifferential operator p, if certain conditions are satisfied. This will then lead to a regularity theorem for solutions of the equation $pu = 0$, if we apply the theory from chapter III. In this section we assume that the principal part p_m of p is homogeneous of order m in a conic neighborhood of (x^0, ξ^0) and that it vanishes precisely of order s on a real analytic non-radial homogeneous involutive variety Σ for which $(x^0, \xi^0) \in \Sigma$. We also assume that p_m is real-analytic and associate with p_m its localization p^T to $N\Sigma$, which is, as in section 4.4, its Taylor expansion of order s considered for the points from Σ.

Further we assume that in $N\Sigma$ we are given a real-analytic bihomogeneous manifold Λ of codimension q which is regular involutive with respect to the relative 2-form on $N\Sigma$ in a bineighborhood of (x^0, ξ^0, σ^0), $(x^0, \xi^0, \sigma^0) \in \Lambda$, in the sense of section 4.7. We assume that p^T vanishes precisely of some order s' on Λ.

2. In the next part of the section we shall show that in the conditions above we can find an analytic canonical change of coordinates in which p has a very simple trihomogeneous principal part which is elliptic in a suitable bineighborhood of (x^0, ξ^0, σ^0) if p^T is "transversally elliptic" to Λ, in a sense which will be specified later on. In fact, we observe at first that we can always find a conic neighborhood V of (x^0, ξ^0) and a

canonical change of coordinates κ in V such that in the new coordinates Σ is given by $\{\xi' = 0\}$. Identifying T^*X with $N\Sigma$ in these coordinates, we will have

$$(p_m \circ \kappa)(x, \xi) = p^T(x, \xi'', \xi') + O(|\xi'|^{s+1}|\xi|^{m-s-1}), \tag{4.10.1}$$

where now, (since we have replaced p by $p \circ \kappa$)

$$p^T(x, \xi'', \xi') = \sum_{|\alpha|=s} \partial_{\xi'}^{\alpha}(p_m \circ \kappa)(x, 0, \xi'')\xi'^{\alpha}/\alpha!.$$

Actually more than (4.10.1) is true in that we can also estimate derivatives of the remainder term $p_m \circ \kappa - p^T$:

$$|\partial_x^{\alpha}\partial_{\xi}^{\beta}(p_m \circ \kappa - p^T)(x, \xi)| \leq c^{|\alpha|+|\beta|+1}\alpha!\,\beta!\,|\xi'|^{s+1-|\beta|}|\xi|^{m-s-1}. \tag{4.10.2}$$

At this moment, we use the assumptions on Λ and write

$$\Lambda = \{(x, \xi'', \sigma); f_j(x, \xi'', \sigma) = 0, j \leq q\},$$

where the f_j are real-analytic, $(1,1)$–bihomogeneous, satisfy $\{f_j, f_k\}_r = 0$, $\forall j, k$, and are such that the $\bar{d}f_j$ and $\sum' \xi_i dx_i$ are linearly independent and such that

$$\text{rank}(\partial f_i/\partial \xi_j)_{i \leq q, j \leq d} = q.$$

Since p^T vanishes of order s' on Λ we can then write that

$$p^T(x, \xi'', \xi') = \sum_{|\beta|=s'} a_{\beta}(x, \xi'', \xi')f^{\beta},$$

where the a_{β} are defined in a bineighborhood of (x^0, ξ^0, σ^0) and are bihomogeneous of bidegree $(m - s', s - s')$.

We can then repeat the constructions from section 4.4 and introduce the Taylor expansion p^{TT} of order s' of p^T at Λ. Arguing as there we obtain that p^{TT} gives a well-defined function on $N\Lambda = T_{\Lambda}N\Sigma/T\Lambda$. In fact, if $w \in T_{(x,\xi'',\sigma)}N\Sigma$ for some $(x, \xi'', \sigma) \in \Lambda$, then we have

$$p^{TT}(x, \xi'', \sigma, w) = \sum_{|\gamma|=s'} \partial^{\gamma}p^T(x, \xi'', \sigma)w^{\gamma}/\gamma!, \tag{4.10.3}$$

but actually in (4.10.3), p^{TT} depends only on the class of w in $T_{(x,\xi'',\sigma)}N\Sigma/T_{(x,\xi'',\sigma)}\Lambda$.

Definition 4.10.1. *We say that p^T is transversally elliptic to Λ if*

$$p^{TT}(x, \xi'', \sigma, w) \neq 0 \text{ for all } (x, \xi'', \sigma) \in \Lambda \text{ and all } w \in T_{(x,\xi'',\sigma)}N\Sigma/T_{(x,\xi'',\sigma)}\Lambda.$$

It is clear from the very definition that this has an invariant meaning in $N\Lambda$.

Assuming transversal ellipticity it now follows, taking into account the tri-homogeneity of p^{TT}, that

$$c|\xi''|^{m-s}|\sigma|^{s-s'}|w|^{s'} \leq |p^{TT}(x,\xi'',\sigma,w)| \leq c'|\xi''|^{m-s}|\sigma|^{s-s'}|w|^{s'},$$

if (x,ξ'',σ) is in a bineighborhood of $(x^0,(0,\xi^{0''}),\sigma^0)$ and $w \neq 0$.

To see what this gives for p, we consider a homogeneous real-analytic canonical change of coordinates $\chi^*(y,\eta) = (x,\xi)$ defined in a bineighborhood of (y^0,η^0,τ^0), $\eta^{0'} = 0$, of the type considered in section 4.9, for which

$$(f_j \circ \chi^*)(y,\eta) = \tau_j + O(|\eta'|^2/|\eta''|), \qquad j \leq q,$$

respectively

$$(f_j \circ \chi)(y,\eta) = \tau_j, \qquad j \leq q,$$

where χ is the bihomogeneous approximation of χ^*. Here we have denoted η_j by τ_j if $j \leq q$ and have assumed that

$$\chi(y^0,\eta^0,\tau^0) = (x^0,\xi^0,\sigma^0).$$

It follows from this that we have

$$(p^T \circ \chi^*)(y,\eta) = \sum_{|\beta|=s'} (a_\beta \circ \chi^*)(y,\eta)\tau^\beta + O(|\tau|^{s'-1}|\eta'|^{s-s'+2}|\eta''|^{m-s-1})$$

on a bineighborhood of (y^0,η^0,τ^0). Since the same is true also for complex (y,η) if y is close to y^0 and $|\operatorname{Im}\eta| < c|\operatorname{Re}\eta'|$, we obtain from Cauchy's inequalities

$$|\partial_y^\alpha \partial_\eta^\beta [(p^T \circ \chi^*)(y,\eta) - \sum a_\beta \circ \chi^*(y,\eta)\tau^\beta]| \leq c^{|\alpha|+|\beta|+1}\alpha!\,\beta!$$
$$|\tau|^{s'-1-|\beta|}|\eta'|^{s-s'+2}|\eta''|^{m-s}.$$

This is all we can achieve as a simplification of p if we use canonical transformations alone. A further simplification can be achieved however if we compute the tri-homogenous principal part of $\sum_{|\beta|=s}(a_\beta \circ \chi^*)(y,\eta)\tau^\beta$. To do so, let us replace first χ^* with its bihomogeneous approximation χ. We can then argue as in section 4.9 to obtain

$$|(a_\beta \circ \chi^* - a_\beta \circ \chi)(y,\eta)| \leq c|\eta'|^{s-s'+1}|\eta''|^{m-s-1}.$$

Indeed, since the latter remains true also for complex (y,η) as long as $|y - y^0| < c$, $|\operatorname{Im}\eta| < c|\operatorname{Re}\eta'|$, it follows that actually

$$|\partial_y^\alpha \partial_\eta^\beta (a_\beta \circ \chi^* - a_\beta \circ \chi)(y,\eta)| \leq c^{|\alpha|+|\beta|+1}\alpha!\,\beta!\,|\eta'|^{s-s'+1-|\beta|}|\eta''|^{m-s-1}. \qquad (4.10.4)$$

Next we observe that $a_\beta \circ \chi$ is $(m-s', s-s')$ bihomogeneous, but is not tri-homogeneous. We shall therefore replace it by the tri-homogeneous

$$b_\beta(y, \sigma'', \eta'') = (a_\beta \circ \chi)(y, (0, \sigma''), \eta'')$$

where we have written $\sigma = (\tau, \sigma'')$, but note that b_β actually does not depend on τ. Thus, b_β is just the tri-homogeneous principal part of $a_\beta \circ \chi$ in that it is immediately seen that we have

$$(a_\beta \circ \chi)(y, \eta) - b_\beta(y, \eta) = O(|\tau||\eta'|^{s-s'-1}|\eta|^{m-s}).$$

It also follows from this that

$$p^T \circ \chi - \sum_{|\beta|=s'} b_\beta \tau^\beta = O(|\tau|^{s'+1}|\eta'|^{s-s'+1}|\eta|^{m-s}).$$

Since after the change of coordinates $(y, \eta'', \sigma) \to \chi(y, \sigma, \eta'')$, Λ has the defining equations $\{\tau = 0\}$ it is now clear that $\sum b_\beta(y, \eta)\tau^\beta$ is just the Taylor expansion of $p^T \circ \chi$ of order s' computed for points in Λ in the coordinates (y, η'', σ). This gives $\sum b_\beta(y, (0, \sigma''), \eta'')\tau^\beta$ a meaning in $N\Lambda$ and also shows that

$$|\partial_y^\alpha \partial_\eta^\beta (p^T \circ \chi - \sum b_\beta \tau^\beta)(y, \eta)| \leq c^{|\alpha|+|\beta|+1}\alpha!\,\beta!\,|\tau|^{s'+1-|\beta|}|\eta'|^{s-s'+1}|\eta|^{m-s}.$$

It is important to note that

$$|\sum_{|\beta|=s'}(a_\beta \circ \chi)(y, \eta)\tau^\beta| \geq c|\tau|^{s'}|\eta'|^{s-s'}|\eta''|^{m-s}$$

on a bineighborhood of (y^0, η^0, τ^0). If this bineighborhood is small, it will follow then that

$$|\sum b_\beta(y, \eta)\tau^\beta| \geq c|\tau|^{s'}|\eta'|^{s-s'}|\eta''|^{m-s}. \tag{4.10.5}$$

To conclude our discussion, we have thus succeded (and this was one of our main aims in this part of the notes) in finding b_β, bihomogeneous of bidegree $(m-s', s-s')$, with (4.10.5), and so that

$$\begin{aligned}
|\partial_y^\alpha \partial_\eta^\beta [(p_m \circ \kappa \circ \chi^*)(y, \eta) - \Sigma b_\beta(y, \eta)\tau^\beta]| &\leq c^{|\alpha|+|\beta|+1}\alpha!\,\beta!\,[|\tau|^{s'+1-|\beta|}|\eta'|^{s-s'-1}|\eta|^{m-s}) \\
&\quad + |\tau|^{s'-1-|\beta|}|\eta'|^{s-s'+2}|\eta''|^{m-s-1} \\
&\quad + |\tau|^{s-1-|\beta|}|\eta'|^{s-s'+1}|\eta''|^{m-s-1}].
\end{aligned} \tag{4.10.6}$$

3. All the above referred to the principal part p_m of our initial operator p and shows that p is an tri-microlocally elliptic operator, once (x^0, ξ^0, σ^0) has been fixed in Λ. We conclude from the theory of pseudodifferential operators in chapter III that we have the following theorem:

Theorem 4.10.2. *Assume p_m is as before and assume p is of form $p = p_m + R$ where R is a classical analytic pseudodifferential operator of order m-1 defined in a conic neighborhood of (x^0, ξ^0). Also let u be a solution of $p(x, D)u = 0$ near (x^0, ξ^0), i.e. assume that $(x^0, \xi^0) \notin WF_A p(x, D)u$. Finally choose coordinates as before, consider $w^{0\prime} \in R^q_{w'}$ and denote $w^0 = (w^{0\prime}, 0)$. Then it follows that*

$$(y^0, \eta^0, \tau^0, w^0) \notin WF^3_{A,s} u.$$

4.11   Canonical transformations and estimates for symbols

1. Let S be a nondegenerate real analytic (1,1)-bihomogeneous phase function on a bineighborhood of (x^0, η^0, τ^0) and let $F^*, F, G^*, G, H^*, H, I^*, I, \chi^*$ and χ be associated with S as in the preceding section. We denote by $y^0 = (S_{\eta'}(x^0, \tau^0, \eta^{0\prime\prime}), x^{0\prime\prime})$ and let $(x^0, \xi^0, \sigma^0) = \chi(y^0, \eta^0, \tau^0)$. (See section 4.1.) To prepare for the calculus of F.I.O. we shall now consider symbols a^*, b^*, c^*, d^* of form

$$a^*(x, \eta) = a(x, F^*(x, \eta)), \ b^*(y, \eta) = b(H^*(y, \eta), \eta), \ c^*(x, \xi) = c(x, I^*(x, \xi)),$$

$$d^*(x, \eta) = d(G^*(x, \eta), \eta),$$

for some previously fixed symbols a, b, c, d. Our goal is here, roughly speaking, to show that a^*, b^*, c^*, d^* can be regarded as symbols in two-microlocalization on appropriate sets, if the same is true for a, b, c, d. This must be made, of course, more precise, in that we should state explicitly which are the domains of definition, fix some explicit symbol classes, etc. We do this here for the case of a^* to show how the argument goes. The other cases can be discussed in a similar fashion, so we shall omit details. We mention however that we shall say a few words on the domains of definition in these cases in section 5.5.

2. Assume then that a is defined on a set of form

$$E = \{(x, \xi); |x - x^0| < c, \xi \in \Gamma, |\xi'| > c'(1 + |\xi|^\delta)\}$$

where Γ is some bineighborhood of (ξ^0, σ^0). The first thing to note here is that $|\eta'| > c_1(1 + |\eta|^\delta)$, together with $|x - x^0| < c_2$ will imply $|F^*_{\xi'}(x, \eta)| > c'(1 + |F^*(x, \eta)|^\delta)$ and that we can find a biheighborhood Γ_1 of (η^0, τ^0) and c_2 so that $F^*(x, \eta) \subset\subset \Gamma$ if $\eta \in \Gamma_1$, $|x - x^0| < c_2$. All this follows from the sections 4.4 and 4.8. We conclude that a^* is well-defined on

$$E' = \{(x, \eta); |x - x^0| < c_3, \eta \in \Gamma_1, |\eta'| > c_1(1 + |\eta|^\delta)\},$$

if c_3 is small enough. It would remain to estimate the derivatives of a^*, assuming that a itself satisfies the estimates for a symbol. An explicit calculation of the derivatives of a^* as a function of (x, η) is here of course unpleasant, but we can avoid this calculation if we use Cauchy's inequalities. Here we need of course the complex-analytic versions of the results from section 4.8. Actually it is easy to see that a^* can be extended analytically to a set of form

$$E'' = \{(z, \eta) \in C^{2n}; \ |z - x^0| < c_4, \ \operatorname{Re} \eta \in \Gamma_2, |\operatorname{Im} \eta| < c_5| \operatorname{Re} \eta'|, \ |\eta'| > c_6(1 + |\eta|^6)\}.$$

As for the estimate of $|a^*|$ in terms of η, we may simply start from the estimate of $|a|$ and use that

$$|F^*_{(\xi')}(x, \eta)| \sim |\eta'|, \ |F^*_{(\xi'')}(x, \eta)| \sim |\eta''|.$$

We omit further details.

Chapter 5

Fourier Integral Operators

1. The main purpose of this part of these notes is to develop a calculus of F.I.O.'s adapted to the study of the second analytic wave front set. Similar results are probably true for F.I.O.'s related to higher order wave front sets. The respective theory would depend on a multi-symplectic geometry similar to the bi-symplectic geometry from chapter IV, but we have not tried to work out what really happens. For second microlocalization we shall consider essentially two classes of operators. The first consists of classical F.I.O.'s A defined in a full conic neighborhood of $(x^0, \eta^0) \in \Sigma$, where Σ is a homogeneous analytic regular involutive manifold, and we shall assume initially that $\Sigma = \{\xi' = 0\}$. When the canonical transformation underlying A leaves Σ invariant, then A operates well on WF_A^2: cf. section 5.1. (Here of course, second mirolocalization is with respect to the bihomogeneous structure $M_0 = R^n$, $M_1 = \{\xi; \xi'' = 0\}$.) In fact, this result allows for an extension of the definition of WF_A^2 to an invariantly defined notion $WF_{A,\Sigma}^2$, where Σ is in arbitrary position, and standard F.I.O.'s operate well on $WF_{A,\Sigma}^2$, too. (Cf. section 5.2.) We should mention here that a theory of contact transformations in relation with second microlocalization has also been developed in Laurent [1]. The theory from there is however not directly applicable to distributions, in that the constructions to which it refers are in spaces which are even larger than the space of hyperfunctions. In particular we hope that the definitions which we use here will be easier to understand for people who are more familiar with the C^∞-theory than with analytic microlocalization.

Also for the second class of F.I.O.'s we shall assume that $\Sigma = \{\xi' = 0\}$, but the phases and symbols are not necessarily defined on full conic neighborhoods of (x^0, η^0). In fact, in this second case we shall fix $(x^0, \eta^0, \tau^0) \in \dot{N}\Sigma$ and phases and symbols will be defined only on suitable bineighborhoods of (x^0, η^0, τ^0). To compensate for the potentially singular behaviour of the phases and symbols we shall however assume that the phase has the special form from section 4.8, an assumption which is also motivated by the results from section 4.10, which shows that the geometry of the canonical map underlying our operator relates well to the bicanonical structure on $\dot{N}\Sigma$. It is then no surprize that we shall prove (in section 5.3) that the F.I.O.'s under consideration transform WF_A^2 according to the bicanonical approximations of the respective canonical maps.

2. Results on the action of F.I.O.'s on wave front sets are only useful if they are supplemented by a calculus of composition of F.I.O.'s among themselves. For the operators from the first class the relevant results follow directly from the theory of standard analytic F.I.O.'s of Sato-Kawai-Kashiwara [1], but we should perhaps mention that almost all what we need also follows from Liess-Rodino [2] for the choice of weight function $\varphi = |\xi|$. This is particularily convenient, since for the operators from the second class we shall rely in part on the results from Liess-Rodino [2] anyway, since, as explained above, k-microlocalization is closely related to a form of G_φ-microlocalization.

Here we should add, that one could probably incorporate both classes of F.I.O.'s above into one single class of operators which operates between two regular involutive analytic manifolds Σ and Σ' of the same codimension and which are in arbitrary position. The reason why we have not tried to argue in this way is that then we could not have used the existing results on standard F.I.O.'s, or on F.I.O.'s in G_φ-classes. Moreover, we believe that the theory of two-microlocal F.I.O.'s becomes more transparent if one splits off from it everything what can already be achieved with standard F.I.O.'s. Indeed, it is this what one will try to do anyway in practical applications: simplify a given situation as much as possible using standard F.I.O.'s and then try a further simplification with the aid of the two-microlocal calculus in special coordinates.

5.1   Classical F.I.O.'s which leave $\xi' = 0$ invariant.

1. Consider X open in R^n, denote by $\Sigma = \{(y, \eta) \in T^*X, \eta' = 0\}$ and let $(y^0, \eta^0) \in \Sigma$, $(x^0, \xi^0) \in \Sigma$. Let moreover ω be a real analytic homogeneous nondegenerate phase function defined in a conic neighborhood of (x^0, η^0) and denote by χ the canonical transformation associated with ω. We assume that $\chi(y^0, \eta^0) = (x^0, \xi^0)$ and that Σ is left invariant under χ. The latter means in terms of ω that

$$\omega_{x'}(x, 0, \eta'') = 0 \text{ for all } x, \eta'' \text{ under consideration.}$$

We study in this section classical analytic F.I.O.'s (defined on microfunctions) associated with ω. Explicitly we assume that $a \in S_{|\xi|}^m(\Gamma)$ is a standard analytic symbol and consider an analytic F.I.O. associated with a with the aid of the formula

$$Au(x) = \int e^{i\omega(x,\eta)}a(x,\eta)h(\eta)\hat{u}(\eta)d\eta, u \in C_0^\infty(R^n), \qquad (5.1.1)$$

where $h \in C^\infty(R^n)$ is identically one if η is in some small conic neighborhood V of η^0 and $|\eta| \geq 1$, vanishes outside some other cone V', and has bounded derivatives of any order. Here V' is chosen so that for some suitable neighborhood U of x^0 we have $U \times V' \subset\subset \Gamma$, but other restrictions shall be put on V' in the sequel. Later on we shall also sometimes call A a F.I.O. of the "first kind". Such operators should be seen in relation to what we shall call F.I.O. of the "second kind" associated with a and ω, which we shall define by

$$\mathcal{F}(A^*v)(\eta) = (2\pi)^n h(\eta) \int e^{-i\omega(x,\eta)}a(x,\eta)v(x)\,dx$$

where h is as before. The factor $(2\pi)^n$ is for later convenience, in that it gives formally

$$A^*v(y) = \int\int h(\eta)e^{-i\omega(x,\eta) + i\langle y,\eta\rangle}a(x,\eta)v(x)\,dx\,d\eta.$$

We should mention that the need to distinguish between operators of the first and of the second kind appears since we shall not consider general F.I.O.'s of the form

$$\int e^{i\omega(x,y,\eta)}a(x,y,\eta)h(\eta)u(y)\,dy d\eta. \qquad (5.1.2)$$

Actually, in the present section, this does not seem to be a loss of generality, since for first order analytic microlocalization it is probably known that the operators from (5.1.2) can always be realized as operators of the first kind (5.1.1). (For C^∞-F.I.O.'s this is proved in Hörmander [1].)

Note that we have defined A in (5.1.1) only on C^∞- functions with compact support. What we gain in doing so is that the integral in (5.1.1) has a direct meaning, but the case when u is a general distribution with compact support is not much more difficult to treat. To make this clear let us recall briefly from Hörmander [1] how one would have to proceed in the general case. In fact, the best thing to do, is to introduce first the distribution kernel $K \in D'(U \times R^n)$ of A (formally) by

$$K(f) = \int e^{i\omega(x,\eta) - i\langle y,\eta\rangle}a(x,\eta)h(\eta)f(x,y)dx\,dy\,d\eta , f \in C_0^\infty(U \times R^n). \qquad (5.1.3)$$

Of course, integration in η may pose problems and the triple integral from (5.1.3) is to be regarded as an oscillatory integral in the sense from Hörmander [1]. By this we mean that

$$K(f) = \lim_{\varepsilon \to 0} K_\varepsilon(f),\ K_\varepsilon(f) = \int e^{i\omega(x,\eta) - i\langle y,\eta\rangle}a(x,\eta)\psi(\varepsilon\eta)h(\eta)f(x,y)dx\,dy\,d\eta,$$

$$(5.1.4)$$

where ψ is any function in $S(R^n)$ with $\psi(0) = 1$. It is not difficult to show that $K(f)$, as defined formally in (5.1.4), exists and has a well-defined meaning, in that it does not depend on the choice of ψ. In fact, both assertions follow from an alternative way to regularize the integral in (5.1.3), which is also described in Hörmander [1]. What one does is to consider, for some given $\kappa \in C_0^\infty$, which is identically one in a neighborhood of zero, the partial differential operator L

$$Lf = \kappa f + (1 - \kappa(\eta))b[\Sigma|\eta|^2((\partial\omega/\partial\eta_j) - y_j)(\partial/\partial\eta_j) + \Sigma(\partial\omega/\partial x_j)(\partial/\partial x_j) - \Sigma\eta_j(\partial/\partial y_j)]f,$$

where

$$b = i[|\eta|^2\Sigma((\partial\omega/\partial\eta_j) - y_j)^2 + \Sigma(\partial\omega/\partial x_j)^2 + |\eta|^2)]^{-1}.$$

It is clear that L has the property

$$L(e^{i\omega(x,\eta) - i\langle y,\eta\rangle}) = e^{i\omega(x,\eta) - i\langle y,\eta\rangle},$$

so that

$$K_\epsilon(f) = \int e^{i\omega(x,\eta) - i\langle y,\eta\rangle}(^tL)^k[\psi(\epsilon\eta)a(x,\eta)h(\eta)f(x,y)]dx\,dy\,d\eta,$$

for any natural number k. Fact is then that if k is large enough, then

$$\int e^{i\omega(x,\eta) - i\langle y,\eta\rangle}(^tL)^k[a(x,\eta)h(\eta)f(x,y)]dx\,dy\,d\eta \qquad (5.1.5)$$

is absolutely convergent and equal to $\lim_{\epsilon \to 0} K_\epsilon(f)$. In particular (5.1.4) das not depend on ψ and (5.1.5) does not depend on k if k is large enough. We have thus given a meaning to K from (5.1.3), and may associate an operator $A : C_0^\infty(R^n) \to D'(U)$ with it. Of course, A is just the operator given by (5.1.1). It is moreover interesting to note that A maps $C_0^\infty(R^n)$ actually into $C^\infty(U)$ and that it admits a natural extension to an operator $A : E'(R^n) \to D'(U)$. From the above it is now also clear that the best way to study A as an operator between $E'(R^n)$ and $D'(U)$ is via its distribution kernel K. Here we avoid this problem altogether by restricting our attention to the case of C_0^∞-functions. This is justified by the fact that in many cases in which one would like to apply the results from this chapter, one will dispose of alternative machineries in the case of the underlying C^∞- problem. What we gain in restricting our attention to (5.1.1) is greater transparency in the proofs and simpler notations. (Anyway, the arguments which one would need to treat the case of $u \in E'(R^n)$ are not fundamentally different from the ones employed here.)

To return now to our discussion of (5.1.1), let us note that it is not difficult to see (cf. e.g. Liess- Rodino [2]) that if h, h' are two functions as before, then

$$(x^0, \xi^0) \notin WF_A \int e^{i\omega(x,\eta)}a(x,\eta)(h - h')(\eta)\hat{u}(\eta)d\eta$$

and that

$$(y^0, \eta^0) \notin WF_A u \text{ implies } (x^0, \xi^0) \notin WF_A Au.$$

This shows that A determines a well-defined operator on microfunctions. It is the main content of this section that A operates well also on WF_A^2. Our first remark in this direction is that if A^1 and A^2 are two analytic F.I.O.'s which are both associated with some canonical transformations which leave Σ invariant, then so is $B = A^1 \circ A^2$. Moreover, it follows from the theory of standard analytic F.I.O.'s that $(x^0, \xi^0) \notin WF_A(B - A^1 \circ A^2)u$, $\forall u$. In particular

$$(x^0, \xi^0, \sigma^0) \notin WF_A^2(B - A^1 \circ A^2)u,$$

for any choice of σ^0. Our next result is

Proposition 5.1.1. *Let A be of form (5.1.1), assume that χ leaves $\eta' = 0$ invariant, consider $(y^0, \eta^0, \tau^0) \in N\Sigma$ and denote by χ' the tangent map $\chi' : N\Sigma \to N\Sigma$ associated with χ. If $(y^0, \eta^0, \tau^0) \notin WF_A^2 u$, it follows that $\chi'(y^0, \eta^0, \tau^0) \notin WF_A^2 Au$.*

We recall here that in the definition of WF_A^2 we had the possibility to make up our mind as to which class of sublinear functions we take for the conditions $|\xi'| \geq f(|\xi|)$. In the present chapter we shall make (to simplify terminology) the choice "$|\xi'| \geq c(1 + |\xi|^\delta)$" for some c and δ.

We see in particular in proposition 5.1.1 that the result is expressed in terms of the tangent map χ' of χ. Since χ itself is defined in terms of ω, it is useful to obtain information on χ' directly in terms of ω. We shall use the special coordinates and recall that

$$\chi'(y, \eta, \tau) = \begin{pmatrix} \chi_{(x)}(y, \eta) \\ \chi_{(\xi'')}(y, \eta) \\ \langle \text{grad}_{\eta'} \chi_{(\xi')}(y, \eta), \tau \rangle \end{pmatrix} \text{ if } (y, \eta) \in \Sigma, \tau \in N_{(y,\eta)}\Sigma.$$

Here in fact

$$\chi_{(\xi')}(y, \eta) = \omega_{x'}(x, \eta) = \omega_{x'}(\chi_{(x)}(y, \eta), \eta) \text{ if } (x, \xi) = \chi(y, \eta).$$

Since $\omega_{x'}(x, \eta) = 0$ if $\eta' = 0$ we conclude that

$$\text{grad}_{\eta'} \chi_{(\xi')}(y, \eta) = \omega_{x'\eta'}(x, \eta) \text{ if } \eta' = 0.$$

It is also useful to note that ω will have to be assumed to be nondegenerate, i.e. we shall assume that

$$\det \omega_{x,\eta}(x^0, \eta^0) \neq 0.$$

If $\omega_{x'}(x, 0, \eta'') = 0$ for all $(x, 0, \eta'') \in \Sigma$, we will have $\omega_{x'\eta''}(x, 0, \eta'') = 0$, so we must have

$$\det \omega_{x'\eta'}(x^0, \eta^0) \neq 0, \det \omega_{x''\eta''}(x^0, \eta^0) \neq 0.$$

The first of these conditions corresponds of course to the nondegeneracy of S in section 4.8. We also mention the following corollary of proposition 5.1.1:

Corollary 5.1.2. *If a is elliptic, then $(y^0, \eta^0, \tau^0) \notin WF_A^2 u$ is equivalent to $\chi'(y^0, \eta^0, \tau^0) \notin WF_A^2 Au$.*

(If a is elliptic, the operator A is called an elliptic F.I.O. It is standard that the inverse of an analytic elliptic F.I.O. is an analytic F.I.O. of the type from before associated with χ^{-1}. What is new in the corollary when compared with the proposition, therefore follows from an application of the proposition to the operator A^{-1}. At this moment we should note that in Liess- Rodino [2] it is only proved that A^{-1} can be realized as a F.I.O. of the second kind. If one does not want to go here beyond the theory of analytic F.I.O.'s as presented in Liess-Rodino loc. cit., it is at this moment necessary to supplement proposition 5.1.1 with the corresponding result for F.I.O.'s of the second kind. We shall do a related thing, for other reasons, in section 5.3.)

2. We prepare the proof of proposition 5.1.1 with three lemmas.

Lemma 5.1.3. *Denote $\sigma^0 = \omega_{x'\eta'}(x^0, 0, \eta^{0''})\tau^0$, $\xi^0 = \omega_x(x^0, 0, \eta^{0''})$, so that $\xi^{0'} = 0$. There is an open cone $\tilde{G}$ containing η^0 with the following properties:*

a) Let G^0 be an open cone containing $\eta^0 = (0, \eta^{0''})$. Then we can find $\varepsilon > 0$, $c > 0$ and a bineighborhood Λ of (ξ^0, σ^0) such that

$$|\xi - \omega_x(x, \eta)| \geq c(|\eta| + |\xi|) \text{ if } \xi \in \Lambda, \, \eta \in \tilde{G} \setminus G^0, \, |x - x^0| < \varepsilon. \tag{5.1.6}$$

b) Let G^1 be an open cone in R^d containing τ^0. Then we can find $\varepsilon > 0$, $c > 0$, an open cone G^0 containing η^0, and a bineighborhood Λ of (ξ^0, σ^0) such that

$$|\xi' - \omega_{x'}(x, \eta)| \geq c(|\eta'| + |\xi'|), \text{ if } \xi \in \Lambda, \eta \in G^0, \eta' \notin G^1, |x - x^0| < \varepsilon. \tag{5.1.7}$$

Proof of lemma 5.1.3. a) We denote by Z the halfray $\{t\eta^0; \, t > 0\}$. For fixed G^0 we can then find c_1, c_2, so that $\eta \notin G^0, \tilde{\eta} \in Z$, implies that either $|\eta'' - \tilde{\eta}''| \geq c_1(|\eta''| + |\tilde{\eta}''|)$, or at least $|\eta'| \geq c_2|\eta''|$. It is convenient to fix another constant c_3 and to distinguish the following two cases:

case I : $|\eta'| \leq c_3|\eta''|$, but $|\eta'' - \tilde{\eta}''| \geq c_1(|\eta''| + |\tilde{\eta}''|)$, $\forall \tilde{\eta} \in Z$, case II : $|\eta'| > c_3|\eta''|$.

It follows from the above that if $c_3 \leq c_2$, $\eta \notin G^0$, then we are always either in case I or in case II. We shall show that, shrinking c_3 possibly still further, (5.1.6) will hold in both cases.

Case I. We write for some $\tilde{\eta} \in Z$, which we determine later on,

$$|\xi'' - \omega_{x''}(x, \eta)| \geq -|\xi'' - \omega_{x''}(x^0, 0, \tilde{\eta}'')| + |\omega_{x''}(x^0, 0, \tilde{\eta}'') - \omega_{x''}(x^0, 0, \eta'')| -$$
$$-|\omega_{x''}(x^0, 0, \eta'') - \omega_{x''}(x^0, \eta)| - |\omega_{x''}(x^0, \eta) - \omega_{x''}(x, \eta)|.$$

Here we shall now have that

$$|\omega_{x''}(x^0, 0, \tilde{\eta}'') - \omega_{x''}(x^0, 0, \eta'')| \geq c_4|\eta'' - \tilde{\eta}''| \geq c_1 c_4(|\eta''| + |\tilde{\eta}''|)$$

for some c_4. Moreover, if Λ is chosen small, ξ is in Λ and $\tilde{\eta}$ is chosen suitably (of form $t\eta^0$, as it must be, since it is in Z), we obtain

$$|\xi'' - \omega_{x''}(x^0, 0, \tilde{\eta}'')| \leq c_1 c_4|\tilde{\eta}''|/4 \,, \; |\xi''| \sim |\tilde{\eta}''|.$$

Further, $|\omega_{x''}(x^0, 0, \eta'') - \omega_{x''}(x^0, \eta)| \leq c_5|\eta'| \leq c_5 c_3|\eta''| \leq c_1 c_4|\eta''|/4$ if c_3 is chosen small compared with c_1. Finally, $|\omega_{x''}(x^0, \eta) - \omega_{x''}(x, \eta)| \leq c_1 c_4|\eta''|/4$ if $|x - x^0| < \varepsilon$ and ε is small. Combination of these estimates solves case I.

Case II. We now have $|\eta'| \geq c_3|\eta''|$ for some c_3 fixed so that it fulfills the needs of case I. On the other hand, we have $|\omega_{x''}(x, \eta)| \leq c_6|\eta|$ and $|\omega_{x'}(x, \eta)| = |\omega_{x'}(x, \eta) - \omega_{x'}(x, 0, \eta'')| \geq c_7|\eta'|$ if $\eta \in \tilde{G}$, provided we choose $\tilde{G}$ suitably. It follows for $\eta \in \tilde{G}$ in this case, that $|\omega_{x'}(x, \eta)| \geq c_8|\omega_{x''}(x, \eta)|$. On the other hand, we may assume that $\xi \in \Lambda$ implies $|\xi'| \leq (c_8/2)|\xi''|$. We can conclude that $\xi \in \Lambda$, $|\eta'| \geq c_3|\eta''|$ implies $|\xi - \omega_x(x, \eta)| \geq c_9(|\xi| + |\omega_x(x, \eta)|)$, for some c_9. (Note that due to the inequalities for ξ and ω_x we cannot have $|\xi' - \omega_{x'}(x, \eta)| \leq c(|\xi'| + |\omega_{x'}(x, \eta)|)$ and $|\xi'' - \omega_{x''}(x, \eta)| \leq c(|\xi''| + |\omega_{x''}(x, \eta)|)$ simultaneously if c is very small, etc.)

Proof of lemma 5.1.3 b). We denote $Z' = \{\tilde{\eta}; \tilde{\eta} = t\eta^0 + t'\sigma^0\}$, $t > 0, t' > 0$, and write, for $\tilde{\eta} \in Z'$ to be determined later on,

$$|\xi' - \omega_{x'}(x, \eta)| \geq |\omega_{x'\eta'}(x^0, 0, \tilde{\eta}'')(\eta' - \tilde{\eta}')| - |\xi' - \omega_{x'\eta'}(x^0, 0, \tilde{\eta}'')\tilde{\eta}'| \qquad (5.1.8)$$
$$-|\omega_{x'\eta'}(x^0, 0, \tilde{\eta}'') - \omega_{x'\eta'}(x, 0, \eta'')||\eta'| - |\omega_{x'}(x, \eta) - \omega_{x'\eta'}(x, 0, \eta'')\eta'|$$

and estimate the terms which appear here one by one.

From the assumption that $\tilde{\eta} \in Z'$ it will follow that $|\omega_{x'\eta'}(x^0, 0, \tilde{\eta}'')(\eta' - \tilde{\eta}')| \geq c_{10}|\eta' - \tilde{\eta}'|$ for some c_{10}. Moreover, if G^1 is fixed, then $\eta' \notin G^1$ will imply $|\eta' - \tilde{\eta}'| \geq c_{11}(|\tilde{\eta}'| + |\eta'|)$, so we obtain

$$|\omega_{x'\eta'}(x^0, 0, \tilde{\eta}'')(\eta' - \tilde{\eta}')| \geq c_{12}(|\eta'| + |\tilde{\eta}'|).$$

Next we shrink Λ until we have

$$|\xi' - \omega_{x'\eta'}(x^0, 0, \tilde{\eta}'')\tilde{\eta}'| \leq (c_{12}/4)(|\tilde{\eta}'| + |\xi'|),$$

if $\tilde{\eta} \in Z'$ is suitable.

These are the conditions which we need in this part of the proof for Λ and $\tilde{\eta}$ and we also observe that if c_{12} has been chosen small, then $|\xi'| \sim |\tilde{\eta}'|$.

To treat our third term from the right hand side of (5.1.8), we shall now shrink ε and G^0 in the η'' variables around $\eta^{0\prime\prime}$. We may in fact assume that $\eta \in G^0$ implies $|\eta''/|\eta''| - \eta^{0\prime\prime}/|\eta^{0\prime\prime}||\leq c_{13}$, where c_{13} and ε have been fixed previously so that

$$|\omega_{x'\eta'}(x^0, 0, \tilde{\eta}'') - \omega_{x'\eta'}(x, 0, \eta'')| \leq c_{12}/4, \text{ if } |x - x^0| \leq \varepsilon \text{ and } |\eta''/|\eta''| - \tilde{\eta}''/|\tilde{\eta}''|| \leq \varepsilon.$$

Finally, to treat the last term from the left hand side of (5.1.8), we shrink G^0 around $\eta' = 0$ (in the sense that we assume that $|\eta'|/|\eta''|$ becomes suitably small if $\eta \in G^0$). From Taylor expansion of order one in the variables η' we obtain then

$$|\omega_{x'}(x, \eta) - \omega_{x'\eta'}(x, 0, \eta'')\eta'| \leq c_{14}|\eta'|^2 \max | \operatorname{grad}_{\eta'}\omega_{x'\eta'}(x, \theta', \eta'')|,$$

where the max is for θ' in the segement $[0, \eta']$. If G^0 is sufficiently narrow around $\eta' = 0$ we will then have that $|(\theta', \eta'')| \sim |\eta|$, and shrinking it still further around $\eta' = 0$ (if necessary) it follows that

$$c_{14}|\eta'|^2 \max | \operatorname{grad}_{\eta'}\omega_{x'\eta'}(x, \theta', \eta'')| \leq c_{15}|\eta'|^2/|\eta| \leq c_{12}|\eta'|/4.$$

Combination of all these estimates, gives (5.1.7).

3. Our next lemma is similar to lemma 4.1.3, but refers to a geometrically simpler situation.

Lemma 5.1.4. *Let K be compact in R_x^n. Then there is a conic neighborhood V' of η^0 and constants $C_i > 0$ so that $\eta \in V'$ together with $|\xi'| \leq C_1|\eta'|$, or together with $|\xi'| > C_2|\eta'|$, implies*

$$|\xi' - \omega_{x'}(x, \eta)| \geq C_3(|\xi'| + |\eta'|), \text{ for all } x \in K. \tag{5.1.9}$$

Proof. We may assume that $\eta' \in V'$ implies $C_4|\eta'| \leq |\omega_{x'}(x, \eta)| \leq C_5|\eta'|$ for $x \in K$. It follows that $|\xi' - \omega_{x'}(x, \eta)| \geq |\omega_{x'}(x, \eta)| - |\xi'| \geq C_4|\eta'| - |\xi'|$, which gives (5.1.9) in the case $|\xi'| < C_1|\eta'|$ if C_1 is small. In a similar way $|\xi' - \omega_{x'}(x, \eta)| \geq |\xi'| - |\omega_{x'}(x, \eta)| \geq |\xi'| - C_5|\eta'|$ gives (5.1.9) when $|\xi'| > C_2|\eta'|$ and C_2 is large.

Our final lemma is:

Lemma 5.1.5. *Let $y^0 = \omega_\eta(x^0, 0, \eta^{0\prime\prime})$ and fix $\tilde{c}$. Then there is $\varepsilon > 0$ and an open cone $G^0 \subset R^n$ which contains η^0 so that*

$$|y^0 - \omega_\eta(x, \eta)| \leq \tilde{c}, \quad \text{if } |x - x^0| < \varepsilon, \ \eta \in G^0.$$

Indeed,

$$
\begin{aligned}
|y^0 - \omega_\eta(x, \eta)| \ \leq \ & |\omega_\eta(x^0, 0, \eta^{0\prime\prime}) - \omega_\eta(x^0, 0, \eta'')| + \qquad\qquad (5.1.10) \\
& |\omega_\eta(x^0, 0, \eta'') - \omega_\eta(x, 0, \eta'')| + |\omega_\eta(x, 0, \eta'') - \omega_\eta(x, \eta)|.
\end{aligned}
$$

The first term from the right hand side of (5.1.10) becomes here smaller than $\tilde{c}/3$ for $\eta \in G^0$ if we shrink G^0 in the variables η'' around $\eta^{0\prime\prime}$. The second term is smaller than $C|x - x^0|$ and becomes therefore smaller than $\tilde{c}/3$ if $|x - x^0| < \varepsilon$ and ε is sufficiently small. Finally, the last term can be estimated by $C|\eta'|/|\eta|$, so also this term becomes less than $\tilde{c}/3$ if we shrink G^0 in the η' variables.

4. Proof of proposition 5.1.1. We choose $\delta > 0$ and open cones Γ^0, Γ^1 so that $\eta^0 \in \Gamma^0 \subset R^n$, $\tau^0 \in \Gamma^1 \subset R^d$, and so that we can find a bounded sequence $u_k \in E'(R^n)$ with the following properties:

a) $\operatorname{supp} u_k \subset \{y; |y - y^0| \leq 2\delta\}$, $\quad u_k = u$ for $|y - y^0| < \delta$,

b) there are $c, c', \delta < 1$ so that

$$|\hat{u}_k(\eta)| \leq c(ck/|\eta'|)^k \ \text{ if } \eta \in \Gamma^0, \eta' \in \Gamma^1, |\eta'| > c'(1 + |\eta|^\delta).$$

We then set $(x^0, \xi^0, \sigma^0) = \chi'(y^0, \eta^0, \tau^0)$, i.e. ,

$$\xi^0 = \omega_x(x^0, 0, \eta^{0\prime\prime}),$$

$$\sigma^0 = \omega_{x'\eta'}(x^0, 0, \eta^{0\prime\prime})\tau^0,$$

$$y^0 = \omega_\eta(x^0, 0, \eta^{0\prime\prime}).$$

What we want to show is that we can find a bounded sequence $v_k \in E'(R^n)$ and a bineighborhood Λ of (ξ^0, σ^0) so that

c) $v_k = Au$ for $|x - x^0| < c_1$,

d) $|\hat{v}_k(\xi)| \leq c_2(c_2 k/|\xi'|)^{k-c_3}$ if $\xi \in \Lambda$, $|\xi'| \geq c_3(1 + |\xi|^\delta)$.

As is customary, we shall look for v_k in the form $v_k = f_k Au$ for some C^∞ functions which satisfy

$$|D^\alpha f_k(x)| \leq c_4^{|\alpha|+1} k^{|\alpha|} \text{ for } |\alpha| \leq k \qquad (5.1.11)$$

and which vanish for all x with $|x - x^0| > c_5$ for some constant c_5. It is no loss of generality if we assume $c_5 = 2c_1$. The main restriction on the f_k is then formulated in terms of c_1.

Let us now also fix k and choose a finite number of functions $\rho_i \in C_0^\infty$, $i = 0, ..., N$, with the following properties

e) supp $\rho_0 \subset \{y; |y - y^0| < \delta\}$,

f) $y \in$ supp ρ_i, $i > 0$, implies $|y - y^0| > 2\delta/3$,

g) $\Sigma \rho_i \equiv 1$ on supp u,

h) $|D^\alpha \rho_i(y)| \leq c_6^{|\alpha|+1} k^{|\alpha|}$ for $|\alpha| \leq k$,

for some constant c_6 which does not depend on k. (However, it is clear that the ρ_i depend on k and that c_6 depends on δ.) From b) it will then follow that

$$|\mathcal{F}(\rho_0 u)(\eta)| \leq c_7(c_7 k/|\eta'|)^k, \text{ if } \eta \in \Gamma^0, \eta' \in \Gamma^1, |\eta'| > c'(1 + |\eta|^\delta) \, .$$

It is now convenient to write $\mathcal{F}(f_k Au)$ in the form $\sum_{i \geq 0} I_i$, where $I_i = \mathcal{F}(f_k A\rho_i u)$. Explicitly we thus have that

$$I_i(\xi) = \int e^{-i\langle x, \xi\rangle + i\omega(x,\eta) - i\langle y,\eta\rangle} f_k(x)a(x,\eta) \, h(\eta)\rho_i(y)u(y)dyd\eta dx,$$

as an oscillatory integral: we assume for simplicity that u is a function. Here h is a C^∞ cut-off function as in the beginning of this section and we may assume that $\eta \in$ supp h implies $\eta \in V'$ where V' is the cone from lemma 5.1.4.

To estimate the I_i, we shall now distinguish a number of cases, according to which i we have and where η lies. In the following discussion, ξ will be fixed with $|\xi'| \geq c_3(1 + |\xi|^\delta)$ and will lie in some suitable bineighborhood Λ of (ξ^0, σ^0). In fact, while estimating the I_i, several conditions on c_1 and Λ will be imposed. Note then e.g., that shrinking c_1 will increase the constant in (5.1.11). However, the choices of c_1, Λ, will depend only on geometric conditions, whereas the value of the constant c_4 is not changing the geometry of our problem.

α) In this part of the proof we study the contribution of the regions $C_1|\eta'|/2 \geq |\xi'|$, respectively $2C_2|\eta'| \leq |\xi'|$, where C_1, C_2 are from lemma 5.1.4, which we apply for $K = \{x; |x - x^0| \leq 2c_5\}$. Note that K contains supp f_k. The idea is here to integrate partially in x'. Since in other regions we shall also have to use partial integrations in η, cut-off to this region must be done carefully. In fact, we use some function $h_k \in C^\infty(R^n)$ so that

$$|\partial^\alpha h_k(\eta)| \leq c_8^{|\alpha|+1} k^{|\alpha|}/(|\eta'| + 1)^{|\alpha|} \text{ if } |\alpha| \leq k,$$

(where c_8 must not depend on k),

$$h_k(\eta) = 1 \text{ if } |\xi'| > 2C_2|\eta'| + 1 \text{ or } |\xi'| + 1 \leq C_1|\eta'|/2,$$

and

$$h_k(\eta) = 0 \text{ if } |\xi'| > C_1|\eta'| \text{ or } |\xi'| < C_2|\eta'|.$$

Denote then by

$$I_i' = \int \int \int e^{-i\langle x, \xi \rangle + i\omega(x,\eta) - i\langle y, \eta \rangle} f_k(x)a(x,\eta)h(\eta)h_k(\eta)\rho_i(y)u(y)dy d\eta dx.$$

As mentioned above we can estimate I_i' (for all i) by partial integration in the x'- variables. We recall here briefly the rather classical argument of how this is done. Starting point is that the operator

$$L = | - \xi' + \omega_{x'}(x,\eta)|^{-2} \langle -\xi' + \omega_{x'}(x,\eta), grad_{x'} \rangle$$

satisfies

$$Le^{-i\langle x, \xi \rangle + i\omega(x,\eta)} = e^{-i\langle x, \xi \rangle + i\omega(x,\eta)}.$$

We also note that here $| - \xi' + \omega_{x'}(x,\eta) | \geq (|\xi'| + |\eta'|)$ as a consequence of lemma 5.1.4. It follows now at first that

$$I_i' = \int e^{-i\langle x, \xi \rangle + i\omega(x,\eta) - i\langle y, \eta \rangle} (^tL)^k(f_k(x)a(x,\eta)h(\eta)h_k(\eta)\rho_i(y)u(y))dy d\eta dx$$

where ^tL is the formal adjoint of L. The main point is that

$$|(^tL)^k(f_k(x)a(x,\eta))| \leq c_9^{k+1} k^k (|\xi'| + |\eta'|)^{-k} \tag{5.1.12}$$

for some constant c_9 which does not depend on k. The technical trick how to prove (5.1.12) is explained in Hörmander [2] and we omit the details. (A situation very similar to the preceding one also appears in Liess-Rodino [2]). It is now clear that the desired estimate for I_i' follows after an integration.

In view of the estimate above we may assume in all what follows that $|\xi'| \sim |\eta'|$. The next case is then very easy.

β) $i = 0$, $\eta \in \Gamma^0$, $\eta' \in \Gamma^1$. Here we have the estimates for $\mathcal{F}(\rho_0 u)$. Since $|\xi'| \sim |\eta'|$ this immediately gives an estimate of the desired type.

In the remaining part of the proof we shall always use partial integrations in a way similar to that in part α). The situation is particularily easy in case

γ) $i = 0, \eta \in \Gamma^0, \eta' \notin \Gamma^1$. In this case we can base the argument on lemma 5.1.3 and argue exactly as in part α) using partial integrations in x'. (Here we may need to choose c_1 small.)

δ) In this part of the proof we assume $i > 0$ and fix some conic neighborhood G^0 of η^0 so that lemma 5.1.5 becomes valid. Roughly speaking we want to study here the contributions from $\eta \in G^0$. To do so, we consider cut-off functions h'_k so that $h'_k(\eta) = 1$, if $|\eta| \geq 2$ lies in some still smaller conic neighborhood G'^0 of η^0, so that supp $h'_k \subset G^0$, and so that

$$|D^\alpha h'_k(\eta)| \leq c_{11}^{|\alpha|+1} k^{|\alpha|} / |\eta|^{|\alpha|}, \text{ if } |\alpha| \leq k.$$

We may assume that $h'_k h = h'_k$ and want to estimate

$$I''_i = \int e^{-i\langle x,\xi\rangle + i\omega(x,\eta) - i\langle y,\eta\rangle} f_k(x) a(x,\eta) h(\eta) h_k(\eta) h'_k(\eta)$$
$$\rho_i(y) u(y) dy d\eta dx.$$

Here we apply lemma 5.1.5 : if $|x - x^0| < \delta$ for some small δ, then it will follow that

$$|y - \omega_\eta(x,\eta)| \geq \varepsilon/2 \text{ if } y \in \text{ supp } \rho_i, i > 0, \eta \in G^0.$$

We can then integrate partially in η, starting from the operator

$$L' = |y - \omega_\eta(x,\eta)|^{-2}\langle y - \omega_\eta(x,\eta), grad \eta\rangle,$$

and continue as in part α). We omit further details.

ε) The last case is when $i > 0$ and $\eta \notin G'^0$, or $i = 0$ and $\eta \notin \Gamma^0$. Here we integrate partially in x, using lemma 5.1.3. The situation is very similar to that in part α), respectively γ) and we omit the details.

5.2   Invariant definition of $WF^2_{A,\Sigma}$

1. Let Σ be a regular homogeneous involutive analytic manifold defined in a conic neighborhood Γ of (y^0, η^0). We can then find an analytic homogeneous canonical change of coordinates χ, which is defined in a conic neighborhood of (y^0, η^0), $\chi(y^0, \eta^0) = (x^0, \xi^0)$,

such that $\chi(\Sigma) = \Sigma'$ is given by $\xi' = 0$. We may then consider a standard analytic elliptic F.I.O. Q associated with χ. (Cf. section 5.1.) Let us also denote by χ' the normal map $\chi' : N\Sigma \to N\Sigma'$ and consider $(y^0, \eta^0, \tau^0) \in N\Sigma$.

Definition 5.2.1. *We shall say by definition that* $(y^0, \eta^0, \tau^0) \notin WF_{A,\Sigma}^2 u$, *if and only if*

$$\chi'(y^0, \eta^0, \tau^0) \notin WF_A^2 Qu.$$

The subset $WF_{A,\Sigma}^2$ *in* $N\Sigma$ *defined in this way shall be called the second analytic wave front set with respect to* Σ.

2. In order to show that $WF_{A,\Sigma}^2$ is well-defined, we must check that $(y^0, \eta^0, \tau^0) \notin WF_{A,\Sigma}^2 u$ does not depend on which canonical transformation χ we choose to obtain $\chi(\Sigma) \subset \{\xi' = 0\}$ and which elliptic analytic F.I.O. Q associated with χ is chosen to test if $(x^0, \xi^0, \sigma^0) \in WF_A^2 Qu$ or not. This is essentially a consequence of results from section 5.1, but in order to make clear that we also use the theory of standard analytic F.I.O.'s, we explain it in some detail. Also note that if we can prove that $WF_{A,\Sigma}^2$ is well-defined in the above sense, then it follows that if Σ is already in the form $\{\xi; \xi' = 0\}$, then $WF_{A,\Sigma}^2$ coincides with the second wave front set WF_A^2 defined as in section 2.1 with respect to the bihomogeneous structure given by $M_0 = R^n$, $M_1 = \{\xi; \xi_{d+1} = \cdots = \xi_n = 0\}$.

3. The problem is then this: Let χ and κ be two homogeneous analytic canonical transformations defined in a conic neighborhood Γ of (y^0, η^0) so that for some conic neighborhood V of (x^0, ξ^0),

$$\chi(\Sigma \cap \Gamma) \cap V = \{\xi' = 0\} \cap V,$$

$$\kappa(\Sigma \cap \Gamma) \cap V = \{\xi' = 0\} \cap V,$$

and let Q and R be elliptic analytic F.I.O.'s associated with χ and κ respectively. Also assume that $\chi(y^0, \eta^0) = \kappa(y^0, \eta^0) = (x^0, \xi^0)$. If we denote Qu by v and Ru by w, we must then show that

$$\chi'(y^0, \eta^0, \tau^0) \notin WF_A^2 v \text{ is equivalent with } \kappa'(y^0, \eta^0, \tau^0) \notin WF_A^2 w. \tag{5.2.1}$$

To show this, let us look at the analytic F.I.O. $T = Q \circ R^{-1}$, which is associated with the canonical transformation $\mu = \chi \circ \kappa^{-1}$ and for which we have $v = Tw$ in the microfunction sense. We denote the variables in $\chi(\Sigma \cap \Gamma)$ by (x, ξ) and those in $\kappa(\Sigma \cap \Gamma)$ by (z, θ). It follows that μ transforms (z, θ) into (x, ξ) and maps Σ' into Σ'. Accordingly we write $\chi'(y^0, \eta^0, \tau^0) = (x^0, \xi^0, \sigma^0)$, $\kappa'(y^0, \eta^0, \tau^0) = (z^0, \theta^0, \nu^0)$ and we have $(x^0, \xi^0, \sigma^0) = \mu'(z^0, \theta^0, \nu^0)$. Moreover we can write

$$Tu = \int e^{i\omega(x,\theta)} a(x,\theta) h(\theta) \hat{u}(\theta) d\theta, \tag{5.2.2}$$

for some analytic nondegenerate homogeneous phase function ω, some elliptic analytic symbol a and some C^∞ function h which is identically one in a conic neighborhood of θ^0 and has bounded derivatives of any order. Until here we have used the standard theory of analytic F.I.O.'s, i.e., Q, R, R^{-1}, T are to be regarded as operators in microfunctions defined near (y^0, η^0) and (z^0, θ^0) respectively. Explicitely, as a relation in distributions, $T = Q \circ R^{-1}$ means then that

$$(x^0, \xi^0) \notin WF_A(Q \circ R^{-1} - T)u,$$

which gives in particular

$$(x^0, \xi^0, \sigma^0) \notin WF_A^2(Q \circ R^{-1})u.$$

The relation (5.2.1) thus comes to the fact that $(z^0, \theta^0, \nu^0) \notin WF_A^2 w$ is equivalent to $\mu'(z^0, \theta^0, \nu^0) \notin WF_A^2 T w$. That this is indeed the case, is a consequence of section 5.1.

4. At this moment it is now possible to give a number of "invariant" statements concerning WF_A^2. We only give here two examples.

Definition 5.2.2. *Consider Σ as before and consider $u \in D'(U)$. We call 2-microsupport of u the set*

$$sing\ supp_A^2 u = \{(x, \xi) \in \Sigma;\ there\ is\ \sigma \in N\Sigma_{(x,\xi)}\ so\ that\ (x, \xi, \sigma) \in WF_A^2 u\}.$$

Thus, the second singular support is by definition the projection on Σ of the second analytic wave front.

Proposition 5.2.3. *(Invariant form of the micro-Holmgren theorem.) Let V be an open connected set in a leaf of the foliation of Σ and assume that*

$$sing\ supp_A^2 u \cap V = \emptyset.$$

It follows that $WF_A u \cap V$ is either equal to V or empty.

5.3   F.I.O. associated with phase functions which live on bineighborhoods

1. In section 5.1 we have described how classical analytic F.I.O.'s which leave $\{\eta' = 0\}$ invariant operate on the second analytic wave front set. A result similar to proposition 5.1.1 is valid for F.I.O. which are associated with phase functions which are defined only

on some bineighborhood of $(x^0, \eta^0, \tau^0) \in N\Sigma$, but which are of the form from section 4.8. We assume then as in that section that ω is of form

$$\omega(x, \eta) = S(x, \eta) + \langle x'', \eta'' \rangle,$$

where S is a real-analytic, real-valued function defined on a bineighborhood $\Gamma = U \times G$ of (x^0, η^0, τ^0), where U is a neighborhood of x^0 and $G \subset R^n_\xi$ is an open bineighborhood of (η^0, τ^0), $\eta^{0\prime} = 0$. Moreover we assume that S is positively (1,1)-bihomogeneous and that it satisfies the conditions from section 4.8. Note that starting from second derivatives, ω behaves like a symbol in $S^1_{|\eta'|}$, i.e.,

$$|\partial_x^\alpha \partial_\eta^\beta \omega(x, \eta)| \le c^{|\alpha|+|\beta|+1} \alpha! \beta! \, |\eta'|^{-|\beta|} \text{ if } |\alpha| + |\beta| \ge 2,$$

provided $(x, \eta) \in U' \times G'$, where $U' \subset\subset U$ and G' is an open bineighborhood of (η^0, τ^0) with $G' \subset\subset G$. If $a \in S^m_{|\eta'|}(U \times [G \cap \{\eta; |\eta'| > c(1 + |\eta|^\delta)\}])$ is given we shall now introduce a F.I.O. by

$$Au = \int e^{i\omega(x, \eta)} a(x, \eta) h(\eta) \hat{u}(\eta) d\eta, \quad u \in C_0^\infty. \tag{5.3.1}$$

Here $h(\eta) = 1$ for $\eta \in G'$, $|\eta'| \ge 3c(1 + |\eta|^\delta)$, respectively $h(\eta) = 0$ if $\eta \notin G''$ or if $|\eta'| \le 2c(1 + |\eta|^\delta)$, for some previously fixed bineighborhoods $G' \subset\subset G'' \subset\subset G$ of (η^0, τ^0). We also assume that all derivatives of h are bounded.

As in Liess-Rodino [2] or section 5.1 we shall call A a F.I.O. of the first kind. We may also associate with a and ω a F.I.O. A^* of the "second kind", which is almost the formal adjoint of A. Explicitly, the Fourier transform of A^* is defined by

$$\mathcal{F}(A^*v)(\eta) = (2\pi)^n h(\eta) \int e^{-i\omega(x, \eta)} a(x, \eta) v(x) dx, \text{ if } v \in C_0^\infty(U), \tag{5.3.2}$$

where h is as before.

2. The operators A and A^* have all the properties which one expects some F.I.O.'s to have. In fact, these operators are very close to the F.I.O.'s in G_φ-classes, considered under appropriate conditions for ω and φ in Liess-Rodino [2], provided one takes $\varphi = |\eta'|$ there. Moreover, even though one of the conditions which relate ω with φ in Liess-Rodino [2] is not formally satisfied, it is still possible to obtain the main properties of the operators in (5.3.1),(5.3.2) from the corresponding results in Liess-Rodino, loc.cit. As far as the mapping properties of our operators are concerned, we shall not rely however on results from Liess-Rodino, loc.cit., since the situation is also very close to the one from section 5.1.

Proposition 5.3.1. *Let χ^* be the canonical transformation associated with $\omega = S + \langle x'', \eta'' \rangle$ and let χ be its bihomogeneous approximation. Consider $(y^0, \eta^0, \tau^0) \in N\Sigma$ and* *denote by* $(x^0, \xi^0, \sigma^0) = \chi(y^0, \eta^0, \tau^0)$. *Let further A be a F.I.O. of the first kind and A^* a F.I.O. operator of the second kind associated with ω. Then*

$$(y^0, \eta^0, \tau^0) \notin WF_A^2 u \text{ implies } (x^0, \xi^0, \sigma^0) \notin WF_A^2 Au, \qquad (5.3.3)$$

and

$$(x^0, \xi^0, \sigma^0) \notin WF_A^2 v \text{ implies } (y^0, \eta^0, \tau^0) \notin WF_A^2 A^* v. \qquad (5.3.4)$$

3. As stated above, the proof of proposition 5.3.1 is very similar to that of proposition 5.1.1. We start by deducing results of a geometric nature which are similar to the lemmata 5.1.3, 5.1.4, 5.1.5.

Lemma 5.3.2. *Consider $K \subset U$ compact and let G' be an open bineighborhood of (η^0, τ^0). If G' is chosen small enough, then we can find $c > 0$ so that*

$$|\omega_x(x, \eta^1) - \omega_x(x, \eta^2)| \geq c|\eta^1 - \eta^2|, \text{ if } \eta^i \in G. \qquad (5.3.5)$$

Proof. Using Taylor expansion and assuming that G' is convex we obtain that

$$|\omega_{x'}(x, \eta^1) - \omega_{x'}(x, \eta^2)| \geq c' \sum_{j \leq d} [|S_{x_j \eta'}(x, \tilde{\eta}^j)(\eta^{1'} - \eta^{2'}) - S_{x_j \eta''}(x, \tilde{\eta}^j)(\eta^{1''} - \eta^{2''})|], \quad (5.3.6)$$

for some $\tilde{\eta}^j \in [\eta^1, \eta^2]$, respectively that

$$\omega_{x''}(x, \eta^1) - \omega_{x''}(x, \eta^2) = \eta^{1''} - \eta^{2''} + S_{x''}(x, \eta^1) - S_{x''}(x, \eta^2) = \qquad (5.3.7)$$

$$\eta^{1''} - \eta^{2''} + O(|\eta^{1'} - \eta^{2'}|) + O(\max(|\tilde{\eta}'|/|\tilde{\eta}''|)|\eta^{1''} - \eta^{2''}|),$$

where the maximum is for all $\tilde{\eta}$ in the segement $[\eta^1, \eta^2]$. Shrinking G' if necessary, it follows from 5.3.7 that

$$|\omega_{x''}(x, \eta^1) - \omega_{x''}(x, \eta^2)| \geq c''|\eta^{1''} - \eta^{2''}|$$

for some $c'' > 0$ if $|\eta^{1''} - \eta^{2''}| \geq c'''|\eta^{1'} - \eta^{2'}|$ and c''' is large. The proof will therefore come to an end if we can show that

$$|\omega_{x'}(x, \eta^1) - \omega_{x'}(x, \eta^2)| \geq c|\eta^{1'} - \eta^{2'}| \qquad (5.3.8)$$

if we assume that $|\eta^{1''} - \eta^{2''}| \leq c'''|\eta^{1'} - \eta^{2'}|$ and that G' is small. To prove this we use (5.3.6) and the fact that

$$S_{x_j \eta''}(x, \tilde{\eta}^j)(\eta^{1''} - \eta^{2''}) = O(|\tilde{\eta}'^j|/|\tilde{\eta}''^j|)|\eta^{1''} - \eta^{2''}| = O(|\tilde{\eta}'^j|/|\tilde{\eta}^{j''}|)|\eta^{1'} - \eta^{2'}|$$

for the η^1, η^2 for which we argue now and $j \leq d$. Here of course $|\tilde{\eta}^{j'}|/|\tilde{\eta}^{j''}|$ will become as small as we please if G' is shrinked suitably. Moreover, we can argue as in section 4.8 and obtain that

$$\sum_{j \leq d} |S_{x_j \eta'}(x, \tilde{\eta}^j)(\eta^{1'} - \eta^{2'})| \geq c|\eta^{1'} - \eta^{2'}|,$$

and (5.3.8) follows.

Lemma 5.3.3. *a) Let G^0 be an open cone containing $\eta^0 = (0, \eta^{0\prime\prime})$ and consider a bineighborhood $\tilde{G} \subset\subset G$ of (η^0, τ^0). Then we can find $\varepsilon > 0, c > 0$ and a bineighborhood Λ of (ξ^0, σ^0) such that*

$$|\xi - \omega_x(x, \eta)| \geq c(|\eta| + |\xi|) \text{ if } \xi \in \Lambda, \eta \in \tilde{G} \setminus G^0, |x - x^0| < \varepsilon. \tag{5.3.9}$$

b) Let G^1 be an open cone in R^d containing τ^0. Then we can find $\varepsilon > 0$, $c > 0$, an open cone G^0 containing η^0 and a bineighborhood Λ of (ξ^0, σ^0) such that

$$|\xi - \omega_x(x, \eta)| \geq c(|\eta'| + |\xi'|), \text{ if } \xi \in \Lambda, \eta \in G^0, \eta' \notin G^1, |x - x^0| < \varepsilon. \tag{5.3.10}$$

Proof. We fix a small bineighborhood Γ of (η^0, τ^0). It follows from section 4.8 that $\omega_x(x^0, \Gamma)$ contains a bineighborhood of (ξ^0, σ^0). (Cf. section 4.8.) It is then no loss of generality to assume that $\Lambda \subset \omega_x(x^0, \Gamma)$. We may then assume that $\xi = \omega_x(x^0, \tilde{\eta})$ for some $\tilde{\eta} \in \Gamma$. Also recall from section 4.8 that $|\xi| \sim |\tilde{\eta}|$, $|\xi'| \sim |\tilde{\eta}'|$. The main thing is then to estimate $|\omega_x(x^0, \tilde{\eta}) - \omega_x(x, \eta)|$ from below when $\tilde{\eta}$ is in a small bineighborhood of (η^0, τ^0), $|x - x^0| < \varepsilon$ and $\eta \notin G^0$ in part a), respectively $\eta \in G^0$ but $\eta' \notin G^1$ in part b).
Here we observe that $|\omega_x(x^0, \tilde{\eta}) - \omega_x(x, \tilde{\eta})| \leq c|x - x^0||\tilde{\eta}'|$, which will be a negligible term if $|x - x^0| \leq \varepsilon$ and ε is small. Let us next observe that in view of lemma 5.3.2 $|\omega_x(x, \tilde{\eta}) - \omega_x(x, \eta)| \geq c_1|\tilde{\eta} - \eta|$ for some c_1. It remains now to observe that $\tilde{\eta} \in \Gamma$, $\eta \notin G^0$ will imply $|\eta - \tilde{\eta}| \geq c_2(|\eta| + |\tilde{\eta}|)$, whereas $\tilde{\eta} \in \Gamma$, $\eta \in G^0$, $\eta' \notin G^1$ will imply $|\eta' - \tilde{\eta}'| \geq c_3(|\eta'| + |\tilde{\eta}'|)$ for some c_2, c_3 if Λ and Γ have been chosen small. This gives the lemma.

Lemma 5.3.4. *Let $\omega = S(x, \eta) + \langle x'', \eta'' \rangle$ and let χ be the bihomogeneous approximation of χ^*. Denote by $(x^0, \xi^0, \sigma^0) = \chi(y^0, \eta^0, \tau^0)$. Also fix c_1. Then we can find $\varepsilon > 0$ and a bineighborhood Γ of (η^0, τ^0) so that*

$$|y^0 - \omega_\eta(x, \eta)| \leq c_1 \text{ if } \eta \in \Gamma, |x - x^0| \leq \varepsilon.$$

Proof. The relation between (x^0, ξ^0, σ^0) and (y^0, η^0, τ^0) is (cf. the construction of bihomogeneous approximations)

$$y^{0\prime} = S_{\eta'}(x^0, \tau^0, \eta^{0\prime\prime}), y^{0\prime\prime} = x^{0\prime\prime}, \sigma^0 = S_{x'}(x^0, \tau^0, \eta^{0\prime\prime}), \xi^{0\prime\prime} = \eta^{0\prime\prime}.$$

It follows that

$$\begin{aligned}
|y^0 - \omega_\eta(x, \eta)| &\leq |y^0 - \omega_\eta(x^0, \eta)| + |\omega_\eta(x, \eta) - \omega_\eta(x^0, \eta)| \\
&\leq |S_{\eta'}(x^0, \tau^0, \eta^{0\prime\prime}) - S_{\eta'}(x^0, \eta)| + |S_{\eta''}(x^0, \eta)| + c|x - x^0| \\
&\leq |S_{\eta'}(x^0, \tau^0, \eta^{0\prime\prime}) - S_{\eta'}(x^0, \eta)| + c|\eta'|/|\eta''| + c|x - x^0|.
\end{aligned}$$

Here $c|\eta'|/|\eta''|$ is for $\eta \in \Gamma$ smaller than $c_1/3$, if Γ is shrinked around $\eta' = 0$, $c|x - x^0| <$ $c_1/3$ if $\varepsilon < c_1/(3c)$, and $S_{\eta'}$ is bihomogeneous of bidegree $(0,0)$ and continuous. If Γ is shrinked, we can therefore also make $|S_{\eta'}(x^0, \tau^0, \eta^{0''}) - S_{\eta'}(x^0, \eta)|$ smaller than $c_1/3$.

Lemma 5.3.5. *Let K be a compact in U. There are constants $C_i > 0$ so that $|\eta'| <$ $C_1|\eta''|$ together with $|\xi'| < C_2|\eta'|$, or together with $|\xi'| > C_3|\eta'|$, implies*

$$|\xi' - \omega_{x'}(x,\eta)| \geq C_4(|\xi'| + |\eta'|), \forall x \in K. \tag{5.3.11}$$

The proof is identical with that of lemma 5.1.4.

Proof of proposition 5.3.1

(5.3.3). The proof of (5.3.3) is almost a repetition of the proof of proposition 5.1.1. The main differences are the following:

a) Starting from second derivatives, ω behaves like a function in $S^1_{|\eta'|}$. (This simplifies the situation.)

b) $a \in S^m_{|\eta'|}$ instead of $a \in S^m_{|\eta|}$.

c) The domain on which we integrate in η is smaller. Accordingly, there is less space for cut-offs. This affects only the parts of the argument when we integrate partially in η. In fact, here the cut-off functions h_k will only satisfy the estimates

$$|D^\alpha h_k(\eta)| \leq c^{|\alpha|+1} k^{|\alpha|} |\eta'|^{-|\alpha|} \text{ for } |\alpha| \leq k.$$

We omit further details.

Proof of (5.3.4). One may obtain (5.3.4) from (5.3.3) if one also uses the results from section 5.4 below. Another possibility is to argue in a way similar to the arguments from (5.3.3). It may then be convenient to study the kernel of A^* first: this is the way in which the statement corresponding to (4.3.4) in general G_φ-classes is proved in Liess-Rodino [2]. Here we describe an argument based on proposition 2.1.15 which takes advantage of the fact that in the definition of A^*v we actually arrive first at the Fourier transform of A^*v and compute from this A^* itself. We content ourselves with a brief description of the idea of the argument. The first thing is then to consider a partition of unity on supp v: we shall in fact fix k and consider $f_i \in C_0^\infty, i = 0, ..., s$, so that

$(\Sigma f_i^2)v = v,$

$f_0 \equiv 1$ in a neighborhood of $x^0,$

$x \in$ supp f_i implies $|x - x^0| > c_1$ if $i > 0$, and

$$|\mathcal{F}(f_0 v)(\xi)| \le c(ck/|\xi'|)^k, \text{ if } \xi \text{ is in a bineighborhood } \Lambda \text{ of } (\xi^0, \sigma^0) \text{ and } |\xi'| > c(1 + |\xi|^\delta),$$

$$(5.3.12)$$

$$\text{and } |D^\alpha f_0(x)| \le c_2^{|\alpha|+1} k^{|\alpha|} \text{ for } |\alpha| \le k.$$

We must also assume that the supports of the f_i are of sufficiently small diameter, as will be specified towards the end of the proof. The constants c, c_1, c_2 must not depend on k here.

We can then write $\mathcal{F}(A^* v) = \sum_i \mathcal{F}(A^*(f_i^2 v))$ and $\mathcal{F}(A^*(f_0^2 v))$ can be estimated quite easily from:

$$\mathcal{F}(A^*(f_0^2 v))(\eta) = (2\pi)^n h(\eta) \int e^{-i\omega(x, \eta) + i\langle x, \xi \rangle} f_0(x) a(x, \eta) \mathcal{F}(f_0 v)(\xi) d\xi \, dx.$$

We are in fact interested to estimate this for η in a small bineighborhood G' of (η^0, τ^0). We may assume that $x \in$ supp f_0 together with $\eta \in G'$ implies $\omega_x(x, \eta) \in \Lambda'' \subset\subset \Lambda$. When $\xi \in \Lambda$ we can estimate $\mathcal{F}(f_0 v)$ by (4.3.5), and for the remaining ξ we can apply partial integration in x.

As for $w_i = A^* f_i^2 v$, $i > 0$, we shall base the proof of $(y^0, \eta^0, \tau^0) \notin W F_A^2 w_i$ on proposition 2.1.15. In fact, $\mathcal{F}(A^* f_i^2 v)$ is clearly in $S'(R^n)$ and it is also obvious that it has an holomorphic extension to Re $\eta \in G$, $c'(1 + |\eta|)^\delta + |Im \eta| \le c''|Re \eta'|$ for some c', c''. It suffices then to find a compact K so that

a) $y^0 \notin K$,

and so that for suitable constants c''', k,

b) $|\mathcal{F}(A^* f_i^2 v)(\eta)| \le c'''(1 + |\eta|)^k e^{H_K(Im \eta)}$, if $Re \eta \in G$, $c'(1 + |\eta|)^\delta + |Im \eta| \le c''|Re \eta'|$.

Actually, K is not difficult to find. We shall fix $z \in$ supp f_i and $\tilde{\eta} \in G$ and define

$$K = \{y; |y' - S_{\eta'}(z, \tilde{\eta})| + |y'' - z''| < \varepsilon\}$$

for some ε to be specified in a moment. To have a) we must in fact choose ε so that

$$|y^{0\prime} - S_{\eta'}(z, \tilde{\eta})| + |y^{0\prime\prime} - z''| > \varepsilon,$$

if G is sufficiently narrow.

That this is possible is essentially a consequence of the fact that $z \in$ supp f_i, $i > 0$ implies $|z - x^0| > \varepsilon$; the actual proof is similar to that of lemma 5.3.4 and we shall omit the details. (Cf. anyway section 5.4 below.)

To estimate $\mathcal{F}(A^* f_i^2 v)(\eta)$ it suffices to show that

$$\sup_{x \in \text{supp } f_i} Im\,\omega(x,\eta) \le H_K(Im\,\eta),$$

for all $\eta \in C^n$ with $Re\,\eta \in G$, $|Im\,\eta| < c''|Re\,\eta'|$. ($a(x,\eta)$ is estimated as a symbol if δ, c'' are suitable.) Here we observe that it follows from Taylor expansion, using also the bihomogeneity of S, that

$$Im\,\omega(x,\eta) = \langle x'', Im\,\eta'' \rangle + \langle S_{\eta'}(x, Re\,\eta), Im\,\eta' \rangle + O(|Im\,\eta||Re\,\eta'|/|Re\,\eta| + |Im\,\eta|^2/|Re\,\eta'|).$$

Shrinking c'' and G if necessary we can achieve that

$$Im\,\omega(x,\eta) \le \langle x'', Im\,\eta'' \rangle + \langle S_{\eta'}(x, Re\,\eta), Im\,\eta' \rangle + (\varepsilon/2)|Im\,\eta|.$$

To conclude the argument it suffices then to observe that we will have

$$|x'' - z''| + |S_{\eta'}(x, Re\,\eta) - S_{\eta'}(z, \tilde{\eta})| < (\varepsilon/2)$$

if $z, x \in \text{supp } f_i$, $Re\,\eta \in G$ and if supp f_i and G are sufficiently small.

5.4 Phase functions associated with Lipschitzian weight functions

1. In section 5.3 we have sketched a direct proof for the mapping properties of F.I.O.s which are associated with phase functions of form $S(x,\eta) + \langle x'', \eta'' \rangle$. When we want to derive the main composition rules for these operators it seems more convenient to use the results and arguments on F.I.O.s associated with general weight functions from Liess-Rodino [2]. For the convenience of the reader we recall here at first the set-up from Liess-Rodino for such operators.

Let us assume then that a Lipschitzian weight function φ is fixed on $G \subset R^n$ and consider X open in R^n. As we have seen in section 2.2, φ is then actually a Lipschitzian weight function on a set of type $G_{c\varphi}$ for some $c > 0$. We shall now also assume that ω is real-analytic and that we have

$$|\partial_x^\alpha \partial_\eta^\beta \omega(x,\eta)| \le c^{|\alpha|+|\beta|+1} \alpha! \beta! \varphi(\eta)^{1-|\beta|} \text{ if } x \in X, \eta \in G_{c\varphi}, |\alpha| + |\beta| \ge 2, \qquad (5.4.1)$$

i.e. that (speaking in a sloppy way), $\omega \in S_\varphi^1$, starting from second derivatives.

Next we assume that ω is nondegenerate. By this we mean that the following conditions are satisfied:

$$\forall X' \subset\subset X , \exists c' \text{ such that } |\omega_x(x,\eta)|/\varphi(\eta) + |\omega_\eta(x,\eta)| \ge c', \forall x \in X', \forall \eta \in G, \qquad (5.4.2)$$

$$|\det \omega_{x\eta}(x,\eta)| \geq c', \forall x \in X', \forall \eta \in G. \tag{5.4.3}$$

2. A complete calculus of F.I.O.'s in general G_φ classes is somewhat unpleasant to state because of the lack of a fiber structure associated with the weight function φ. A number of conditions relating φ to ω must therefore be fulfilled before we can arrive at a reasonable calculus at all. The main conditions are the conditions B,C,D below. (We follow here the numerotation for conditions from Liess-Rodino [2]. In that paper there is also a condition A (on the relation between symbols from classes of type S_φ^m and $\tilde{S}_\varphi^m$), which does not refer directly to ω. The fact that the respective condition is always satisfied in the conditions from higher analytic microlocalization is the content of proposition 3.4.3.)

Condition B. We shall say that condition B is valid if

$$\varphi(\omega_x(x,\eta)) \sim \varphi(\eta) \text{ on } X \times G. \tag{5.4.4}$$

The geometric interpretation of this condition is the following: $\varphi(\eta)$ and $\varphi(\omega_x(x,\eta))$ are meant to measure the specific length of η and of the phase component of the image of (y,η) under the canonical transformation associated with the generating function $\omega(x,\eta) - \langle y, \eta \rangle$. The condition then tells us that these quantities remain comparable.

Remark 5.4.1. *(Cf. Liess-Rodino [2]) If (5.4.4) is valid and X' is fixed with $X' \subset\subset X$, then we can find c so that*

$$\varphi(\omega_x(x,\eta)) \sim \varphi(\eta) \quad on \ X' \times G_{c\varphi}. \tag{5.4.5}$$

Our next condition is meant to say that points (x,η^1), (x,η^2) which stay apart initially, do also stay apart -in a quantitative way- when mapped by ω_x:

Condition C. We shall say that ω satisfies (or has) property C (on (X,G)) if: $\forall X' \subset\subset X$, $\forall c_1$, $\exists c_2$ such that $\eta^1, \eta^2 \in G$, $|\eta^1 - \eta^2| \geq c_1 \varphi(\eta^1)$, implies

$$|\omega_x(x,\eta^1) - \omega_x(x,\eta^2)| \geq c_2(\varphi(\eta^1) + \varphi(\eta^2)), \text{ if } x \in X', \eta^1, \eta^2 \in G.$$

A related condition in the x -variables is

Condition D. ω is said to satisfy condition D if : $\forall X' \subset\subset X$, $\forall c_1 > 0$, $\exists c_2 > 0$ such that

$$|\omega_\eta(x^1,\eta) - \omega_\eta(x^2,\eta)| \geq c_2 \text{ if } |x^1 - x^2| > c_1, x^1, x^2 \in X', \eta \in G.$$

Remark 5.4.2. *Remarks similar to remark 5.4.1 apply. (Cf.Liess-Rodino, loc.cit.)*

Remark 5.4.3. $|\eta^1 - \eta^2| \geq c_1\varphi(\eta^1)$ *implies* $|\eta^1 - \eta^2| \geq c_2\varphi(\eta^2)$. *In view of this condition C, can also be written as:*

$$|\eta^1 - \eta^2| \geq c(\varphi(\eta^1) + \varphi(\eta^2)) \quad \Rightarrow \quad |\omega(x,\eta^1) - \omega_x(x,\eta^2)| \geq c_2(\varphi(\eta^1) + \varphi(\eta^2)), \quad \text{if } x \in X',$$
$$\eta^1, \eta^2 \in G.$$

3. It is interesting to observe that all these conditions are verified in the situation from second microlocalization when the phase function is as in section 5.3. To be more precise, assume that $\omega(x,\eta) = S(x,\eta) + \langle x'', \eta'' \rangle$ for some S which is positively (1,1)-bihomogeneous and satisfies the properties (4.8.1), (4.8.3), (4.8.4), on a bineighborhood of (x^0, η^0, τ^0). As for φ, we will have $\varphi(\eta) = |\eta'|$, which is a Lipschitzian weight function if we restrict our attention to a set of form $|\eta'| > c|\eta|^\delta$ for some $\delta < 1$. (It will be in such regions that all important action takes place.)

Observe then at first that, ω being real-analytic, (5.4.1) is a consequence of the bihomogeneity. Moreover, ω is nondegenerate. In fact, (5.4.3) follows from the condition on S in section 4.8 (we have observed this already) and to check (5.4.2) we may observe that

$$|\omega_{x'}(x,\eta)| = |S_{x'}(x,\eta)| = |\langle S_{x'\eta'}(x,\eta), \eta' \rangle| \geq c|\eta'|.$$

Next, to check that B holds, we note that

$$\varphi(\omega_x(x,\eta)) = |\omega_{x'}(x,\eta)| = |S_{x'}(x,\eta)| = O(|\eta'|),$$

since $S_{x'}$ is (1,1)-bihomogeneous. It is then clear that B must hold since, as remarked while checking non-degeneracy, $|S_{x'}(x,\eta)| \geq c|\eta'|$. Let us also mention in passing that condition B is also satisfied in the situation from section 5.1.

Then we observe that much more is true, than what is required in condition C, in that lemma 5.3.3 says that $|\omega_x(x,\eta^1) - \omega_x(x,\eta^2)| \geq c|\eta^1 - \eta^2|$. Finally, also D is valid in a strengthened form, as is clear from the following result

Lemma 5.4.4. *Suppose X is convex. If G is small enough and X' is fixed with $X' \subset\subset X$, we can find c so that*

$$|\omega_\eta(x^1,\eta) - \omega_\eta(x^2,\eta)| \geq c|x^1 - x^2|, \quad \text{if } x^1, x^2 \in X', \eta \in G.$$

Proof of lemma 4.4.4. Using Taylor expansion we obtain that

$$\omega_{\eta''}(x^1,\eta) - \omega_{\eta''}(x^2,\eta) = x^{1''} - x^{2''} + O(|\eta'|/|\eta''|)|x^1 - x^2|.$$

This already gives the desired inequality in the case when $|x^{1\prime} - x^{2\prime}| < c'|x^{1\prime\prime} - x^{2\prime\prime}|$. It suffices therefore to show that

$$|\omega_{\eta'}(x^1, \eta) - \omega_{\eta'}(x^2, \eta)| \geq c''|x^{1\prime} - x^{2\prime}|,$$

in the case that $|x^{1\prime} - x^{2\prime}| \geq c'|x^{1\prime\prime} - x^{2\prime\prime}|$.

Here we apply once more Taylor expansion to write, for $j \leq d$, that

$$|\omega_{\eta_j}(x^1, \eta) - \omega_{\eta_j}(x^2, \eta)| = S_{\eta_j x'}(\tilde{x}^j, \eta)(x^{1\prime} - x^{2\prime}) + O(|\eta'|/|\eta''|)|x^{1\prime\prime} - x^{2\prime\prime}|$$

for some $\tilde{x}^j$ which lie on the segment $[x^1, x^2]$. The proof then comes to an end if we show that

$$\sum |S_{\eta_j x'}(\tilde{x}^j, \eta)(x^{1\prime} - x^{2\prime})| \geq c_1|x^{1\prime} - x^{2\prime}|.$$

This essentially follows from the nondegeneracy condition for S and we have encountered similar situations several times in section 4.8. We omit further details.

5.5 Composition of F.I.O.

1. We fix $1 \leq d < n$ and consider $x^0 \in R^n$, $\eta^0 \in \dot{R}^n$ for which $\eta^{0\prime} = 0$ and $\tau^0 \in R^d$. Further consider X open in R^n, $x^0 \in X$, G an open bineighborhood of (η^0, τ^0) and let S be a real analytic phase function on $X \times G$ with the properties from section 4.8. As in section 5.3 we denote by $\omega(x, \eta) = S(x, \eta) + \langle x'', \eta'' \rangle$, $\xi^0 = (0, \eta^{0\prime\prime})$, $\sigma^0 = S_{x'\eta'}(x^0, \tau^0, \eta^{0\prime\prime})$, and by $y^0 = (S_{\eta'}(x^0, \tau^0, \eta^{0\prime\prime}), x^{0\prime\prime})$. If χ is the bihomogeneous approximation of the canonical map χ^* associated with ω, we thus have $(x^0, \xi^0, \sigma^0) = \chi(y^0, \eta^0, \tau^0)$. (Cf. relation (4.1.2).)

Let us now also fix some bineighborhood Γ of (x^0, ξ^0, σ^0). If $X' \subset\subset X$, X' open, $x^0 \in X'$ and a bineighborhood G' of (η^0, τ^0) are fixed sufficiently small, it will follow that $(x, \omega_x(x, \eta)) \in \Gamma$ for $x \in X'$, $\eta \in G'$. Moreover, we will have $|\omega_{x'}(x, \eta)| \geq c'(1 + |\omega_{x''}(x, \eta)|^\delta)$ if $|\eta'| \geq c(1 + |\eta|^\delta)$ and c was large enough. All this follows from section 4.8. We then consider $a \in S^m_{|\eta'|}(X \times [G \cap \{|\eta'| > c(1 + |\eta|^\delta)\}])$, $p \in S^\mu_{|\xi'|}(\{(x, \xi) \in \Gamma, |\xi'| > c'(1 + |\xi|^\delta)\})$ and denote by A and by P some F.I.O., respectively pseudodifferential operator, associated with a and p. If X', G' and c are suitable, we may regard $(x, \eta) \to p(x, \omega_x(x, \eta))$ as a symbol in $S^\mu_{|\eta'|}(X' \times G' \cap \{|\eta'| > c(1 + |\eta|^\delta)\})$. (Cf. section 4.9.) We fix X, G initially small enough to have a reasonable theory for A. Finally fix $g \in C_0^\infty(X)$, $g \equiv 1$ in a neighborhood of x^0.

Proposition 5.5.1. *If X', G' are small enough, then we can find $b \in S_{|\eta'|}^{m+\mu}(X' \times [(G' \cap \{|\eta'| > c''(1 + |\eta|^\delta)\}])$, so that for the F.I.O. B associated with b we have*

$$(x^0, \xi^0, \sigma^0) \notin WF_A^2(PgA - B)u, \forall u \in C_0^\infty. \tag{5.5.1}$$

Moreover, modulo lower order terms,

$$b \sim p(x, \omega_x(x, \eta))a(x, \eta). \tag{5.5.2}$$

Remark 5.5.2. *a) It is of course essential to note here that $p(x, \omega_x(x, \eta))$ is a two-microlocal symbol. Cf. the discussion above.*

b) Thus, $p(x, \omega_x(x, \eta))a(x, \eta)$ is the principal part of b in the sense from section 3.3. Since p and a are not supposed to be finite sums of bihomogeneous terms it does not make sense to consider principal parts in the sense of section 3.6.

c) Of course, once the choices for P, g, A, B are fixed, more is true than (5.5.1) in that

$$WF_{|\eta'|}(PgA - B)u \cap \Lambda = \emptyset$$

for some small bineighborhood Λ of (x^0, ξ^0, σ^0). Similar remarks will apply for all propositions later on in this section, but we do not state them explicitly.

2. Next we consider $a \in S_{|\eta'|}^m(X, [G \cap \{|\eta'| > c(1 + |\eta|^\delta)\}])$, $b \in S_{|\eta'|}^\mu(X, [G \cap \{|\eta'| > c(1 + |\eta|^\delta)\}])$, and denote by A and B^* F.I.O.s of the first, respectively second, kind associated with a and b. Thus explicitly

$$Au(x) = \int e^{i\omega(x,\eta)} h(\eta) a(x, \eta) \hat{u}(\eta) d\eta,$$

$$(B^*v)(y) = \int \int e^{i\langle y, \eta \rangle - i\omega(x,\eta)} h'(\eta) b(x, \eta) v(x) dx \, d\eta,$$

where h and h' are as in section 5.1. If $g \in C_0^\infty(X)$, we may now consider the composition B^*gA. We want to show that when $g \equiv 1$ in a neighborhood of x^0, then B^*gA is equivalent, two-microlocally on $Y \times G$, to a pseudodifferential operator, where Y is a small neighborhood of y^0 and G is a small bineighborhood of (η^0, τ^0). Moreover, as in the classical theory, we want the principal symbol of B^*gA to be

$$a(\omega_\eta^{-1}(y, \eta), \eta) \, b(\omega_\eta^{-1}(y, \eta), \eta) \, |\det \omega_{x\eta}(\omega_\eta^{-1}(y, \eta), \eta)|^{-1}. \tag{5.5.3}$$

Here ω_η^{-1} is the inverse of $x \to \omega_\eta(x, \eta)$ for fixed η, and for η we must assume that $|\eta'| > c(1 + |\eta|^\delta)$. (In section 4.8 we have denoted ω_η^{-1} by H^*. Here we shall use somewhat sloppier notations.)

Of course, for this to make sense we need to know that $(\omega_\eta^{-1}(y,\eta),\eta) \in X \times G$ for $(y,\eta) \in Y \times G^1$. It follows from section 4.8 that this will be the case if $Y \times G^1$ is a small bineighborhood of (y^0,η^0,τ^0). Moreover we must show that the expression from (5.5.3) is a symbol. Except for the factor $D^* = |\det \omega_{x\eta}(\omega_\eta^{-1}(y,\eta),\eta)|^{-1}$ this has already been established in section 4.9. To see what happens with D^*, we first observe that the bihomogeneous principal part of $|\det \omega_{x\eta}(x,\eta)|^{-1}$ is $\det S_{x'\eta'}$, which is an elliptic symbol of bidegree (0,0). We can then invert it and use the argument from section 4.9 to compose it with the map $(y,\eta) \to (\omega_\eta^{-1}(y,\eta),\eta)$.

Proposition 5.5.3. *If X,G,Y,G^1 are suitable, then we can find $q \in S_{|\eta'|}^{m+\mu}(Y,G^1 \cap \{|\eta'| > c(1+|\eta|^\delta)\})$ so that*

$$(y^0,\eta^0,\tau^0) \notin WF_A^2(B^* gA - Q)u, \forall u \in C_0^\infty(Y) \tag{5.5.4}$$

where Q is the pseudodifferential operator associated with q and where we have assumed that $g \in C_0^\infty(X)$ is identically one in a neighborhood of x^0. Moreover, the principal part of q is given by (5.5.3).

3. In a similar way we may try to compose $A B^*$. Also this should be a pseudodifferential operator and the symbol should be this time

$$(2\pi)^n a(x,\omega_x^{-1}(x,\xi)) \, b(x,\omega_x^{-1}(x,\xi)) \, |\det \omega_{x\eta}(x,\omega_x^{-1}(x,\xi))|^{-1}, \tag{5.5.5}$$

where $\omega_x^{-1}(x,\xi) = \eta$ is the solution of $\omega_x(x,\eta) = \xi$ for fixed x. Note that this will exist for (x,ξ) in a small bineighborhood of (x^0,ξ^0,σ^0). (Cf. once more section 4.8.) In addition we must also know that $|\eta'| > c'(1+|\eta|^\delta)$ if $|\xi'| > c(1+|\xi|)^\delta$ and c is large enough. Also this follows from section 4.8. Moreover, we see as before that the expression (5.5.5) defines a symbol.

Proposition 5.5.4. *Consider X,G small. There is a bineighborhood Γ of (x^0,ξ^0,σ^0) and $q \in S_{|\xi'|}^{m+\mu}(\Gamma \cap |\xi'| > c(1+|\xi|)^\delta)$ such that*

$$(x^0,\xi^0,\sigma^0) \notin WF_A^2(AB^* - Q)v, \forall v \in C_0^\infty(X),$$

where Q is the pseudodifferential operator associated with q. Moreover, the principal symbol of q is given by (5.5.5).

4. Finally, consider $a \in S_{|\eta'|}^m(X,G \cap \{|\eta'| > c(1+|\eta|)^\delta\})$, $p \in S_{|\eta'|}^\mu(Y,G \cap \{|\eta'| > c(1+|\eta|)^\delta\})$. If $g \in C_0^\infty(Y)$ is identically one in a neighborhood of y^0, we can consider

the composition AgP where A and P are the F.I.O. and pseudodifferential operator associated with a and p respectively. We want to write AgP as a F.I.O. B associated with a symbol b which has principal part

$$a(x, \eta) \, p(\omega_\eta(x, \eta), \eta). \tag{5.5.6}$$

We must make sure then that $\omega_\eta(x, \eta) \in Y$ and this will be the case if (x, η) is in a small bineighborhood of (x^0, η^0, τ^0). We also mention that (5.5.6) indeed defines a symbol.

Proposition 5.5.5. *Let* $g \in C_0^\infty(Y)$ *be identically one in a neighborhood of* y^0. *We can then find* Y *open with* $y^0 \in Y$, *a bineighborhood* Γ *of* (x^0, η^0, τ^0), *and a symbol* $b \in S_{|\eta'|}^{m+\mu}(\Gamma \cap \{|\eta'| > c'(1 + |\eta|)^\delta\})$ *such that for the F.I.O. of the first kind* B *associated with* b *the following relation holds:*

$$(x^0, \xi^0, \sigma^0) \notin WF_A^2(AgP - B)u, \forall u \in C_0^\infty(Y).$$

Moreover, the principal part of b *is given by (5.5.6).*

5. As in first order microlocalization a useful consequence of the results from before is that F.I.O.'s which are associated with elliptic symbols are elliptic, in the sense that the statement from proposition 5.3.1 can be reversed.

Proposition 5.5.6. *Let all notations be as in proposition 5.3.1 and assume in addition that* a *is invertible in* $S_{|\eta'|}^m(U \times [G \cap \{\eta; |\eta'| > c(1 + |\eta|^\delta)\}])$. *Then we have*

$$(y^0, \eta^0, \tau^0) \notin WF_A^2 u \text{ if and only if } (x^0, \xi^0, \sigma^0) \notin WF_A^2 Au,$$

and

$$(x^0, \xi^0, \sigma^0) \notin WF_A^2 v \text{ if and only if } (y^0, \eta^0, \tau^0) \notin WF_A^2 A^* v.$$

This is indeed a consequence of the propositions 5.5.3, 5.5.4 and of the regularity theorems for pseudodifferential operators from chapter III.

On a similar vein, we can also solve the equations

$$v = Au, \qquad w = A^* v$$

modulo error terms which have no second wave front set at (y^0, η^0, τ^0), respectively (x^0, ξ^0, σ^0), if a is elliptic. What we mean by this is that if, e.g., v is given, then we can find $u \in C_0^\infty(Y)$, $g' \in C_0^\infty(X)$ and $v' \in C_0^\infty(X)$ so that $v = g'Au + v'$ and $(y^0, \eta^0, \tau^0) \notin WF_A^2 v'$.

The preceding remarks are, once more as in standard microlocalization, the basis of the reduction of the study of two-microregularity questions for the equation $P(x, D)u = 0$ to simpler models. Indeed, if A, B^* are elliptic F.I.O. associated with some phase function S, then we may consider the operator

$$Q = B^* g P g' A$$

where g and g' lie in C_0^∞, have sufficiently small supports and are identically one on some still smaller neighborhoods of x^0. It follows from the results above that there is a pseudodifferential operator $\tilde{Q}$ with principal symbol

$$b(\omega_\eta^{-1}(y, \eta), \eta) p(\omega_\eta^{-1}(y, \eta), \omega_x(\omega_\eta^{-1}(y, \eta), \eta)) a(\omega_\eta^{-1}(y, \eta), \eta) \det \omega_{x\eta}(\omega_\eta^{-1}(y, \eta), \eta)|^{-1}$$

$$(5.5.7)$$

and so that

$$(y^0, \eta^0, \tau^0) \notin WF_A^2(B^* g P g' A - \tilde{Q})w, \ \forall w \in C_0^\infty(Y).$$

It follows that

$$(y^0, \eta^0, \tau^0) \notin WF_A^2 \tilde{Q}u, \ \forall u \in C_0^\infty(Y)$$

is equivalent with

$$(x^0, \xi^0, \sigma^0) \notin WF_A^2 Pv, \ \forall v \in C_0^\infty(X).$$

Indeed, we may assume that v is of form $v' + g'Au$, with $(y^0, \eta^0, \tau^0) \notin WF_A^2 v'$ and then have the following chain of equivalences

$$(x^0, \xi^0, \sigma^0) \notin WF_A^2 Pv \Leftrightarrow (x^0, \xi^0, \sigma^0) \notin WF_A^2 Pg'Au \Leftrightarrow$$

$$(y^0, \eta^0, \tau^0) \notin WF_A^2 B^* g P g' Au \Leftrightarrow (y^0, \eta^0, \tau^0) \notin WF_A^2 \tilde{Q}u.$$

The interesting thing here now is that

$$(x, \eta) \to (\omega_\eta^{-1}(y, \eta), \omega_x(\omega_\eta^{-1}(y, \eta), \eta))$$

is precisely the canonical transformation χ^* associated with S in section 4.8. Indeed, in the notations from that section, $\omega_\eta^{-1} = H^*$ and $\omega_x = F^*$. Apart from an elliptic factor, the principal symbol of $\tilde{Q}$ is then just $p \circ \chi^*$.

6. The propositions from above are thus the standard results on the calculus of F.I.O. which one wants to have in order to use canonical transformations at ease. We want to show that they are consequences of arguments from Liess-Rodino [2, part III], although they do not follow formally from the statements there. To explain what happens, let us introduce the following condition:

Let $x^0 \in X$, $y^0 \in R^n$ and $\Lambda \subset R^n_\eta$ be given. We say that (x^0, y^0, Λ) is $\omega-$ compatible if

$$\forall c, \exists c' \text{ such that } |y^0 - \omega_\eta(x^0, \eta)| < c \text{ if } \eta \in \Lambda \text{ and } |\eta| > c'. \tag{5.5.8}$$

Note that (5.5.8) roughly says that

$$(x^0, \omega_x(x^0, \Lambda)) \text{ and } (\omega_\eta(\eta^0, \Lambda), \Lambda)$$

are related by the canonical transformation χ associated with ω in the sense that

$$(x^0, \omega_x(x^0, \Lambda)) \sim \chi(y^0, \Lambda),$$

and it is a condition meant to compensate for the lack of a fiber structure associated with φ. Indeed, if the conditions B,C,D are fulfilled, and if A is a F.I.O. associated on some set $X \times G$, $\Lambda \subset G$ with ω and if (x^0, y^0, Λ) is $\omega-$compatible, then it is proved in Liess-Rodino [2] that

$$(y^0, \Lambda) \cap WF_\varphi u = \emptyset \text{ implies } (x^0, \omega_x(x^0, \Lambda)) \cap WF_\varphi Au = \emptyset.$$

Unfortunately, $\omega-$compatibility does not hold in the situation from second microlocalization as considered here, although in fact not much is missing:

if $(x^0, \xi^0, \sigma^0) = \chi(y^0, \eta^0, \tau^0)$ and c_1 is given, then we may choose L of form $\{\eta; \eta = t\eta^0 + t'(\tau^0, 0), t > 0, t' > 0, t' < t\}$ so that

$$|y^0 - \omega_\eta(x^0, \eta)| \leq c_1 \text{ if } \eta \in L.$$

For all practical purposes this could replace $\omega-$ compatibility, but there is another problem with a direct application of results from Liess-Rodino [2], which comes from the fact that $|\eta'|-$ neighborhoods of sets of type L are not bineighborhoods of (η^0, τ^0). One may try then to apply the results from Liess-Rodino, loc.cit., simultaneously for families of generating elements which sweep out a full bineighborhoods of the points under consideration. In the case of the results from section 5.3 this approach did not seem justified. In fact, for one thing, we could model our arguments on the results from section 5.1 and on the other hand, had we worked e.g. with generating elements different from those generated by (η^0, τ^0), then the corresponding x^0 would also have been changed. The situation is different in the case of the results presented in this section, and the reasons to rely on the results and arguments from Liess-Rodino, loc.cit. are more compelling. In fact, independent proofs of these results are quite long, and then significant portions of the proofs from that paper can be taken over with no change. Indeed, the proofs of the results on composition of F.I.O. from Liess-Rodino [2] essentially all start with a part in which, after a preliminary transformation of the integrals which give PgA, B^*gA, AB^*, AgP, the formal expansion (at the level of symbols in $SF^{m+\mu}$) for the symbols from the respective proposition are found. In a second part of the proof these formal expansions are estimated and only in the third part the statements on the wave front sets are proved. The proofs of the first two parts now carry over from Liess-Rodino, loc. cit. to our present situation or have been settled in chapter IV. The reason is that in these parts of the argument only the conditions B,C,D are used effectively and these conditions are also valid here. In particular, this will give all the statements on the form of the principal parts of the symbols in $S^{m+\mu}_{|\eta'|}$ which we are trying to find. The situation is unfortunately slightly different in the third part of the argument, since there $\omega-$compatibility was used in Liess-Rodino, loc.cit. However, even in this part we are closer to the situation from that paper than in section 5.3, in that now it suffices to vary generating elements either only on the (x, ξ, σ) side or on the (y, η, τ) side. To see what happens, let us consider e.g. the case of proposition 5.5.1. We can then argue roughly as follows.

We fix $x, \eta, |\eta'| > c(1 + |\eta|)^\delta$ in a small bineighborhood Γ of (x^0, η^0, τ^0) and denote by $\xi = \omega_x(x, \eta)$, $y = \omega_\eta(x, \eta)$. It follows in a trivial way that (x, y, η) is $\omega-$compatible. Arguing as in Liess-Rodino [2], we can now conclude that

$$\{(x, \xi)\} \cap WF_{|\xi'|}(PgA - Q)u = \emptyset, \tag{5.5.9}$$

in an (and this is the main part of the argument) "uniform way", provided that Γ was small enough. In fact that (5.5.9) with no condition on uniformity is valid, is a triviality, since (x, ξ) is an one point set. To explain what we mean by uniformity, note that, formally (5.5.9) says, for fixed (x, ξ), that we can find a constant c_1 so that if $f_k \in C_0^\infty(X)$ is a sequence of functions which satisfy supp $f_k \subset \{z; |z-x| < c_1\}, f_k(z) = 1$ if $|z - x| < c_1/2$, $|D^\alpha f_k(x)| \le c_2^{|\alpha|+1} k^{|\alpha|}$ for $|\alpha| \le k$, then

$$|\mathcal{F}(f_k(PgA - Q)u)(\theta)| \le c_3(c_3 k/|\theta'|)^{k-c_4}, \text{ if } |\xi - \theta| < c_5|\xi'| , \tag{5.5.10}$$

where c_3, c_4 and c_5 do not depend on k. While this is trivial for a fixed (x, ξ), the "uniformity" from above now should mean that, once we have fixed c_1 suitably and once we have fixed a sequence f_k as before, the inequalities from (5.5.10) are valid with some constants c_3, c_4, c_5 which do not depend on (x, ξ). Since (x, ξ) sweeps out (by section 4.8) a bineighborhood of (x^0, ξ^0, σ^0) when (x, η) runs through a bineighborhood of (x^0, η^0, τ^0) this is precisely what is needed to prove (5.5.1). We omit further details.

5.6   Invariant meaning and proof of theorem 1.1.3

1. We assume that $X, G, p, \Sigma, p^T, \Lambda, W, B, u$ are as in the statement of theorem 1.1.3 and that homogeneous analytic canonical coordinates are chosen so that for a conic neighborhood V of (x^0, ξ^0)

$$\Sigma \cap V = \{(x, \xi); \xi' = 0\}. \tag{5.6.1}$$

Actually we have already studied the invariant meaning of the notions which appear in theorem 1.1.3. We know in particular that the fact that Λ is regularly involutive for the relative symplectic structure on $N\Sigma$ does not depend on the way we reduce Σ to the form (5.6.1), that the projections of the leaves of the relative bicharacteristic foliation of Λ onto Σ are well-defined, and even the fact that the condition that p is transversally elliptic to Λ has an invariant meaning in $N\Lambda$.

2. As for the proof of theorem 1.1.3, the main intermediate step is to show, under the assumptions and notations from the statement, that

$$\{((x', x^{0\prime\prime}), \xi^0, \sigma);\ x' \in W, \sigma \in \dot{R}^d \} \cap WF_A^2 u = \emptyset. \tag{5.6.2}$$

In fact, once this is proved, we know from the micro-Holmgren theorem that we have the following alternative:

either we have $\{(x', x^{0\prime\prime}, \xi^0);\ x' \in W\} \subset WF_A u$

or else $\{(x', x^{0\prime\prime}, \xi^0);\ x' \in W\} \cap WF_A u = \emptyset$.

From the assumption in the theorem it now follows that we can find points $x' \in W$ so that $\{(x', x^{0\prime\prime}, \xi^0) \notin WF_A u$. Of the two possibilities in the alternative, it must be therefore the second one which actually holds.

3. In order to prove (5.6.2) we shall now at first observe that we already know from section 3.8 that

$$WF_A^2 u \subset \{(\lambda, \sigma) \in N\Sigma;\ p^T(\lambda, \sigma) = 0 \}.$$

The only problem in (5.6.2) can therefore appear for the points $(\lambda, \sigma) \in \Lambda$. To study these points we shall use once more the micro-Holmgren theorem to show that also here we have an alternative similar to the one for $WF_A u$.

Proposition 5.6.1. *Let all the assumptions on p be as before and consider $(\lambda^0, \sigma^0) \in \Lambda$. Let L be the leaf of the relative bicharacteristic foliation of Λ which contains (λ^0, σ^0). If u is a solution of $WF_{A,\Sigma}^2 p(x, D)\tilde{u} \cap \Gamma = \emptyset$, then the following alternative holds:*

either we have $L \subset WF^2_{A,\Sigma}\tilde{u}$,

or else $L \cap WF^2_{A,\Sigma}\tilde{u} = \emptyset$.

Note that when we apply this alternative for the solution u of $p(x, D)u = 0$ which appears in the statement of theorem 1.1.3, then one of the assumptions on u is that L is not a subset of $WF^2_{A,\Sigma}u$. (Here we use that there is no second wave front set where there is no first wave front set.) The proof of proposition 5.6.1 will therefore conclude the proof of theorem 1.1.3.

3. The statement of proposition 5.6.1 is two-microlocal: it suffices to check it when we intersect L with some suitably chosen bineighborhood G of (λ^0, σ^0). The important thing is now that proposition 5.6.1 refers to any solution $\tilde{u}$ of $WF^2_{A,\Sigma}p(x, D)\tilde{u} \cap G = \emptyset$ and not just to the solutions of the type considered in theorem 1.1.3, i.e. for those solution for which we have some information far away from (λ^0, σ^0). We can then argue as in section 5.5 and transform the operator p two-microlocally to a pseudodifferential operator $\tilde{q}$ with principal symbol $p \circ \chi^*$, where χ^* is a canonical transformation of the type considered in section 4.8. When we choose χ^* so that Λ has the form $\{\tau = 0\}$ on a small bineighborhood G' of (λ^0, σ^0), then we are in the situation from theorem 4.10.2. In our fixed coordinates we can then conclude that $(\lambda, \sigma, w) \notin WF^3_{A,s}v$ whatever solution v of $WF^2_A\tilde{q}v \cap G' = \emptyset$ and whatever $(\lambda, \sigma) \in L \cap G$ and $w \in \dot{R}^{d'}$ we fix. We then have the alternative from proposition 5.6.1 on G' as a consequence of proposition 2.1.12. This concludes the proof.

Chapter 6

Conical refraction, hyperbolicity and slowness surfaces

A particularily important case where propagation phenomena occur is that of hyperbolic operators. One big advantage when dealing with hyperbolic operators is that one can study them often starting from a parametrix for the Cauchy problem. Results on propagation of singularities are in particular related to the singularities of the parametrix, which singularities in turn can often be read off from the explicit form of that parametrix, when such a form is known. An important notion which appears in this context is that of an influence domain. One of the ideas underlying the results on conical refraction in these notes, has been that it basically sufficed to have good information about various localizations of the operator under study. It is therefore a natural idea to see what happens if one assumes for an operator which is as in the beginning of section 1.1, that its localization p^T along Σ is hyperbolic. This is what is effectively done in Melrose-Uhlmann [1], respectively Laubin [1], with the additional condition that p^T is second order. The result which is obtained in Melrose-Uhlmann (in the C^∞ case) and Laubin, loc.cit., is then similar to the one from theorem 1.1.3, but is expressed in terms of influence domains for p^T. We establish the relation with our results in section 6.2, but before we can do so, it is necessary to analyze the relation between influence domains and conditions of the type which appear in theorem 1.1.3 in more detail. In section 6.3, on the other hand, we shall establish a result related to section 3.9, but assuming now only that $p_{m,2}$ is hyperbolic.

We continue the chapter then with a discussion of how the singularities of the characteristic surface make themselves felt in conical refraction. We also discuss in some detail two of the most important cases where conical refraction is known to occur: the system of crystal optics and the system of elasticity for cubic crystals.

6.1   Influence domains and bicharacteristics

1. We collect in this section a number of remarks on second order hyperbolic operators which should be wellknown, but are anyhow of independent interest. Let U be open in $R^{n+1} = R_x^n \times R_t$ which contains 0 and denote the variables from R^{n+1} by $z = (x, t)$. The Fourier-dual variables shall be denoted by $\lambda = (\xi, \tau)$. Also consider a symmetric matrix $A = (a_{ij})_{i,j=1}^n$ of real-valued real analytic functions a_{ij} on U such that

$$A(z) = (a_{ij}(z))_{i,j=1}^n \text{ is positive definite for any } z \in U. \qquad (6.1.1)$$

To simplify notations we shall assume $0 \in U$ and shall also denote by $B = (b_{ij})_{i,j=0}^n$ the following symmetric matrix:

$$b_{00} = 1, b_{0i} = b_{i0} = 0, \text{ if } i \geq 1, b_{ij} = -a_{ij} \text{ if } i, j \geq 1.$$

More generally, we could start from the very beginning from a symmetric nondegenerate $(n + 1) \times (n + 1)$ matrix B of positivity 1, but this would only have complicated the notations, so we shall stick to the case considered above.

With A we now associate the second order operator

$$p(x, t, D_x, D_t) = D_t^2 - \sum_{i,j=1}^n a_{ij}(x, t) D_{x_i} D_{x_j},$$

which is, in view of (6.1.1), strictly hyperbolic with respect to the hyperplanes $t = const$. Next we recall that after a change of variables we may assume that

$$A(0) = Id,$$

where Id is the identity matrix, an assumption which makes our calculations more transparent.

As we are generally looking into propagation phenomena, we are interested in the bicharacteristic curves associated with p. In our present case, these are the curves $s \to (z(s), \lambda(s))$ which satisfy

$$\frac{d}{ds} t(s) = 2\tau(s), \qquad (6.1.2)$$

$$\frac{d}{ds} x_i(s) = -2 \sum_j a_{ij}(z(s)) \xi_j(s), \qquad (6.1.3)$$

$$\frac{d}{ds} \lambda(s) = \sum_{i,j} \text{grad}_z a_{ij}(z(s)) \xi_i(s) \xi_j(s). \qquad (6.1.4)$$

To the system (6.1.2), (6.1.3), (6.1.4) we normally shall associate some initial conditions $z(0) = z^0$, $\lambda(0) = \lambda^0$ and s shall be considered on the maximal interval containing 0 on which we can solve the system. In general, the length of this interval may depend on the initial conditions (z^0, λ^0), but if $U' \subset\subset U$ is fixed, the solution can not break down at s^0 as long as $z(s) \in U'$ for $s < s^0$. Moreover, a rough estimate of the existence interval suffices here and is easy to obtain from homogeneity. In fact, p being homogeneous of order 2 in λ, it is immediate that if $(z(s), \lambda(s))$ is a solution of the system with $z(0) = z^0$, $\lambda(0) = \lambda^0$, then $\tilde{z}(s) = z(\sigma s)$, $\tilde{\lambda}(s) = \sigma\lambda(\sigma s)$ solves the system with initial conditions $\tilde{z}(0) = z^0$, $\tilde{\lambda}(0) = \sigma\lambda^0$. Practically this means that we can normalize the situation to $|\lambda^0| = 1$ or $\tau^0 = 1$. It follows that if z^0 is in a small neighborhood of zero, then we can find $c > 0$ so that

the solution of (6.1.2), (6.1.3), (6.1.4) with $z(0) = z^0$, $\lambda(0) = \lambda^0$ exists for $|s| < c/|\lambda^0|$.

$$(6.1.5)$$

In particular, bicharacteristics dispose of more time when initially $|\lambda^0|$ is small. We next recall that a bicharacteristic is called "null" if $p(z^0, \lambda^0) = 0$ and that $p(z(s), \lambda(s)) = 0$ for all s along a null-bicharacteristic. Here we can now use that $p(z, \lambda) = 0$ implies

$$|\xi(s)| \leq c'|\tau(s)| \text{ for some } c' \qquad (6.1.6)$$

if $z(s)$ lies in some previously fixed neighborhood U' of the origin. It follows that $\xi(s) = 0$ if $\tau(s) = 0$. On the other hand, the only solution of the system (6.1.2), (6.1.3), (6.1.4) for which $\lambda^0 = 0$ is $z = const$, $\lambda = 0$. We conclude that on a null-bicharacteristic $\tau(s)$ can never change sign. Note that it follows from 6.1.2 that t is a natural parameter on the bicharacteristic. If then e.g. $\tau^0 > 0$, we shall call the associated bicharacteristic "future- oriented". The estimate (6.1.6) also shows when combined with the system, that $|\dot{x}(s)| \leq c''|\dot{t}(s)|$ on bicharacteristics, if $z(s) \in U'$. This shows that the speed of the projection of a bicharactistic into physical space is finite. Since we know already that a bicharacteristic can not blow up as long as $z(s) \in U'$ it is immediate that we must have the following result:

Proposition 6.1.1. *If a neighborhood U'' of the origin in R^{n+1} and $\delta > 0$ are fixed suitably small, it follows that any null bicharacteristic $z(s), \lambda(s)$ for which $z(0) \in U''$, $t(0) < \delta$, must eventually achieve $t(s) \geq \delta$.*

Remark 6.1.2. *It is useful sometimes to observe that not only does $\tau(s)$ never vanish on a bicharacteristic, but also the speed with which it can decrease can not be too big. This is based on the following lemma:*

Lemma 6.1.3. *Let c, c' be given. Then we can find c'' so that for every continuously differentiable function $\tau : [0, c] \to R_+$ which satisfies*

$$\left| \frac{d}{ds} \tau(s) \right| \leq c' |\tau(s)|^2, \quad \tau(0) = 1,$$

it follows that

$$\tau(s) \geq 1/2 \ if \ s \in [0, c''].$$

Proof. We denote by I the set of points $\sigma \in [0, c]$ for which $\tau(\sigma) \leq 1$. We have that

$$\tau(s) = \int_0^s \dot{\tau}(\sigma) \, d\sigma \geq 1 - \int_{I \cap [0,s]} |\dot{\tau}(\sigma)| \, d\sigma \geq 1 - \int_{I \cap [0,s]} c' \, d\sigma \geq 1 - sc'.$$

The assertion from the lemma is then true for $c'' = 1/(2c')$.

Remark 6.1.4. *Note here that the τ-component of a bicharacteristic satisfies an estimate of the form $|\dot{\tau}(s)| \leq c' |\tau(s)|^2$ as long as $z(s) \in U'$. Since $\dot{t}(s) = 2\tau(s)$ the lemma then gives an estimate from below for t.*

2. Our next goal in this section is to look into what we shall call "propagative sets" associated with p. The terminology is here based on

Definition 6.1.5. *A closed set $K \subset U$ is called locally propagative for p at $z^0 \in U$ if it has the following properties:*

a) $z^0 \in K$,

b) *there is $\delta > 0$ so that for every $\tilde{z} \in K \cap \{z; t < t^0 + \delta\}$ and every $\tilde{\lambda}$ with $p(\tilde{z}, \tilde{\lambda}) = 0$ and $\tilde{\tau} > 0$, the (x, t) component of the bicharacteristic which starts at $(\tilde{z}, \tilde{\lambda})$ stays in K as long as $t(s) \leq t^0 + \delta$,*

c) *any bicharacteristic of the type from b) eventually achieves $t(s) \geq t^0 + \delta$.*

Remark 6.1.6. *One of the reasons why we are interested in propagative sets is the following: if u is a solution of $p(x, t, D_x, D_t)u = 0$ which is real analytic in $K \cap \{z; t > t^0 + \delta'\}$ then u is automatically also real analytic for $z \in K \cap \{z; t \leq t^0 + \delta'\}$ if δ' has been fixed small enough.*

3. We shall now show how one can explicitly construct a locally propagative set, which is related to the construction of a local parametrix for the Cauchy problem for p. To

prepare for this construction, we need some preliminaries concerning the projection of the bicharacteristic flow on the z-space.

To fix notations, we shall denote by $(\varphi(s,\lambda), \psi(s,\lambda))$ the solution of the system (6.1.2), (6.1.3),(6.1.4) (i.e., $\varphi \sim z, \psi \sim \lambda$) with initial conditions

$$\varphi(0,\lambda) = 0, \tag{6.1.7}$$

$$\psi(0,\lambda) = \lambda. \tag{6.1.8}$$

An interesting example is when $p = D_t^2 - \sum_j (\partial/\partial x_j)^2$, in which case

$$\varphi(s,\lambda) = 2s \begin{pmatrix} \tau \\ -\xi \end{pmatrix} , \quad \psi(s,\lambda) = \lambda.$$

Returning to the general case, it is a well-known property of systems of ordinary differential equations that the φ and ψ are real analytic functions of s and λ and we want to obtain further information on φ as a function of λ. Of course we have

$$\varphi(s,0) = 0 \text{ for all } s,$$

the homogeneity gives

$$\varphi(s,\sigma\lambda) = \varphi(\sigma s, \lambda), \psi(s,\sigma\lambda) = \sigma\psi(\sigma s, \lambda), \text{ if } \sigma \in R, \tag{6.1.9}$$

and derivating the initial conditions, we obtain

$$\varphi_\lambda(0,\lambda) = 0 , \quad \psi_\lambda(0,\lambda) = Id. \tag{6.1.10}$$

Moreover, derivating the system we get

$$\frac{\partial}{\partial s}\frac{\partial}{\partial \lambda_j}\varphi_i(s,\lambda) = \sum_k \frac{\partial^2}{\partial z_k \partial \lambda_i}p(\varphi(s,\lambda), \psi(s,\lambda))\frac{\partial}{\partial \lambda_j}\varphi_k(s,\lambda)$$

$$+ \sum_k \frac{\partial^2}{\partial \lambda_k \partial \lambda_i}p(\varphi(s,\lambda), \psi(s,\lambda))\frac{\partial}{\partial \lambda_j}\psi_k(s,\lambda).$$

Inserting (6.1.10) into this, gives

$$\frac{\partial}{\partial s}\frac{\partial}{\partial \lambda_j}\varphi_i(0,\lambda) = \sum \frac{\partial^2}{\partial \lambda_k \partial \lambda_i}p(0,\lambda)\delta_{kj} = \frac{\partial^2}{\partial \lambda_j \partial \lambda_i}p(0,\lambda) = 2b_{ij}(0).$$

When, for example, $|\lambda| \leq 2$, we conclude from all this that

$$\frac{\partial}{\partial \lambda}\varphi(s,\lambda) = 2sB(0) + O(s^2)$$

for small s. Here and later on in this section, we shall write $O(s^2)$ for matrix functions which can be estimated in the form

$$|O(s^2)| \le c|s^2|$$

for some c which does not depend on the independent variables of $O(s^2)$, there where these variables are considered. It follows in particular that

$$\det \left(\frac{\partial}{\partial \lambda} \varphi(s, \lambda) \right) \ne 0, \text{ if } |\lambda| \le 2 \text{ and } s \ne 0 \text{ is small.}$$

We can then conclude with the aid of the inverse mapping theorem that we have

Proposition 6.1.7. *If s is fixed small enough, we can find two neighborhoods $U \subset R_\lambda^{n+1}$ and $V \subset R_z^{n+1}$ of the origin in R_λ^{n+1}, respectively R_z^{n+1}, so that*

$$\lambda \to \varphi(s, \lambda)$$

maps U diffeomorphically onto V.

Remark 6.1.8. *The size of U may depend here on s, but we can play on (6.1.9) to increase the size of U at the expense of s. In particular we may assume for $s < \varepsilon$ that U contains $|\lambda| < 3/2$.*

Let us next denote by $\Gamma^\pm = \{\lambda \in R^{n+1}; p(0, \lambda) > 0, \pm\tau > 0\}$, which in fact, due to our assumption that $A(0) = Id$, comes to

$$\Gamma^\pm = \{\lambda \in R^{n+1}; \pm\tau > |\xi|\}.$$

Further we consider

$$Q_\varepsilon^\pm = \{z \in R^{n+1}; \exists \lambda \in \Gamma^\pm, \exists s > 0, \text{ with } |\lambda| + s < \varepsilon \text{ such that } z = \varphi(s, \lambda)\}.$$

Actually it is sometimes convenient to observe that for $\varepsilon \le 1$

$$Q_\varepsilon^+ \subset \tilde{Q}_\varepsilon^+, \tag{6.1.11}$$

where $\tilde{Q}_\varepsilon^+ = \{z \in R^{n+1}; z = \varphi(\varepsilon, \lambda) \text{ for some suitable } \lambda \in \Gamma^+ \text{ with } |\lambda| < \varepsilon\}$. (To establish (6.1.11) we note that for $s \le \varepsilon$ we can write $\varphi(s, \lambda) = \varphi(\varepsilon, (s/\varepsilon)\lambda)$.)

It is now also important to observe that we have

Lemma 6.1.9. *If ε is small enough, we can find a neighborhood of the origin W such that*

$$\partial Q_\varepsilon^+ \cap W \subset \{z; \text{ there is } \lambda \in \partial\Gamma^+ \text{ such that } z = \varphi(\varepsilon, \lambda) \text{ and } |\lambda| < \varepsilon\}.$$

(∂Q_ε^+ is the boundary of Q_ε^+, $\partial \Gamma^+$ that of Γ^+.) This is indeed a consequence of the fact that φ is a diffeomorphism near 0.

Conversely, if δ and a neighborhood W' of the origen are fixed suitably, then

$$\{z; \text{ there is } \lambda \in \partial \Gamma^+ \text{ such that } z = \varphi(\delta, \lambda) \text{ and } |\lambda| < \delta\} \cap W' \text{ is contained in } \partial Q_\varepsilon^+.$$

What we want to prove next is that Q_ε^+ is locally propagative. Before we can show this, we need some additional information on the boundary of Q_ε^+. We start our study with the following result:

Lemma 6.1.10.

$$\langle \psi(s, \lambda), \frac{\partial}{\partial \lambda_j} \varphi(s, \lambda) \rangle = s \frac{\partial}{\partial \lambda_j} p(0, \lambda). \tag{6.1.12}$$

Proof. (Cf. e.g. Laubin [1].) When $s = 0$ this follows from $(\partial/\partial \lambda_j)\varphi(0, \lambda) = 0$. Next we observe that

$$\frac{\partial}{\partial s} \langle \psi(s, \lambda), \frac{\partial}{\partial \lambda_j} \varphi(s, \lambda) \rangle = -\langle p_z(\varphi(s, \lambda), \psi(s, \lambda)), \frac{\partial}{\partial \lambda_j} \varphi(s, \lambda) \rangle$$

$$+ \langle \psi(s, \lambda), \frac{\partial}{\partial \lambda_j} [p_\lambda(\varphi(s, \lambda), \psi(s, \lambda))] \rangle =$$

$$- \langle p_z(\varphi, \psi), \frac{\partial}{\partial \lambda_j} \varphi \rangle + \frac{\partial}{\partial \lambda_j} \langle \psi, p_\lambda(\varphi, \psi) \rangle - \langle \frac{\partial}{\partial \lambda_j} \psi, p_\lambda(\varphi, \psi) \rangle =$$

$$- \langle p_z(\varphi, \psi), \frac{\partial}{\partial \lambda_j} \varphi \rangle + 2 \frac{\partial}{\partial \lambda_j} [p(\varphi, \psi)] - \langle \frac{\partial}{\partial \lambda_j} \psi, p_\lambda(\varphi, \psi) \rangle = \frac{\partial}{\partial \lambda_j} [p(\varphi, \psi)].$$

By Euler's relation for homogeneouse functions, the latter is equal to

$$(\partial/\partial \lambda_j)[p(\varphi(s, \lambda), \psi(s, \lambda))],$$

which is also equal to $(\partial/\partial \lambda_j)[p(0, \lambda)]$ since p is constant on bicharacteristics.

Lemma 6.1.11. *Consider ω^0 with $|\omega^0| = 1$. The vector $\psi(s^0, (\omega^0, 1))$ has the direction of the inward normal to ∂Q_ε^+ at $\varphi(s^0, (\omega^0, 1))$ if s^0 is small.*

(Also cf. Laubin [1] for a related statement.)

Proof. The inward normal to $\{\lambda; |\tau| = |\xi|, \tau > 0\}$ at some point (τ^0, ξ^0) is $(\tau^0/2, -\xi^0/2)$. We denote by φ_* the cotangent map to $\lambda \to \varphi(s^0, \lambda)$ at $\lambda^0 = (\omega^0, 1)$. The tangent map in the fiber over $(\omega^0, 1)$ is given by multiplication with the matrix

$$\frac{\partial \varphi}{\partial \lambda}(s^0, \lambda^0).$$

We need then verify that

$$\frac{{}^t\partial\varphi}{\partial\lambda}(s^0,\lambda^0)\psi(s^0,\lambda^0) = a\begin{pmatrix} \tau^0 \\ -\xi^0 \end{pmatrix}$$

for some positive a. In view of lemma 6.1.10 we now have that

$$\left[\frac{{}^t\partial\varphi}{\partial\lambda}(s^0,\lambda^0)\psi(s^0,\lambda^0)\right]_i = \sum_j \frac{\partial}{\partial\lambda_i}\varphi_j(s^0,\lambda^0)\psi_j(s^0,\lambda^0) = \langle\psi,\frac{\partial\varphi}{\partial\lambda_i}\rangle = s\frac{\partial}{\partial\lambda_i}(\tau^2 - |\xi|^2)(\lambda^0).$$

Our next result is:

Proposition 6.1.12. *Let $\tilde{z} \in \partial Q_\varepsilon^+$ and consider $\tilde{\lambda}$ so that $p(\tilde{z},\tilde{\lambda}) = 0$, $\tilde{\tau} > 0$. Also denote by*

$$L_{\tilde{z},\tilde{\lambda}} = (z(s),\lambda(s))$$

the null-bicharacteristic of p which starts at $(\tilde{z},\tilde{\lambda})$. Then $z(s) \in Q_\varepsilon^+$ for small positive ε.

Proof. We have seen that the inward normal at $z^0 = \varphi(s^0,(\omega^0,1))$ to ∂Q_ε^+ is parallel to $\psi(s^0,(\omega^0,1))$. It suffices then to check that

$$\langle\psi(s^0,(\omega^0,1)), \dot{z}(0)\rangle > 0.$$

Since $\dot{z}(0) = p_\lambda(\tilde{z},\tilde{\lambda})$ this comes to

$$\langle\psi(s^0,(\omega^0,1)), p_\lambda(\tilde{z},\tilde{\lambda})\rangle > 0. \tag{6.1.13}$$

Here we recall that $\psi(s^0,(\omega^0,1))$ satisfies $p(z^0,\psi(s^0,(\omega^0,1))) = 0$ and we may also assume that $\psi_{n+1}(s^0,(\omega^0,1)) > 0$. ($\psi_{n+1}$ is the τ component of ψ.) The relation (6.1.13) is therefore a consequence of the calculations made in section 1.3.

We can conclude from all this discussion that we have

Corollary 6.1.13. *The set Q_ε^+ is locally propagative for small ε for $z^0 = 0$.*

Remark 6.1.14. *Our notion of propagative set is closely related to the influence domains of Leray [1], Duistermaat [1] or Wakabayashi [3],[5] but we shall not study the explicit relation with those notions here.*

6.2   The canonical form of operators when p_m^T is second order and hyperbolic.

1. To establish the relation between theorem 1.1.3 and the results from Melrose-Uhlmann [1], respectively Laubin [1], we have to recall here a construction which is made in these

papers. The assumption is then that a classical analytic pseudodifferential operator $p(x, D)$ is given on a set $X \times G$, G some open conic neighborhood of ξ^0, such that for some point x^0 we have

a) $p_m(x^0, \xi^0) = 0$, $\operatorname{grad} p_m(x^0, \xi^0) = 0$,

b) $\Sigma = \{(x, \xi) \in \Gamma; p_m(x, \xi) = 0, \operatorname{grad} p_m(x, \xi) = 0\}$ is a real analytic, nonradial involutive manifold of codimension k, $k \geq 2$,

c) $\operatorname{Hess} p_m(x, \xi)$ is of rank k and positivity 1 if $(x, \xi) \in \Sigma$.

Here p_m is the principal part of p and $\Gamma = X \times G$.

Under these assumption the form of p_m can be simplified significantly. We recall here, how this is done, since the constructions which we need later on are easier to understand for the simplified form of p_m.

Proposition 6.2.1. *(Laubin [1]) There is a homogeneous real analytic canonical transformation* χ *of a conic neighborhood* U *of* (x^0, ξ^0) *to a conic neighborhood* V *of* $\chi(x^0, \xi^0)$ *such that*

$$\chi(U \cap \Sigma) = \{(x, \xi) = U; \xi_1 = \cdots = \xi_{n-k} = 0\},$$

$$(p_m \circ \chi^{-1})(x, \xi) = \alpha(x, \xi)(\xi_1^2 - \sum_{i,j=2}^{n-k} a_{ij}(x, \xi)\xi_i \xi_j),$$

where $\alpha(x^0, \xi^0) \neq 0$, *the matrix* $a_{ij}(x^0, \xi^0)$ *is positive definite,* α *is positively homogeneous in* ξ *of degree* $m-2$ *and the* a_{ij} *are positively homogeneous of degree 0.*

Remark 6.2.2. *Results related to this proposition also appear in Melrose-Uhlmann [1] and in Lascar [1].)*

Proof of proposition 6.2.1. (Cf. Laubin [1].) After a preliminary real analytic canonical change of coordinates we can assume that $\Sigma = \{\xi_i = 0, i = 1, ..., n - k\}$. Since p_m vanishes of order two on Σ it follows that

$$p_m(x, \xi) = \langle A(x, \xi)\xi', \xi' \rangle$$

where A is real analytic and has homogeneous entries of degree $m - 2$. (The variables ξ are once more written as $\xi = (\xi', \xi'')$ where this time $\xi' = (\xi_1, ..., \xi_{n-k})$.) In view of the assumptions, A is of rank k and positivity 1 at (x^0, ξ^0). We may in particular choose variables so that $A(x^0, \xi^0) = \operatorname{diag}(1, -1, ..., -1)$ so that also $(\partial/\partial \xi_1)^2 p_m(x^0, \xi^0) \neq 0$. From the Weierstrass preparation theorem we can then write

$$p_m(x, \xi) = \alpha(x, \xi)(\xi_1^2 + \tilde{A}(x, \xi_2, ..., \xi_n)\xi_1 + B(x, \xi_2, ..., \xi_n)) = \alpha(x, \xi)((\xi_1 - a(x, \xi_2, ..., \xi_n))^2$$

$$+ \, b(x, \xi_2, ..., \xi_n)),$$

where a and b are real analytic, homogeneous of order one, respectively two, and where

$$a(x^0, \xi_2^0, ..., \xi_n^0) = b(x^0, \xi_2^0, ..., \xi_n^0) = 0, \alpha(x^0, \xi^0) \neq 0.$$

Next we use that p_m vanishes of order two on Σ and obtain

$$a(x, (0, \xi'')) \equiv b(x, (0, \xi'')) \equiv 0, \ \operatorname{grad}_{\xi'} b(x, (0, \xi'')) \equiv 0.$$

We conclude in particular that $b(x, \xi_2, ..., \xi_n) = \sum_{i,j=2}^{n-k} b_{ij}(x, \xi_2, ..., \xi_n) \xi_i \xi_j$ for some symmetric positive definite real analytic matrix (b_{ij}).

We must still get rid of a. This shall be done with the aid of an additional canonical transformation, which we shall choose of form

$$(x, \xi) \rightarrow (x_1, \varphi(x, \xi_2, ..., \xi_n), \xi_1 - a(x, \xi_2, ..., \xi_n), \psi(x, \xi_2, ..., \xi_n))$$

for some vector-valued functions $\varphi = (\varphi_2, ..., \varphi_n)$ and $\psi = (\psi_2, ..., \psi_n)$ still to be defined. In order for this to be canonical, we need in particular

$$\{\xi_1 - a(x, \xi_2, ..., \xi_n), \varphi_j\} = 0, \forall j = 2, ..., n, \{\xi_1 - a(x, \xi_2, ..., \xi_n), \psi_j\} = 0, \forall j = 2, ..., n,$$

where $\{,\}$ is the standard Poisson bracket. This comes to

$$\frac{d}{dt} \varphi_j = \{a, \varphi_j\}', \frac{d}{dt} \psi_j = \{a, \psi_j\}',$$

where $\{,\}'$ is the Poisson bracket in the variables $x_2, ..., x_n$ and where $t = x_1$. We also assume $x_1^0 = 0$. To the previous system we add the initial conditions

$$\varphi_j(0, x_2, ..., x_n, \xi_2, ..., \xi_n) = x_j, \psi_j(0, x_2, ..., x_n, \xi_2, ..., \xi_n) = \xi_j.$$

The remaining Poisson brackets are then automatically in the desired relation. We check this e.g. for φ_j, φ_k. In fact, $c_{jk} = \{\varphi_j, \varphi_k\}'$ satisfies $c_{jk|t=0} = 0$ and $(d/dt)c_{jk}$ satisfies the first order partial differential equation

$$\frac{d}{dt} c_{jk} = \{\{a, \varphi_j\}', \varphi_k\}' + \{\varphi_j, \{a, \varphi_k\}'\}' = \{a, \{\varphi_j, \varphi_k\}'\}' = \{a, c_{jk}\}'.$$

The only solution of this problem is $c_{jk} = 0$ Moreover, the set $\xi' = 0$ is left invariant under the transformation, in that

$$\psi(x, 0, \xi'') = (0, \xi''). \tag{6.2.1}$$

To see this, we note that indeed $\psi(0, x_2, ..., x_n, 0, \xi'') = (0, \xi'')$ by our choice of initial conditions and that $\psi(x, 0, \xi'')$ also satisfies the system

$$\frac{\partial}{\partial x_1} \psi(x, 0, \xi'') = \{a, \psi\}'_{|\xi'=0} = \sum_{j=2}^{n-k} \frac{\partial}{\partial \xi_j} a(x, 0, \xi'') \frac{\partial}{\partial x_j} \psi(x, 0, \xi'')$$

in view of $a(x, 0, \xi'') = 0$. That $\psi(x, 0, \xi'') = (0, \xi'')$ follows therefore from Holmgrens uniqueness theorem. This concludes the proof of the proposition.

2. After an application of the proposition, it is now clear that the Taylor expansion p_m^T of order two of p_m at the points from Σ is

$$\tau^2 - \langle A(x, 0, \xi'')\sigma, \sigma \rangle$$

where we have set $\xi' = (\tau, \sigma)$. For any (x'', ξ'') fixed on a leaf of the canonical foliation of Σ this is a symbol of a second order hyperbolic differential operator of the type discussed in the preceding section. We can then associate with these operators the propagative sets Q_ϵ^+ constructed in corollary 6.1.13. Theorem 1.1.3 gives then the following result

Theorem 6.2.3. *Assume that coordinates have been chosen as described above and that* u *is a solution of* $p(x, D)u = 0$ *on* $X \times G$. *If we can find* ϵ *and a neighborhood* W *of* (x^0, ξ^0) *such that*

$$WF_A u \cap W \cap (Q_\epsilon^+ \setminus (x^0, \xi^0)) = \emptyset$$

then it follows that $(x^0, \xi^0) \notin WF_A u$.

This is essentially the result from Laubin [2], stated in special coordinates. For the C^∞ case it corresponds to the result of Melrose-Uhlmann [1]. That the statement has an invariant meaning is here a consequence of the constructions from chapter IV and the preceding section. A completely different argument is given in Laubin loc. cit., respectively Melrose-Uhlmann loc. cit.

6.3 The case when $p_{m,k}$ is hyperbolic

1. We mention explicitly a result along the lines from the sections 3.8, 3.9 in the case when $p_{m,k}$ is hyperbolic with respect to some direction N and when p is constant coefficient. For simplicity we shall assume k=2, although we think that this is not essential. We start by recalling some elementary inequalities concerning hyperbolic polynomials. Let us in fact assume that q is a homogeneous hyperbolic polynomial on R^d which is hyperbolic with respect to the direction $N = (1, 0, ..., 0)$. Explicitly this means that

$$q(N) \neq 0 \text{ and } q(\sigma + itN) \notin 0, \forall \sigma \in R^d, \forall t \in \dot{R}.$$

We shall write $\sigma = (\tau, \eta)$, $\tau \in R$, $\eta \in R^{d-1}$, for the variables from R^d in the phase space, respectively $s = (t, y)$, $t \in R$, $y \in R^{d-1}$ for their pre-dual variables in the base space.

Associated with q and N is the hyperbolicity cone Γ. We recall that this is the connected component of the set $\{\sigma \in R^d; q(\sigma) \neq 0\}$ which contains N; it is a remarkable fact that (cf. e.g. Hörmander [5])

$$q(\sigma) \neq 0 \text{ if Im } \sigma \in \Gamma, \text{ whatever } \sigma \in C^d \text{ is }. \tag{6.3.1}$$

Also denote Γ^0 the polar cone of Γ:

$$\Gamma^0 = \{s \in R^d; \langle s, \sigma \rangle > 0, \forall \sigma \in \Gamma\}.$$

Γ^0 will play an essential role in the propagation phenomena which we shall analyze. Technically this is based on:

Lemma 6.3.1. *Let K be a convex compact in R^{d-1} and denote by G the cone*

$$G = \{s = (t, y) \in R^d; t > 0, y/t \in K\}.$$

In particular, $y \in K$ implies $(1, y) \in G$. Also let G^0 be the polar of G. If H_K is the supporting function of K, then

$$\sigma \in G^0 \text{ precisely if } \tau > H_K(-\eta).$$

Proof. Consider $\sigma \in G^0$. Due to the definition of G this means that

$$\langle \sigma, (1, y) \rangle > 0, \forall y \in K, \text{ i.e. } \tau > \langle -\eta, y \rangle, \ \forall y \in K.$$

This gives $\tau > \max_{y \in K} \langle -\eta, y \rangle = H_K(-\eta)$. Conversely, if $\tau > \max_y \langle -\eta, y \rangle$, then $\langle \sigma, s \rangle > 0$, $\forall s \in G$, so $\sigma \in G^0$.

We want to apply lemma 6.3.1 for the case when $G = \Gamma^0$, with Γ the hyperbolicity cone of q with respect to N. We must then set

$$K = \{y \in R^{d-1}; (1, y) \in \Gamma^0\}$$

and will have $G^0 = \Gamma$ since $\Gamma^{00} = \Gamma$. It follows that $\sigma \in \Gamma$, $\tau > 0$, precisely if $\tau > H_K(-\eta)$. If now $q(\sigma) = 0$ for some $\sigma \in C^d$ with Im $\tau > 0$, we cannot have, as seen above, Im $\sigma \in \Gamma$, so we must necessarily have

$$\text{Im } \tau < H_K(-\text{Im } \eta).$$

We also observe that $0 \in K$ since $(1, 0, ..., 0) \in \Gamma$. It follows that $H_K(\xi) \geq 0$, so we obtain for all solutions of $q(\sigma) = 0$

$$\operatorname{Im} \tau_+ < H_K(-\operatorname{Im} \eta). \tag{6.3.2}$$

2. After these preliminaries, we return to the situation from section 3.7. We assume that an homogeneous polynomial p_m is given on R^n, consider ξ^0 with $p_m(\xi^0) = 0$ and associate with it a chain

$$\xi^0, \xi^1, ..., \xi^{k-1}, \ p_{m,0}, p_{m,1}, ..., p_{m,k}, \ s(0), s(1), ..., s(k-1),$$

which satisfies the conditions a),b),c) from section 3.7. As there we shall assume that coordinates are chosen with $\xi^j = (0, ..., 0, 1, 0, ..., 0)$ with the "1" on position $n - j$. Later on we shall also sometimes regard $p_{m,k}$ as a polynomial on R^{n-k}, the latter being identified with the subspace $\{\xi; \xi_{n-k+1} = \cdots \xi_n = 0\}$ of R^n.

As for the condition that $p_{m,k}$ be elliptic on R^{n-k}, called condition d) in section 1.1, we replace it here by the condition

e) $p_{m,k}$ is hyperbolic with respect to the direction $N = (1, 0, ..., 0) \in R^{n-k}$.

It should be noted that an important case when e) is valid is when p itself is hyperbolic with respect to N. Indeed, it is wellknown (cf. e.g. Hörmander [5], vol.I) that if some polynomial is hyperbolic with respect to some direction, then all localizations are also hyperbolic with respect to that direction. Actually less is needed in that it suffices to assume that p_m is microhyperbolic with respect to the direction N. We mention this explicitly since microhyperbolicity is clearly related to some of our conditions later on in these notes. Let us then at first recall the following definition due to Andersson [1] and Kawai-Kashiwara [2]. (The present formulation is from Hörmander [5])

Definition 6.3.2. *A real-analytic function f in the open set $X \subset R^n$, $0 \in X$, is called microhyperbolic with respect to $\theta \in \dot{R}^n$ at 0 if there is a neighborhood U of 0 and a positive continuous function $t(x)$ on U such that*

$$f(x + it\theta) \neq 0, \ if \ 0 < t < t(x), x \in U.$$

It is wellknown that if f is microhyperbolic with respect to θ and if f^T is its localization polynomial at 0, then f^T is hyperbolic with respect to θ.

Our last assumption on our total operator p is that it is of form

$$p = p_m + R,$$

where R is an arbitrary lower order linear partial differential operator with constant coefficients. Our main result from this section is:

Proposition 6.3.3. *Let $x^0 = (1, 0, ..., 0)$ and assume that the above conditions are valid for the operator p with $k=2$. Denote by $\Gamma \subset R^{n-2}$ the hyperbolicity cone of $p_{m,2}$ and by*

$$K \subset R^{n-3}, \ K = \{(x_2, ..., x_{n-2}); (1, x_2, ..., x_{n-2}) \in \Gamma^0\}.$$

Consider further a solution u of $p(D)u = 0$ which is defined on a neighborhood of the convex hull of the set

$$\{(0, x_2, ..., x_{n-2}, 0, 0) \in R^n; (x_2, ..., x_{n-2}) \in K\} \cup \{x^0\}$$

and for which

$$((0, y, 0, 0), \xi^0, \xi^1) \notin WF_A^2 u, \ \forall y \in K.$$

Then it follows that

$$(x^0, \xi^0, \xi^1) \notin WF_A^2 u.$$

Remark 6.3.4. *a) Under the stronger assumption that p itself is hyperbolic with respect to N essentially the same conclusion is obtained in Laubin-Esser [1] with completely different methods.*

b) A similar result is probably true for general k, provided one works with the wave front set $WF_{A,s}^k$.

c) The statement of proposition 6.3.3 is of "conical refraction" type, at the level of WF_A^2. Arguing as in the last part of the proof of theorem 1.1.6 we can obtain from it statements on conical refraction for WF_A. As in the related case of proposition 3.10.2 no new ideas are involved in the passage to results for WF_A, so we do not state any such result explicitly.

3. The proof of proposition 6.3.3 is closely related to that of proposition 3.10.2, although it is easier at one essential point. To simplify the situation we shall assume that p is homogeneous, i.e. that $R \equiv 0$. (We do not make this assumption in lemma 6.3.5, to make clear what kind of complications appear in the general case.) The first remark is to observe that it suffices to show that

$$((0, y, 0, 0), \xi^0, \xi^1, \mu) \notin WF_{A,s}^3 u, \ \forall y \in K, \ \forall \mu \tag{6.3.3}$$

implies

$$((1 - t, ty, 0, 0), \xi^0, \xi^1, \mu) \notin WF_{A,s}^3 u, \ \forall y \in K, \ \forall \mu, \forall 0 \leq t \leq 1. \tag{6.3.4}$$

In fact, if we can show this, then proposition 6.3.3 itself follows from the micro-Holmgren theorem, applied at the level of WF_A^2. The fact that $WF_{A,s}^3$ comes in in this proof, is of course no surprize, since the assumptions involve $p_{m,2}$, which is an entity computed for the purpose of tri-microlocalization. We also observe that homogeneity reasons show that it suffices to prove (6.3.4) for t=0. We shall later on use the notations $\zeta' = (\zeta_2, ..., \zeta_n)$, $\tau = \zeta_1$, $\zeta'' = (\zeta_2, ..., \zeta_{n-1})$.

4. The first thing which we need is an inequality along the lines of the inequalities from section 3.9. What we get using (6.3.2), rather than the assumption $\bar{d}$) from that section is (arguing as there):

Lemma 6.3.5. *Let all the assumptions be as in proposition 6.3.3. We can then find $c, c', c'', M, \delta < 1, \beta'$, and a bineighborhood Γ of ξ^0, ξ^1 such that*

$$Im\,\tau_+ \le H_K(\,Im\,\zeta'') + c''(1 + |\zeta|^\delta) + c|\zeta'|^{1+\beta'}/|\zeta|^{\beta'} + c|\zeta''|^{1+\beta'}/|\zeta'|^{\beta'}, \qquad (6.3.5)$$

$$if\ |p(-\zeta)| \le c'(1 + |\zeta|)^{-M}, \quad Re\,\zeta \in \Gamma.$$

Moreover, when $p = p_m$, we may take here $c'' = 0$, i.e. the term $c'' \ln(1 + |\zeta|^\delta)$ is not needed in (6.3.5).

Remark 6.3.6. *The situation in proposition 6.3.3 is considerably easier (we discuss this later on) if instead of (6.3.5) we have the stronger inequality*

$$Im\,\tau_+ \le H_K(\,Im\,\zeta'') + c(1 + |\zeta|^\delta)\ if\ |p(-\zeta)| \le c'(1 + |\zeta|)^{-M}, \quad Re\,\zeta \in \Gamma.$$

Note that this is the case when p itself is hyperbolic.

5. We shall use arguments from the proof of the micro-Holmgren theorem. The main step in the argument is the following proposition:

Proposition 6.3.7. *Under the assumptions made in proposition 6.3.3 the following is true: $\exists \beta$, $\exists \beta'$, $\forall d'$, $\forall r' \in \mathcal{R}(\xi^0, \xi^1, \beta, \delta = 0)$, $\forall \varepsilon'$, $\forall b'$, $\exists d$, $\exists r \in \mathcal{R}(\xi^0, \xi^1, \beta', 0)$, $\exists \varepsilon$, $\exists b$, $\exists c$ such that every $h \in A(C^n)$ which satisfies*

$$|h(\lambda)| \le exp\,[d\,r(-Re\,\zeta) + Im\,\tau_+ + \varepsilon|Im\,\lambda| + b\,ln(1 + |\zeta|)]$$

can be decomposed in the form

$$h = h_1 + h_2 + {}^t p(-\zeta)h_3, \qquad (6.3.6)$$

with $h_i \in A(C^n)$ satisfying

$$|h_1(\zeta)| \le c\ exp\,[d'r'(-Re\,\zeta) + H_K(Im\,\zeta'') + \varepsilon'|Im\,\zeta| + b'ln(1 + |\zeta|)], \qquad (6.3.7)$$

$$|h_2(\zeta)| \le c\ exp\,[max\,(Im\,\tau_+, H_K(Im\,\zeta'')) + \varepsilon'|Im\,\zeta| + b'ln(1 + |\zeta|)]. \qquad (6.3.8)$$

To see that this suffices to prove that (6.3.3) implies (6.3.4) (with $t = 0$), let us fix $\beta, c_1, d'', \varepsilon'', b'', r'' \in \mathcal{R}(\xi^0, \xi^1, \beta, 0)$ such that $w \in E'(R^n)$ together with

$$|\hat{w}(\zeta)| \leq \exp\left[d''r''(-\operatorname{Re}\zeta) + H_K(\operatorname{Im}\zeta'') + \varepsilon''|\operatorname{Im}\zeta| + b''\ln(1 + |\zeta|)\right]$$

implies $|u(w)| \leq c_1$. Such β, c_1, d'', ε'', b'', r'' do exist precisely since

$$((0, y, 0, 0), \xi^0, \xi^1, \mu) \notin WF^3_{A,s}\, u, \forall \mu, \forall y \in K.$$

We now apply proposition 6.3.7 for $d' = d''/2$, $r'' = r'$, $\varepsilon' = \varepsilon''/2$ and $b'' << b'$ and get d, r, ε, b. If then $v \in E'(R^n)$ satisfies

$$|\hat{v}(\zeta)| \leq \exp\left[dr(-\operatorname{Re}\zeta) + \operatorname{Im}\tau_+ + \varepsilon|\operatorname{Im}\zeta| + b\ln(1 + |\zeta|)\right], \qquad (6.3.9)$$

then we can decompose $\hat{v}$ as in proposition 6.3.7. It follows with $v^i = F^{-1}h_i$ that

$$v = v^1 + v^2 + {}^t p(D)v^3 \qquad (6.3.10)$$

where

$$|\hat{v}^1(\zeta)| \leq c\exp\left[d''r''(-\operatorname{Re}\zeta)/2 + H_K(\operatorname{Im}\zeta'') + \varepsilon''|\operatorname{Im}\zeta|/2 + b''\ln(1 + |\zeta|)\right], \quad (6.3.11)$$

$$|\hat{v}^2(\zeta)| \leq c\exp\left[\max\left(\operatorname{Im}\tau_+, H_K(\operatorname{Im}\zeta'')\right) + \varepsilon''|\operatorname{Im}\zeta|/2 + b''\ln(1 + |\zeta|)\right]. \qquad (6.3.12)$$

Here (6.3.10) almost gives

$$u(v) = u(v^1) + u(v^2) \qquad (6.3.13)$$

and (6.3.11) almost leads to $|u(v^1)| \leq c_2$. We can then ideally hope to obtain that $|u(v)| \leq C$ for any v as in (6.3.9). This of course would give that $(tx^0, \xi^0, \xi^1) \notin WF^2_A u$ for any t between zero and one. The only problem with (6.3.13) and $|u(v^1)| \leq c_2$ is that v^1 and v^3 are actually analytic functionals and not, e.g., C_0^∞ functions. Since $|u(v^2)| \leq c_3$ if b'' is chosen suitably (i.e., large negative), it then essentially remains to give a meaning to the two statements made just before. The way out is here to approximate the v^i with distributions v^i_j, $j = 1, 2, 3, ...$, with compact support, in such a way that the

$$v^1_j + v^2_j + {}^t p(D)v^3_j$$

approximate

$$v = v^1 + v^2 + {}^t p(D)v^3$$

well enough. Actually, what we want is that $\hat{v}^1$ be approximated by $\hat{v}^1_j$ in the "class" of estimates (6.3.11), $\hat{v}^2$ by $\hat{v}^2_j$ in the class of estimates (6.3.12), etc. The argument is here similar to the one in section 2.3 and we omit further details. (We should mention however, that an additional complication comes from the fact that we have also to consider terms of form $p(-\zeta)\hat{v}^3$. How to adapt the argument from section 2.3 to this situation is clear, e.g., when one compares with the situation in section 2.2 from Liess [2]. Also cf. the proof of proposition 6.3.7 itself.)

6. We are essentially left with the proof of proposition 6.3.7. The situation is similar to that in the proof of theorem 2.1.12. We would like to argue as in the proof of lemma 2.3.1, but we see that we have an additional term $p(-\zeta)h_3$. The reason why we need such a term is that a decomposition of h into a sum $h = h_1 + h_2$ with h_1, h_2 satisfying the inequalities (6.3.11), (6.3.12) can only be performed in a neighborhood of $V_p = \{\zeta \in C^n; \ p(-\zeta) = 0\}$. The relation between decompositions of type

$$h = h^1 + h^2 \text{ in some special neighborhood of } V_p \tag{6.3.14}$$

and decompositions of type

$$h = h_1 + h_2 + p(-\zeta)h_3 \text{ on } C^n \tag{6.3.15}$$

is given by the following lemma, which is a consequence of the fundamental principle. (Cf. here section 3.11.)

Lemma 6.3.8. *Let $c > 0, M \geq 0$ be given. For every $d'' > 0$, $\varepsilon'' > 0$, $b'' \in R$, $r \in \mathcal{R}(\xi^0, \xi^1, \beta, 0)$, we can find $c' > 0$, $d' > 0$, $\varepsilon' > 0$, $b' \in R$, c'', M', so that for every*

$$h^i \in A(\zeta \in C^n; \ |p(-\zeta)| < c(1 + |\zeta|^2)^{-M}), i = 1, 2,$$

which satisfy

$$|h^1(\zeta)| \leq \ exp\ [d'r(-Re\,\zeta) + H_K(Im\,\zeta'') + \varepsilon'|Im\,\zeta| + b'ln(1 + |\zeta|)],$$

respectively

$$|h^2(\zeta)| \leq \ exp\ [max\ (Im\,\tau_+, H_K(Im\,\zeta'')) + \varepsilon'|Im\,\zeta| + b'ln(1 + |\zeta|)],$$

on their domains of definition, we can find $h_1, h_2 \in A(C^n)$ so that

$$|h_1(\zeta)| \leq c' \ exp\ [d''r(-Re\,\zeta) + H_K(Im\,\zeta'') + \varepsilon''|Im\,\zeta| + b''ln(1 + |\zeta|)],$$

respectively

$$|h_2(\zeta)| \leq c' \ exp\ [max\ (Im\,\tau_+, H_K(Im\,\zeta'')) + \varepsilon''|Im\,\zeta| + b''ln(1 + |\zeta|)],$$

and so that for some $h'' \in A(\zeta \in C^n; \ |p(-\zeta)| \leq c''(1 + |\zeta|^2)^{-M'})$ we have

$$h^i - h_i = p(-\zeta)h^{i'}, i = 1, 2, \ if\ |p(-\zeta)| \leq c''(1 + |\zeta|^2)^{-M'}.$$

Moreover, c'', M' can here be chosen actually independent of $d'', \varepsilon'', b'', r$.

(In the preceding lemma, h^1 and h^2 are not directly related one to the other.)

It follows in particular that if $h \in A(C^n)$ and

$$h = h^1 + h^2 \text{ on } |p(-\zeta)| < c(1 + |\zeta|^2)^{-M}$$

and if h_1, h_2 are associated with h^1, h^2 as in lemma 6.3.8, then we can find $h_3 \in A(C^n)$ such that

$$h = h_1 + h_2 + p(-\zeta)h_3,$$

so that we can indeed pass from a decomposition of type (6.3.14) to one of type (6.3.15).

Remark 6.3.9. *It appears from this argument that rather than working on a full neighborhood of V, one should work directly on V, using analytic functions defined on the "multiplicity variety" associated with $p(-\zeta)$. (For definitions cf. Ehrenpreis [1], Palamodov [1] and also section 3.11. The terminology is that of Ehrenpreis.) Direct work on V is however unpleasant because of the singularities V may have. (Cf. anyway Liess [2] for a related situation where it was necessary to work much more on V than it is here.) The reason now why we work on sets of type $|p(-\zeta)| < c(1 + |\zeta|^2)^{-M}$ is that these sets are domains of holomorphy, if M is rational (cf. proposition 6.3.10 below) and that they are not excessively narrow at infinity. (Cf. proposition 6.3.12 below.)*

7. Having made up our mind that we want to work on subsets of $\{\zeta; |p(-\zeta)| \leq c(1 + |\zeta|^2)^{-M}\}$, we are forced to look into the situation from lemma 2.3.1 on such sets. In particular this means that we have to solve $\bar{\partial}$- systems on sets of this type. Since $\bar{\partial}$- equations are solved best on domains of holomorphy, it is useful to observe that we have:

Proposition 6.3.10. *a) Let $M > 0$ be a rational number and let $c > 0$ be given. Then*

$$U = \{\zeta \in C^n; |p(-\zeta)| < c(1 + |\zeta|^2)^{-M}\}$$

is a domain of holomorphy.
b) For any c there is c' and a domain of holomorphy U so that

$$\{\zeta \in C^n; c' + |Im\,\zeta| < |Re\,\zeta|/c'\} \subset U \subset \{\zeta \in C^n; c + |Im\,\zeta| < |Re\,\zeta|/c\}.$$

Proof. a) If $\gamma = (\gamma_1, ..., \gamma_n), \gamma_i \in C, |\gamma_i| = 1$, is given and $M = M'/M''$ for some positive integers M', M'', then

$$U_\gamma = \{\zeta \in C^n; |p(-\zeta)|^{2M''}|(1 + \sum(\gamma_i\zeta_j)^2)|^{2M'} < c\}$$

is a domain of holomorphy and we have

$$U = \cap_\gamma U_\gamma \ ,$$

where the intersection is over all γ as above.

b) It suffices to show that the set

$$V = \{\zeta \in C^n; c_1 + |Im\,\zeta|^2 < |Re\,\zeta|^2, |Im\,\zeta|^2 < c_2 |Re\,\zeta|^2\},$$

is a domain of holomorphy for any choice of c_1, provided $c_2 \leq 1$. To check that V is a domain of holomorphy, we observe that so is $V' = \{\zeta \in C^n; c_1 + |Im\,\zeta|^2 < |Re\,\zeta|^2\}$ since it is the preimage of the set $\{z \in C; Re\,z < 0\}$ under the holomorphic map $\zeta \to z = -c_1 + \sum_{j=1}^n \zeta_j^2$. It suffices then to observe that the domain $V'' = \{\zeta \in C^n; |Im\,\zeta|^2 < c_2 |Re\,\zeta|^2\}$ is given by the equation $\rho(\zeta) < 0$, where $\rho(\zeta) = |Im\,\zeta|^2 - c_2 |Re\,\zeta|^2$ is plurisubharmonic when $c_2 \leq 1$, is C^∞ for $\zeta \neq 0$ and satisfies $\mathrm{grad}_{\zeta,\bar\zeta}\rho(\zeta) \neq 0$ when $\rho = 0$, $\zeta \neq 0$.

8. When solving $\bar\partial$-equations on domains of holomorphy, we shall essentially rely on Hörmanders theory from [5]. Actually, the theory from that paper refers to L^2-estimates, whereas here we work with point-wise, i.e. in fact sup-norm, estimates. Fortunately we are not interested in optimal estimates and may also shrink domains of definition. What one can then do, is to pass from some initial sup-norm estimates to related (but somewhat weaker) L^2-estimates, solve the $\bar\partial-$system on U, in the frame of L^2-estimates, and recover sup-norm estimates on a smaller domain for the solution using the ellipticity of the $\bar\partial-$system. Explicitly this leads to the following result, which we recall from Liess [2].

Proposition 6.3.11. *Let $U \subset C^n$ be a domain of holomorphy, consider a plurisubharmonic function $\varphi : U \to R$ and assume $\psi : U \to R$ is a function such that $\varphi(\zeta) \leq \psi(\zeta)$, $\forall \zeta \in U$, and such that we can find $\rho > 0$ and $c' > 0$ for which $|\psi(\zeta^1) - \psi(\zeta^2)| \leq c'$ whenever $|\zeta^1 - \zeta^2| \leq \rho$. Then there are constants $c > 0$ and $\mu \geq 0$ such that the system*

$$\bar\partial g = W$$

has a solution $g \in C^\infty(U)$ with

$$|g(\zeta)| \leq c(1 + (dist\,[\zeta, (C^n \setminus U)])^{-n})\, exp\,[\psi(\zeta) + \mu\,ln(1 + |\zeta|)] \tag{6.3.16}$$

for any $(0,1)$ form $w \in C^\infty_{(0,1)}(U)$ with $\bar\partial w = 0$ which satisfies $|w(\zeta)| \leq exp\,\varphi(\zeta)$.

9. We see in particular from the preceding proposition that estimates become bad near the boundary of U. In the present context this is not disturbing, since we can stay away

from the boundary of U by shrinking the domain on which we consider the solution g of $\bar{\partial}g = w$. That while doing so, we still remain in the class of domains of type $\{\zeta; |p(-\zeta)| < c(1 + |\zeta|^2)^{-M}\}$, is a consequence of the following result:

Proposition 6.3.12. *Let c', M' be given and denote by*

$$U' = \{\zeta; |p(-\zeta)| < c'(1 + |\zeta|^2)^{-M'}\}.$$

If $c'' > 0$ and $M'' > 0$ are suitable, we can find $c > 0$ and $M > 0$ so that

$$|p(-\zeta)| < c''(1 + |\zeta|)^{-M''} \text{ implies } \text{dist}\,(\zeta, C^n \setminus U') > c(1 + |\zeta|^2)^{-M}.$$

Remark 6.3.13. *The bad term* $\text{dist}\,(\zeta, (C^n \setminus U))$ *in (6.3.16) can thus be estimated by* $C(1 + |\zeta|)^C$ *if we only look at the ζ with $|p(-\zeta)| \le \tilde{c}(1 + |\zeta|)^{-\tilde{M}}$, $\tilde{c}, \tilde{M}$ suitable.*

Proof. It is a well-known consequence of the Tarski-Seidenberg principle (cf. Palamodov [1]) that we can find $c_1, c_2, \beta', \beta'', \mu', \mu''$ so that

$$|p(-\zeta)| \le c_1 \, \text{dist}\,(\zeta, V_p)^{\beta'}(1 + |\zeta|^2)^{\mu'}, \forall \zeta \in C^n,$$

respectively,

$$\text{dist}\,(\zeta, V_p) \le c_2|p(-\zeta)|^{\beta''}(1 + |\zeta|^2)^{\mu''}, \forall \zeta \in C^n.$$

Consider then μ with $|p(-\mu)| \ge c'(1 + |\mu|^2)^{-M'}$ and λ with $|p(-\lambda)| \le c''(1 + |\lambda|^2)^{-M''}$, for some c'' and M'' still to be determined. We must show that if c'', M'' are suitable, then we can find c, M so that under the above assumptions

$$|\zeta - \mu| \ge c(1 + |\zeta|^2)^{-M}.$$

This is only a problem if $|\zeta - \mu| \le 1$ (say) and our initial problem is anyway easy for $|\zeta|$ small. We may therefore assume without loss of generality that $|\lambda| \sim |\mu|$. We now have

$$\text{dist}\,(\mu, V_p) \ge (1/c_1)^{1/\beta'}|p(-\mu)|^{1/\beta'}(1 + |\mu|^2)^{-\mu'/\beta'} \ge c_3(1 + |\mu|^2)^{-M'/\beta'-\mu'/\beta'}$$

and

$$\text{dist}\,(\zeta, V_p) \le c_2|p(-\zeta)|^{\beta''}(1 + |\zeta|^2)^{\mu''} \le c_4(1 + |\zeta|^2)^{-M''\beta''+\mu''}.$$

Assuming that $-M'/\beta' - \mu'/\beta' > M''\beta'' + \mu''$, the desired inequality now follows if we use that $|\zeta - \mu| \ge \text{dist}\,(\mu, V_p) - \text{dist}\,(\zeta, V_p)$.

10. Our next remark is that actually we need not necessarily perform the decomposition $h = h^1 + h^2$ on all of $|p(-\zeta)| < c(1 + |\zeta|^2)^{-M}$, in that it suffices to prove

Proposition 6.3.14. *Let $c > 0$ be given. $\exists \tilde{c}$, $\forall \beta > 0, \forall c', \forall M', \forall d'', \forall r'' \in \mathcal{R}(\xi^0, \xi^1, \beta, 0)$, $\forall \varepsilon'', \forall b'', \exists \beta', \exists d', \exists r' \in \mathcal{R}(\xi^0, \xi^1, \beta', 0), \exists \varepsilon', \exists c'', \exists M', \exists b', \exists c_1$ such that if*

$$h \in A(\zeta \in C^n; |p(-\zeta)| < c'(1 + |\zeta|^2)^{-M'}, c + |Im\,\zeta| < |Re\,\zeta''|/c)$$

satisfies

$$|h(\zeta)| \leq exp\,[d'r'(-Re\,\zeta) + Im\,\tau_+ + \varepsilon'|Im\,\zeta| + b'ln(1 + |\zeta|)],$$

then we can find

$$h'_i \in A(\zeta \in C^n; |p(-\zeta)| < c''(1 + |\zeta|^2)^{-M''}, \tilde{c} + |Im\,\zeta| < |Re\,\zeta''|/\tilde{c})$$

with

$$h = h'_1 + h'_2,$$

$$|h'_1(\zeta)| \leq c_1\,exp\,[d''r''(-Re\,\zeta) + H_K(Im\,\zeta'') + \varepsilon''|Im\,\zeta| + b''ln(1 + |\zeta|)],$$

$$|h'_2(\zeta)| \leq c_1\,exp\,[max\,(Im\,\tau_+, H_K(Im\,\zeta'')) + \varepsilon''|Im\,\zeta| + b''ln(1 + |\zeta|)].$$

Indeed, once proposition 6.3.12 is proved, we can obtain a decomposition on a set of form $\{\zeta \in C^n; |p(-\zeta)| < c'(1 + |\zeta|^2)^{-M'}\}$ if we also use:

Proposition 6.3.15. *Let $c_1 > 0$ be given.$\forall r \in \mathcal{R}(\xi^0, \xi^1, \beta, 0), \forall \varepsilon''' > 0, \forall b''', \forall c, \forall c_1, \forall M$, $\exists d', \exists \varepsilon', \exists b', \exists c', \exists M', \exists c_2$, such that for any*

$$h'_1 \in A(\zeta \in C^n; |p(-\zeta)| < c(1 + |\zeta|^2)^{-M}, c_1 + |Im\,\zeta| < |Re\,\zeta''|/c_1)$$

which satisfies

$$|h'_1(\zeta)| \leq exp\,[d'r(-Re\,\zeta) + H_K(Im\,\zeta'') + \varepsilon'|Im\,\zeta| + b'ln(1 + |\zeta|)],$$

there exist

$$h_1 \in A(\zeta \in C^n; |p(-\zeta)| < c'(1 + |\zeta|^2)^{-M'})$$

and

$$f_1 \in C^\infty(\{\zeta \in C^n; |p(-\zeta)| < c'(1 + |\zeta|^2)^{-M'}, 2c_1 + |Im\,\zeta| < |\,Re\,\zeta''|/(2c_1)\})$$

such that

$$h'_1 = h_1 + f_1,$$

$$|h_1(\zeta)| \leq c_2\,exp\,[d'r(-Re\,\zeta) + H_K(Im\,\zeta'') + \varepsilon'''|Im\,\zeta| + b'''ln(1 + |\zeta|)],$$

$$|f_1(\zeta)| \leq c_2\,exp\,[H_K(Im\,\zeta'') + \varepsilon'''|Im\,\zeta| + b'''ln(1 + |\zeta|)].$$

Note in particular that in the part

$$\{\zeta \in C^n;\, |p(-\zeta)| < c'(1 + |\zeta|^2)^{-M'},\, 2c_1 + |\operatorname{Im}\zeta| \geq |\operatorname{Re}\zeta''|/(2c_1)\},$$

which is essentially what we "gain" in proposition 6.3.15, we have the estimate

$$|h_1(\zeta)| \leq c_2 \exp\left[(2d'c_1 + \varepsilon''')|\operatorname{Im}\zeta| + H_K(\operatorname{Im}\zeta'') + b'''ln(1 + |\zeta|)\right].$$

(Recall here that $r(\xi) \leq |\xi''|$.) When combining the propositions 6.3.12 and 6.3.15, we can write any $h \in A(\zeta \in C^n;\, |p(-\zeta)| < c(1 + |\zeta|)^{-M})$ which satisfies

$$|h(\zeta)| \leq \exp\left[dr(-\operatorname{Re}\zeta) + \operatorname{Im}\tau_+ + \varepsilon|\operatorname{Im}\zeta| + b\,ln(1 + |\zeta|)\right]$$

for suitable d, r, ε, b in the form

$$h = h_1' + h_2' = h_1 + (h_2' + f_1).$$

The function h_1 is then of the type desired in proposition 6.3.11 and the same is also true for $h_2 = h_2' + f_1$, or rather for its holomorphic extension to $|p(-\zeta)| < c'(1 + |\zeta|)^{-M'}$. In fact, the holomorphic extension is just $h - h_1$ and the estimates for $h - h_1$ are of the desired type if we have started from suitable $d''', \varepsilon''', b''', r$, in view of the remark on h_1 and a similar remark for h.

Proof of proposition 6.3.15. (Sketch) The proof is rather simple. As above we observe that

$$|h_1'(\zeta)| \leq c'' \exp\left[(2c_1d' + \varepsilon')|\operatorname{Im}\zeta''| + H_K(\operatorname{Im}\zeta'') + b'\,ln(1 + |\zeta|)\right]$$

on the set $\{\zeta \in C^n;\, 2c_1 + |\operatorname{Im}\zeta| > |\operatorname{Re}\zeta|/(2c_1)\}$. A C^∞-correction of the desired type can therefore be found with the aid of a $\bar\partial$ argument of the type used below. We omit details.

11. We are left with the proof of proposition 6.3.12. To understand how we want to argue, we assume for a moment that, rather than having the inequality (6.3.5), the inequality

$$\operatorname{Im}\tau_+ \leq H_K(\operatorname{Im}\zeta'') \text{ if } |p(-\zeta)| \leq c(1 + |\zeta|)^{-M},\ c + |\operatorname{Im}\zeta| < |\operatorname{Re}\zeta''|/c \qquad (6.3.17)$$

holds, if $c > 0$ and $M \geq 0$ are suitably. It would follow for h as in proposition 6.3.12 that

$$|h(\zeta)| \leq c' \exp\left[dr(-\operatorname{Re}\zeta) + H_K(\operatorname{Im}\zeta'') + \varepsilon|\operatorname{Im}\zeta| + b\,ln(1 + |\zeta|)\right],$$

so we simply could set $h_1' = h, h_2' = 0$ and were done. Unfortunately, (6.3.5) is weaker than (6.3.17) in two respects:

- the set on which the inequality (6.3.5) is valid is smaller, in that it refers only to a bineighborhood of (ξ^0, ξ^1),

- the inequality itself is (there where it is valid) weaker.

To overcome these difficulties, we shall effectively use true decompositions of h and, implicitely, also the fact that in the definition of $WF_{A,s}^3$ we are interested anyway mainly in domains of form $|\xi''| > c'|\xi'|^{1+\beta}/|\xi_n|^\beta$.

To make clear how this method works we shall now at first assume that h satisfies an inequality which is somewhat stronger than what we effectively know and show later how one can reduce the general case to the one we can already treat.

Proposition 6.3.16. *Let $\beta < \beta'$, where β' is the constant from lemma 6.3.5. Consider $d'', r'' \in \mathcal{R}(\xi^0, \xi^1, \beta, 0), \varepsilon'', c', M', b'', c_1$.*
Then we can find $\varepsilon', b', c'', M'', c_2, c_3$ such that for every d_1, d_2 there is a bineighborhood Γ of (ξ^0, ξ^1) with the following property: if $\Gamma' \subset\subset \Gamma$ and if $h \in A(\zeta \in C^n; |p(-\zeta)| < c'(1 + |\zeta|)^{-M'}, c_1 + |Im\,\zeta| < |Re\,\zeta''|/c_1)$ satisfies

$$|h(\zeta)| \leq \; exp \; [d'' r''_{\Gamma''}(-Re\,\zeta) + Im\,\tau_+ - d_1|Re\,\zeta'|_\Gamma^{1+\beta}/|Re\,\zeta|^\beta - d_2|Re\,\zeta''|_\Gamma^{1+\beta}/|Re\,\zeta'|^\beta$$
$$+ \; \varepsilon'|Im\,\zeta| + b'ln(1 + |\zeta|)].$$

then we can find $h_i \in A(\zeta \in C^n; |p(-\zeta)| < c''(1 + |\zeta|)^{-M''}, c_2 + |Im\,\zeta| < |Re\,\zeta''|/c_2)$. $i = 1, 2$, such that

$$h = h_1 + h_2,$$

$$|h_1(\zeta)| \leq c_3 \; exp \; [d'' r''(-Re\,\zeta) + H_K(Im\,\zeta'') + \varepsilon''|Im\,\zeta| + b''ln(1 + |\zeta|)],$$

$$|h_2(\zeta)| \leq c_3 \; exp \; [max \; (Im\,\tau_+, H_K(Im\,\zeta'')) + \varepsilon''|Im\,\zeta| + b''ln(1 + |\zeta|)].$$

To simplify the notations we shall denote henceforth in this section for $g : R^n \to R$ by g_Γ the function $g_\Gamma(\xi) = g(\xi)$ if $\xi \in \Gamma$, $g_\Gamma(\xi) = 0$ if $\xi \notin \Gamma$.

Proof of proposition 6.3.16. Let d_1, d_2 be given. We choose Γ so that

$$Im\,\tau_+ \leq H_K(Im\,\zeta'') + d_1|Re\,\zeta'|^{1+\beta}/|Re\,\zeta|^\beta + d_2|Re\,\zeta''|^{1+\beta}/|Re\,\zeta'|^\beta.$$

if $|p(-\zeta)| < c'(1 + |\zeta|)^{-M'}, Re\,\zeta \in \Gamma$. The existence of such a Γ follows from lemma 6.3.5 if we note that

$$|Re\,\zeta''|^{\beta'-\beta}/|Re\,\zeta'|^{\beta'-\beta} \cdot |Re\,\zeta'|^{\beta'-\beta}/|Re\,\zeta|^{\beta'-\beta}.$$

will be arbitrarily small in Γ, provided we shrink Γ sufficiently. Since $d'' r''_{\Gamma''}(-Re\,\zeta) = 0$ for $Re\,\zeta \notin -\Gamma''$ we can therefore apply the following lemma. in which we denote for fixed c, c_1, M by $U(c, c_1, M)$ the set

$$U(c, c_1, M) = \{\zeta \in C^n; |p(-\zeta)| < c(1 + |\zeta|)^{-M}, c_1 + |Im\,\zeta| < |Re\,\zeta|/c_1\}.$$

Lemma 6.3.17. *Let c, c_1, M and bicones $\Gamma' \subset\subset \Gamma$ be given and let*

$$\varphi_i : U(c, c_1, M) \to R, \ i = 1, 2 \ and \ \psi : U(c, c_1, M) \to R$$

be functions such that

$$\varphi_2(\zeta) \le \psi(\zeta) \ if \ Re \ \zeta \in \Gamma,$$

$$\varphi_1(\zeta) = 0 \ if \ Re\zeta \notin \Gamma',$$

ψ *is Lipschitz continuous and plurisubharmonic.*

Then we can find c', c_1', M' such that for every ε'', b'' there are ε', b' with the following property: every $h \in A(U(c, c_1, M))$ which satisfies

$$|h(\zeta)| \le \ exp \ [\varphi_1(\zeta) + \varphi_2(\zeta) + \varepsilon'|Im\zeta| + b'ln(1 + |\zeta|)], \zeta \in U(c, c_1, M)$$

can be decomposed in the form

$$h = h_1 + h_2$$

for some $h_1, h_2 \in A(U(c', c_1', M'))$ which satisfy

$$|h_1(\zeta)| \le c_2 \ exp[\varphi_1(\zeta) + \psi(\zeta) + \varepsilon''|Im\zeta| + b''ln(1 + |\zeta|)], \zeta \in U(c', c_1', M'),$$

$$|h_2(\zeta)| \le c_2 \ exp[max \ (\varphi_2(\zeta), \psi(\zeta)) + \varepsilon''|Im\zeta| + b''ln(1 + |\zeta|)], \zeta \in U(c', c_1', M').$$

Proof of lemma 6.3.17. The proof is straightforward. Using a C^∞ partition of unity associated with the two sets

$$\{\zeta \in U(c, c_1, M); Re \ \zeta \in \Gamma\}, \{\zeta \in U(c, c_1, M); Re \ \zeta \notin \Gamma'\},$$

we can find $f_1, f_2 \in C^\infty(U(c, c_1, M))$ such that

$$|f_1(\zeta)| + |f_2(\zeta)| \le |h(\zeta)|,$$

$$|f_1(\zeta)| \le \ exp[\varphi_1(\zeta) + \psi(\zeta) + \varepsilon'|Im \ \zeta| + b'ln(1 + |\zeta|)],$$

$$|f_2(\zeta)| \le \ exp[\varphi_2(\zeta) + \varepsilon'|Im \ \zeta| + b'ln(1 + |\zeta|)]$$

and such that

$$|\bar{\partial} f_1(\zeta)| \le c_2 \exp[\psi(\zeta) + \varepsilon'|Im \ \zeta| + b'ln(1 + |\zeta|)].$$

From our discussion above on the solvability of the $\bar{\partial}$-system it follows that we can find c', c_1' and $g \in C^\infty(U(c', c_1', M))$ such that $\bar{\partial} f_1 = \bar{\partial} g$ and such that

$$|g(\zeta)| \le c_3 \exp[\psi(\zeta) + \varepsilon''|Im \ \zeta| + b''ln(1 + |\zeta|)].$$

We may then set $h_1 = f_1 - g, h_2 = f_2 + g$.

It remains to explain how we can reduce the general case in proposition 6.3.12 to the case considered in the preceding result. This is the content of the following two propositions.

Proposition 6.3.18. *Let* $\Gamma, \Gamma', \Gamma''$ *be fixed bineighborhoods of* (ξ^0, ξ^1) *with* $\Gamma'' \subset\subset \Gamma' \subset\subset$ Γ. $\forall d'', \forall r'', \forall \varepsilon'', \forall b_2, \forall c', \forall M', \forall c, \forall c_1 \; \exists d', \; \exists r', \; \exists \varepsilon', \; \exists b_1, \; \exists c'', \; \exists c_2, \; \exists M'', \; \exists c,$ *such that* $\forall d''', \; \exists d, \; \exists c_3$ *with the following property: for any*

$$h \in A(\zeta \in C^n; |p(-\zeta)| < c'(1 + |\zeta|)^{-M'}, |Im\,\zeta| < |Re\,\zeta''|/c_1)$$

which satisfies

$$|h(\zeta)| \le \; exp \; [d'r'(-Re\,\zeta) + Im\,\tau_+ + \varepsilon'|Im\,\zeta| - d'''|Re\,\zeta'|_\Gamma^{1+\beta}/|Re\,\zeta|^\beta + b_1 ln(1 + |\zeta|)],$$

we can find $h_1, h_2 \in A(\zeta \in C^n; |p(-\zeta)| < c''(1 + |\zeta|)^{-M''}, c_2 + |Im\,\zeta| < |Re\,\zeta''|/c_2)$ *with*

$$h = h_1 + h_2,$$

$$|h_1(\zeta)| \le c_3 \; exp \; [d''r''_{\Gamma''}(-Re\,\zeta) + Im\,\tau_+ + \varepsilon''|Im\,\zeta| - d|Re\,\zeta'|_{\Gamma'}^{1+\beta}/|Re\,\zeta|^\beta$$
$$-d|Re\,\zeta''|_{\Gamma'}^{1+\beta}/|Re\,\zeta'|^\beta + b_2 ln(1 + |\zeta|)],$$

$$|h_2(\zeta)| \le c_3 \; exp \; [Im\,\tau_+ + \varepsilon''|Im\,\zeta| + b_2 ln(1 + |\zeta|)].$$

Remark 6.3.19. *One of the main features is here that* r''. $|Re\,\zeta'|$, *and* $|Re\,\zeta''|$ *are cut off to different cones. In this way we get a transition region where* r'' *is not felt anymore but where* $-d|Re\,\zeta''|^{1+\beta}/|Re\,\zeta'|^\beta$, *respectively* $-d|Re\,\zeta'|^{1+\beta}/|Re\,\zeta|^\beta$ *can still be used in estimates. Note that we have a situation of the same kind already in the assumption:* r' *is chosen later than* Γ, *so it may have a support which is not large enough to compensate the term involving* $-d'''$ *using* r'. *(The term would become useless, if compensated by* $d'r'$.)

For the proof of proposition 6.3.18 we need two lemmas.

Lemma 6.3.20. *Let* $r'' \in \mathcal{R}(\xi^0, \xi^1, \beta, 0)$ *be given. Then there is a bineighborhood* Γ_1 *of* (ξ^0, ξ^1) *so that*

$$|\xi''|^{1+\beta}/|\xi'|^\beta \le c_1|\xi'|^{1+\beta}/|\xi|^\beta + cr''(-\xi) \; if \; \xi \in \Gamma_1. \tag{6.3.18}$$

Moreover, we can shrink c_1 *as much as we please, if we shrink* Γ_1 *accordingly.*

Proof. From the definition of $\mathcal{R}(\xi^0, \xi^1, \beta, 0)$ it follows that we can find a bineighborhood Γ_1 of (ξ^0, ξ^1) and C_2, C_3 so that

$$|\xi''| \le C_3 r''(-\xi) \; if \; \xi \in \Gamma_1 \; and \; |\xi''| > C_2|\xi'|^{1+\beta}/|\xi|^\beta$$

which gives (6.3.18) in the region $|\xi''| \ge C_2|\xi'|^{1+\beta}/|\xi|^\beta$. If, on the other hand, $|\xi''| \le$ $C_2|\xi'|^{1+\beta}/|\xi|^\beta$, then (6.3.18) is valid automatically.

Our second lemma is a reformulation of a result from section 2.5.

Lemma 6.3.21. *Let $d''' > 0, \varepsilon'', \tilde{c}$ and a bineighborhood $\Gamma' \subset\subset \Gamma$ of (ξ^0, ξ^1) be given. Then there is $c, d^{IV} > 0$ and a plurisubharmonic function ψ on $|Im\,\zeta| < \tilde{c}|Re\,\zeta'|$ so that*

$$-d'''|Re\,\zeta'|_{\Gamma'}^{1+\beta}/|Re\,\zeta|^{\beta} < \psi(\zeta) < -2d^{IV}|Re\,\zeta'|_{\Gamma'}^{1+\beta}/|Re\,\zeta|^{\beta} + (\varepsilon''/4)|Im\,\zeta| + c.$$

Proof of proposition 6.3.18. Let Γ_1 be as in lemma 6.3.20 and fix $d'', d''', d^{IV}, d''' > d^{IV}$. It follows in particular that if d is small enough then

$$d|Re\,\zeta''|_{\Gamma_1}^{1+\beta}/|Re\,\zeta'|^{\beta} \le (d''/2)r_{\Gamma_1}''(-Re\,\zeta) + d^{IV}|Re\,\zeta'|_{\Gamma_1}^{1+\beta}/|Re\,\zeta|^{\beta} + c_4. \qquad (6.3.19)$$

Next we fix $\Gamma_3 \subset\subset \Gamma_2 \subset\subset \Gamma'' \cap \Gamma_1$ and a plurisubharmonic function φ on $|Im\,\zeta| < c|Re\,\zeta''|$ so that

$$\varphi(\zeta) \ge 0 \text{ for } Re\,\zeta \in \Gamma_2,$$

$$\varphi(\zeta) \le -d|Re\,\zeta''|^{1+\beta}/|Re\,\zeta'|^{\beta} + (\varepsilon''/4)|Im\,\zeta| + c_5 \text{ if } Re\,\zeta \in \Gamma \setminus \Gamma'',$$

$$\varphi(\zeta) \le (\varepsilon''/4)|Im\,\zeta| + c_5 \text{ if } Re\,\zeta \notin \Gamma \text{ or } Re\,\zeta \in \Gamma''.$$

(For a similar result, see proposition 2.5.8.) In order to find such a function φ we may of course have to shrink d. We may then also assume supp $r' \subset \Gamma_2$.
We now choose two C^∞-functions g_1, g_2 such that $h = g_1 + g_2$,

$$|g_1(\zeta)| + |g_2(\zeta)| \le |h(\zeta)|,$$

$$g_1(\zeta) = h(\zeta) \text{ for } Re\,\zeta \in \Gamma_3,$$

$$g_2(\zeta) = h(\zeta) \text{ for } Re\,\zeta \notin \Gamma_2,$$

$$|\bar{\partial}g_j(\zeta)| \le c \exp\left[Im\,\tau_+ - d'''|Re\,\zeta'|_{\Gamma'}^{\beta}/|Re\,\zeta|^{\beta} + \varepsilon'|Im\,\zeta| + b_1 ln(1 + |\zeta|)\right].$$

We can then solve $\bar{\partial}u = \bar{\partial}g_1$ on $\{\zeta; |p(-\zeta)| < c''(1 + |\zeta|)^{-M''}, c_2 + |Im\,\zeta| < |Re\,\zeta''|/c_2\}$ so that

$$|u(\zeta)| \le c \exp[Im\,\tau_+ - 2d^{IV}|Re\,\zeta'|_{\Gamma'}^{\beta}/|Re\,\zeta|^{\beta} - d|Re\,\zeta''|_{\Gamma' \setminus \Gamma''}^{1+\beta}/|Re\,\zeta'|^{\beta} + \varepsilon''|Im\,\zeta| + b_2 ln(1+|\zeta|)].$$

($c'', c_2, M'', \varepsilon', b_1, c$ are assumed suitable.) The functions $h_1 = g_1 - u, h_2 = g_2 + u$ are then analytic. Moreover it is clear that h_2 satisfies the desired estimates. As for h_1, the "part" u satisfies the inequalities required for h_1 explicitly for $Re\,\zeta \notin \Gamma''$ and g_1 vanishes for $Re\,\zeta \notin \Gamma''$. On the other hand, when $Re\,\zeta \in \Gamma''$, $d|Re\,\zeta''|^{1+\beta}/|Re\,\zeta'|^{\beta}$ is dominated by

$$(d''/2)r''(-Re\,\zeta) + d^{IV}|Re\,\zeta'|^{1+\beta}/|Re\,\zeta|^{\beta}$$

in view of (6.3.19), so also in this case the inequalities are valid.

Proposition 6.3.22. *Let $\beta > 0$ be fixed and denote by $\beta' = 2\beta + \beta^2$. $\exists \Gamma$, $\forall d_1$, $\forall \varepsilon_1$, $\forall c''$, $\forall M''$, $\forall c_2$, $\forall b_1$, $\forall r_1 \in \mathcal{R}(\xi^0, \xi^1, \beta', 0)$, $\exists d_2$, $\exists d_3$, $\exists \varepsilon_2$, $\exists c'$, $\exists M'$, $\exists c_1$, $\exists b_2$, $\exists r_2 \in \mathcal{R}(\xi^0, \xi^1, \beta, 0)$, $\exists c_3$ with the property that*

$$\forall h \in A(\zeta \in C^n; |p(-\zeta)| < c''(1+|\zeta|)^{-M''}, c_2 + |Im\,\zeta| < |Re\,\zeta''|/c_2)$$

which satisfies

$$|h(\zeta)| \leq \ exp\left[d_2 r_2(-Re\,\zeta) + Im\,\tau_+ + \varepsilon_2|Im\,\zeta| + b_2 ln(1+|\zeta|)\right], \tag{6.3.20}$$

$\exists f_1, f_2 \in A(\zeta \in C^n; |p(-\zeta)| < c'(1+|\zeta|)^{-M'}, c_1+|Im\,\zeta| < |Re\,\zeta''|/c_1)$ *such that* $h = f_1+f_2$,

$$|f_1(\zeta)| \leq c_3 \ exp\left[d_1 r_1(-Re\,\zeta) + Im\,\tau_+ + \varepsilon_1|Im\,\zeta| - d_3|Re\,\zeta'|_\Gamma^{1+\beta'}/|Re\,\zeta|^{\beta'} + b_1 ln(1+|\zeta|)\right].$$

$$|f_2(\zeta)| \leq c_3 \ exp\left[\varepsilon_1|Im\,\zeta| + b_1 ln(1+|\zeta|)\right].$$

Before we can prove the proposition 6.3.22 we need a lemma.

Lemma 6.3.23. *Let $\beta > 0, c \geq 1$ and a small open bineighborhood Γ of (ξ^0, ξ^1) be given. Denote $2\beta + \beta^2$ by β'. Then there are c_1, c_2, c_3 and for every $\Gamma'' \subset\subset \Gamma' \subset\subset \Gamma$ a plurisubharmonic function φ on $|Im\,\zeta| \leq c_1|Re\,\zeta''|$ so that*

$$\varphi(\zeta) \geq 0 \ for \ Re\,\zeta \in \Gamma'' \cap \{\xi \in R^n; |\xi''| > c|\xi'|^{1+\beta}/|\xi|^\beta\},$$

$$\varphi(\zeta) \leq -|Re\,\zeta'|^{1+\beta'}/|Re\,\zeta|^{\beta'} + c_2|Im\,\zeta| + c_3 \ if \ Re\,\zeta \in \Gamma \setminus \Gamma' \ or$$

$$Re\,\zeta \in \Gamma', |Re\,\zeta''| < (c/2)|Re\,\zeta'|^{1+\beta}/|Re\,\zeta|^\beta.$$

$$\varphi(\zeta) \leq c_2|Im\,\zeta| + c_3, \ if \ Re\,\zeta \notin \Gamma.$$

Proof. The proof is similar to that of proposition 2.5.8. We fix c' so that $\theta \in \Gamma''$ together with $\xi \notin \Gamma'$ implies

$$|\xi' - \theta'| \geq c'|\xi'|,$$

and fix

$$\theta \in \Gamma'' \ with \ |\theta''| > c|\theta'|^{1+\beta}/|\theta|^\beta. \tag{6.3.21}$$

With θ we associate (as a preparation,) the function $f_\theta : C^n \to R_+$

$$f_\theta(\zeta) = |\zeta - \theta\langle\zeta, \theta\rangle/|\theta|^{-2}|,$$

i.e., f_θ is the distance from ζ to the line generated by θ. It is clear that

$$f_\theta(\zeta) \sim [|Im\,\zeta| + |\frac{Re\,\zeta}{|Re\,\zeta|} - \frac{\theta}{|\theta|}|\,|Re\,\zeta|], \tag{6.3.22}$$

if $\operatorname{Re}\zeta \in \Gamma^0$, provided Γ^0 is a sharp cone around ξ^0. It follows therefore that

$$f_\theta(\zeta) \geq c_5|\operatorname{Im}\zeta| + c'|\operatorname{Re}\zeta'| \text{ if } \operatorname{Re}\zeta \notin \Gamma^0 \setminus \Gamma'.$$

We also claim that we have

$$f_\theta(\zeta) \geq c_6|\operatorname{Re}\zeta'|^{1+\beta}/|\operatorname{Re}\zeta|^\beta \text{ if } \operatorname{Re}\zeta \in \Gamma^3 \text{ and } |\operatorname{Re}\zeta''| < (c/2)|\operatorname{Re}\zeta'|^{1+\beta}/|\operatorname{Re}\zeta|^\beta. \quad (6.3.23)$$

Proof of (6.3.23). We may assume that $|\theta| = |\operatorname{Re}\zeta|$. Furthermore, we choose $c_7 > 0$ so that $|\operatorname{Re}\zeta' - \theta'| \leq c_7|\operatorname{Re}\zeta'|$ implies $|\theta'| \geq (3/4)|\operatorname{Re}\zeta'|$. It follows then for such ζ that

$$f_\theta(\zeta) \geq |\theta'' - \operatorname{Re}\zeta''| \geq c|\theta'|^{1+\beta}/|\theta|^\beta - (c/2)|\operatorname{Re}\zeta'|^{1+\beta}/|\operatorname{Re}\zeta|^\beta$$

$$\geq ((3c/4) - c/2)|\operatorname{Re}\zeta'|^{1+\beta}/|\operatorname{Re}\zeta|^\beta$$

In the opposite case, when $|\operatorname{Re}\zeta' - \theta'| \geq c_7|\operatorname{Re}\zeta'|$, we have on the other hand that

$$f_\theta(\zeta) \geq |\operatorname{Re}\zeta' - \theta'| \geq c_7|\operatorname{Re}\zeta'| \geq c_6|\operatorname{Re}\zeta'|^{1+\beta}/|\operatorname{Re}\zeta|^\beta.$$

We have now proved (6.3.23) and add to the estimates for f_θ already obtained the trivial estimate

$$f_\theta(t\theta) = 0 \text{ if } t > 0.$$

Furthermore we consider for each θ as in (6.3.21) a plurisubharmonic function φ_θ on $\{\zeta; |\operatorname{Im}\zeta| < c_1|\operatorname{Re}\zeta''|\}$ so that

$$\varphi_\theta(\zeta) \leq -f_\theta(\zeta) + c_{10}|\operatorname{Im}\zeta| + c_{11}, \text{ if } \operatorname{Re}\zeta \in \Gamma^0,$$

$$\varphi_\theta(\zeta) \geq -c_{12}f_\theta(\zeta)$$

$$\varphi_\theta(\zeta) \leq c_{10}|\operatorname{Im}\zeta| + c_{11}, \text{ if } \operatorname{Re}\zeta \notin \Gamma^0.$$

The existence of such functions is guaranteed by proposition 2.5.6. It follows in particular that $\varphi_\theta(t\theta) \geq 0$ if $t \in R$. A function φ as desired is then the upper regularization of

$$\psi(\zeta) = \sup_\theta \varphi_\theta(\zeta).$$

Proof of proposition 6.3.22. We fix c_4, Γ_1, so that

$$r_1(\xi) = |\xi''| \text{ if } \xi \in \Gamma_1, |\xi''| > c_4|\xi'|^{1+\beta}/|\xi|^\beta$$

and denote $\Gamma' = \Gamma_1 \cap \Gamma$, Γ from the preceding lemma. It is no loss of generality to assume that $c_4 \geq 1$. We also fix $\Gamma_2 \subset\subset \Gamma'' \subset\subset \Gamma'$. Finally we consider $r_2 \in \mathcal{R}(\xi^0, \xi^1, \beta, 0)$ so that

$$r_2(\xi) = 0 \text{ if } |\xi''| \leq 3c_4|\xi'|^{1+\beta}/|\xi|^\beta \text{ or } \xi \notin \Gamma_2.$$

If h satisfies (6.3.20) then we can find $g_1, g_2 \in C^\infty$ so that $h = g_1 + g_2$ on $|p(-\zeta)| < c''(1 + |\zeta|)^{-M''}$, $c_2 + |\text{Im}\,\zeta| < |\text{Re}\,\zeta''|/c_2$, and which have the following properties:

$$|g_1(\zeta)| + |g_2(\zeta)| \leq |h(\zeta)|,$$

$$\text{supp}\ g_1 \subset \{\zeta; |\text{Re}\,\zeta''| \geq 2c_4 |\text{Re}\,\zeta'|^{1+\beta}/|\text{Re}\,\zeta|^\beta, \text{Re}\,\zeta \in \Gamma''\},$$

$$\text{supp}\ g_2 \subset \{\zeta; |\text{Re}\,\zeta''| \leq 3c_4 |\text{Re}\,\zeta'|^{1+\beta}/|\text{Re}\,\zeta|^\beta, \text{or Re}\,\zeta \notin \Gamma_2\},$$

$$\text{supp}\ \bar\partial g_1 \subset \{\zeta; 2c_4 |\text{Re}\,\zeta'|^{1+\beta}/|\text{Re}\,\zeta|^\beta \leq |\text{Re}\,\zeta''| \leq 3c_4 |\text{Re}\,\zeta'|^{1+\beta}/|\text{Re}\,\zeta|^\beta, \text{Re}\,\zeta \in \Gamma_2 \text{ or }$$

$$|\text{Re}\,\zeta''| \geq 2c_4 |\text{Re}\,\zeta'|^{1+\beta}/|\text{Re}\,\zeta''|^\beta \text{ and Re}\,\zeta \in \Gamma'' \setminus \Gamma_2\},$$

$$|\bar\partial g_1(\zeta)| \leq c_5 \exp\left[\text{Im}\,\tau_+ + \varepsilon_2 |\text{Im}\,\zeta| + b_2 ln(1 + |\zeta|)\right].$$

Note that the condition on supp $\bar\partial g_1$ gives in particular that $\bar\partial g_1 = 0$, if $\text{Re}\,\zeta \in \Gamma_2$ and $(3/2)c_4 |\text{Re}\,\zeta'|^{1+\beta}/|\text{Re}\,\zeta|^\beta \geq |\text{Re}\,\zeta''|$ or if $\text{Re}\,\zeta \notin \Gamma''$ and therefore $\bar\partial g_1$ satisfies any inequality of type $|\bar\partial g_1(\zeta)| \leq \exp \psi(\zeta)$ on that set. If $c', c_1, M', d_3, \varepsilon_2, b_2, c_6, c_7$ are suitable, we can then solve (using the lemma and results from above on the theory of the $\bar\partial$ system) $\bar\partial u = \bar\partial g_1$ on

$$\{\zeta \in C^n; |p(-\zeta)| < c'(1 + |\zeta|)^{-M'}, c_1 + |\text{Im}\,\zeta| < |\text{Re}\,\zeta''|/c_1\}$$

so that

$$|u(\zeta)| \leq c_6 \exp\left[\text{Im}\,\tau_+ - d_3 |\text{Re}\,\zeta'|^{1+\beta'}/|\text{Re}\,\zeta|^{\beta'} + \varepsilon_1 |\text{Im}\,\zeta| + b_1 ln(1 + |\zeta|)\right]$$

if $|\text{Re}\,\zeta''| < c_4 |\text{Re}\,\zeta'|^{1+\beta}/|\text{Re}\,\zeta|^\beta$, $\text{Re}\,\zeta \in \Gamma'$ or $\text{Re}\,\zeta \in \Gamma \setminus \Gamma'$, respectively

$$|u(\zeta)| \leq c_7 \exp\left[\text{Im}\,\tau_+ + \varepsilon_1 |\text{Im}\,\zeta| + b_1 ln(1 + |\zeta|)\right]$$

if $|\text{Re}\,\zeta''| > c_4 |\text{Re}\,\zeta'|^{1+\beta}/|\text{Re}\,\zeta|^\beta$ and $\text{Re}\,\zeta \in \Gamma'$ or $\text{Re}\,\zeta \notin \Gamma$.

For d_2, r_2, d_3 small, the desired decomposition is

$$f_1 = g_1 - u, f_2 = g_2 + u.$$

The only problem is the estimation of f_1 with the term $-d_3 |\text{Re}\,\zeta'|^{1+\beta'}_\Gamma/|\text{Re}\,\zeta|^{\beta'}$. Here we recall that g_1 vanishes for $|\text{Re}\,\zeta''| < 2c_4 |\text{Re}\,\zeta'|^{1+\beta}/|\text{Re}\,\zeta|^\beta$ or $\text{Re}\,\zeta \notin \Gamma''$ and on $|\text{Re}\,\zeta''| < c_4 |\text{Re}\,\zeta'|^{1+\beta}/|\text{Re}\,\zeta|^\beta$, $\text{Re}\,\zeta \in \Gamma'$ or $\text{Re}\,\zeta \in \Gamma \setminus \Gamma'$ u is good. The desired estimates will therefore hold when $|\text{Re}\,\zeta''| < 2c_4 |\text{Re}\,\zeta'|^{1+\beta}/|\text{Re}\,\zeta|^\beta$, $\text{Re}\,\zeta \in \Gamma'$ or when $\text{Re}\,\zeta \in \Gamma \setminus \Gamma'$. When, on the other hand $|\text{Re}\,\zeta''| > c_4 |\text{Re}\,\zeta'|^{1+\beta}/|\text{Re}\,\zeta|^\beta$ and $\text{Re}\,\zeta \in \Gamma_1$, then we can dominate $d_3 |\text{Re}\,\zeta'|^{1+\beta'}/|\text{Re}\,\zeta|^{\beta'}$ for small d_3 by $(d_1/2)r_1(-\text{Re}\,\zeta)$. (Note that the situation was very similar to that from the last part of the proof of lemma 6.3.16.)

6.4   Singular points on surfaces

1. One of the major technical sources for difficulties in dealing with operators with characteristics of variable multiplicity is that the characteristic surface for such operators is not smooth. It seems then interesting to analyze the singularities which can appear in some detail from a geometrical point of view, and we do so in this section in some elementary cases, relying on some definitions taken from Sommerville [1]. From a different point of view and with different goals, a (much) deeper analysis of related phenomena is made in a number of papers and books on higher order extensions of the Morse lemma: cf. e.g. the book of Arnold-Gusein/Zade-Varchenko [1].

2. Let f be a real-valued C^∞ function defined on a neighborhood U of $\xi^0 \in R_\xi^n$ and assume that

$$\partial_\xi^\alpha f(\xi^0) = 0 \text{ for } |\alpha| < s, \sum_{|\alpha|=s} |(\partial_\xi^\alpha f)(\xi^0)| \neq 0.$$

To simplify notations we shall almost always assume in this section that $\xi^0 = 0$. We consider the localization f^T of f at 0. Thus

$$f^T(v) = \sum_{|\alpha|=s} \partial_\xi^\alpha f(0) v^\alpha / \alpha!.$$

It is clear from this in particular that

$$(\partial/\partial v)^s f^T(0) = (\partial/\partial v)^s f(0), \text{ for any } v. \tag{6.4.1}$$

We are interested in the relation between the sets $V = \{\xi; f(\xi) = 0\}$ and $W = \{\xi; f^T(\xi) = 0\}$. Of course, when, e.g., $V = \{0\}$, this is not very interesting. In any case, V will not be in general a smooth submanifold of U. It is therefore interesting to note that very often V admits "natural" tangent vectors at 0. Here we shall use the following definition:

Definition 6.4.1. *A vector $v \in \dot{R}^n$ is called tangent to V at 0, if we can find $\varepsilon > 0$ and a C^1 curve $\gamma : (-\varepsilon, \varepsilon) \to R^n$ so that*
α) $\gamma(0) = 0$,
β) $\gamma(t) \in V, \forall t$,
γ) $\dot{\gamma}(0) = v$.

In particular we see that the present definition coincides with the standard definition of a tangent vector, if V is smooth at 0. Of course the set of tangent vectors is void when $V = \{0\}$, so we are interested in conditions which lead to a large set of tangent vectors. The following proposition, which we also needed in section 1.2, is a result in this direction.

Proposition 6.4.2. *Consider $v \in \dot{R}^n$. If v is tangent to V at 0, then $f^T(v) = 0$. Conversely, if f is real analytic and if $v \in \dot{R}^n$ is a solution of $f^T(v) = 0$ so that*

$$\operatorname{grad} f^T(v) \neq 0, \tag{6.4.2}$$

then v is tangent to V at 0.

Remark 6.4.3. *The importance of conditions of type (6.4.2) is well-established in the theory of critical points of functions. (Cf. Arnold-Gusein/Zade-Varchenko [1].)*

Proof of proposition 6.4.2. If v is tangent to V, then we can find a smooth curve γ : $(-\varepsilon, \varepsilon) \to R^n$ so that $\gamma(0) = 0$, $\dot{\gamma}(0) = v$, $f(\gamma(t)) \equiv 0$. It follows that $(d/dt)^s f(\gamma(0)) = 0$ and it remains to observe that

$$(d/dt)^s f(\gamma(0)) = f^T(v).$$

Conversely, assume f is real analytic and let v be a nontrivial solution of $f^T(v) = 0$, for which $\operatorname{grad} f^T(v) \neq 0$. We shall look for a real analytic γ with $\gamma(0) = 0$, $\dot{\gamma}(0) = v$ and $f(\gamma(t)) \equiv 0$. We want then at first to find a formal solution

$$\gamma(t) = \sum_{i \geq 0} v_i t^i, v_i \in R^n, \tag{6.4.3}$$

of

$$f(\gamma(t)) = O(t^j), \forall j. \tag{6.4.4}$$

Since $\gamma(0) = 0$ we shall set $v_0 = 0$, and from $\dot{\gamma}(0) = v$ it follows that we will have $v_1 = v$. Actually, if $v_1 = v$, then it is clear that

$$f(\gamma(t)) = O(t^{s+1}).$$

Our next remark is, that if we write formally

$$f(\gamma(t)) = \sum_{j \geq s} a_i t^i, \tag{6.4.5}$$

then a term $v_i t^i$ from the formal expansion of f comes in in the expansion (6.4.5) for the first time when we compute a_{s+i-1}. This makes it possible to determine the v_i inductively. Assume in fact that $v_1, ..., v_{j-1}$ have already been found so that $a_i = 0$ for $i \leq s + j - 2$. We shall then determine v_j so that also $a_{s+j-1} = 0$. To do so, we observe that

$$f(\gamma(t)) = f^T(v_1 t + v_j t^j) + g t^{s+j-1} + O(t^{s+j}),$$

where g depends only on the choice of $v_1, ..., v_{j-1}$ and may therefore be assumed known. Here we observe that

$$f^T(v_1 t + v_j t^j) = \langle \operatorname{grad} f^T(v_1), v_j \rangle t^{s+j-1} + O(t^{s+j}),$$

so it suffices to choose v_j so that

$$\langle \operatorname{grad} f^T(v_1), v_j \rangle + g = 0.$$

This is of course possible under the assumption that $\operatorname{grad} f^T(v_1) = \operatorname{grad} f^T(v) \neq 0$.

We have thus proved the existence of a formal solution of (6.4.4). A real analytic solution for (6.4.4) will then also exist in view of the following result from Artin [1]:

Theorem 6.4.4. *Consider a system of analytic equations*

$$f(x, y) = 0, \tag{6.4.6}$$

where $f(x, y) = (f_1(x, y), ..., f_m(x, y))$ are convergent series in the variables $x = (x_1, ..., x_n)$, $y = (y_1, ..., y_N)$. Suppose that $\bar{y}(x) = (\bar{y}_1(x), ..., \bar{y}_N(x))$ is a formal power series solution of (6.4.6) without constant term. Let c be some positive integer. Then there is a convergent series solution $y(x) = (y_1(x), ..., y_N(x))$ of (6.4.6) such that $\partial_x^\alpha y(0) = \partial_x^\alpha \bar{y}(0)$ for $|\alpha| \leq c$.

(The result is true both in the real and in the complex setting.)

Proposition 6.4.2 deals with a case when it is easy to construct tangent vectors to a surface at a singular point. Contrary to what happens at regular points, the set of tangent vectors will not have an affine structure in the general case. It is therefore interesting to mention that two-dimensional imbedded planes appear in a situation closely related to microcharacteristic directions. Let us in fact assume e.g. that p is a homogeneous polynomial of degree m in the variables ξ which vanishes of order s at $\xi^0 = (0, ..., 0, 1)$, denote by p_1 the localization of p at ξ^0 and consider $\sigma^0 \in \dot{R}^{n-1}$ so that $p_1(\sigma^0) = 0$. Assume that $\operatorname{grad} p_1(\sigma^0) \neq 0$. Since we have

$$p(\xi) = \xi_n^{m-s} p_1(\xi') + \sum_{j=0}^{m-s-1} \xi_n^j Q_j(\xi'),$$

for some polynomials Q_j of degree $m - j$, we conclude that p_1 is also the localization of $p(\xi', 1)$ at $\xi' = 0$. From proposition 6.4.2 it follows that we can find a curve $t \to \gamma(t) \in R^{n-1}$ so that $\gamma(0) = 0$, $p(\gamma(t), 1) \equiv 0$, $\dot{\gamma}(0) = \sigma^0$. We also fix a function $t \to \mu(t)$ so that $\mu(0) = 1$, and denote by $F(t)$ the curve $t \to (\mu(t)\gamma(t), \mu(t))$. It is clear that $p(F(t)) \equiv 0$

and that $F(0) = \xi^0$. Moreover, we have that $\dot{F}(0) = (\dot{\gamma}(0), \dot{\mu}(0))$. Since we can here choose $\dot{\mu}(0)$ as we please and since we can change the speed on F, we have now proved the following result:

Proposition 6.4.5. *Assume that* p, ξ^0, σ^0 *are as before and fix* $t > 0, t' > 0$. *Denote by* $\theta = t\xi^0 + t'(\sigma^0, 0)$. *Then* θ *is a tangent vector to* $V = \{\xi; p(\xi) = 0\}$.

3. After these general remarks on tangent vectors, we turn our attention to the case when $s = 2$. In this case, the localization of f at 0 is simply half the quadratic form associated with the Hessian $H_{\xi\xi}f$ of f at 0. We want also to make sure that V is a true surface near 0. Of course, we could, for example, assume that $H_{\xi\xi}f(0)$ were nondegenerate, so it would follow from the Morse lemma that, in suitably chosen coordinates, f had the form

$$f = h(\xi) \left(\sum_{i=1}^{n} a_i \xi_i^2 \right), \quad h(\xi) \neq 0,$$

where all $a_i \in R$ are of absolute value 1. If the quadratic form $\xi \to \sum_i a_i \xi_i^2$ were nondegenerate, $\sum_i a_i \xi_i^2$ would be the equation of a conic surface then. We shall return to this situation at the end of this section. In what follows, we need to consider however also cases when $H_{\xi\xi}f(0)$ is degenerate, so we shall make stronger assumptions of a geometric nature. Our assumptions on f and on V will be in fact the following:

a) $f(0) = 0$, grad $f(0) = 0$,

b) there is $\varepsilon > 0$ so that $\xi \in V \setminus \{0\}$, $|\xi| < \varepsilon$, implies grad $f(\xi) \neq 0$,

c) for a special choice of linear coordinates,

$$(\partial f / \partial \xi_1)^2 f(0) \neq 0,$$

and for the same choice of coordinates

d) there is $\varepsilon > 0$ so that for every $|\xi'| < \varepsilon$ there is at least one solution ξ_1 of $f(\xi_1, \xi') = 0$.

Remark 6.4.6. *i) As a consequence of b), V is smooth at all points* $\xi \in V \setminus \{0\}$ *which are close to* 0. *A singularity cannot be excluded at* 0 *however and will effectively appear in the generic case.*

ii) The condition d) expresses the fact that the point 0 *lies in the "interior" of the surface* V.

4. It is elementary to observe that, as a consequence of c) and d), $f(\xi_1, \xi') = 0$ has precisely two smooth roots ξ_1 for every fixed ξ', provided $|\xi'|$ is small. Indeed, it follows

from the Malgrange preparation theorem (but the situation is quite elementary here and f will be analytic in all applications anyway, so simpler results would suffice), that

$$f(\xi) = h(\xi)g(\xi), g(\xi) = \xi_1^2 + a(\xi')\xi_1 + b(\xi'), \tag{6.4.7}$$

for two $C^\infty-$ functions a and b defined in a neighborhood of $0 \in R^{n-1}$ and a $C^\infty-$ function h defined in a neighborhood of $0 \in R^n$ which satisfy

$$h(0) \neq 0, \ a(0) = 0, \ b(0) = 0 \ .$$

When $|\xi|$ is small, the zeros of $f(\xi_1, \xi') = 0$ must be of form

$$\xi_1 = (-a(\xi') \pm \sqrt{\Delta(\xi')})/2, \tag{6.4.8}$$

where

$$\Delta(\xi') = a^2(\xi') - 4b(\xi') \tag{6.4.9}$$

is the discriminant of g in the variable ξ_1. The assumption d) tells us that at least one of the values ξ_1 given by (6.4.8) must be real, but so is then also automatically the other. We also observe for later use that this happens because then necessarily $\Delta(\xi') \geq 0$. Actually, more is true, in that we have that $\Delta(\xi') \neq 0$ for small $|\xi'|$. Indeed, if we had $\Delta(\tilde{\xi}') = 0$ we could conclude that we also had $\operatorname{grad}\Delta(\tilde{\xi}') = 0$, which would leed to

$$a(\tilde{\xi}') \operatorname{grad}_{\xi'}a(\tilde{\xi}') - 2 \operatorname{grad}_{\xi'}b(\tilde{\xi}') = 0.$$

On the other hand, we would have that

$$\operatorname{grad}_\xi(\tilde{\xi}_1{}^2 + a(\tilde{\xi}')\tilde{\xi}_1 + b(\tilde{\xi}')) = (2\tilde{\xi}_1 + a(\tilde{\xi}'), \ \operatorname{grad}_{\xi'}a(\tilde{\xi}')\tilde{\xi}_1 + \ \operatorname{grad}_{\xi'}b(\tilde{\xi}')) = 0,$$

since $\tilde{\xi}_1{}' = -a(\tilde{\xi}')/2$, which would contradict assumption b). Since we must of course have $\Delta(0) = 0$, and, as a consequence of $\operatorname{grad} f(0) = 0$, that $\operatorname{grad}_{\xi'}\Delta(0) = 0$, it follows that we have in fact

$$\partial_{\xi'}^\alpha \Delta(0) = 0 \text{ for } |\alpha| < 2, \ H_{\xi'\xi'}\Delta(0) \geq 0.$$

5. In much of what follows, conditions on $H_{\xi'\xi'}\Delta(0)$ play an important role. It is therefore interesting to note that, while the explicit calculation of Δ depends on a precise knowledge of the decomposition of f into the form (6.4.7), the calculation of $H_{\xi'\xi'}\Delta(0)$ does not. This is based on the fact that the Hessian of f at 0 is $h(0) H_{\xi\xi}g(\xi)$, if f has the form (6.4.7). Now

$$\langle H_{\xi\xi}(\xi_1^2 + a(\xi')\xi_1 + b(\xi'))v, \ v\rangle = 2v_1^2 + 2\langle \operatorname{grad}_{\xi'}a(0), v'\rangle v_1 + \langle H_{\xi'\xi'}b(0)v', v'\rangle,$$

so the discriminant of $\langle H_{\xi\xi}g(0)v, v\rangle$ in v_1 is

$$4\langle \text{grad}_{\xi'}a(0), v'\rangle^2 - 8\langle H_{\xi'\xi'}b(0)v', v'\rangle.$$

On the other hand we obtain, if we also use $a(0) = 0$, that

$$\langle H_{\xi'\xi'}\Delta(0)v', v'\rangle = 2\langle \text{grad}_{\xi'}a(0), v'\rangle^2 - 4\langle H_{\xi'\xi'}b(0)v', v'\rangle. \qquad (6.4.10)$$

It follows that $\langle H_{\xi'\xi'}\Delta(0)v', v'\rangle$ is a multiple of the discriminant of the localization polynomial f^T of f at 0. Since $\langle H_{\xi'\xi'}\Delta(0)v', v'\rangle/2$ itself is the localization of Δ when $H_{\xi'\xi'}\Delta(0) \neq 0$, we have then proved the following lemma:

Lemma 6.4.7 *Assume that the localization of f at x^0 is a second order polynomial and that $H_{\xi'\xi'}\Delta(0) \neq 0$. Then the localization of the discriminant is proportional to the discriminant of the localization and both have the same sign.*

6. In the sequel we are interested in the relation between conditions on the Hessian $H_{\xi\xi}f(0)$ of f at 0 and conditions on $H_{\xi'\xi'}\Delta(0)$. We use once more that $H_{\xi\xi}f(0)$ is a multiple of $H_{\xi\xi}g(0)$. It is also convenient to write g in the form

$$g(\xi) = (\xi_1 + a(\xi')/2)^2 - \Delta(\xi')/4 \qquad (6.4.11)$$

where Δ is, as before, the discriminant of f. This suggests to perform, at least temporarily, the change of variables $\xi \to \eta$ given by

$$\eta_1 = \xi_1 + a(\xi'), \ \eta_i = \xi_i \text{ if } i \geq 2.$$

In the new variables, g then has the form $g(\eta) = \eta_1^2 - \Delta(\eta')/4$, with Δ unchanged. We can read off from this the following result:

Lemma 6.4.8. *There are equivalent:*
i) $H_{\xi\xi}f(0)$ has rank $k + 1$,
ii) $H_{\xi'\xi'}\Delta(0)$ has rank k.

Moreover, when $H_{\xi'\xi'}\Delta(0)$ has rank k, then $H_{\xi\xi}f(0)$ has precisely one (strictly) positive and precisely k (strictly) negative eigenvalues, the other eigenvalues being zero.

(Here we use of course also that $H_{\xi'\xi'}\Delta(0) \geq 0$.)

7. We now turn to a discussion of the set of tangent vectors to V at 0, assuming that f satisfies a),b),c),d). According to what we have established in the beginning of this section, tangent vectors must satisfy the relation

$$\langle H_{\xi\xi}f(0)v, v\rangle = 0. \qquad (6.4.12)$$

Actually, as in proposition 6.4.2, if $v \neq 0$ satisfies (6.4.12), then it is tangent to V at 0. To see this, let us consider a vector v which satisfies (6.4.12), i.e. for which

$$2v_1^2 + 2v_1 \langle \operatorname{grad}_{\xi'} a(0), v' \rangle + \langle H_{\xi'\xi'} b(0) v', v' \rangle = 0.$$

We assume, to make a choice, that

$$v_1 = [-\langle \operatorname{grad}_{\xi'} a(0), v' \rangle + (\langle \operatorname{grad}_{\xi'} a(0), v' \rangle^2 - 2\langle H_{\xi'\xi'} b(0) v', v' \rangle)^{1/2}]/2.$$

Next, we define a curve $\gamma : (-\varepsilon, \varepsilon) \to R^n$ by $\gamma_i(t) = v_i t$, if $i \geq 2$, respectively, with the notation $\tilde{\gamma}(t) = (\gamma_2(t), ..., \gamma_n(t))$,

$$\gamma_1(t) = [-a(\tilde{\gamma}(t)) + \sqrt{\Delta(\tilde{\gamma}(t))}]/2 \text{ for } t \geq 0,$$

and

$$\gamma_1(t) = [-a(\tilde{\gamma}(t)) - \sqrt{\Delta(\tilde{\gamma}(t))}]/2 \text{ for } t < 0.$$

We claim that the function γ_1 is C^1. Here we observe that the only problems can appear at $t = 0$. Actually, we must have that $\Delta(\tilde{\gamma}(t)) = ct^{2j} + O(t^{2j+1})$ for some natural number j and some positive constant c, so the statement follows. We must then only show that $\dot{\gamma}(0) = v$. This follows from

$$(d/dt)\sqrt{\Delta(\tilde{\gamma}(t))}_{|t=0} = (\langle H_{\xi'\xi'} \Delta(0) v', v' \rangle)^{1/2}/\sqrt{2}.$$

Indeed, we have $\Delta(\tilde{\gamma}(t)) = t^2 \langle H_{\xi'\xi'} \Delta(0) v', v' \rangle/2 + O(t^3)$. Also cf. here (6.4.10).

8. We have now checked that the tangent vectors to V at 0 are precisely the solutions of (6.4.12), which is a quadratic equation. Moreover, as seen above, the quadratic form $v \to \langle H_{\xi\xi} f(0) v, v \rangle$ is never definite, so the solutions of (6.4.12) form a nontrivial cone. However, this cone can degenerate, so there are several geometric situations which can occur. Two cases, which are in some sense extreme, are here of special interest to us.

a) We assume at first that rank $H_{\xi\xi} f(0) = n$. In this case, V is in suitably chosen coordinates of form $\xi_1^2 - \xi_2^2 - \cdots - \xi_n^2 = 0$. In particular, we have thus a true singularity at 0. We shall call 0 a conical singularity then.

b) The other case in which we are directly interested is when rank $H_{\xi\xi} f(0) = 1$. The equation (6.4.12) degenerates then completely to the equation of an hyperplane, counted twice. To find this hyperplane explicitly, note that in this case $H_{\xi'\xi'} \Delta(0) \equiv 0$. We will then have that

$$H_{\xi\xi} f(0) = \begin{pmatrix} 2 & {}^t\operatorname{grad}_{\xi'} a(0) \\ \operatorname{grad}_{\xi'} a(0) & H_{\xi'\xi'} a^2(0)/4 \end{pmatrix}.$$

Since $a(0) = 0$, the equation $\langle H_{\xi\xi} f(0)v, v \rangle = 0$ comes to

$$2v_1^2 + 2\langle \mathrm{grad}_{\xi'} a(0), v' \rangle + \langle \mathrm{grad}_{\xi'} a(0), v' \rangle^2 / 2 = (1/2)(2v_1 + \langle \mathrm{grad}_{\xi'} a(0), v' \rangle)^2.$$

The equation (6.4.12) gives thus explicitly

$$2v_1 + \langle \mathrm{grad}_{\xi'} a(0), v' \rangle = 0. \tag{6.4.13}$$

It is also worthwhile to analyze the regularity class of V for a moment. In fact, since now $H_{\xi'\xi'} \Delta(0) = 0$, we obtain in combination with $\Delta(\xi') \geq 0$ for small $|\xi'|$, that

$$\partial_{\xi'}^\alpha \Delta(0) = 0 \text{ for } |\alpha| \leq 3.$$

It follows in particular that $\xi' \to \sqrt{\Delta(\xi')}$ is a C^1 map at 0. Since away from 0 the map is (in view of $\Delta(\xi') \neq 0$) anyway C^∞, it is C^1 in a full neighborhood of 0. We can then write $V = V_+ \cup V_-$, where $V_\pm$ are the two hypersurfaces given by the equations

$$f_\pm(\xi) = 2\xi_1 + a(\xi') \mp \sqrt{\Delta(\xi')} = 0.$$

Both functions $f_\pm$ are by the preceding C^1, so their tangent spaces are given by the equations

$$\langle df_\pm(0), v \rangle = 0, \ v \in \dot{R}^n.$$

Here we have in fact $df_+(0) = df_-(0) = 2d\xi_1 + \langle \mathrm{grad}_{\xi'} a(0), d\xi' \rangle$, so explicitly the last equality comes to (6.4.13): the tangent vectors defined in definition 6.4.1 coincide with the tangent space defined as in standard differential geometry for each of the $V_\pm$. (The main point is here of course the fact that the $V_\pm$ are C^1 regular and have the same tangent space. That this is precisely the set of tangent vectors from definition 6.4.1 comes then as no surprize.)

We conclude the section with the following

Definition 6.4.9. *(Sommerville [1]) Let f satisfy the conditions a),b),c),d). Then we shall say that 0 is an unode for V (or that it is a singularity of unodal type for V) if the set of tangent vectors is a hyperspace.*

According to the above, this is equivalent with requiring that rank $H_{\xi\xi} f(0) = 1$ or that $H_{\xi'\xi'} \Delta(0) \equiv 0$. We also mention that in Sommerville [1] the word "unode" is an abbreviation for "uniplanar node". (In the book of Sommerville only the case $n = 3$ is considered, so tangent spaces are two-dimensional planes. In the present day literature, related definitions seem to appear rather in books on algebraic than on differential geometry.)

6.5   Remarks on the velocity, the slowness and the wave surface

1. In the general theory of linear partial differential equations of the last decades, the analysis is centered strongly on the characteristic surface of the equation under consideration. In constant coefficient elasticity theory and in crystal optics (and perhaps elsewhere), other surfaces have been considered traditionally along with the characteristic surface and are preferred in applications often even nowadays. For the convenience of the reader, and in part, as a preparation for the next section, we describe here how these surfaces are defined and what their physical signification is. We shall assume that a homogeneous polynomial F of degree m is given on $R_\lambda^{n+1} = R_\xi^n \times R_\tau$ and that F has the following properties A),B),C) below, but we should mention that additional conditions are needed if effects which appear only at the level of higher microlocalization are to be considered.

A) $F(0, \tau) = 0$ implies $\tau = 0$, and $F(D)$ is hyperbolic with respect to the plane $t = 0$.

The second part gives here that the equation $F(\xi, \tau) = 0$ has for every $\xi \in R^n$ precisely m real roots τ, when multiplicities are also counted.

B) $F(\xi, 0) = 0$ implies $\xi = 0$.

The interpretation of this condition is that plane wave solutions of form

$$u = \exp\left(i\langle x, \xi\rangle + it\tau\right)$$

of $F(D)u = 0$ are never stationary. (This will be discussed in more detail later on.)

C) $\mathrm{grad}_\lambda F(\lambda^0) \neq 0$ implies $(\partial/\partial\tau)F(\lambda^0) \neq 0$.

Geometrically this means that on the regular part of the characteristic surface $V = \{\lambda; F(\lambda) = 0\}$, the equation $F(\lambda) = 0$ is equivalent locally to an equation of form $\tau = \tau(\xi)$ for some smooth function $\xi \to \tau(\xi)$, but there is also a physical interpretation of this condition which will be discussed in nr.6 below. (The interpretation is roughly that the speed of propagation on bicharacteristics is finite.)

We shall now associate with F the following three surfaces:

Definition 6.5.1. *Consider the surfaces X, S and W defined by*

$$X = \{\xi \in R^n; F(\xi/|\xi|, |\xi|) = 0\},$$

$$S = \{\xi \in R^n; F(\xi, 1) = 0\},$$

whereas W is the envelope of the family of hyperplanes

$$Z_\lambda = \{x; \langle x, \xi \rangle + \tau = 0, F(\xi, \tau) = 0\}.$$

X is called the "velocity", S the "slowness" and W the "wave" surface of F.

3. Before we justify the terminology fixed in definition 6.5.1, we make a number of elementary remarks on the relation between X, S, W and V. Actually, it is obvious that S is just the projection onto the ξ-space of the intersection of V with $\tau = 1$. Moreover, it is clear that $\theta \in X$, precisely when $\theta/|\theta|^2 \in S$, so that X and S are inverse to each other with respect to the unit sphere. We shall see later on that if $\theta \in X$ is given, then $|\theta|$ has a natural interpretation as a speed. For the point $\xi = \theta/|\theta|^2$ we have $|\xi| = 1/|\theta|$, so it is natural to call $|\xi|$ a "slowness". This will justify the name of S if we can justify that of X. Next we turn our attention to W. The terminology refers here of course to the standard procedure of Ch. Huygens to construct wave fronts as envelopes of families of more elementary waves, in this case plane waves of a rather particular form. We shall however see that W contains also explicitly information relative to an important notion of velocity in wave motion. (We distinguish in this section between "speeds" and "velocities". A "speed" is, in general, the absolute value of a "velocity".) We start here by observing that technically the equation of W is obtained by eliminating the parameter ξ from

$$\langle x, \xi \rangle + \tau(\xi) = 0, F(\xi, \tau(\xi)) = 0, \mathrm{grad}_\xi(\langle x, \xi \rangle + \tau(\xi)) = 0,$$

assuming that $\xi \to \tau(\xi)$ is a local C^1 solution (there where such a solution exists) of $F(\xi, \tau) = 0$. Actually, all equations are homogeneous, so we may restrict our attention to the ξ with $\tau(\xi) = 1$. ξ is then a generic point on the slowness surface. Moreover, $\mathrm{grad}(\langle x, \xi \rangle + \tau(\xi)) = 0$ implies $\langle x, \xi \rangle + \tau(\xi) = 0$, by Euler's relation for homogeneous functions, so, technically, the condition $\langle x, \xi \rangle + \tau(\xi) = 0$ is superfluous. Finally, $\mathrm{grad}(\langle x, \xi \rangle + \tau(\xi)) = 0$ gives $x = -\mathrm{grad}\ \tau(\xi)$, i.e., x has the direction of the normal to the surface $\tau(\xi) = 1$. The relation

$$x = -\mathrm{grad}\ \tau(\xi),\ \xi \in S,$$

is thus a parametric description of W and the radius vector of a point in W is the normal to S at some associated point in S. It is well known from differential geometry that the converse must then also hold. (The discussion is not complete of course, since we say nothing on the singular parts of W or S. It should illustrate however well enough what is going on.)

4. Let us now also explain why X is called "velocity" surface. This is related to a particular notion of velocity associated with wave motion for special solutions of $F(D)u = 0$. A "plane wave solution" of this equation is by definition a solution of form

$$u = f(\langle x, \xi \rangle + t\tau),$$

where f is in $C^\infty(R)$ (say). To have $F(D)u = 0$, we need of course that $F(\xi, \tau) = 0$, i.e. that (ξ, τ) lies on the characteristic surface of F. We also recall that in the case $n = 1$ it is customary to call ξ the "wave number" of the solution u, whereas $\tau/(2\pi)$ is its "circular frequency". In the (x, t)-space the planes $\langle x, \xi \rangle + t\tau = c$ are for fixed c the hyperplanes of equal phase. For an observer in the physical x-space it is perhaps however more interesting to consider here for each fixed time t the projection of $\langle x, \xi \rangle + t\tau = c$ onto R^n_x. What we obtain in this way is then a family of planes

$$\langle x, \xi \rangle = c - t\tau$$

of equal phase, parametrized by the time t. The normal to these planes is just ξ, so the perception of the observer will be that of a movement of these planes in the direction of the normal. If the projection was $\langle x, \xi \rangle = c$ at time $t = 0$, it will be $\langle x, \xi \rangle = c - \tau$ after one unit of time. The speed of the propagation of the family of planes, measured along the normal direction, is then $-\tau/|\xi|$. It is called "phase velocity" in classical mechanics. Since $F(\xi, \tau) = 0$ implies $F(\xi/|\xi|, \tau/|\xi|) = 0$, we see that $\tau\xi/|\xi|^2$ is a point of X. The surface X contains then all information on phase velocities in a rather explicit form.

5. Phase velocity is a very intuitive concept of velocity in wave propagation, but it refers only to plane waves and is therefore not very interesting from a physical point of view. Another concept of velocity has been introduced by Stokes under the name "group velocity" and refers to wave packets grouped around certain frequencies. It was later on recognized by Raileigh that group velocity is related to the way energy is transported in the wave motion, but here we shall remain basically within the reach of the construction of Stokes. (We follow here Sommerfeld [1], to which we also refer for a more thorough discussion of the considerations of Raileigh.) The basic idea in the construction of Stokes, can already be seen if one looks at two plane waves which have frequencies which are infinitessimally close. Moreover, we may even assume that these plane waves have a very special form. Assume in fact that u^1, u^2 are two plane wave solutions of $F(D)u = 0$ of form

$$u^i = \sin\left(\langle x, \xi^i \rangle + t\tau^i\right), \; i = 1, 2.$$

Thus the u^i are both of amplitude 1, but for the frequencies we assume that $(\xi^1, \tau^1) \neq$

(ξ^2, τ^2). We may then consider the solution $u = u^1 + u^2$ of $F(D)u = 0$ and have

$$u = 2 \cos \left((\langle x, \xi^1 - \xi^2\rangle + t(\tau^1 - \tau^2))/2\right)\sin \left((\langle x, \xi^1 + \xi^2\rangle + t(\tau^1 + \tau^2))/2\right).$$

If $|\xi^1 - \xi^2| + |\tau^1 - \tau^2|$ is sufficiently small, then the factor $\sin \left((\langle x, \xi^1 + \xi^2\rangle + t(\tau^1 + \tau^2))/2\right)$ will roughly behave like u^1 or u^2 itself. However, an additional oscillation is imposed on this factor by the factor $\cos \left((\langle x, \xi^1 - \xi^2\rangle + t(\tau^1 - \tau^2))/2\right)$. From the point of view of oscillations as a function in (x, t), this is slowly oscillating, since $|\lambda^1 - \lambda^2|$ is small. However, the phase velocity of the latter factor, as observed in the x-space, is

$$- (\tau^1 - \tau^2)/|\xi^1 - \xi^2| \tag{6.5.1}$$

and it points in the direction $(\xi^1 - \xi^2)/|\xi^1 - \xi^2|$. We assume that (ξ^1, τ^1) is a regular point on the characteristic surface of F, i.e. that we can parametrize the characteristic surface near (ξ^1, τ^1) by $(\xi, \tau(\xi))$, where $\tau(\xi)$ is a locally defined smooth function. In particular we may then assume that $\tau^i = \tau(\xi^i)$ for $i = 1, 2$. The limit for $\xi^2 \to \xi^1$ in (6.5.1) will then exist, e.g. if $\xi^2 = \xi^1 + se^i$, where $s \in R_+$, and $e^i = (0, ... 0.1, 0, ..., 0)$ with "1" on position i. We will then have in fact that

$$\lim_{s \to 0, s > 0} -(\tau^1 - \tau^2)/|\xi^1 - \xi^2| = -(\partial/\partial\xi_i)\tau(\xi^1).$$

The map $\xi \to -\mathrm{grad}_\xi \tau(\xi)$ contains thus all information on these limits and we shall call $-\mathrm{grad}_\xi \tau(\xi^1)$ the "group velocity" of the solutions of $F(D)u = 0$ at frequency ξ^1. In particular this gives W the meaning of being the surface of group velocities.

6. It is here interesting to relate the group velocity of Stokes to the velocity of propagation of wave front sets in microlocal analysis. Let us then assume that u is an arbitrary solution of $F(D)u = 0$, defined in a neighborhood U of $0 \in R^{n+1}$, say, and assume that $(0, \lambda^0) \in WF u$ (or $WF_A u$). Here $WF u$ is the C^∞ wave front set of u. We assume moreover that $\mathrm{grad}_\lambda F(\lambda^0) \neq 0$, so it follows from the general theory that the connected component of

$$\{(s\, \mathrm{grad}_\lambda F(\lambda^0), \lambda^0); \ s \in R, \ s\, \mathrm{grad}_\lambda F(\lambda^0) \in U\}$$

which contains $(0, \lambda^0)$ lies in $WF u$ (respectively $WF_A u$). On the space-time part of the bicharacteristic $s \to (s\, \mathrm{grad}F(\lambda^0), \lambda^0)$, we can now introduce the time as a natural parameter. It is then clear that the velocity of propagation in the x-space along the bicharacteristic, when t is moving, is

$$\mathrm{grad}_\xi F(\lambda^0)/((\partial/\partial\tau)F(\lambda^0)).$$

(Here we recall that in view of our assumption C), $(\partial/\partial\tau)F(\lambda^0) \neq 0$ when $\mathrm{grad}_\lambda F(\lambda^0) \neq 0$.) From $F(\xi, \tau(\xi)) \equiv 0$ we now obtain that

$$\mathrm{grad}_\xi F(\lambda^0)/(\partial/\partial\tau)F(\lambda^0)) = -\,\mathrm{grad}_\xi \tau(\xi),$$

if $\xi \to \tau(\xi)$ is a locally defined smooth solution of $F(\xi, \tau) = 0$ with $\tau(\xi^0) = \tau^0$. We see therefore that the group velocity is precisely the velocity of propagation on the associated bicharacteristic curves. Since bicharacteristic curves correspond to propagation of particles in classical mechanics, we have also an heuristic relation between wave propagation and particle propagation. (If such a relation is interesting, as is e.g. the case in optics.) W is then the surface reached by the particles after one unit of time.

7. We have thus seen that there is more than one possibility to introduce velocities associated with wave motion. Although for some few cases these notions will leed to the same quantities for all frequencies (the simplest example is the standard wave equation), these notions are essentially different in the general case for generic frequencies. It is then comforting to note that the directions of these velocities can at least not be orthogonal to each other. This is a consequence of the assumption B) and of Euler's relation for homogeneous functions. We have in fact that $\tau(\xi) = \langle \xi, \mathrm{grad}_\xi \tau(\xi) \rangle$, so $\langle \xi, \mathrm{grad}_\xi \tau(\xi) \rangle = 0$ would leed to $\tau(\xi) = 0$.

6.6   Singular points on the slowness surface and conical refraction

1. Let F be a homogeneous polynomial of order m in the variables $\lambda = (\xi, \tau)$. We assume that F is as in section 6.5. In addition we assume that F has no multiple irreducible factors and that $F(\xi, \tau) = F(\xi, -\tau)$. In particular it follows from the last assumption that the degree m of F is even. While the second of these additional assumptions is mainly to give the slowness surface a nicer shape, the first is purely technical and serves to simplify notations. (In other words, both assumptions are for convenience.) We want to see how one can recognize from information on the singularities of the slowness surface S associated with F the type of conical refraction which will occur.

We start our argument with a few simple minded remarks on S. Let us then fix $\xi \in S^{n-1} = \{\xi \in R^n; |\xi| = 1\}$. Since we can find at least one and at most $m/2$ positive roots τ of $F(\xi, \tau) = 0$, it follows that there are between one and $m/2$ positive solutions ρ of $F(\rho\xi, 1) = 0$. We conclude that for each fixed direction $\xi \in S^{n-1}$ we can find between one and $m/2$ points of form $\rho\xi$, $\rho > 0$, in S. Moreover, from the assumptions on F it follows immediately that there are two positive constants c, c' so that

$$\xi \in S \text{ implies } c \leq |\xi| \leq c'.$$

An observer placed at $0 \in R^n$ will then perceive S as an at most $m/2$ sheeted surface which stays, in some sense, over S^{n-1}. Of course the various sheets may intersect and at points of intersection ξ^0 we will have necessarily that

$$\text{grad}_\xi f(\xi^0) = 0 \text{ if } f(\xi) = F(\xi, 1).$$

It is standard here to observe (but we do not need this statement later on) that the innermost sheet of S must be convex. In fact, its homogeneous flowout in the (ξ, τ)-space is just the hyperbolicity cone of F, which, as observed several times already, is known to be convex. At an heuristic level, the statement comes of course from the fact that if this innermost sheet were not convex, then we could find a line which had more than m points of intersection with S. (Cf. Courant-Hilbert [1] or section 1.3 for a closely related situation.)

2. Since we want to relate the singularities of S to conical refraction, we shall now analyze to what extent the assumptions made in section 6.4 hold for the defining function f of the slowness surface. Let us then fix $\xi^0 \in S$ and denote by $\lambda^0 = (\xi^0, 1)$, so that $F(\lambda^0) = 0$. We shall assume that

I. $\text{grad}_\lambda F(\lambda^0) = 0$,
II. $(\partial/\partial\tau)F(\lambda) \neq 0$ for any $\lambda \in \{\theta \in R^{n+1}; F(\theta) = 0\}$ which lies in a sufficiently small conic neighborhood of λ^0 and is not proportional to λ^0,
III. $(\partial/\partial\tau)^2 F(\lambda^0) \neq 0$.

It follows from these assumptions that
I)$'$ $\text{grad}\, f(\xi^0) = 0$,
II)$'$ $\langle \xi,\ \text{grad}_\xi f(\xi) \rangle \neq 0$ if $\xi \in S$ is close to ξ^0.
III)$'$ $\langle \xi^0,\ \text{grad}_\xi \rangle^2 f(\xi^0) \neq 0$.

In fact, I)$'$ is obvious and II)$'$ follows from Euler's relation for homogeneous functions, in that we have for $\xi \in S$ that

$$\sum (\xi_j \partial/\partial\xi_j) F(\xi, 1) + (\partial/\partial\tau) F(\xi, 1) = 0.$$

Finally, to check that III)$'$ is valid, we observe that, with our standard notations for localizations,

$$\langle \xi^0,\ \text{grad}_\xi \rangle^2 f(\xi^0) = \langle H_{\xi\xi} f(\xi^0)\xi^0, \xi^0 \rangle = 2f_{\xi^0}^T(-\xi^0) = 2F_{\lambda^0}^T((0,1)) = (\partial^2/\partial\tau^2) F(\lambda^0).$$

The third equality is here a consequence of relation (1.1.12).

3. After these preparations we can now check that the assumptions a), b), c), d) from nr. 3 in section 6.4 are satisfied for f at ξ^0. In fact, this is obvious for a) and b) and c) will hold in polar coordinates (ρ, ω) near ξ^0. Here $\rho = |\xi|$ and ω are some local coordinates on S^{n-1} near $\xi^0/|\xi^0|$. Indeed, we will have that $(\partial/\partial\rho)^2$ is proportional to $\langle \xi^0, \text{grad}_\xi \rangle^2$, so c) follows from III)'. Moreover, d) is also satisfied, since if $\xi^0 \in S$ and if $\xi/|\xi|$ is close to $\xi^0/|\xi^0|$, then we can find a root τ close to one of $F(\xi, \tau) = 0$ and it will follow that $\xi/\tau \in S$.

4. Let us now assume that some (classical) analytic pseudodifferential operator $p(x, t, D_x, D_t)$ is given on some conic neighborhood Γ of $(0, \lambda^0)$ and that our F from before is the principal part of p. The characteristic variety of the operator is then

$$\{(x, t, \lambda) \in \Gamma; F(\lambda) = 0\},$$

but we shall argue, to simplify notations, as if we had $\Gamma = R_{x,t}^{n+1} \times R_\lambda^{n+1}$. When grad $F(\lambda^0) \neq 0$, it is wellknown that

$$(0, \lambda^0) \notin WF_A u, p(x, t, D_x, D_t)u = 0 \text{ implies } (s \text{ grad} F(\lambda^0), \lambda^0) \notin WF_A u, \forall s \in R,$$

so no conical refraction occurs. This is the reason why we shall assume in fact that $\text{grad}_\lambda F(\lambda^0) = 0$. We shall assume now, in addition, that the properties I), II), III) considered in nr.2 of this section, hold. We shall here only look into the two extreme cases considered in section 6.4, namely when rank $H_{\xi\xi} f(\xi^0) = n$ and when rank $H_{\xi\xi} f(\xi^0) = 1$. In the first case, ξ^0 is a conical singularity of S and we have seen in section 6.4 that the signature of $H_{\xi\xi} f(\xi^0)$ is $+ - \cdots -$. Since in view of (1.1.12)

$$p_{m,1}(v, w) = F_{\lambda^0}^T(v, w) = \langle H_{\xi\xi} f(\xi^0)(v - w\xi^0), v - w\xi^0 \rangle,$$

we can conclude that the localization $p_{m,1}$ of the principal part p_m of the operator p at ξ^0 is nonelliptic. Actually, it is a quadratic form which is of rank n. This is a situation which we have effectively encountered in section 1.3 and we have seen that "true" conical refraction occurs. In the other case, when rank$H_{\xi\xi} f(\xi^0) = 1$, ξ^0 is an unode. We will thus have that $\langle H_{\xi\xi} f(\xi^0)\sigma, \sigma \rangle = \langle x^0, \sigma \rangle^2$ for some $x^0 \in R^n$, which is in fact proportional to the normal to the sheets of S at ξ^0. For $p_{m,1}$ we obtain, once more in view of (1.1.12),

$$p_{m,1}(v, w) = \langle x^0, v - w\xi^0 \rangle^2 = \langle (x^0, -\langle x^0, \xi^0 \rangle), (v, w) \rangle^2.$$

As we have seen in section 1.3, this means that singularities propagate in the direction $(x^0, -\langle x^0, \xi^0 \rangle)$ and no conical refraction occurs.

5. In problems which are models for physical phenomena, the speed of propagation of disturbances should always be finite. It is then interesting to note that when the $p_{m,1}$ above is associated with a slowness surface which has the properties introduced in the beginning of this section, then the speed on the relative bicharacteristics associated with p_m has this property. We prove this assuming, to simplify notations, that the rank of $H_{\xi\xi} f(\xi^0)$ is n. Consider in fact some vector $\theta^0 \neq 0$, orthogonal to λ^0, such that $p_{m,1}(\theta^0) = 0$ and denote $\Sigma = \{(z, \lambda); \lambda \text{ is proportional to } \lambda^0\}$. Thus, e.g., $(0, \lambda^0, \theta^0)$ is microcharacteristic. The relative bicharacteristic of p_m associated with $(0, \lambda^0, \theta^0)$ is then

$$(s \operatorname{grad} p_{m,1}(\theta^0), \lambda^0, \theta^0), \quad s \in R,$$

where $(s \operatorname{grad} p_{m,1}(\theta^0), \lambda^0, \theta^0)$ is viewed as a point in $N\Sigma$ for each fixed s. (We recall here that it is related to the propagation of disturbances e.g. by the fact that the second wave front set is propagated along it.)

The speed on this bicharacteristic will now be finite if

$$(\partial/\partial w)p_{m,1}(\theta^0) \neq 0.$$

That this is the case in the situation at hand is the content of the following result:

Proposition 6.6.1. *Assume that F satisfies the conditions I,II,III, denote by $f(\xi) = F(\xi, 1)$ and assume that rank $H_{\xi\xi} f(\xi^0) = n$, $n \geq 3$. Let $(v, w) \in R^{n+1}$ be a solution of*

$$\langle H_{\xi\xi} f(\xi^0)(v - w\xi^0), v - w\xi^0 \rangle = 0.$$

Then it follows that

$$\frac{\partial}{\partial w} \langle H_{\xi\xi} f(\xi^0)(v - w\xi^0), v - w\xi^0 \rangle \neq 0.$$

Before we start the argument, we observe that if the conclusion of the proposition were false, then the vector $N = (0, 1) \in R^{n+1}$ would be tangent to the surface

$$W = \langle H_{\xi\xi} f(\xi^0)(v - w\xi^0), v - w\xi^0 \rangle = 0.$$

This justifies why we are interested first in the following remark on tangent vectors to cones of second order:

Lemma 6.6.2. *Let $Q(\eta) = \sum a_{ij}\eta_i\eta_j$ be a nondegenerate quadratic form of signature $+ - \cdots -$ in $n \geq 3$ variables and denote $\Gamma = \{\eta; Q(\eta) \geq 0\}$. If $|\eta^0| = 1$ is a given direction then η^0 is tangent to Γ at some point $\mu^0 \in \partial\Gamma$ if and only if $Q(\eta^0) \leq 0$. In the case of two variables, the condition of tangency is $Q(\eta^0) = 0$.*

Proof. We may assume that $Q(\eta) = \eta_n^2 - \eta_1^2 - \cdots - \eta_{n-1}^2$ and it is then no essential restriction to assume in addition that $|\mu^{0\prime}| = |\mu_n^0| = 1$, $\mu^{0\prime} = (\mu_1^0, ..., \mu_{n-1}^0)$. The condition $Q(\eta^0) \leq 0$ is then equivalent to $|\eta^{0\prime}| \geq |\eta_n^0|$ and η^0 is tangent to $\partial\Gamma$ at μ^0 precisely if

$$0 = \langle \eta^0, \ \text{grad} \ Q(\mu^0)\rangle/2 = \langle \eta^{0\prime}, \mu^{0\prime}\rangle - \eta_n^0\mu_n^0.$$

Assume first that $|\eta^{0\prime}| < |\eta_n^0|$. It follows that $|\langle \eta^{0\prime}, \mu^{0\prime}\rangle| \leq |\eta^{0\prime}| < |\eta_n^0|$, so we cannot have $\langle \eta^{0\prime}, \mu^{0\prime}\rangle = \eta_n^0\mu_n^0$ and therefore η^0 cannot be tangent to $\partial\Gamma$ at μ^0. If, conversely, $|\eta_n^0| \leq |\eta^{0\prime}|$, and if we set $\mu_n^0 = 1$, then it is clearly possible to find $\mu^{0\prime} \in R^{n-1}$ with $|\mu^{0\prime}| = 1$ and $\langle \eta^{0\prime}, \mu^{0\prime}\rangle = \eta_n^0$, provided $n - 1 \geq 2$. In the case $n = 2$, the situation is in so far different that we have $\langle \eta^{0\prime}, \mu^{0\prime}\rangle = \eta_1^0\mu_1^0$ and we can have tangency only if $|\eta_1^0| = |\eta_2^0|$.

Proof of proposition 6.6.1. We may assume that $(\partial^2/\partial\tau^2)F(\lambda^0) > 0$. It follows then that $\langle \xi^0, \ \text{grad}_\xi\rangle^2 f(\xi^0) > 0$, so that the signature of $\langle H_{\xi\xi}f(\xi^0)\xi, \xi\rangle$ is $+ - \cdots -$. (See lemma 6.4.8.) Next we observe that

$$p_{m,1}(v, w) = \langle H_{\xi\xi}f(\xi^0)(v - w\xi^0), v - w\xi^0\rangle.$$

It is therefore clear that the signature of $p_{m,1}$ as a quadratic form on the space of variables orthogonal to λ^0 is also $+ - \cdots -$. According to the preceding lemma it suffices then to show that $p_{m,1}(N) > 0$. This is clear however, since $p_{m,1}(N) = (\partial/\partial\tau)^2 F(\lambda^0)/2$.

6.7   Conical refraction in free space for the system of crystal optics

1. In the present section we shall study conical refraction (still in free space) for the system of (linear) crystal optics. Since the situation from crystal optics is covered by most papers on conical refraction, our exposition shall be brief. One particularity is here that we study a system rather than a scalar equation. Actually, the passage from the scalar case to square systems, does not involve in many cases any additional effort, since the components of the solution of the system will satisfy usually some scalar equation associated with some notion of determinant of the system. The system of crystal optics is just Maxwell's system specialized to the case of homogeneous media. Explicitly it is thus (cf. e.g. Courant-Hilbert [1])

$$\partial_t(\varepsilon E) - \text{curl} \ H = 0, \tag{6.7.1}$$

$$\partial_t(\mu H) + \text{curl} \ E = 0, \tag{6.7.2}$$

$$\text{div} \ (\varepsilon E) = 0, \text{div} \ (\mu H) = 0, \tag{6.7.3}$$

where E and H are the electric and magnetic components of some electromagnetic field, ε is the dielectric tensor and μ is the tensor of magnetic permeability. (Both ε and μ are 3×3 matrices with constant entries.) Actually, choosing coordinates suitably, we may assume (in crystal theory) that

$$\varepsilon = \begin{pmatrix} \varepsilon_1 & & 0 \\ & \epsilon_2 & \\ 0 & & \varepsilon_3 \end{pmatrix}, \quad \mu = \begin{pmatrix} \nu & & 0 \\ & \nu & \\ 0 & & \nu \end{pmatrix} \tag{6.7.4}$$

where the ε_i and ν are positive constants. Note that this means in particular that the crystal is magnetically isotropic, but there is still some anisotropy in the electric components. (For the fact that crystals may be assumed magnetically isotropic, cf. Born-Wolf [1] or Sommerfeld [1]. To assume magnetic isotropy will simplify here the notations, but the system is not effectively more complicated in the general case. For a discussion of this, cf. Kline-Kay [1].) The crystal is optically isotropic if $\varepsilon_1 = \varepsilon_2 = \varepsilon_3$. It is called uniaxial if two of the three ε_i coincide and differ from the third (Iceland spar is an example) and it is called biaxial if $\varepsilon_i \neq \varepsilon_j$ whenever $i \neq j$. (An example is aragonite.) The case of isotropic crystals is of no particular interest in the present situation. For uniaxial crystals, we will have rather simple phenomena of conical (in fact double) refraction and we shall not discuss them here at all. We shall here only consider the case of biaxial crystal. To make a choice we shall assume then that

$$\varepsilon_3 < \varepsilon_2 < \varepsilon_1.$$

2. The system (6.7.1), (6.7.2) is a determined square system of partial differential equations, so the full system (6.7.1), (6.7.2), (6.7.3) is overdetermined. The equations (6.7.3) are called "constraint equations". While they play in general an important rôle in Maxwell's theory, they are not really essential in the present section. We should therefore mention that the constraint equations become important when we shall consider conical refraction for transmission problems in chapter VIII. It is now also convenient to write the system (6.7.1), (6.7.2) in the form

$$P(D)u = 0, \tag{6.7.5}$$

where $u = (u', u'')$, $u' = E$, $u'' = H$ and

$$P(D) = \begin{pmatrix} \varepsilon\, \partial_t & -curl \\ curl & \mu\, \partial_t \end{pmatrix}.$$

The characteristic variety of P is then by definition the surface $\{\lambda \in R^4; \det P(\lambda) = 0\}$, where $P(\lambda)$ is the matrix which is obtained if we replace ∂_t by τ and ∂_{x_i} by ξ_i in $P(D)$.

After some calculations (cf. Courant-Hilbert [1]) and using the notations

$$d_j = 1/(\mu \varepsilon_j),$$

$$\psi(\xi) = (d_2 + d_3)\xi_1^2 + (d_3 + d_1)\xi_2^2 + (d_1 + d_2)\xi_3^2,$$

$$\varphi(\xi) = d_2 d_3 \xi_1^2 + d_3 d_1 \xi_2^2 + d_1 d_2 \xi_3^2,$$

we arrive at the following expression for $\det P$:

$$\det P(\lambda) = \frac{\tau^2}{d_1 d_2 d_3}(\tau^4 - \psi(\xi)\tau^2 + \varphi(\xi)|\xi|^2). \tag{6.7.6}$$

The relevance of $\det P$ for solutions of $P(D)u = 0$ comes from the fact that if we assume that u is a solution of $P(D)u = 0$, then it automatically follows that the components u_i of u (there are of course six of them) satisfy $(\det P)(D)u_i = 0$. In view of this, we can then obtain results on conical refraction for the solutions of the system (6.7.1), (6.7.2) if we study conical refraction for the scalar equation $\det P(D)v = 0$. Later on we shall write

$$p(\lambda) = \tau^4 - \psi(\xi)\tau^2 + \varphi(\xi)|\xi|^2$$

so that $\det P(\lambda) = \tau^2 p(\lambda)$. It is interesting to note that for solutions u of the full Maxwell system which are compactly supported in x we can prove that actually $p(D)u_i = 0$. (Cf. in fact the following chapter where we effectively need this result.) Here we content ourselves with observing that the factor τ^2 will be elliptic for the characteristic points which we consider, so that it can be ignored for our problems of conical refraction anyway.

3. Our main goal in this section is to study the characteristic surface and the slowness surface of $p(D)$ from the point of view of their singularities. Let us observe then that the roots τ of $p(\xi, \tau) = 0$ are of form

$$\tau = \pm \left(\frac{\psi(\xi) \pm \sqrt{\Delta(\xi)}}{2} \right)^{1/2} \quad , \quad \Delta(\xi) = \psi^2 - 4\varphi(\xi)|\xi|^2,$$

and it is easy to see (cf. Courant-Hilbert [1]) that Δ can be written as

$$((\sqrt{d_3 - d_2}\xi_1 - \sqrt{d_2 - d_1}\xi_3)^2 + (d_3 - d_1)\xi_2^2)((\sqrt{d_3 - d_2}\xi_1 + \sqrt{d_2 - d_1}\xi_3)^2 + (d_3 - d_1)\xi_2^2) \ge 0.$$

We observe next that the relation $\varepsilon_3 < \varepsilon_2 < \varepsilon_1$ gives $d_1 < d_2 < d_3$. It follows that for each fixed ξ the equation $p(\xi, \tau) = 0$ has 4 real roots, not more than two of them coinciding when $\xi \neq 0$ (since then by necessity $\psi^2 > \psi^2 - 4\varphi(\xi)|\xi|^2$). Moreover, double roots can only occur when $\xi_2 = 0$ and

$$\frac{\xi_1}{\xi_3} = \pm \sqrt{\frac{d_2 - d_1}{d_3 - d_2}}. \tag{6.7.7}$$

(Asymetries in the roles of the ξ_i are here and later on due to the fact that we only look at real roots.)

4. It is now easy to study the singularity types of the slowness surface or characteristic surface of $p(D)$ at the singular points. The equation of the slowness surface is

$$1 - \psi(\xi)\tau^2 + \varphi(\xi)|\xi|^2 = 0,$$

and the surface itself is called "Fresnel's" surface in crystal optics. Corresponding to (6.7.7) we have four singular points on it. Let us consider such a point, e.g.,

$$\xi^0 = \left(\sqrt{\frac{1}{d_2}\frac{d_2 - d_1}{d_3 - d_1}} \, , \, 0 \, , \, \sqrt{\frac{1}{d_2}\frac{d_3 - d_2}{d_3 - d_1}} \right)$$

and denote by $\lambda^0 = (\xi^0, 1)$. It is easily seen from the explicit form of the discriminant that the Hessian of the discriminant has rank two at ξ^0. It is therefore clear that ξ^0 is a conical singularity for the slowness surface of p. We have seen in the introduction and section 6.4 that this leeds to effective conical refraction for solutions of $p(D)v = 0$. One can also compute the localization polynomial p^T of p at λ^0, explicitly. What one obtains is (cf. Esser [1]):

$$\begin{aligned}
p^T &= 8((-\tau + \frac{1}{2\sqrt{d_2}}(d_2 + d_3)\sqrt{\frac{d_2 - d_1}{d_3 - d_1}}\xi_1 + \frac{1}{2\sqrt{d_2}}(d_1 + d_2)\sqrt{\frac{d_3 - d_2}{d_3 - d_1}}\xi_3)^2 \\
&\quad -(\frac{1}{2\sqrt{d_2}}(d_3 - d_2)\sqrt{\frac{d_2 - d_1}{d_3 - d_1}}\xi_1 + \frac{1}{2\sqrt{d_2}}(d_1 - d_2)\sqrt{\frac{d_3 - d_2}{d_3 - d_1}}\xi_3)^2 \\
&\quad - \frac{1}{4d_2}(d_2 - d_1)(d_3 - d_2)\xi_2^2).
\end{aligned}$$

6.8 Singular points on the slowness surface of the system of elasticity for cubic crystals

1. The system of linear elasticity for crystals is of form

$$\frac{\partial^2 u_i}{\partial t^2} = \sum_{j,p,q=1}^{3} c_{ijpq}\frac{\partial^2 u_j}{\partial x_p \partial x_q}, i = 1, 2, 3. \tag{6.8.1}$$

The system thus depends on the $3^4 = 81$ constants c_{ijpq}, but these constants are not all independent, in that we will always have the relations

$$c_{ijpq} = c_{jipq} = c_{ijqp} = c_{pqij}.$$

It follows that the number of independent constants is in fact 21. An additional restriction on the range of values for the elasticity constants c_{ijpq} comes from the condition that the quadratic form of strain energy:

$$(1/2) \sum c_{ijpq} e_{ij} e_{pq}$$

is positiv definit on symmetric tensors (e_{ij}). As a consequence the system (6.8.1) is hyperbolic in the time direction.

If the crystal under consideration exhibits some symmetries, the number of independent constants is reduced still further. From the point of view of their symmetries, crystals are grouped in crystal classes. Whereas in crystal optics, we essentially had only three different types of crystals (homogeneous, uniaxial and biaxial crystals), there are essentially six classes as far as acoustic properties are concerned: isotropic, hexagonal, cubic, tetragonal, rhomboedric and trigonal ones. (Cf. e.g. Sommerfeld [1] or Born-Wolf [1].) The easiest case is of course the isotropic one. In this case we are left with essentially two constants, the Lamé constants of elasticity theory. The slowness surface of the system of elasticity in the isotropic case is the union of two spheres. The next simple case is, mathematically speaking, the hexagonal class. Hexagonal crystals are transversally isotropic, so the difficulties are essentially those of the two space variables case. The slowness surface is some special fourth order algebraic curve in the plane which is rotated around some symmetry axis. In the case of hexagonal crystals there are essentially five free constants. The simplest interesting case is that of cubic crystals. The number of free constants is three, which is only one more than in the isotropic case and two less than in the hexagonal case, but the slowness surfacs is more interesting than it is in the hexagonal case. The interest in cubic crystals comes in part from the fact that many precious or valuable crystals are cubic. (Examples of cubic crystals are gold and salt.) After a number of renotations for constants, the characteristic equation for the system (6.8.1) is, in the cubic case and if coordinates are chosen suitably, written in what is called "Kelvin's form" (cf. Musgrave [1]),

$$\frac{b\xi_1^2}{\tau^2 - c|\xi|^2 + (b-a)\xi_1^2} + \frac{b\xi_2^2}{\tau^2 - c|\xi|^2 + (b-a)\xi_2^2} + \frac{b\xi_3^2}{\tau^2 - c|\xi|^2 + (b-a)\xi_3^2} = 1. \quad (6.8.2)$$

Here $b - a$ is, in some sense, a measure of the anisotropy of the crystal: $b - a = 0$ would correspond to the isotropic case. We shall therefore always assume in the sequel that $b - a \neq 0$. Note that the equation (6.8.2) is left unchanged if we permute the ξ_i among themselves, which is of course a consequence of the cubic symmetry of the crystal. We should also note here that if we require the same kind of invariantness for the equation of Fresnel from crystal optics (cf. the last section), then the surface is automatically the one from the optically isotropic situation. Indeed, it is in fact well known that cubic crystals are optically isotropic.

2. The meaning of (6.8.2) is of course that

$$\sum_i b\xi_i^2(\tau^2 - c|\xi|^2 + (b-a)\xi_{i+1}^2)(\tau^2 - c|\xi|^2 + (b-a)\xi_{i+2}^2) = \prod_j (\tau^2 - c|\xi|^2 + (b-a)\xi_j^2),$$

where in the sum indices are computed modulo 3. Corresponding to (6.8.2), and with a similar convention, we have for the equation of the slowness surface

$$\frac{b\xi_1^2}{1 - c|\xi|^2 + (b-a)\xi_1^2} + \frac{b\xi_2^2}{1 - c|\xi|^2 + (b-a)\xi_2^2} + \frac{b\xi_3^2}{1 - c|\xi|^2 + (b-a)\xi_3^2} = 1. \qquad (6.8.3)$$

The following result is wellknown in crystal theory.

Proposition 6.8.1. *Assume $b \neq 0$, $b \neq a$, $b \neq -a$, $c > 0$, $3c - b + a > 0$. Then the only double points of the surface from (6.8.2) with $\xi \neq 0$ are those for which either*
a) $\xi_1 = \pm\xi_2 = \pm\xi_3$,
or else
b) $\xi_i = \xi_j = 0$ for two indices in $\{1, 2, 3\}$.

Remark 6.8.2. *When $b = 0$, then we will have a triple root at $\xi_1^2 = \xi_2^2 = \xi_3^2$ and double roots at $\xi_1^2 = \xi_2^2$, etc. We have not made a similar analysis for the other exceptional values from the statement of proposition 6.8.1, but, as far as this author knows, there are no examples of crystals in the literature when triple roots appear effectively. The conditions $c > 0$ and $3c - b + a > 0$ are needed to obtain real roots for specific values of ξ. For crystals they follow therefore from the positivity of the energy form.*

Before we prove proposition 6.8.1 we make the following remark which seems standard in this context (cf. Duff [1]):

Lemma 6.8.3. *Let $a_1 \leq a_2 \leq a_3 \in R$ and $b_i > 0$, $i = 1, 2, 3$ be given. Denote by*

$$g(t) = \prod_{i=1}^{3} (t - a_i) - \sum_{i=1}^{3} b_i(t - a_{i+1})(t - a_{i+2}).$$

Then the three roots of $g(t) = 0$ are all real and if $a_1 < a_2 < a_3$, in each of the open intervals (a_1, a_2), (a_2, a_3), (a_3, ∞) we have precisely one such root. Moreover, double roots can only appear when $a_1 = a_2 = a_3$. A similar statement is valid in the case when the b_i are strictly negative.

Proof. Assume first that $a_1 < a_2 < a_3$. We have

$$\lim_{t \to \pm\infty} g(t) = \pm\infty,$$

and

$$\text{sign } g(a_i) = (-1)^i.$$

We have therefore three changes of sign on $(-\infty, \infty)$ and they occur precisely in the intervals of the statement. If now, e.g., $a_1 = a_2$, then one of the roots of $g(t) = 0$ is $t = a_1$ and the other two satisfy

$$g'(t) = (t - a_1)(t - a_3) - (b_1 + b_2)(t - a_1)(t - a_3) = 0.$$

When $a_1 \neq a_3$, we can argue as before and see that the last equation has two roots which are located in the open intervals (a_1, a_3), (a_3, ∞). No double roots can appear in this situation. Conversely, if $a_1 = a_3$, then $t = a_1$ is of course a double root of $g(t) = 0$, the third root being $t = a_1 + 3\sum_i b_i$.

Proof of proposition 6.8.1.   When $\xi_1^2 = \xi_2^2 = \xi_3^2$, then $\tau^2 = (3c - b + a)\xi_1^2$ is a double root. Here τ is real precisely if we assume that $3c - b + a > 0$, which is one of the conditions from the assumption. Similarly, if $\xi_i = \xi_j = 0$ for two indices i and j, then $\tau^2 = c\xi_k^2$ is a double root, where k is the remaining index.

Assume now conversely that a double root appears, for some fixed ξ.

a) We assume at first that $\xi_1 \xi_2 \xi_3 \neq 0$ and want to show that $\xi_1^2 = \xi_2^2 = \xi_3^2$ then. Denote $a_i = c|\xi|^2 - (b - a)\xi_i^2$, $b_i = b\xi_i^2$, for $i = 1, 2, 3$. It is no loss of generality to assume $\xi_3^2 \leq \xi_2^2 \leq \xi_1^2$, so $a_1 \leq a_2 \leq a_3$ if $b - a > 0$, as we will assume to make a choice. Further, we shall assume, once more to make a choice, $b > 0$. From remark 6.8.3, applied to the function g associated with the present values of the a_i, b_i, it follows that a double root can only appear if $a_1 = a_2 = a_3$, which means $\xi_1^2 = \xi_2^2 = \xi_3^2$.

b) It remains to consider the case $\xi_1 \xi_2 \xi_3 = 0$. We may assume e.g. that $\xi_3 = 0$ and want to show that if a double root appears, then we must also have $\xi_1 \xi_2 = 0$. In any case, the fact that $\xi_3 = 0$ already gives us that one of the roots must be $\tau^2 = c(\xi_1^2 + \xi_2^2)$. We use the notations a_i and b_i with their meaning from part a) of the proof and denote by g'' the function

$$g''(t) = (t - a_1)(t - a_2) - b_1(t - a_2) - b_2(t - a_1) = 0.$$

If $a_1 \neq a_2$ and $\xi_1 \xi_2 \neq 0$ the two roots of $g''(t) = 0$ will be distinct and a double root can only appear if $g''(c(\xi_1^2 + \xi_2^2)) = 0$. This will only be the case if $b + a = 0$, but we have excluded this in the assumption. The situation is similar when $a_1 = a_2$ and we omit further details.

We have now seen where the double roots of the characteristic surface are located and want to see what kind of singularities the slowness surface has at the corresponding points. The first result is

Proposition 6.8.4. *Assume* $3c - b + a > 0$. *Then the slowness surface S has a conical singularity at the points $\tilde{\xi}$ where $\tilde{\xi}_1^2 = \tilde{\xi}_2^2 = \tilde{\xi}_3^2$.*

Proof. The equation of the slowness surface is $f(\xi) = 0$, where

$$f(\xi) = \sum_i b\xi_i^2(1 - c|\xi|^2 + (b-a)\xi_{i+1}^2)(1 - c|\xi|^2 + (b-a)\xi_{i+2}^2) - \prod_j (1 - c|\xi|^2 + (b-a)\xi_j^2).$$

We must show that the Hessian is nondegenerate at $\tilde{\xi}_1^2 = \tilde{\xi}_2^2 = \tilde{\xi}_3^2 = \sigma^2$ where σ satisfies $1 - (3c - b + a)\sigma^2 = 0$. At first we observe that the term $\prod_j (\tau^2 - c|\xi|^2 + (b-a)\xi_j^2)$, vanishes of third order at $\tilde{\xi}$, and will therefore not give any contribution to the Hessian there. Moreover, a nontrivial contribution to a second derivative of a term

$$b\xi_i^2(\tau^2 - c|\xi|^2 + (b-a)\xi_{i+1}^2)(\tau^2 - c|\xi|^2 + (b-a)\xi_{i+2}^2)$$

at $\tilde{\xi}$ will only appear in situations when we leave $b\xi_i^2$ underivated and derivate each of the factors

$$(1 - c|\xi|^2 + (b-a)\xi_j^2),\ j = i+1, j = i+2$$

once. Thus we will have that actually,

$$H_{\xi\xi}f(\tilde{\xi}) = b\sigma^2 H_{\xi\xi}[\sum_i (1 - c|\xi|^2 + (b-a)\xi_{i+1}^2)(1 - c|\xi|^2 + (b-a)\xi_{i+2}^2)].$$

Here it is convenient to perform two coordinate transforms. The first is $\eta_i = \xi_i^2$. It suffices then to show that the Hessian of

$$G(\eta) = \sum_i (1 - c(\eta_1 + \eta_2 + \eta_3) + (b-a)\eta_{i+1})(1 - c(\eta_1 + \eta_2 + \eta_3) + (b-a)\eta_{i+2})$$

is nondegenerate at $(\sigma^2, \sigma^2, \sigma^2)$. This is the same with the Hessian at 0 of

$$H(\eta) = \sum_i (-c(\eta_1 + \eta_2 + \eta_3) + (b-a)\eta_{i+1})(-c(\eta_1 + \eta_2 + \eta_3) + (b-a)\eta_{i+2}).$$

We can now change coordinates a second time, setting

$$(b-a)\eta_i - c(\eta_1 + \eta_2 + \eta_3) = \theta_i,\ i = 1, 2, 3,$$

which is indeed a coordinate transform if $3c \neq b - a$. In the new coordinates we now have $H(\eta(\theta)) = \theta_1\theta_2 + \theta_2\theta_3 + \theta_3\theta_1$, and the desired nondegeneracy can be read off directly.

We now turn our attention to the remaining singular points of the slowness surface.

Proposition 6.8.5. *Consider* $\tilde{\xi} = (1/\sqrt{c}, 0, 0)$. *Then* $\tilde{\xi}$ *is an unode of the slowness surface. The same holds of course also for all other points of the slowness surface which lie on the coordinate axes.*

Proof. We must show that rank $H_{\xi\xi}f(\tilde{\xi}) = 1$, where f is the function introduced in the last proof. We denote by $Q_i = 1 - c|\xi|^2 + (b-a)\xi_i^2$ and have that Q_2 and Q_3 vanish of first order at $\tilde{\xi}$. It is then clear that $H_{\xi\xi}f(\tilde{\xi}) = H_{\xi\xi}\tilde{f}(\tilde{\xi})$ where

$$\tilde{f}(\xi) = \prod_i Q_i - b\xi_1^2 Q_2 Q_3 = Q_2 Q_3 (Q_1 - b\xi_1^2).$$

Here $(Q_1 - b\xi_1^2)(\tilde{\xi}) = -a\tilde{\xi}^2 \neq 0$ and $Q_2 Q_3$ vanishes of order two at $\tilde{\xi}$. It follows that in fact $H_{\xi\xi}f(\tilde{\xi}) = -(a/c)H_{\xi\xi}(Q_2 Q_3)(\tilde{\xi})$. Further we note that $(\partial/\partial\xi_i)Q_j(\tilde{\xi}) = 0$ if $i, j \in \{2, 3\}$. The only second order derivative of $Q_2 Q_3$ which is different from 0 at $\tilde{\xi}$ is then $(\partial/\partial\xi_1)^2 Q_2 Q_3(\tilde{\xi}) = -4c^2|\tilde{\xi}|^2 = -4$. The desired assertion now follows.

3. Having studied the nature of the singular points on the charactersitic surface, it is now clear from the theory developed above that the singular points $\pm\xi_1 = \pm\xi_2 = \pm\xi_3$ (all sign combinations are to be taken) give rise to conical refraction, whereas the remaining singularities do not. A deeper question is however that of the physical relevance of the results one so obtains. Indeed, microlocalization is, in some sense, high frequency asymptotics and waves of very high frequencies are perhaps not realistic phenomena of elasticity theory. One would then need results similar to the ones above for the case of medium-high frequencies. Needless to say, the first problem is with the choice of appropriate definitions, etc. In this situation it is therefore comforting to note that conical refraction in crystal acoustics has been studied in the physical literature and that arguments of geometrical optics are used in this context. (See Musgrave [1] and the literature cited there.)

Chapter 7

Propagation of regularity up to the boundary

1. Results on propagation of singularities in free space do not suffice in general for physical applications, since boundary effects will appear. It is therefore an interesting question to analyze to what extent information on propagation of singularities in free space is relevant for results on propagation of singularities for boundary problems. We shall address this question in chapter VIII for "transversal" real analytic boundaries, making however the rather strong assumption that the operator under consideration has constant coefficients. The main tool which we need is a result on propagation of regularity up to the boundary which will be established in the present chapter. On the other hand, application of this result is only meaningful if combined with a result on microlocal solvability of Cauchy problems for operators which are microlocally hyperbolic: cf. proposition 7.2.3 below. (Also cf. here the comments in the introduction to chapter VII.) We should note that in the frame of hyperfunctions many results on propagation of regularity up to the boundary are known and that in particular one can obtain the results which we effectively apply in chapter VIII from that theory (when combined with results on conical refraction in free space in hyperfunctional set-up): cf. e.g. Schapira [2], [3], Kataoka [2], [3]. However, (apart from the fact that the present approach is perhaps sometimes simpler,) only part of the results on conical refraction from these notes have also been studied in hyperfunctions, and moreover, we think that our approach can sometimes be extended to situations in which one can not remain in the analytic category.

7.1   Interior regularity when the traces are smooth

1. Later on it will be convenient to establish explicit relations between the analytic wavefront set of a solution u of a boundary problem and the regularity of the Cauchy traces u_j of u. The following result is then useful:

Theorem 7.1.1. *Let $p(x, D_x)$ be a linear partial differential operator of order m with analytic coefficients defined in a neighborhood of $0 \in R^n_x$ for which the surface $S = \{x; x_n = 0\}$ is noncharacteristic and let u be a distribution which is defined in a neighborhood of 0 and satisfies*

$$p(x, D)u = 0 \text{ in } x_n > 0 \text{ near } 0.$$

Assume that $(0, \xi'^0) \notin WF_A(\partial/\partial x_n)^j u_{|x_n=0}$ for $j = 0, ..., m-1$. Then there is $\varepsilon > 0$ such that

$$(x, \xi) \notin WF_A u \text{ if } 0 < x_n < \varepsilon \text{ and } |x'| + |\xi' - \xi'^0| < \varepsilon. \tag{7.1.1}$$

Here $x' = (x_1, ..., x_{n-1}), \xi' = (\xi_1, ..., \xi_{n-1})$.

Moreover, when $p(x, D)u = 0$ in a full neighborhood of 0, then we will have $(x, \xi) \notin WF_A u$ for all (x, ξ) such that $|x| + |\xi' - \xi'^0| < \varepsilon$.

Theorem 7.1.1 is due to Schapira [1] and is also implicit in Kataoka [1]. (For constant coefficient operators a closely related result is considered in Liess [6].) Also cf. Hörmander [5] and Liess [7] (where a variant in G_φ- classes of the theorem -which covers in particular the analytic case- is given).

Note that for twosided solutions the converse to theorem 7.1.1 is also true: if u is a solution of $p(x, D)u = 0$ in a neighborhood of zero such that

$$(0, (\xi'^0, \nu)) \notin WF_A u, \text{ for all } \nu \text{ such that } p_m(0, (\xi'^0, \nu)) = 0, \tag{7.1.2}$$

then

$$(0, \xi'^0) \notin WF_A(\partial/\partial x_n)^j u_{|x_n=0}. \tag{7.1.3}$$

Indeed, if (7.1.2) holds, then there is no vector of form $(0, (\xi'^0, \xi_n))$ in $WF_A u$. (In fact, $WF_A u$ is contained near 0 in the characteristic variety of p and characteristic vectors of the form $(0, (\xi'^0, \xi_n))$ are excluded by our assumption.) The conclusion follows then from the following wellknown result of Hörmander-Sato on the regularity of a restriction:

Theorem 7.1.2. *Let u be a distribution defined in a neighborhood of 0 such that $(0, (0, ..., 0, \pm 1)) \notin WF_A u$. Denote by w the restriction of u to $x_n = 0$ (which is well defined near 0 in view of the assumptions). Also consider $\xi'^0 \in \dot{R}^{n-1}$. If $\{(0, (\xi'^0, \nu)); \nu \in R\} \cap WF_A u = \emptyset$ it follows then that $(0, \xi'^0) \notin WF_A w$.*

7.2   Extension across hypersurfaces of solutions of constant coefficient microlocally hyperbolic operators

1. Let $p(D)$ be a constant coefficient linear partial differential operator for which the hypersurface $x_n = 0$ is noncharacteristic. Also fix some open cone Γ in $R_{\xi'}^{n-1}$, $\xi' = (\xi_1, ..., \xi_{n-1})$. We shall say that $p(D)$ is microlocally hyperbolic on Γ, if we can find a constant c so that

$$(\xi', \tau) \in \Gamma \times C, p(\xi', \tau) = 0 \text{ implies } |Im\,\tau| \le c. \tag{7.2.1}$$

The main result of this section is

Proposition 7.2.1. *Assume that p satisfies (7.2.1) and that $x_n = 0$ is noncharacteristic for $p(D)$. Let $u \in D'(x; |x| < \varepsilon)$ be a solution of*

$$p(D)u = 0, \quad on \ \{x; |x| < \varepsilon,\ x_n > 0\},$$

and fix $\varepsilon' < \varepsilon$ and $\Gamma' \subset\subset \Gamma$. We can then find $v \in D'(x; |x| < \varepsilon')$ such that

$$p(D)v = 0 \ on \ \{x; |x| < \varepsilon'\}$$

and such that

$$(\{x'; |x'| < \varepsilon'\}, \Gamma') \cap WF_A((\partial/\partial x_n)^j(u-v)_{|x_n=0_+}) = \emptyset, \ j = 0, ..., m-1. \tag{7.2.2}$$

b) When we combine proposition 7.2.1 with theorem 7.1.1 it follows in addition if $\Gamma'' \subset\subset \Gamma'$ is fixed, that there is $\varepsilon'' > 0$ so that

$$(x, (\xi', \tau)) \notin WF_A(u-v), \ if\ x_n > 0,\ |x| < \varepsilon'',\ \tau \in R, \xi' \in \Gamma''. \tag{7.2.3}$$

Here $w_{|x_n=0_+}$ is, for some distribution $w \in D'(U \cap V)$, the (distributional) trace from the side $x_n > 0$ of w to $x_n = 0$. ($u - v$ being defined on both sides of $x_n = 0$, distributional traces of $(\partial/\partial x_n)^j(u-v)$ exist in view of Peetre's theorem. Since we are in the analytic category, it is perhaps worth mentioning that hyperfunctional traces would make sense, by a result of P.Schapira and H.Komatsu, also if u were defined only in the region $U \cap \{x; x_n > 0\}$. The result is not surprizing in fact, since u admits a hyperfunctional extension to a full neighborhood of 0 since the sheaf of hyperfunctions is flabby.)

Remark 7.2.2. *a) At the end of the section we shall see that the theorem remains true also when we replace the surface $x_n = 0$ by some more general surface H which has the property that the tangent plane to H remains locally in the region where the solution is defined.*

b) It is not difficult to see that (7.2.1) gives the seemingly stronger condition : if $\Gamma' \subset\subset \Gamma$, then

$$p(\zeta', \tau) = 0, \ \ Re\,\zeta' \in \Gamma' \text{ implies } |Im\,\tau| \leq c'(1 + |Im\,\zeta'|). \tag{7.2.4}$$

In fact, the function

$$f(\zeta') = \sup_{\tau, p(\zeta', \tau) = 0} |Im\,\tau|$$

is plurisubharmonic and satisfies $f(\zeta') \leq c''(1 + |\zeta'|)$ for some constant c''. The assertion follows therefore from the Phragmén-Lindelöf principle. (For a suitable form of this principle, cf. Hörmander [7].) We shall later on sometimes work directly with (7.2.4), rather than with (7.2.1).

2. Proposition 7.2.1 is a direct consequence of the following

Proposition 7.2.3. *Let $p(D)$ be as in proposition 7.2.1. Let further $f_0, ..., f_{m-1} \in D'(x'; |x'| < \varepsilon)$. If $\varepsilon' < \varepsilon$ is fixed, we can find $f \in D'(x; |x| < \varepsilon')$ such that*

$$p(D)f = 0, \tag{7.2.5}$$

$$(\{x'; |x'| < \varepsilon'\}, \Gamma') \cap WF_A(f_j - (\partial/\partial x_n)^j f_{|x_n = 0_+}) = \emptyset, \ j = 0, ..., m - 1. \tag{7.2.6}$$

In fact, if u is given as in proposition 7.2.1, then a distribution as needed in the conclusion of that proposition is just the f associated with $f_j = (\partial/\partial x_n)^j u_{|x_n = 0_+}$, $j = 0, ..., m - 1$.

3. **Proof of proposition 7.2.3.** We denote by

$$T : \quad \begin{matrix} A(C^n) \\ \times \\ \prod_{j=0}^{m-1} A(C^{n-1}) \end{matrix} \quad \longrightarrow A(C^n)$$

the map which associates with $g \in A(C^n)$, $g_j \in A(C^{n-1})$, $j = 0, ..., m - 1$, the solution h of the Cauchy problem

$$p(D)h = g \text{ on } R^n, \tag{7.2.7}$$

$$(i\partial/\partial x_n)^j h_{|x_n = 0_+} = g_j, j = 0, ..., m - 1. \tag{7.2.8}$$

(The map T is well defined in view of the global version of the Cauchy-Kowalewska theorem.) Let

$$
{}^tT : A'(C^n) \longrightarrow
\begin{array}{c}
A'(C^n) \\
\times \\
\prod_{j=0}^{m-1} A'(C^{n-1})
\end{array}
$$

be the dual map (which acts thus between spaces of analytic functionals). Explicitly, if $v \in A'(C^n)$ is given, then v is related to

$$
{}^tTv = \begin{pmatrix} w \\ w_i, i \le m-1 \end{pmatrix}
$$

by the fact that

$$
v = {}^t p(D)w + \sum_{j=0}^{m-1} w_j \otimes (-i\partial/\partial x_n)^j \delta_{x_n}, \tag{7.2.9}
$$

where δ_{x_n} is the Dirac distribution in the variable x_n at $x_n = 0$. (7.2.9) is of course a relation in analytic functionals. Actually, (7.2.7), (7.2.8) and (7.2.9) are related by

$$
v(h) = w(g) + \sum_j w_j(g_j).
$$

The interesting thing is now that from (7.2.9) we obtain a rather explicit form for the Fourier-Borel transform of w and of the w_j. (That this is so, is of course well-known. Cf. e.g. Kiselman [1].) In fact, if we take the Fourier-Borel transform of (7.2.9), we get

$$
\hat v(\zeta) = p(-\zeta)\hat w(\zeta) + \sum_{j=0}^{m-1} \hat w_j(\zeta')\zeta_n^j, \tag{7.2.10}
$$

which shows that $\hat w$ and the $\hat w_j$ are just the quotient and remainder terms in a Weierstrass-type decomposition of $\hat v$. One can compute $\hat w$ explicitly from (7.2.10) using contour integration formulas. Note in fact that it follows from (7.2.10) that

$$
\frac{1}{2\pi i}\frac{\hat w(\zeta', \zeta_n + \sigma)}{\sigma} = \frac{1}{2\pi i}\frac{\hat v(\zeta', \zeta_n + \sigma)}{\sigma p(-\zeta', -\zeta_n - \sigma)} - \frac{1}{2\pi i}\sum_{j=0}^{m-1} \hat w_j(\zeta')\frac{(\zeta_n + \sigma)^j}{\sigma p(-\zeta', -\zeta_n - \sigma)},
$$

whenever σ and $p(-\zeta', -\zeta_n - \sigma)$ are different from zero. Integrating this in the complex σ- plane over a contour of form $|\sigma| = c(1 + |\zeta'|)$ for some sufficiently large c, (we consider contours in relation with contour integrals always with counterclockwise orientation,) we obtain

$$
\hat w(\zeta', \zeta_n) = \frac{1}{2\pi i}\int_{|\sigma|=c(1+|\zeta'|)}\frac{\hat v(\zeta', \zeta_n + \sigma)}{\sigma p(-\zeta', -\zeta_n - \sigma)}d\sigma, \tag{7.2.11}
$$

since

$$
\int_{|\sigma|=c(1+|\zeta'|)}\frac{(\zeta_n + \sigma)^k}{\sigma p(-\zeta', -\zeta_n - \sigma)}d\sigma = 0, \text{ for } k \le m-1, \tag{7.2.12}
$$

if c is large enough. (We can apply the residuum theorem at ∞, or else just estimate the integral for $c \to \infty$. The value of the integral is constant in c for large c, since the roots of $p(\zeta', \tau) = 0$ will all satisfy an estimate of form $|\tau| \leq c_1(1 + |\zeta'|)$ for some c_1. On the other hand it is also immediate that the integral will go to zero for $c \to \infty$. All this is of course standard. Cf. e.g. Kiselman [1].) Inserting (7.2.11) back into (7.2.10) we also get for suitable polynomials Q_j which do not depend on v (cf. once more Kiselman, loc. cit.) that

$$\hat{w}_j(\zeta') = \int_{\Gamma_{\zeta'}} \frac{Q_j(\zeta', \tau)\hat{v}(\zeta', \tau)}{p(-\tau, -\zeta')} d\tau$$

where $\Gamma_{\zeta'}$ is the contour $\{\sigma; |\sigma| = \bar{c}(1 + |\zeta'|)\}$ for some large enough $\bar{c}$ and has counterclockwise orientation. It is moreover important to observe that the contour $\Gamma_{\zeta'}$ can be deformed to any other piecewise smooth contour $\gamma_{\zeta'}$ as long as we do not cross the roots of $p(-\zeta', -\tau) = 0$ in the process of deformation.

The main point is now that if $k \geq 0$ is fixed, then we can find ε', k' so that if $v \in E'(R^n)$ satisfies

$$|\hat{v}(\zeta)| \leq (1 + |\zeta|)^{-k'} e^{\varepsilon'|Im\,\zeta|}, \text{ for } \zeta \in C^n,$$

then

$$|\hat{w}_j(\xi')| \leq c'(1 + |\xi'|)^{-k}, \text{ if } \xi' \in \Gamma.$$

This is a consequence of our assumption of microlocal hyperbolicity and is in fact seen from an easy contour deformation in the integrals defining the $\hat{w}_j$'s. We shall encounter a somewhat more complicated situation of the same type in section 7.7, where we shall give more details. Here we fix k so that $(1 + |\xi'|)^k \mathcal{F}^{-1}(\varphi f_j) \in L^1(R^{n-1})$, where $\varphi \in C_0^\infty$ is fixed with $\varphi \equiv 1$ in a neighborhood of 0. We also assume of course that the support of φ lies in the common domain of definition of the f_j.

We can then define a distribution f on $|x| < \varepsilon'$ by setting

$$f(v) = (2\pi)^{-n} \int_\Gamma \sum_j \hat{w}_j(-\xi') \mathcal{F}(\varphi f_j)(\xi')\, d\xi', \text{ if } v \in C_0^\infty(x; |x| < \varepsilon'). \tag{7.2.13}$$

It is clear that f is well-defined and it is immediate that $p(D)f = 0$. Indeed, we need to check that $f({}^t p(D)\tilde{v}) = 0$ if $\tilde{v} \in C_0^\infty$, but this follows from the remark that

$${}^t T({}^t p(D)\tilde{v}) = \begin{pmatrix} \tilde{v} \\ 0 \\ \cdot \\ \cdot \\ \cdot \\ 0 \end{pmatrix},$$

so that the w_j associated with $v = {}^t p(D)\tilde{v}$ all vanish. Moreover,

$$(-i\partial/\partial x_n)^j f_{|x_n=0} = \mathcal{F}^{-1}(\mathcal{F}(\varphi f_j)\chi_\Gamma), \qquad (7.2.14)$$

where χ_Γ is the characteristic function of Γ. In fact, to compute $(-i\partial/\partial x_n)^j f_{|x_n=0}$, we observe that

$$(-i\partial/\partial x_n)^j f_{|x_n=0}(g) = f(g \otimes (i\partial/\partial x_n)^j \delta_{x_n}) = (2\pi)^{-n} \int_\Gamma \hat{g}(-\xi')\mathcal{F}(\varphi f_j)(\xi')\, d\xi',$$

since

$$ {}^t T(g \otimes (i\partial/\partial x_n)^j \delta_{x_n}) = {}^t(0,\dots,0,g,0,\dots,0),$$

where g is on position $j+1$.

From (7.2.14) it is now also clear that

$$(\{(x',0);|x'| < \varepsilon''\}, \Gamma') \cap WF_A(f_j - (\partial/\partial x_n)^j f_{|x_n=0_+}) = \emptyset, \ j = 0,\dots,m-1,$$

where $\varepsilon'' > 0$ is chosen so that φ is identically one on $|x'| < \varepsilon''$, so the proof is complete.

Remark 7.2.4 . *From the standard geometrical optics construction of solutions of strictly hyperbolic equations, one can prove a result similar to proposition 7.2.3 for operators which satisfy the following conditions:*

   a) p has analytic coefficients,

   b) p_m is of principal type,

   c) $|x| \le \varepsilon$, $\xi' \in \Gamma$, $p_m(x,\xi',\tau) = 0$ implies $\tau \in R$.

The conclusion is then however only that

$$(0, \Gamma' \times R) \cap WF_A p(x,D)f = \emptyset,$$

rather than $p(x,D)f = 0$.

(If we admit hyperfunction extensions, much better results are know.)

4. In concluding this section we want to extend the validity of proposition 7.2.5 to a case of nonflat boundary. We shall then assume that $\rho : U \to R$ is a real analytic function so that grad $\rho(0) = (0,\dots,0,1)$ and so that $x_n > 0$ implies $\rho(x) > 0$. Assume further that $u \in D'(U)$ satisfies the equation

$$p(D)u = 0 \ \text{on} \ U \cap \{x; \rho(x) > 0\}.$$

Then it follows that we can find a solution $v \in D'(V)$ of $p(D)v = 0$ on V so that

$$(0, \Gamma) \cap WF_A((\partial/\partial n)^j(u - v)_{|\rho=0+}) = \emptyset, \ j = 0, ..., m - 1,$$

where "$_{|\rho=0+}$" means restriction to $\rho = 0$ from the part $\rho > 0$. n is here the normal to $\rho = 0$.

Actually this follows immediately from the results above if we combine them with the following result of Kataoka [1] and Lebeau [2] (also cf. Uchida [1]):

Theorem 7.2.5. *Let w satisfy $p(D)w = 0$ on $(\rho > 0 \cup \{x; x_n > 0\}) \cap V$, where ρ is a real analytic function with $\rho(0) = 0$, and grad $\rho = (0, ..., 0, 1)$ and assume that $(0, \xi'^0) \notin WF_A(\partial/\partial x_n)^j w_{|x_n=0+}, \ j = 0, ..., m - 1$. Then it follows that*

$$(0, \xi'^0) \notin WF_A(\partial/\partial n)^j w_{|\rho=0+}.$$

(The result of Kataoka-Lebeau is valid, more generally, for operators with analytic coefficients.)

One of our main goals in this chapter will be to obtain a result which can replace the remark before the statement of theorem 7.2.5 in the case when we do not assume that $x_n > 0$ implies $\rho(x) > 0$.

7.3   The wave front set of a restriction

1.   We want to discuss here some technical lemmas concerning wave front sets of a restriction to an analytic manifold S of form

$$S = \{x; x' \in U, x_n = g(x')\},$$

where $g : U' \to R$ is a real analytic function on an open neighborhood U' of $0 \in R^{n-1}$. We also denote by $\chi : U' \to R^n$ the map

$$\chi_i(y') = y_i \text{ if } i \leq n - 1,$$

$$\chi_n(y') = g(y'),$$

and by $x^0 = \chi(0)$. To simplify notations we shall assume that grad $g(0) = 0$. The cotangent space to S at x^0 can then be identified in a natural way with R^{n-1}. (We use the standard pullback.) We also consider a distribution u defined in a neighborhood Ω of x^0 and are interested to establish wether

$$(x^0, \xi^{0\prime}) \neq WF_A u_{|S} \tag{7.3.1}$$

provided the restriction of u to S has a reasonable meaning. Actually, the meaning of $u_{|S}$ will be given to us later on by Peetre's theorem, since u will satisfy a partial differential equation in the region $x_n > g(x')$ and S will be noncharacteristic for that equation. It seems convenient here to explain in some detail how the result of Peetre will be used and to discuss how various notions of restrictions of distributions to S are defined.

The first remark is that if u is smooth, then the restriction of u to S is defined as a linear functional on $C_0^\infty(S)$ by

$$u_{|S}(h) = \int_S u(x)h(x)d\sigma = \int u(\chi(y'))h(\chi(y'))s(y')\,dy', \ h \in C_0^\infty(S), \qquad (7.3.2)$$

where $d\sigma = s(y')\,dy'$, $s(y') = \sqrt{1 + |\text{grad } g(y')|^2}$, is the surface element on S. It is convenient to write (7.3.2) in the form

$$u_{|S}(h) = (h\,d\sigma)(u), \qquad (7.3.3)$$

where $(h\,d\sigma)(u)$ is the action of the distribution $h\,d\sigma$ on the smooth funtion u. In particular this suggests that it is natural to regard $u_{|S}$ as a functional on densities on S, rather than as a functional on functions. Actually, $(h\,d\sigma)(u)$ may have a natural meaning even in the case when u is not smooth. This is in particular the case when

$$(x, n(x)) \notin WF\,u \text{ and } (x, -n(x)) \notin WF\,u, \ \forall x \in S, \qquad (7.3.4)$$

where $n(x)$ is the normal to S at x. That this is so is a standard result in microlocal analysis (cf. e.g. Hörmander [1]) and it also follows that if $\psi \in C_0^\infty(R^n)$ is chosen with $\int \psi(x)\,dx = 1$, then

$$u_{|S} = \lim_{\varepsilon \to 0}(u * \psi_\varepsilon)_{|S}, \text{ if } \psi_\varepsilon(x) = \varepsilon^{-n}\psi(x/\varepsilon). \qquad (7.3.5)$$

Finally, we consider the case when $p(x, D)u = 0$ in some region of form $U = \{x; x' \in U', g(x') < x_n < g_1(x')\}$, for some linear partial differential operator with smooth coefficients for which S is noncharacteristic and under the additional assumption that u is extendible accross S. If we fix $\eta > 0$ and $U'' \subset U'$ so that $g(x') + \eta < g_1(x')$ for $x' \in U''$, and so that $S_\delta = \{x; x' \in U'', x_n = g(x') + \delta\}$ is still noncharacteristic for p if $0 < \delta < \eta$, then $u_{|S_\delta}$ has a meaning in the sense of microlocal analysis. Moreover, we can identify $D'(S_\delta)$ for each δ with $D'(x; x' \in U'', x_n = g(x'))$ and the limit

$$\lim_{\delta \to 0} u_{|S_\delta}$$

will then exist and be equal to $u_{|S}$. This is in fact part of Peetre's theorem, applied in the coordinates $y = (y', y_n)$, $y' = x'$, $y_n = x_n - g(x')$. Indeed, in these coordinates, S_δ

has the form $\{y' \in U'', y_n = \delta\}$ and Peetre's theorem states that

$$u \in C_0^\infty([0, \varepsilon); D'(U'')).$$

(Cf. Melrose [1].) We omit further details, since the fact that the limit $\lim_{\delta \to 0} u_{|S_\delta}$ exists will be clear anyway in our situation later on. We should observe however in concluding that it also follows that

$$(\partial/\partial n)^j u_{|S_\delta} \to (\partial/\partial n)^j u_{|S} \text{ for } \delta \to 0_+, \text{ for any } j. \tag{7.3.6}$$

2. We have now given a meaning to $u_{|S}$ and return to our discussion of (7.3.1). Since S is parametrized by U' it is convenient to transport $u_{|S}$ to U' and to check (7.3.1) there. We are here free in principle to transport $u_{|S}$ to U' as we please, but it is natural to use the map

$$\kappa : D'(S) \to D'(U'),$$

defined by

$$\kappa(v)(f) = v(f \circ \chi^{-1}), \text{ if } v \in D'(S), f \in C_0^\infty(U'). \tag{7.3.7}$$

We will then have

$$(x^0, \xi^{0'}) \notin WF_A v \Leftrightarrow (0, \xi^{0'}) \notin WF_A \kappa(v). \tag{7.3.8}$$

Explicitly, (7.3.7) means that

$$\kappa(u_{|S})(f) = [(f \circ \chi^{-1}) \, d\sigma](u), \tag{7.3.9}$$

when u is a smooth function, but we have already recalled that $[(f \circ \chi^{-1}) \, d\sigma](u)$ makes sense under the assumption (7.3.4). From (7.3.5) we can then conclude that

$$\kappa(u_{|S})(f) = u_{|S}(f \circ \chi^{-1}) = \lim_{\varepsilon \to 0}(u * \psi_\varepsilon)_{|S}(f \circ \chi^{-1}) = \lim_{\varepsilon \to 0}[(f \circ \chi^{-1}) \, d\sigma](u * \psi_\varepsilon)$$
$$= [(f \circ \chi^{-1}) \, d\sigma](u),$$

so (7.3.9) holds also in this case. If, finally, u satisfies $p(x, D)u = 0$, p as before, then we have with the notations above that

$$\kappa(u_{|S})(f) = \lim_{\delta \to 0_+} [(f \circ \chi_\delta^{-1}) \, d\sigma_\delta](u),$$

where $d\sigma_\delta$ is the surface element on the surface S_δ and $\chi_\delta(x') = (x', g(x') + \delta)$.

It is clear from all this, that if we want to establish (7.3.1), then we need to estimate $((f \circ \chi_\delta^{-1}) d\sigma_\delta)(u)$, where $f \in C_0^\infty(U')$ is so that

$$|\hat{f}(\zeta')| \leq \exp[d|\operatorname{Re} \zeta'|_{\Gamma'} + \varepsilon|\operatorname{Im} \zeta'| + b \ln(1 + |\zeta'|)].$$

We are therefore interested in estimating the Fourier transform of $(f \circ \chi^{-1}) d\sigma_\delta$, if we know how to estimate $\hat{f}$. How this is done, is the content of the following section, where we have of course only considered the case $\delta = 0$.

7.4   Estimates for the Fourier transform of some surface densities

1. We want to compute estimates for $\mathcal{F}[(h \circ \chi^{-1}) \, d\sigma]$ starting from estimates for $\hat{h}$. (All notations are from the preceding section.) Here we observe that

$$\mathcal{F}[(h \circ \chi^{-1}) \, d\sigma](\xi) = [(h \circ \chi^{-1}) \, d\sigma](\exp(-i\langle x, \xi \rangle)) = \int \exp(-i\langle \chi(y'), \xi \rangle)$$
$$h(y')s(y') \, dy'.$$

When $\rho \in C_0^\infty(R^{n-1})$ is identically one on supp h and supp ρ lies in the domain of definition of s, it also follows that

$$\mathcal{F}[(h \circ \chi^{-1}) \, d\sigma](\xi) = (2\pi)^{-n} \int \int \exp(-i\langle \chi(y'), \xi \rangle + i\langle y', \eta' \rangle) \hat{h}(\eta')\rho(y')s(y') \, dy'd\eta'.$$
$$(7.4.1)$$

When estimating $\mathcal{F}(h \circ \chi^{-1})$ we shall choose for ρ functions which have additional regularity properties. To simplify the notations we shall now assume that χ has the following properties for some $\delta > 0$:

i) $\chi_n(y') < 0$ if $|y'| < \delta$,

ii) $\chi_i(y') = y_i$ for $i \leq n - 1$,

iii) $K = \{y; |y'| \leq \delta, \chi_n(y') < y_n \leq 0\}$ is convex.

If ζ is complex we will have then that

$$|\exp(-i\langle \chi(y'), \zeta \rangle)| \leq \exp(H_K(\operatorname{Im}\zeta)) \text{ for } |y'| \leq \delta,$$

where H_K is the supporting function of K.

Our main estimate is

Lemma 7.4.1. *Let* $\chi : U' \to R^n$ *be a real analytic map such that* i), ii), iii) *hold and assume that* $d' > 0$, $\varepsilon' > 0$, $b' \in R$, $c_1 > 0$ *and some open cones* Γ', Γ *are given with* $\Gamma' \subset\subset \Gamma \subset\subset R^{n-1}$. *Then we can find* $d > 0$, $\varepsilon > 0$, $b \in R$, $b'' \in R$, c *such that if* $f \in C_0^\infty(R^{n-1})$ *satisfies*

$$|\hat{f}(\eta')| \leq \exp[(d|\operatorname{Re}\eta'|_{-\Gamma'} + \varepsilon|\operatorname{Im}\eta'| + b\ln(1 + |\eta'|)], \text{ for } \eta' \in C^{n-1}, \qquad (7.4.2)$$

then $I(\zeta)$ *defined by*

$$I(\zeta) = \int \exp(-i\langle \chi(y'), \zeta \rangle) f(y')s(y') \, dy'$$

satisfies

$$|I(\zeta)| \le c \, exp \, [(d'|Re\zeta| + H_K(Im\zeta) + \varepsilon'|Im\zeta| + b' \, ln(1 + |\zeta'|) + b'' ln(1 + |\zeta_n|)] \quad (7.4.3)$$

if $Re\zeta' \in -\Gamma$ *or* $|Re\zeta_n| > c_1|Re\zeta'|$, *respectively*

$$|I(\zeta)| \le c \, exp \, [(H_K(Im\zeta) + \varepsilon'|Im\zeta| + b' \, ln(1 + |\zeta'|) + b'' ln(1 + |\zeta_n|)] \, \text{ in all other cases.}$$
$$(7.4.4)$$

Proof. We argue at first for $|\zeta'| \ge 1$ and perform some preparations needed to arrive at the term $b' ln(1 + |\zeta'|)$ in the exponent of (7.4.3) and (7.4.4). Actually there is no problem with this term when $|\zeta_n| \ge |\zeta'|$, since then $-ln(1 + |\zeta'|) + ln(1 + |\zeta_n|) \ge 0$, so we can always introduce a term of form $b' ln(1 + |\zeta'|)$ with b' arbitrary at the expense of an additional term of form $b'' ln(1 + |\zeta_n|)$, with b'' large positive. To cope with the situation when $|\zeta'| > |\zeta_n|$, we shall perform some preliminary partial integrations in y'. We start by observing that

$$|\text{grad}_{y'}\langle \chi(y'), \zeta\rangle| = |\zeta'| + O(\varepsilon)|\zeta|$$

on the support of f. In this situation we shall have if ε is sufficiently small that

$$|\, \text{grad}_{y'}\langle \chi(y'), \zeta\rangle| \ge (1/2)|\zeta'|.$$

We shall then consider the operator

$$L = |\, \text{grad}_{y'}\langle \chi(y'), \zeta\rangle|^{-2}\langle \text{grad}_{y'}\langle \chi(y'), \zeta\rangle, -i\partial/\partial y'\rangle$$

and write

$$I(\zeta) = \int \, exp \, (-i\langle \chi(y'), \zeta\rangle)(^tL)^k(f(y')s(y'))dy' = \sum_{|\alpha| \le k} I_\alpha(\zeta),$$

$$I_\alpha(\zeta) = \int \, exp \, (-i\langle \chi(y'), \zeta\rangle)(\partial/\partial y')^\alpha f(y')s(y', \zeta)dy'$$

where ^tL is the adjoint of L and

$$(^tL)^k(f(y')s(y')) = \sum_{|\alpha| \le k}((\partial/\partial y')^\alpha f(y'))s_\alpha(y', \zeta), \quad (7.4.5)$$

for some $s_\alpha(y', \zeta)$ chosen so that (7.4.5) is the expansion by the generalized Leibniz rule. It follows in particular that

$$|\partial^\beta s_\alpha(y', \zeta)| \le c^{|\beta|+1}\beta!|\zeta'|^{-k}, \, \text{if } |\zeta'| \ge |\zeta_n|$$

and if ε is small. The number k shall here be fixed larger than $b' + n$. In particular, it is independent of ζ as long as $|\zeta_n| \le |\zeta'|$. It will suffice then to estimate the individual terms I_α. We do so in the remainder of this proof, but mention that this preparation is not needed when either $|\zeta'| \leq 1$ or else when $|\zeta'| \leq |\zeta_n|$ and that we shall leave I in its initial form then. We shall nevertheless always refer in the sequel to the case I_α, since the situation is very similar also when we can work with I in its initial form.

Next we choose some constant c', $0 < c' < 1/4$ so that

$$\theta' \in \Gamma', |\theta' - \tau'| < 2c'(|\theta'| + |\tau'|) \text{ implies } \tau' \in \Gamma'.$$

We can apply lemma 2.3.11 and conclude that it is no loss in generality to assume that $\hat{f}$ satisfies the estimate

$$|\hat{f}(\eta')| \leq \exp\left[d|\mathrm{Re}\,\eta'|_{-\Gamma' \cap \{\eta'; |\eta' - \eta'^0| < c'|\eta'^0|\}} + \varepsilon|\mathrm{Im}\,\eta'| + b\,ln(1 + |\eta'|)\right],$$

for some fixed $\eta'^0 \in -\Gamma'$, rather than the estimate (7.4.2). (After this change, f is perhaps only in $E'(R^{n-1})$ rather than in C^∞.) The fact that complex $\eta' \in C^{n-1}$ come in, is needed to make clear that supp $f \subset \{y'; |y'| \leq \varepsilon\}$, and in particular (7.4.1) will hold if ρ is a function in $C_0^\infty(y'; |y'| \leq 2\varepsilon)$ which is identically one on $|y'| \leq \varepsilon$.

We now also consider a constant $c_2 > 0$ to be specified later on and denote the integer part of $(|\eta'^0| + 1)/c_2$ by j. We can then find a constant c_3, which does not depend on j and a $C_0^\infty(R^{n-1})$ function ρ_j such that

$$\rho_j(y') = 0 \text{ for } |y'| > 2\varepsilon, \rho_j(y') = 1 \text{ for } |y'| \leq \varepsilon, |\partial^\alpha \rho_j(y')| \leq c_3(c_3/\varepsilon)^{|\alpha|}j^{|\alpha|} \text{ if } |\alpha| \leq j.$$

Using this ρ_j, we can argue as in (7.4.1) and write $I_\alpha = I'_\alpha + I''_\alpha$, where

$$I'_\alpha(\zeta) = \int\int_{|\eta' - \eta'^0| > c'|\eta'^0|} \exp\left(-i\langle \chi(y'), \zeta\rangle + i\langle y', \eta'\rangle\right)\hat{f}(\eta')\eta'^\alpha \rho_j(y') s_\alpha(y', \zeta)\, dy' d\eta',$$

$$I''_\alpha(\zeta) = \int\int_{|\eta' - \eta'^0| \leq c'|\eta'^0|} \exp\left(-i\langle \chi(y'), \zeta\rangle + i\langle y', \eta'\rangle\right)\hat{f}(\eta')\eta'^\alpha \rho_j(y') s_\alpha(y', \zeta)\, dy' d\eta'.$$

The estimation of I'_α is rather simple. On the domain of integration we have $|\hat{f}(\eta')| \leq (1 + |\eta'|)^b$, so we will have when $|\zeta'| \geq |\zeta_n|$, $|\zeta'| \geq 1$, that

$$|I'_\alpha(\zeta)| \leq c_4|\zeta'|^{-k} \exp H_K(\mathrm{Im}\,\zeta) \int_{|y'| \leq 2\varepsilon} \int_{|\eta' - \eta'^0| \geq c'|\eta'^0|} (1 + |\eta'|)^b |\eta'|^k\, dy' d\eta'.$$

The integral in η' is here convergent if b is suitable. Our choice of k gives then an estimate of the type from the right hand side of (7.4.3). A similar argument holds for the case $|\zeta_n| \geq |\zeta'|$ or when $|\zeta'| \leq 1$.

We are left with the estimation of I''_α, so we can assume henceforth that $|\eta' - \eta'^0| < c'|\eta'^0|$. In particular we will then have (since $c' < 1/4$) that

$$|\eta'| \leq 2|\eta'^0|, \ |\eta'^0| \leq 2|\eta'|. \tag{7.4.6}$$

As is standard, we will have to consider several regions for the location of η', ζ.

a) $|\eta'^0 - \mathrm{Re}\,\zeta'| < 2c'|\eta'^0|$. We have here $\mathrm{Re}\,\zeta' \in \Gamma$ and $|\eta'| \sim |\mathrm{Re}\,\zeta'|$. From the estimate for $\hat{f}$ we obtain

$$|\hat{f}(\eta')| \leq c_5(1 + |\mathrm{Re}\,\zeta'|)^b \exp d'|\mathrm{Re}\,\zeta'|,$$

if d had been small. It follows that

$$|I'_\alpha{}'(\zeta)| \leq c_6 \exp\left[d'|\mathrm{Re}\,\zeta'| + H_K(\mathrm{Im}\,\zeta)\right](1 + |\mathrm{Re}\,\zeta'|)^b|\zeta'|^{-k+n-1}.$$

(Note here that the volume over which we integrate in the η' space is of the order of magnitude $c|\eta'^0|^{n-1} \sim |\mathrm{Re}\,\zeta'|^{n-1}$.)

We obtain an estimate by the right hand side of (7.4.3) due to our choice of k.

In the remainder of the proof we may now assume that $|\eta'^0 - \mathrm{Re}\,\zeta'| \geq 2c'|\eta'^0|$. Together with $|\eta'^0 - \eta'| < c'|\eta'^0|$ this will give $|\eta' - \mathrm{Re}\,\zeta'| \geq (c'/2)|\eta'|$. We conclude that in fact

$$|\eta' - \mathrm{Re}\,\zeta'| \geq c_7(|\eta'| + |\mathrm{Re}\,\zeta'|).$$

Next we observe that

$$|\mathrm{grad}_{y'}(\langle \chi(y'), \zeta \rangle - \langle y', \eta' \rangle)| \geq |\mathrm{Re}\,\zeta' - \eta'| - c_8\varepsilon|\mathrm{Re}\,\zeta|$$

by our assumption on χ. If $c_1|\mathrm{Re}\,\zeta'| > |\mathrm{Re}\,\zeta_n|$ we can then choose $\varepsilon > 0$ sufficiently small in order to achieve that

$$|\mathrm{grad}_{y'}(\langle \chi(y'), \zeta \rangle - \langle y', \eta' \rangle)| \geq (c_7/2)(|\mathrm{Re}\,\zeta'| + |\eta'|). \tag{7.4.7}$$

The same inequality can moreover be achieved in the case $c_1|\mathrm{Re}\,\zeta'| < |\mathrm{Re}\,\zeta_n|$ provided that $|\eta'^0| > c_9|\mathrm{Re}\,\zeta_n|$ if ε is small. We shall therefore consider as the next case:

b) $|\eta'^0 - \mathrm{Re}\,\zeta'| \geq 2c'|\eta'^0|$ and $c_1|\mathrm{Re}\,\zeta'| > |\mathrm{Re}\,\zeta_n|$ or $|\eta'^0 - \mathrm{Re}\,\zeta'| \geq 2c'|\eta'^0|$, $c_1|\mathrm{Re}\,\zeta'| < |\mathrm{Re}\,\zeta_n|$, but $|\eta'^0| > c_9|\mathrm{Re}\,\zeta_n|$.

The idea is here of course to use (7.4.7) for partial integration in y'. Since it is here where we can finally choose c_2, we will give further details. Denote in fact by

$$\psi(y', \zeta, \eta') = \langle \chi(y'), \zeta \rangle - \langle y', \eta' \rangle$$

and consider the operator

$$F = |\mathrm{grad}_{y'}\psi|^2 \langle \mathrm{grad}_{y'}\psi, -i\partial/\partial y' \rangle.$$

It follows that

$$|I_\alpha''(\zeta)| \le \int \int_{|\eta'-\eta'^0| \le c'|\eta'^0|} \exp\ (i\psi(y',\zeta,\eta'))|\hat{f}(\eta')\eta'^\alpha({}^tF)^j(\rho_j(y')s_\alpha(y',\zeta))|\,dy'\,d\eta'.$$

The integrand can here be estimated by

$$\exp\ (H_K(\operatorname{Im}\zeta) + d|\eta'|)(1 + |\zeta'|)^{-k}(1 + |\eta'|)^b c_{10}(c_{10}j/\varepsilon)^j(|\zeta'| + |\eta'|)^{-j}.$$

Knowing that $|\eta'| \sim |\eta'^0|$ and that $|\eta'|$ is close to $c_2 j$, we can choose for ε fixed, c_2, d so that

$$\exp\ (d|\eta'|)(c_{10}j/\varepsilon)^j|\eta'|^{-j} \le c_{11}.$$

The estimation of I_α'' is now straightforward. We are thus left with the case

c) $|\eta'^0 - \operatorname{Re}\zeta'| \ge 2c'|\eta'^0|$, $c_1|\operatorname{Re}\zeta'| \le |\operatorname{Re}\zeta_n|$, $|\eta'^0| \le c_9|\operatorname{Re}\zeta_n|$. We can estimate $|\hat{f}(\eta')|$ by $\exp\ (d'|\operatorname{Re}\zeta| + b\,ln(1 + |\eta'|))$. The desired estimate follows immediately.

7.5 Boundary values and partial regularity

1. In section 7.3 we have recalled some results from the Melrose-Peetre theory of boundary traces for solutions of partial differential equations to noncharacteristic boundaries. Here we discuss related results for constant coefficient operators expressed in terms of Fourier transforms. We consider Ω a relatively compact convex union of spheres of fixed radius and let $p(D)$ be a constant coefficient linear partial differential operator for which $x_n = 0$ is noncharacteristic. (A relatively compact convex set is a union of spheres precisely when its boundary is of class $C^{1,1}$, i.e. when its boundary is described by functions with Lipschitz continuous derivatives. Cf. Kiselman [2].) A standard way to study the existence and properties of traces to a boundary of form $x_n = 0$ of an extendible solution of the equation $p(D)u = 0$ is to study the partial C^∞ character up to the boundary in the variable x_n of that solution. Although the boundary will be more complicated than $x_n = 0$ in this section, it will be convenient also here to regard our results as results on partial regularity in the x_n variable. For simplicity we shall assume that $u \in D'(R^n)$ and that actually $p(D)u = 0$ on the set

$$\Omega_\delta = \{y \in R^n;\ \exists\, x \in \Omega \text{ s.t. } y' = x', 0 \le y_n - x_n < \delta\}.$$

Our first result is

Lemma 7.5.1. *Let* $b, b' \in R$ *and consider* $h \in A(C^n)$ *such that*

$$|h(\zeta)| \le\ exp\ [H_\Omega(\operatorname{Im}\zeta) + b'\,ln(1 + |\zeta'|) + b\,ln(1 + |\zeta_n|)].$$

Then we can find $h_j \in A(C^n)$ so that

$$\sup_{\zeta} \left[|h_j(\zeta)| / exp \left(H_{\Omega_\delta}(Im\,\zeta) + b'\, ln(1 + |\zeta|) \right) \right] < \infty,$$

and such that

$$\sup_{\zeta} |(h_j - h)(\zeta)| / exp \left(H_{\Omega_\delta}(Im\,\zeta) + b'\, ln(1 + |\zeta'|) + (b + 1)\, ln(1 + |\zeta_n|) \right) \to 0, \;\; if \; j \to \infty.$$
$$(7.5.1)$$

Proof. Denote by $g_j(\zeta) = h(\zeta) \exp\left(-i\zeta_n/j\right)$, so that $h - g_j = h(1 - \exp\left(-i\zeta_n/j\right))$. Here

$$H_{\Omega_\delta}(\xi) = H_\Omega(\xi) + \delta\xi_{n+}$$

and

$$|1 - \exp\left(-i\zeta_n/j\right)| \exp\left(-\delta Im\,\zeta_{n+} - ln(1 + |\zeta|)\right) \to 0 \text{ with } j \to \infty.$$

We conclude from all this that (7.5.1) is valid for h_j replaced by g_j. We shall then set $h_j = g_j\hat{\varphi}_j$, where $\varphi_j(x_n) = j\varphi(jx_n)$ and $\varphi \in C_0^\infty(R)$ satisfies $\int_R \varphi(t)dt = 1$, etc.

Remark 7.5.2. *When we apply the lemma, b' will be a large negative number, so the distribution $\mathcal{F}^{-1}h$ is rather regular in the x'−variables. The lemma gives then a sequence which is regular in addition in the x_n−variable.*

2. Next we prove

Lemma 7.5.3. *Let Ω be a relatively compact convex union of spheres of fixed radius and assume that $u \in D'(R^n)$ satisfies $p(D)u = 0$ on Ω. Also assume that $x_n = 0$ is noncharacteristic for $p(D)$. If $b \in R$ is given, we can find $b' \in R$ and k so that $|u(f)| \le c$ for any $f \in C_0^k(R)$ which satisfies*

$$|\hat{f}(\zeta)| \le exp \left[H_\Omega(Im\,\zeta) + b'\, ln(1 + |\zeta'|) + b\, ln(1 + |\zeta_n|) \right].$$
$$(7.5.2)$$

Proof. If $f \in C_0^k(R^n)$ and supp $f \subset \bar{\Omega}$, then $u({}^tp(D)f)$ will have a natural meaning if k is large enough and we will have $u({}^tp(D)f) = 0$ since $p(D)u = 0$. We also may assume that $|u(f)| \le c'$ for any $f \in C_0^k(R^n)$ for which

$$|\hat{f}(\zeta)| \le \exp\left(H_\Omega(Im\,\zeta) - k\, ln(1 + |\zeta|)\right),$$

and observe that

$$\sup |p(-\zeta)h(\zeta)| / \exp\left(H_\Omega(Im\,\zeta) - k\, ln(1 + |\zeta|)\right) < \infty$$

implies

$$\sup_{\zeta} |h(\zeta)| / \exp\left(H_{\Omega}(\operatorname{Im} \zeta) - k \, ln(1 + |\zeta|)\right) < \infty$$

if $h \in A(C^n)$ (by a theorem of B.Malgrange).

Next we observe that if f satisfies (7.5.2) and b' is suitable, then we can find $f_1, f_2 \in E'(R^n)$ so that

$$f = f_1 + {}^t p(D) f_2 \tag{7.5.3}$$

and

$$|\hat{f}_1(\zeta)| \le c \, \exp\left(H_{\Omega}(\operatorname{Im} \zeta) - k \, ln(1 + |\zeta|)\right). \tag{7.5.4}$$

This follows from the fundamental principle and the following proposition, which we prove at the end of this section

Proposition 7.5.4. *Let $K \subset R^n$ be a compact convex union of spheres of fixed radius and fix $b_1 \in R$. Then we can find $b_2 \in R$, c and a plurisubharmonic function $\rho : C^n \to R$ such that*

$$H_K(\operatorname{Im} \zeta) - b_2 \, ln(1 + |\zeta|) \le \rho(\zeta) \le H_K(\operatorname{Im} \zeta) - b_1 \, ln(1 + |\zeta|) + c, \forall \zeta \in C^n.$$

Since f in (7.5.3) is in C_0^k it will follow from there and (7.5.4) that

$$|\hat{f}_2(\zeta)| \le c'' \, \exp\left(H_{\Omega}(\operatorname{Im} \zeta) - k \, ln(1 + |\zeta|)\right).$$

We conclude that $u({}^t p(D) f_2) = 0$ in a natural way, so that $u(f) = u(f_1)$. This gives the conclusion of lemma 7.5.3.

3. We now return to domains of form Ω_δ and consider $u \in D'(R^n)$ satisfying $p(D)u = 0$ on Ω_δ. If k is large enough, u is defined in a natural way on $C_0^k(\Omega_\delta)$. Also fix b, consider b' and denote by $|\ |_{\Omega,b,b'}$ the seminorm

$$|h|_{\Omega,b,b'} = \sup_{\zeta \in C^n} [|h(\zeta)| / \exp\left(H_{\Omega}(\operatorname{Im} \zeta) + b' \, ln(1 + |\zeta'|) + b \, ln(1 + |\zeta_n|)\right)]$$

and by $A_{\Omega,b,b'}$ the space

$$A_{\Omega,b,b'} = \{ h \in A(C^n); |h|_{\Omega,b,b'} < \infty \}.$$

$A_{\Omega,b,b'}$ is made a normed vector space when endowed with the norm $|\ |_{\Omega,b+1,b'}$, and likewise $\mathcal{F}^{-1}(A_{\Omega,b,b'})$ is considered a normed space by identification with $A_{\Omega,b,b'}$.

If we put the lemmata 7.5.1 and 7.5.3 together, we can now fix b' so that with $u : C_0^k(R^n) \to C$ we can associate a unique continuous linear functional U,

$$U : \mathcal{F}^{-1}(A_{\Omega,b,b'}) \to C$$

for which

$$U(\mathcal{F}^{-1}h) = \lim_{j \to \infty} u(\mathcal{F}^{-1}h_j),$$

where the h_j are associated with h as in lemma 7.5.1. If b' has been chosen suitably, it is also clear that

$$U(\mathcal{F}^{-1}h) = 0, \text{ if } h \text{ is of form } h = p(-\zeta)h'(\zeta),$$

which is roughly speaking the relation $p(D)U = 0$. Later on we shall not distinguish between u and U and simply pretend that u itself defines a functional on the spaces $\mathcal{F}^{-1}(A_{\Omega,b,b'})$. This is justified by the above when b' is suitably chosen. We also observe that the fact that b could be large positive means that u is in some sense regular in the x_n variable, provided we regard it in a weak sense in the x' variables.

4. It remains to give the proof of proposition 7.5.4. We prepare this proof with three lemmas.

Lemma 7.5.5. *There is a plurisubharmonic function* $\psi : C^n \to R \cup \{-\infty\}$ *and* $b > 0$, $c \geq 0$ *so that*

$$\psi(\zeta) \leq |Im\,\zeta| - b\,ln(1 + |\zeta|) + c, \text{ if } \zeta \in C^n, \tag{7.5.5}$$

$$\psi(i\xi) \geq |\xi| - ln(1 + |\xi|) - c, \text{ if } \xi \in R^n. \tag{7.5.6}$$

Proof of lemma 7.5.5. We shall set

$$\psi(\zeta) = ln\,|(\sin \sqrt{\zeta_1^2 + \cdots + \zeta_n^2})/\sqrt{\zeta_1^2 + \cdots + \zeta_n^2}\,|,$$

observing that this is well-defined if the square root is the same in both occurences. It is clear that ψ is plurisubharmonic. When $\xi \in R^n$, then

$$\psi(i\xi) = ln|(e^{|\xi|} - e^{-|\xi|})/|\xi|\,| - ln\,2$$

which gives (7.5.6) immediately.

Our next remark is that

$$|\sin \sqrt{\zeta_1^2 + \cdots + \zeta_n^2}/\sqrt{\zeta_1^2 + \cdots + \zeta_n^2}\,| \leq c'\,exp\,|Im \sqrt{\zeta_1^2 + \cdots + \zeta_n^2}\,|. \tag{7.5.7}$$

Indeed, when $|\sqrt{\zeta_1^2 + \cdots + \zeta_n^2}| \geq 1$ this follows from

$$|\sin \sqrt{\zeta_1^2 + \cdots + \zeta_n^2}| \leq exp\,|Im \sqrt{\zeta_1^2 + \cdots + \zeta_n^2}\,|$$

and for $b = |\sqrt{\zeta_1^2 + \cdots + \zeta_n^2}| \leq 1$, we may use that

$$|\sin a/a| \leq c'' \text{ if } |a| = b.$$

We now fix some constant c_1 to be specified later on. It is then obvious that there is a constant c_2 so that

$$|\sin \sqrt{\zeta_1^2 + \cdots + \zeta_n^2}/\sqrt{\zeta_1^2 + \cdots + \zeta_n^2}| \le c_2(1 + |\zeta|)^{-1} \exp |\operatorname{Im} \zeta| \qquad (7.5.8)$$

if $|\zeta_1^2 + \cdots + \zeta_n^2| \ge c_1(|\operatorname{Re} \zeta| + |\operatorname{Im} \zeta|)^2$. ((7.5.8) follows from $|\operatorname{Im} \sqrt{\zeta_1^2 + \cdots + \zeta_n^2}| \le |\operatorname{Im} \zeta|$. To prove the latter we consider separately the cases $|\operatorname{Re} \zeta| \ge |\operatorname{Im} \zeta|$ and $|\operatorname{Im} \zeta| > |\operatorname{Re} \zeta|$. In the first case,

$$|\operatorname{Im} \sqrt{\zeta_1^2 + \cdots + \zeta_n^2}| = |\operatorname{Im} \sqrt{|\operatorname{Re} \zeta|^2 - |\operatorname{Im} \zeta|^2 + 2i\langle \operatorname{Re} \zeta, \operatorname{Im} \zeta\rangle}|$$

$$\le |\operatorname{Im} \sqrt{|\operatorname{Re} \zeta|^2 - |\operatorname{Im} \zeta|^2 + 2i|\operatorname{Re} \zeta||\operatorname{Im} \zeta|}| = |\operatorname{Im} \zeta|,$$

and in the second case

$$|\operatorname{Im} \sqrt{\zeta_1^2 + \cdots + \zeta_n^2}| \le |\operatorname{Im} \sqrt{\beta^2 - |\operatorname{Im} \zeta|^2 + 2i\beta|\operatorname{Im} \zeta|}| = |\operatorname{Im} \zeta|,$$

where β is chosen so that $|\langle \operatorname{Re} \zeta, \operatorname{Im} \zeta\rangle| = \beta|\operatorname{Im} \zeta|$.)

We have now proved (7.5.8), which gives

$$\psi(\zeta) \le |\operatorname{Im} \zeta| - ln(1 + |\zeta|) - c_2 \text{ if } |\zeta_1^2 + \cdots + \zeta_n^2| \ge c_1(|\operatorname{Re} \zeta| + |\operatorname{Im} \zeta|)^2,$$

with a constant c_2 which depends on c_1. We may then assume in the sequel that

$$|\zeta_1^2 + \cdots + \zeta_n^2| \le c_1(|\operatorname{Re} \zeta| + |\operatorname{Im} \zeta|)^2, \qquad (7.5.9)$$

where c_1 is fixed as pleased. We claim that when c_1 is sufficiently small, then it will follow that

$$|\operatorname{Im} \sqrt{\zeta_1^2 + \cdots + \zeta_n^2}| \le (1/2)|\operatorname{Im} \zeta|. \qquad (7.5.10)$$

Proof of (7.5.10). (7.5.9) means explicitly that

$$||\operatorname{Re} \zeta|^2 - |\operatorname{Im} \zeta|^2 + 2i\langle \operatorname{Re} \zeta, \operatorname{Im} \zeta\rangle| \le c_1(|\operatorname{Re} \zeta| + |\operatorname{Im} \zeta|)^2.$$

We obtain in particular from this that

$$|(|\operatorname{Re} \zeta| - |\operatorname{Im} \zeta|)| \le c_1(|\operatorname{Re} \zeta| + |\operatorname{Im} \zeta|).$$

If $c_3 > 0$ is fixed we can then shrink c_1 until

$$|\operatorname{Im} \zeta| \le c_3|\operatorname{Re} \zeta|, \ |\operatorname{Re} \zeta| \le c_3|\operatorname{Im} \zeta|, \text{ if (7.5.9) holds.} \qquad (7.5.11)$$

This will give $|\operatorname{Im} \sqrt{\zeta_1^2 + \cdots + \zeta_n^2}|^2 \le |\zeta|^2 \le c_1(|\operatorname{Re} \zeta| + |\operatorname{Im} \zeta|)^2 \le c_1(c_3 + 1)^2|\operatorname{Im} \zeta|^2$. It suffices then to shrink c_1 still further until $c_1(c_3 + 1)^2 \le 1/4$.

We have thus checked (7.5.10) and conclude that we may assume that (7.5.9) implies simultaneously (7.5.11) and

$$|\psi(\zeta)| \leq ln\, c' + |\mathrm{Im}\,\sqrt{\zeta_1^2 + \cdots + \zeta_n^2}\,| \leq c_4 + (1/2)|\mathrm{Im}\,\zeta|. \tag{7.5.12}$$

The proof is then completed by observing that

$$(1/2)|\mathrm{Im}\,\zeta| \leq c_5 + |\mathrm{Im}\,\zeta| - ln(1 + |\zeta|), \text{ if (7.5.11) is valid.}$$

Remark 7.5.6. *The holomorphic function* $\sin\sqrt{\zeta_1^2 + \cdots + \zeta_n^2}/\sqrt{\zeta_1^2 + \cdots + \zeta_n^2}$ *on which the preceding argument is based, is closely related to the Fourier transform of the natural density distribution on the unit sphere* $|x| = 1$. *It is of course this idea which stays behind the construction.*

Lemma 7.5.7. *There are constants* $b' > 0$, c' *and a plurisubharmonic function* φ : $C^n \to R$ *so that*

$$|Im\,\zeta| - 2\,ln(1 + |\zeta|) \leq \varphi(\zeta) \leq |Im\,\zeta| - b'\,ln(1 + |\zeta|) + c'.$$

Proof. Let ψ, b, c be as in the conclusion of the preceding lemma. We define for $\theta \in R^n$ by ψ_θ the function

$$\psi_\theta(\zeta) = \psi(\zeta - \theta) - ln(1 + |\theta|).$$

This gives for $\eta \in R^n$

$$\psi_\theta(\theta + i\eta) \geq -ln(1 + |\theta|) - ln(1 + |\eta|) - c + |\eta|$$

$$\psi_\theta(\xi + i\eta) \leq -ln(1 + |\theta|) - b\,ln(1 + \sqrt{|\xi - \theta|^2 + |\eta|^2}) + |\eta| + c \tag{7.5.13}$$

if $\xi \in R^n$. We can now use that

$$-ln(1 + |\theta|) - b\,ln(1 + \sqrt{|\xi - \theta|^2 + |\eta|^2}) \leq -b'\,ln(1 + |\xi| + |\eta|) + c'',$$

so that the upper regularization of $\sup_{\theta \in R^n} \psi_\theta$ is a plurisubharmonic function with the desired properties.

From lemma 7.5.7 we immediately obtain

Lemma 7.5.8. *For every constant* $b' > 0$ *we can find constants* $b > 0$, c *and a plurisubharmonic function* $\chi : C^n \to R$ *so that*

$$|Im\,\zeta| - b\,ln(1 + |\zeta|) \leq \chi(\zeta) \leq |Im\,\zeta| - b'\,ln(1 + |\zeta|) + c.$$

Proof. When $a > 0$ is a fixed constant, we can find $a' \geq 0$ so that

$$|ln\,(1 + at) - ln\,(1 + t)| \leq a', \text{ for } t \geq 0.$$

Indeed, $|ln\,(1 + at) - ln\,(1 + t)| = |\,ln[(1 + at)/(1 + t)]|$ and

$$c_1 \leq \frac{1 + at}{1 + t} \leq c_2, \forall t \geq 0.$$

A function χ with the desired properties is then $\chi(\zeta) = A\varphi(\zeta/A)$, where φ is from the preceding lemma and $A > 0$ is chosen suitably.

Proof of proposition 7.5.4. We write

$$K = \cup_{i \in I} K_i$$

where the K_i are the spheres from the definition of K. We can then for each $i \in I$ construct a plurisubharmonic function χ_i so that

$$H_{K_i}(\mathrm{Im}\ \zeta) - b_2(1 + |\zeta|) \leq \chi_i(\zeta) \leq H_{K_i}(\mathrm{Im}\ \zeta) - b_1(1 + |\zeta|).$$

A function ρ as desired is then the upper regularization of $\sup_{i \in I} \chi_i$.

7.6 Regularity up to the boundary

1. The following is the main result of this chapter

Proposition 7.6.1. *Let $\Gamma \subset \dot{R}^{n-1}$ be an open conic neighborhood of $\xi^{0\prime}$ and assume that*

$$p(\zeta', \tau) = 0 \text{ implies } |Im\,\tau| \leq c(1 + |Im\,\zeta'|), \text{ if } Re\,\zeta' \in \Gamma. \tag{7.6.1}$$

Let further

$$g : \{x' \in R^{n-1}; |x'| < c'\} \rightarrow R_-$$

be a real analytic function with negative values such that

$$grad\,g(0) = 0 \text{ and } \sum_{|\alpha|=2} |\partial_{x'}^\alpha g(x')| \leq c'', \forall |x'| < c'. \tag{7.6.2}$$

Denote by

$$S = \{x \in R^n; |x'| < c', x_n = g(x')\}$$

and consider a distribution u on $\{x \in R^n; |x'| < c'\}$ which satisfies the equation $p(D)u = 0$ in $\{x; |x'| < c', g(x') < x_n < \eta\}$ for some positive η. Then there is a constant C_1, which depends on c, c'', but not effectively on g, with the property that

$$\{(x', \xi^{0\prime}); |x'| < c'\} \cap WF_A(\partial/\partial x_n)^j u_{|x_n=0} = \emptyset, \forall j, \tag{7.6.3}$$

implies

$$((0, g(0)), \xi^{0\prime}) \notin W F_A (\partial/\partial n)^k u_{|S}, \forall k, \tag{7.6.4}$$

provided that

$$|g(0)|/c' \leq C_1, \tag{7.6.5}$$

and provided $c' \leq 1$ itself is small enough. (The size of c' will depend e.g. on the size of c''.)

$((\partial/\partial n)$ is the normal derivative to S and $((0, g(0)), \xi^{0\prime})$ is identified with the cotangent vector to S obtained from the vector $(0, \xi^{0\prime})$ under the pull-back associated with the map $(x', g(x')) \rightarrow x'$.)

Remark 7.6.2. *a) The intuitive meaning of condition (7.6.5) is that $(0, g(0))$ is in the domain of influence of $\{x, |x'| < c', x_n = 0\}$.*

b) A useful remark is that if C_1, C_2 are given constants and if we start from $|g(0)/c'| < C_1'$, then we may assume that simultaneously

$$|g(0)/c'| \leq C_1, \tag{7.6.6}$$

and

$$|\operatorname{grad} g(x')| \leq C_2 \text{ if } |x'| \leq c', \tag{7.6.7}$$

provided c' and C_1' are suitable. Indeed, from (7.6.2) we obtain that

$$|\operatorname{grad} g(x')| \leq (n-1)|x'|c''.$$

Starting from $|g(0)/c'| \leq C_1'$ instead of (7.6.5) we will then obtain (7.6.7) if $(n-1)c'c'' \leq C_2$. For given C_2 this can be achieved by shrinking c' to $c' \leq C_2/(c''(n-1))$. In order to have (7.6.5), we need then have in addition that $C_1'C_2/(c''(n-1)) \leq C_1$. Once C_1 and C_2 are fixed, this is a condition on C_1'.

2. Before we start the proof of proposition 7.6.1 we show that it is no loss in generality to assume that g has the form

$$g(x') = a - \sqrt{r^2 - |x'|^2} \tag{7.6.8}$$

for some $a > 0$ and $r > 0$ so that $a - r = g(0)$. (Since $g(0) < 0$, we must have $r > a$!). This is based on the following lemma

Lemma 7.6.3. *Let g be a negative C^2 function on $\{x'; |x'| < c'\}$ which satisfies the conditions (7.6.2) and (7.6.5) for some given c'', with C_1 still to be specified. Also fix $\check{C}_1$. Then we can find $a > 0$, $r > 0$, C_1, $\tilde{c}' \leq c'$, so that with the notation*

$$\tilde{g}(x') = a - \sqrt{r^2 - |x'|^2},$$

$$\tilde{g}(0) = g(0), \ i.e. \ a - r = g(0), \tag{7.6.9}$$

$$|\tilde{g}(0)| \leq \check{C}_1 \tilde{c}' \tag{7.6.10}$$

$$\tilde{c}' \leq r, \tag{7.6.11}$$

$$\tilde{g}(x') \geq g(x') \ for \ all \ |x'| \leq \tilde{c}', \tag{7.6.12}$$

$$\tilde{g}(x') < 0 \ if \ |x'| \leq \tilde{c}' \tag{7.6.13}$$

$$|\partial^\alpha \tilde{g}(x')| \leq \tilde{c}'' \ if \ |x'| \leq \tilde{c}', |\alpha| = 2. \tag{7.6.14}$$

Moreover, $a, \tilde{c}''$ and r depend here essentially only on c'' and $g(0)$.

Proof of lemma 7.6.3. We write (7.6.12) as

$$- |x'|^2 \geq h(x') \ for \ |x'| < \tilde{c}', \tag{7.6.15}$$

where $h(x') = (g(x') - a)^2 - r^2$, and note that $h(0) = 0$, grad $h(0) = 0$. We also have if $|\alpha| = 2$,

$$|\partial^\alpha h(x')| \leq 2c''|g(x' - a)| + |\text{grad} \, g(x')|^2 \leq 2c''(r + c''|x'|^2) + c''^2|x'|^2.$$

It follows from Taylor expansion that (7.6.15) is valid if $2c''r + 3c''^2|x'|^2$ is small enough. Since c'' is fixed, this gives a limitation for r and $\tilde{c}'$, both in terms of c''. This will be the only condition on r, but $\tilde{c}'$ will have to be shrinked still further. Next we observe that having fixed r, a is defined from (7.6.9). To have (7.6.13), we need then that

$$a = g(0) + r < \sqrt{r^2 - |x'|^2} \ for \ |x'|^2 \leq \tilde{c}'.$$

Explicitly this comes to

$$g^2(0) + 2rg(0) < -\tilde{c}'^2, \ i.e. \ |\tilde{c}'| < |(2r + g(0))g(0)|^{1/2}.$$

Since r was determined in terms of c'' alone, this is a condition expressed in terms of c'' and $g(0)$. We may also shrink $\tilde{c}'$ to have

$$\tilde{c}' \leq \min(r/2, c').$$

The condition $\tilde{c}' \leq c'$ is here needed for (7.6.12) and $\tilde{c}' \leq r/2$ will imply (7.6.14). Finally, to obtain (7.6.10), we shall ask for $C_1 c' \leq \tilde{C}_1 \tilde{c}'$. Once $\tilde{c}'$ is fixed, this can be achieved by choosing C_1 small enough.

3. We have now proved the lemma, which shows that the assumptions in proposition 7.6.1 do also hold for the function $\tilde{g}(x') = a - \sqrt{r^2 - |x'|^2}$. If we can then prove (7.6.4) with g replaced by $\tilde{g}$, then we get the result for our initial g by applying (Lebeau's) theorem 7.2.5.

Remark 7.6.4. *The advantage of working with the function from (7.6.8) is of course that we have transformed our initial problem into a problem of convex analysis.*

The main step in the proof of proposition 7.6.1 is the following lemma, the proof of which shall be given in the following section:

Lemma 7.6.5. *Assume that (7.6.1) is valid for some open cone Γ which contains $\xi^{0\prime}$ and fix some open conic neighborhood $\Gamma' \subset\subset \Gamma$ of $\xi^{0\prime}$. Also assume that c_1, c_2, c_3 are constants for which*

$$c_2 + c_3 \leq 1, 2c_1 + c_2 \leq 1, c_3 + (1 - c_2)c \leq c_1/4 \qquad (7.6.16)$$

(c is the constant from (7.6.1)) and denote by K' the set

$$K' = \{x; |x'| \leq c_3, c_2 - \sqrt{1 - |x'|^2} \leq x_n \leq 0\}.$$

Let finally K be a convex compact set with a C^2 boundary such that

$$\{x; |x'| \leq c_1, c_2 - \sqrt{1 - |x'|^2} \leq x_n \leq 0\} \subset K \subset \{x; |x'| \leq 2c_1, c_2 - \sqrt{1 - |x'|^2} \leq x_n \leq 0\}.$$

There is b''', c_4 such that for all d', b'', b_1, b_2 we can find d, b', c_5 such that if $f \in C_0^\infty(R^n)$ satisfies

$$|\hat{f}(\zeta)| \leq \ exp\left(d|Re\,\zeta| + H_{K'}(Im\,\zeta) + b' \ln(1 + |\zeta'|) + b_2 \ln(1 + |\zeta_n|)\right)$$
$$if \ |Re\,\zeta_n| \geq 2c_4(1 + |Re\,\zeta'|), \qquad (7.6.17)$$

respectively

$$|\hat{f}(\zeta)| \leq \ exp\left(d|Re\,\zeta'|_{-\Gamma'} + H_{K'}(Im\,\zeta) + b' \ln(1 + |\zeta'|) + b_2 \ln(1 + |\zeta_n|)\right) \qquad (7.6.18)$$
$$if \ |Re\,\zeta_n| \leq 2c_4(1 + |Re\,\zeta'|),$$

then there are $h, h' \in \mathcal{F}(E'(R^n)), h_j \in \mathcal{F}(E'(R^{n-1}))$ for which

$$\hat{f}(\zeta) = p(-\zeta)h'(\zeta) + \sum_{j=0}^{m-1} h_j(\zeta')\zeta_n^j + h(\zeta), \qquad (7.6.19)$$

$$|h_j(\zeta')| \le c_5 \ exp \ (d'|Re\zeta'|_{-\Gamma'} + c_1|Im\zeta|/2 + b_1 \ ln(1 + |\zeta'|)), \tag{7.6.20}$$

$$|h(\zeta)| \le c_5 \ exp \ (H_K(Im\,\zeta) + b'' \ ln(1 + |\zeta|)), \tag{7.6.21}$$

$$|h'(\zeta)| \le c_5 \ exp \ (d'|Re\,\zeta| + H_K(Im\,\zeta) + b'' \ ln(1 + |\zeta'|) + b''' \ ln(1 + |\zeta_n|)). \tag{7.6.22}$$

4. Proof of proposition 7.6.1. I. We denote $(0, g(0))$ by x^0 and start with a number of reductions.

a) It suffices to show that

$$(x^0, \xi^{0\prime}) \notin WF_A(\partial/\partial x_n)^k u_{|S}, \forall k, \tag{7.6.23}$$

i.e., to consider derivatives of type $\partial/\partial x_n$ rather than normal derivatives. The reason is that once we know that (7.6.23) is valid, we can conclude that

$$(x^0, \xi^{0\prime}) \notin WF_A(fX^1 \cdots X^r(\partial/\partial x_n)^k u_{|S}), \forall k,$$

for any real analytic function f and any choice of real analytic vector fields X^j which are tangent to S. It suffices then to express $(\partial/\partial n)^r$ for fixed r as a finite sum of terms of form $X^1 \cdots X^i(\partial/\partial x_n)^{r-i}$ with X^i tangent to S, plus a term of form $f(\partial/\partial x_n)^r$ with f real analytic.

b) It suffices to check (7.6.23) for $k = 0$. Indeed, once this is done, we can obtain the general case by observing that $v = (\partial/\partial x_n)^k u$ does also satisfy the assumptions stated for u, so it will follow from what is already proved, that $(x^0, \xi^{0\prime}) \notin WF_A v_{|S}$.

II. To prove (7.6.23) for $k = 0$ we want to apply lemma 7.6.5 in combination with the results from the sections 7.3, 7.4, 7.5. We have already seen that we can assume that S is given by an equation of form $x_n = a - \sqrt{r^2 - |x'|^2}$. After an homothety $x' = ry'$, this comes to $ry_n = (a/r)r - r\sqrt{1 - |y'|^2}$, so we may assume from the very beginning that $r = 1$. The condition $|g(0)| \le c'C_1$ comes then to $1 - a \le c'C_1$. For suitable $c_3 < c'/2$, $2c_1 \le c'/2$, suitable C_1 and $c_2 = a$ we can then assume that the conditions (7.6.16) all hold. Let also K' and K have the meaning from lemma 7.6.5. We note that in view of section 7.4 it will suffice to find d, $\Gamma' \ni \xi^{0\prime}$, c_4 and for every b_2 some b' so that for any $f \in E'(R^n)$ which satisfies the estimates (7.4.3), (7.4.4) it follows that

$$|u(f)| \le c_4.$$

That such choices are possible, will follow from lemma 7.6.5. Indeed, the assumption (7.6.3) shows that we can find $d' > 0$, Γ', b_1, c_5 so that

$$|(\partial/\partial x_n)^j u(x', 0)(G_j)| \le c_5$$

for any $G_j \in E'(R^{n-1})$ which satisfies

$$|\hat{G}_j(\zeta')| \leq \exp(d'|\operatorname{Re}\zeta'|_{-\Gamma'} + (c'/2)|\operatorname{Im}\zeta'| + b_1 \ln(1 + |\zeta'|)). \qquad (7.6.24)$$

Moreover, from the fact that u is extendible we conclude that we can find $c_6, \delta > 0$ and b'' so that

$$|u(G)| \leq c_6$$

for any $G \in E'(R^n)$ so that

$$|\hat{G}(\zeta)| \leq \exp\left(H_K(\operatorname{Im}\zeta) + \delta(\operatorname{Im}\zeta_n)_+ + b'' \ln(1 + |\zeta|)\right). \qquad (7.6.25)$$

We can then write any f which satisfies (7.4.3) and (7.4.4) in the form

$$f = {}^t p(D)G' + G + \sum_{j=0}^{m-1} G_j(x') \otimes (\partial/\partial x_n)^j \delta_{x_n}$$

with the G_j satisfying (7.6.24) and G satisfying (7.6.25).

The proof will now come to an end if we can show that

$$u(f) = u(G) + \sum ((\partial/\partial x_n)^j u_{|x_n=0})(G_j). \qquad (7.6.26)$$

To prove (7.6.26), let us denote by τ_θ, $\theta > 0$, the map $\tau_\theta(f)(x) = f(x', x_n + \theta)$ and consider $\varphi \in C_0^\infty(R^n)$ such that $\int \varphi(x)\, dx = 1$. Also denote $\varepsilon^{-n}\varphi(x/\varepsilon)$ by φ_ε. It follows then from section 7.5 that

$$u(f) = \lim_{\theta \to 0+} u(\tau_\theta f) = \lim_{\theta \to 0+} \lim_{\varepsilon \to 0} u((\tau_\theta f) * \varphi_\varepsilon).$$

Moreover,

$$u((\tau_\theta f) * \varphi_\varepsilon) = u((\tau_\theta G) * \varphi_\varepsilon) + u[\sum (G_j \otimes (\partial/\partial x_n)^j \delta_{x_n=\theta}) * \varphi_\varepsilon]$$

if ε is small compared with θ, since then

$$u({}^t p(D)(\tau_\theta G') * \varphi_\varepsilon) = (p(D)u)((\tau_\theta G') * \varphi_\varepsilon) = 0.$$

It follows for $\varepsilon \to 0$ that

$$u(\tau_\theta f) = u(\tau_\theta G) + \sum ((\partial/\partial x_n)^j u_{|x_n=\theta})(G_j).$$

We obtain (7.6.26) if we let $\theta \to 0$.

7.7 Proof of lemma 7.6.5.

1. All notations are as in the statement of lemma 7.6.5.

I. We start the argument with a preliminary remark, for which we fix a collection $(\partial^1, V^1), \ldots, (\partial^k, V^k)$ of Noetherian operators associated with p. (Cf. section 3.11.) We can then apply the fundamental principle, using also proposition 7.5.4. If b^{IV} and b'' are fixed it follows that we can find b^V, c_6 such that if $H \in A(C^n)$ satisfies

$$|\partial^s H(\zeta)|_{|V^s} \leq \exp\left(H_K(\text{Im}\,\zeta) + b^V \ln(1 + |\zeta'|) + b^{IV} \ln(1 + |\zeta_n|)\right), s = 1, \ldots, k \quad (7.7.1)$$

then there are $h \in A(C^n), h' \in A(C^n)$, such that

$$H(\zeta) = h(\zeta) + p(-\zeta)h'(\zeta),$$

and

$$|h(\zeta)| \leq c_6 \exp\left(H_K(\text{Im}\,\zeta) + b'' \ln(1 + |\zeta|)\right).$$

In view of this, the conclusion of the lemma will follow if we can show that there is b^{IV} such that for any b^V there are c_7 and $h_j \in A(C^n)$ for which (7.6.20) is valid and such that (7.7.1) holds for

$$H = (\hat{f} - \sum_{j=0}^{m-1} h_j(\zeta')\zeta_n^j)/c_7.$$

II. After our preliminary remark, the next step in the proof is to show that we can find $d > 0$ and for every b_2, b_3 some b', c_8, so that if f is given as in the statement of the lemma, then there are $h'_j \in A(C^{n-1})$, $h'' \in A(C^n)$ for which

$$\hat{f}(\zeta) = p(-\zeta)h''(\zeta) + \sum_{j=0}^{m-1} h'_j(\zeta')\zeta_n^j, \quad (7.7.2)$$

$$|h'_j(\zeta')| \leq c_8 \exp\left(d'|\text{Re}\,\zeta'|_{-\Gamma'} + (c_1/2)|\text{Im}\,\zeta'| + b_3 \ln(1 + |\zeta|)\right), \text{ if } \text{Re}\,\zeta' \in -\Gamma. \quad (7.7.3)$$

To see that this is possible, we shall use the contour integration formulas from section 7.2. In fact, using these formulas, we can find $h'_j \in A(C^{n-1})$, $h'' \in A(C^n)$ for which (7.7.2) is valid for all $\zeta \in C^n$. The only problem is then to show that for suitable choices of constants, (7.7.3) is valid. To see under what conditions we can achieve this, let us write at first that

$$H_{K'}(\text{Im}\,\zeta) \leq c_3|\text{Im}\,\zeta'| + (1 - c_2)\text{Im}\,\zeta_{n-}.$$

On the other hand, the assumption on p is that $p(\zeta', \tau) = 0$ implies $|\text{Im}\,\tau| \leq c(1 + |\text{Im}\,\zeta'|)$, if $\text{Re}\,\zeta' \in \Gamma$. With the notations from section 7.2, this suggests to deform the contour $|\tau| = \tilde{c}(1 + |\zeta'|)$ (recall the discussion from that section) to some rectangular contour $\gamma_{\zeta'}$ (in C) which has the following properties:

all zeros σ of $p(-\zeta',\sigma) = 0$ lie inside $\gamma_{\zeta'}$,

$$|p(-\zeta', -\tau)| \geq c_9, \text{ if } \tau \in \gamma_{\zeta'}, \tag{7.7.4}$$

$|\operatorname{Im}\tau| \leq c(1 + |\operatorname{Im}\zeta'|) + c_{10}$ if $\tau \in \gamma_{\zeta'}$ and $\operatorname{Re}\zeta' \in -\Gamma'$,

$|\tau| \leq c_{11}(1 + |\zeta'|)$, if $\tau \in \gamma_{\zeta'}$.

In particular the length of the contour $\gamma_{\zeta'}$ can be estimated by $c_{12}(1 + |\zeta'|)$. Using this contour to define the h'_j and taking into account (7.7.4), it will suffice to show that

$$|\hat{f}(\zeta',\tau)| \leq c_{13} \exp\left(d'|\operatorname{Re}\zeta'|_{-\Gamma'} + (c_1/2)|\operatorname{Im}\zeta'| + (b_3-1)\ln(1+|\zeta'|)\right), \text{ if } \tau \in \gamma_{\zeta'}, \operatorname{Re}\zeta' \in -\Gamma, \tag{7.7.5}$$

provided d, b' are suitable. For c_4 in the conclusion of the lemma we shall now put $c_4 = c_{11}$.

At this moment we use the inequalities (7.6.17), (7.6.18) for $\hat{f}$. The term $H_{K'}(\operatorname{Im}\zeta)$ will not produce any difficulty, since we will have

$$H_{K'}(\operatorname{Im}\zeta) \leq c_3|\operatorname{Im}\zeta'| + (1-c_2)|\operatorname{Im}\zeta_n| \leq (c_3 + (1-c_2)c)|\operatorname{Im}\zeta'| + c_{14} \leq (c_1/4)|\operatorname{Im}\zeta'| + c_{15},$$

$$\text{if } \zeta_n \in \gamma_{\zeta'}, \operatorname{Re}\zeta' \in -\Gamma,$$

as a consequence of (7.6.16). Consider then $\tau \in \gamma_{\zeta'}$. If in addition $|\operatorname{Re}\tau| \leq 2c_4(1+|\operatorname{Re}\zeta'|)$, then we are in the case of the estimate from (7.6.18), so we immediately obtain (7.7.5). (For suitable choices of constants.) Also note that since $|\zeta_n| \leq c_{11}(1 + |\zeta'|)$ for $\zeta_n \in \gamma_{\zeta'}$, we can make

$$b'\ln(1 + |\zeta'|) + \tilde{b}\ln(1 + |\zeta|) \leq (b_2 - 1)\ln(1 + |\zeta'|)$$

by chosing b' suitably.

The other case is when $|\operatorname{Re}\tau| \geq 2c_4(1 + |\operatorname{Re}\zeta'|)$. From $\tau \in \gamma_{\zeta'}$ we know that also $|\tau| \leq c_{11}(1 + |\zeta'|)$. This is only possible when $|\operatorname{Re}\tau| \leq c_{16}(1 + |\operatorname{Im}\zeta'|)$, which, together with $|\operatorname{Re}\tau| \geq 2c_4(1 + |\operatorname{Re}\zeta'|)$, will lead to $|\operatorname{Re}(\zeta',\tau)| \leq c_{17}(1 + |\operatorname{Im}\zeta'|)$. We conclude that the bad term $d|\operatorname{Re}(\zeta',\tau)|$ in (7.6.17) can be estimated by $c_{17}d|\operatorname{Im}\zeta'| + c_{18}$. This is compatible with (7.7.5) if we shrink d until $c_{17}d \leq c_1/4$. We should also observe at this moment that if

$$\sup_{\zeta} |\hat{f}(\zeta)|/\exp\left(H_{K'}(\operatorname{Im}\zeta) + b^{VI}\ln(1 + |\zeta|)\right) < \infty$$

for some b^{VI}, then

$$\sup_{\operatorname{Re}\zeta' \in \Gamma} |h'_j(\zeta)|/\exp\left((c_1/2)|\operatorname{Im}\zeta'| + b^{VII}\ln(1 + |\zeta'|)\right) < \infty. \tag{7.7.6}$$

for some suitable b^{VII}.

This is in fact an easy consequence of the same kind of contour deformation in the definition of h'_j as used before. (The situation is somewhat easier than above in that we do not have a case corresponding to (7.6.17).)

III. After the definition of the h'_j, we now construct $h_j \in A(C^{n-1})$ such that

$$|h'_j(\zeta') - h_j(\zeta')| \le c_{19} \exp\left(3c_1/4\right)|\text{Im } \zeta'| + b_1 \ln(1 + |\zeta'|)), \text{ if } \text{Re } \zeta' \in -\Gamma', \qquad (7.7.7)$$

$$|h_j(\zeta')| \le c_{19} \exp\left((3c_1/4)|\text{Im } \zeta'| + b_1 \ln(1 + |\zeta'|)\right), \text{ if } \text{Re } \zeta' \notin -\Gamma'. \qquad (7.7.8)$$

Such h_j are easy to construct with the aid of a $\bar{\partial}$-argument if b_2 was suitable. Also note that $h_j \in \mathcal{F}(E'(R^{n-1}))$, since h_j satisfies Paley-Wiener type estimates. In fact, when $\text{Re } \zeta \notin -\Gamma'$ this is seen directly from (7.7.8) and for $\text{Re } \zeta \in -\Gamma'$, we note that $h'_j - h_j$ satisfies Paley-Wiener type estimates in view of (7.7.7), whereas h'_j satisfies such estimates in view of (7.7.6). We conclude that the h_j are already as needed in our final conclusion.

IV. To conclude the proof we now want to show that

$$H(\zeta) = f(\zeta) - \sum_{j=0}^{m-1} h_j(\zeta')\zeta_n^j$$

satisfies the inequalities (7.7.1) if divided by some constant c_7. To estimate $\partial^s H$, we shall use for $\text{Re } \zeta' \in -\Gamma'$ that

$$\partial^s(\hat{f} - \sum h_j(\zeta')\zeta_n^j) = \partial_s(\sum(h'_j - h_j)(\zeta')\zeta_n^j),$$

whereas for $\text{Re } \zeta' \notin -\Gamma$ we shall estimate $\partial^s \hat{f}$ and $\partial^s(h_j(\zeta')\zeta_n^j)$, $j \le m-1$ separately. Here we recall that the ∂^s had been of form $(\partial/\partial\theta)^i$ for some i and some direction θ in R^n. We can therefore use Cauchy's inequalities on a sphere of radius one centered at ζ' and use the inequalitities for $h'_j - h_j$. The other cases can be treated in a similar fashion.

7.8   Extension of solutions from the interior

We put here together the propositions 7.6.1 and 7.2.1 to obtain

Proposition 7.8.1. *Let $p(D)$, $\xi^{0'}$, Γ, $g : \{x' \in R^{n-1}; |x'| < c'\} \to R_-$ be as in proposition 7.6.1 and fix $\Gamma' \subset\subset \Gamma$. We can then find $c'' > 2$ and $0 > \delta > g(0)$ so that if $g(x') < \delta$ for $|x'| < c''|g(0) - \delta|$ and if $u \in D'(x; |x'| < c''|g(0) - \delta|)$ is a solution of $p(D)u = 0$ on*

$$\{x \in R^n; |x'| < c''|g(0) - \delta|, g(x') < x_n < \delta\},$$

then we can find $v \in D'(R^n)$ *such that*

a) $p(D)v = 0$, *on* $\{x; |x'| + |x_n - \delta| < c''|g(0) - \delta|/2\}$,

b) $(x', \xi') \notin WF_A((\partial/\partial x_n)^j(u - v)_{|x_n = \delta}$, *if* $|x'| < c''|g(0) - \delta|/2$, $\xi' \in \Gamma'$, $\forall j$,

c) $((0, g(0)), \xi^{0'}) \notin WF_A((\partial/\partial n)^k(u - v)_{|x_n = g(x')})$, $\forall k$.

Remark 7.8.2. *a) In the special case when* $g(x') < g(0)$ *for* $x' \neq 0$, *this result can be replaced in later applications by the much simpler remark following the statement of proposition 7.2.1.*

b) The assumption $c'' > 2$ *is to make sure that* v *is defined in a neighborhood of* $(0, g(0))$.

c) It is clear how to use proposition 7.8.1 to transform a problem with boundary conditions into a problem in free space. In fact, if we have information on the wave front set of the solution u *in the region* $x_n > g(x')$, *say, we can obtain from it information on the wave front set of the Cauchy traces of* u *at* $x_n = \delta$. *From b) in the statement, this gives information on the Cauchy traces of* v *at* $x_n = \delta$, $|x'| < c''|g(0) - \delta|/2$. *Under suitable conditions it will follow from this and results on propagation of singularities in free space that* $((0, g(0)), (\xi^{0'}, \tau)) \notin WF_A v$, *whatever* $\tau \in R$ *is. We can conclude that*

$$((0, g(0)), \xi^{0'}) \notin WF_A((\partial/\partial n)^k v_{|x_n = g(x')}), \forall k.$$

The same will then be true for u *in view of part c) in the statement.*

Chapter 8

Some results on transmission problems

1. In the present chapter we shall apply the results from chapter VII to extend the results on propagation of singularities in free space to cases of boundary problems. The general idea of our approach is in fact very simple: in the situations in which we shall work, it will essentially be possible to extend onesided solutions from one side of the boundary to twosided solutions. Information on the regularity of the solutions in the interior of the domain is carried then to the boundary by our results on propagation of regularity in free space. The picture is finally completed by taking into account the information contained in the boundary conditions. How this is done in some explicit cases will be illustrated in the present chapter in which we shall mainly concentrate on transmission problems. In fact, the methods which we use here can equally well be used for other kinds of boundary problems, but our preference for transmission problems seems justified by the fact that in physical applications conical refraction is certainly associated first of all with the transmission problem in crystal optics. On the other hand, we should also mention that microlocal transmission problems can often be transformed to onesided boundary problems: one side of the domain is reflected with the aid of the transmission interface to the other side, etc. One of the reasons why we have not tried to follow this approach is that a transmission problem for two scalar equations is transformed in this way to a boundary problem for a system for which the intuitive physical meaning of reflected and refracted rays is lost. However, while microlocal transmission problems do not seem to have attracted much attention in the literature (we mention here the papers of Hansen [1], of Nosmas [1] and of Taylor [1,2]), the literature on microlocal mixed problems is more consistent. (Cf. e.g. the papers Ivrii [1,2], Kajitani [1],[2], Kajitani-Wakabayashi [1],[2], and Wakabayashi [1], [2].)

2. The setting in which we work will be rather simple, but it will be general enough to cover some specific situations in crystal optics. What we shall assume is, essentially, that:

a) the operators under consideration are constant coefficient,

b) "rays" and "cones" of propagation hit the interface transversally.

Here the condition a) is effectively needed for the application of the results in the preceding chapter and is, in part, related to our method of proof. There is no doubt that it can be relaxed, if one would use more powerful methods for the study of boundary problems than those developed in section 7.2, as are e.g. the methods from Sjöstrand [2]. Also condition b) is needed when we want to apply the results from the preceding chapter, but the problem with it lies deeper. In fact, that difficulties appear when b) is not valid, is clear already at the level of geometrical optics, if one thinks of the many ways a nontransversal cone can intersect or touch a manifold.

8.1 Formulation of the problem

1. A typical setting for a physical transmission problem is that some waves (e.g. of electromagnetic or acoustic nature) are transmitted from one medium to some other medium through a separating interface. To give a more precise formulation to this, assume that some region X of R^n, $n = 3$, is filled by two media, situated respectively in the subregions X^+ and X^- of X, and that the two media meet along the interface H. If some wave of finite propagation speed is produced, say, in X^-, then it will spread first within X^-, but it may also eventually hit the interface somewhere. At that instance part of the wave (which actually could be all of it in cases of total reflection) will be reflected back into X^-, whereas part of it will be refracted into the region X^+. Of course all this can be very complicated. In the present chapter we shall study such phenomena, placing ourselves in some very simple situations in which, among other things, no boundary effects appear. (When we say "boundary", we mean the boundary of X.) This is often a realistic assumption in that, due to the finite propagation speed, it will anyway take some time before waves produced in the interior of X^- can come under the influence of boundary conditions. We shall then of course also assume that wave propagation in the two regions X^+ and X^- is governed by some linear partial differential equations, p^+ and p^-, respectively. (A typical example is e.g. when the equations p^+ and p^- are just the standard wave equation, but for speeds of light which have a discontinuity along H.

2. We shall now drop our assumption $n = 3$, but still denote by X some open region in R^n which is divided by some smooth interface H into two subregions X^+ and X^-. We assume that we are given two linear partial differential operators $p^+(x, t, D_x, D_t)$ and $p^-(x, t, D_x, D_t)$ which are both defined on $X \times (-a, a)$ for some $a > 0$. Actually it would suffice to assume that the coefficients of p^+ were defined in a neighborhood of $(X^+ \cup H) \times (-a, a)$, with a similar assumption on the coefficients of p^-, but the fact that we assume them defined on $X \times (-a, a)$ is justified (in most cases) by our physical setting, since we could just assume that the medium from X^+ initially filled all of X and has been removed from there at some time, etc. (Actually, there could be problems with the evolution in t of the coefficients. All this is of course not relevant for the situation in which we shall effectively work later on, since then coefficients will be constant.)

Moreover, we assume in this discussion that the surface $H \times (-a, a)$ is noncharacteristic for both operators p^+ and p^- and that the orders of the two operators are the same. As for the regularity of the coefficients, we must assume for the present discussion that they are at least smooth. Our last assumption on the operators is, finally, that they are hyperbolic with respect to the time variable.

Let now u be a distribution on $(X^+ \cup X^-) \times (-a, a)$ so that

$$p^+(x, t, D_x, D_t)u = 0 \text{ for } (x, t) \in X^+ \times (-a, a), \qquad (8.1.1)$$

$$p^-(x, t, D_x, D_t)u = 0 \text{ for } (x, t) \in X^- \times (-a, a), \qquad (8.1.2)$$

and so that the restrictions u^+ of u to $X^+ \times (-a, a)$ and u^- of u to $X^- \times (-a, a)$ admit distributional extensions across $H \times (-a, a)$. It follows from J.Peetre's theorem that we can define the traces v_j^+ of $(\partial/\partial n)^j u^+$ and v_j^- of $(\partial/\partial n)^j u^-$ to $H \times (-a, a)$, where n is the normal to H (for a fixed orientation) and $j = 0, ..., m - 1$. Here m is the common order of p^+ and p^-. A transmission condition in this situation is then a relation between v_j^+, $j = 0, ..., m - 1$ and v_j^-, $j = 0, ..., m - 1$. In the present context we will always assume this relation to be of the form

$$v_i^- = \sum_{j=0}^{m-1} b_{ij} v_j^+, \quad , i = 0, ..., m - 1, \qquad (8.1.3)$$

where $B = (b_{ij})$ is a matrix of pseudodifferential operators on $H \times (-a, a)$. Of course, this requires that we fix the regularity class in which we want to work. Actually later on in this chapter H will be flat or real analytic, so we shall work in the analytic category; correspondingly B will be a matrix of analytic pseudodifferential operators. Conditions of the type of the Lopatinski conditions from boundary value problems should be required for B in the case of the general transmission problem (we shall discuss some geometric conditions in section 8.5) but for most of the results in this chapter, in which we are only interested in passing from X^- to X^+, we only need the pseudolocality of B. We want also to mention explicitly that the conditions from (8.1.3) are the only conditions by which the solutions u^+ and u^- are coupled in the setting of our problem.

3. The transmission problem is not yet completely determined, if we do not fix initial and boundary conditions. For the reasons explained above we do not care much for the boundary conditions. Actually, we shall assume henceforth that $X = R^n$, so no physical boundary appears. Moreover, in physical situations one may often assume that the solutions have compact support in R^n for each fixed t, so in particular no disturbances will come in from infinity. As for the initial conditions, we shall assume that we have knowledge as explicit as needed of the microlocal singularities of all traces of $u^\pm$ at some time t^0 in the past, or of the singularities of $u^\pm$ for some time slice in the past. Actually a very natural condition would be to assume that the $u^\pm$ vanish for large negative times, i.e. before some experiment started. To have nontrivial solutions we would then have to work with inhomogeneous equations of form $p^\pm u^\pm = f^\pm$, rather than with the homogeneous equation $p^\pm u^\pm = 0$. If however the perturbation $f^\pm$ has compact support in the time variable, we still would be left with the homogeneous equations starting from some time on. It is in this later region that we want to place ourselves. Our problem is then to say something on the singularities of $u^\pm$ for such later times.

8.2   Regularity of traces as a consequence of interior regularity, I.

1. Theorem 7.1.1 shows how the regularity of the Cauchy traces leads to interior regularity for linear partial differential operators with analytic coefficients. In this and the following section we consider results in the opposite direction, on how interior regularity can propagate from the interior to the boundary. The main idea in the argument is to transform some initially onesided problem into a twosided problem, so that in fact the desired result is a consequence of our results on conical refraction in free space. Indeed, the argument is very simple and the many assumptions which we make are just to guarantee the simultaneous applicability of the various results on which it is based. We should also mention that our results are somewhat sharper in the case of a flat boundary, so it seemed convenient to discuss flat and curved boundaries in distinct sections.

Remark 8.2.1 *Actually our arguments can be applied in much more general situations.*

2. Before we state the results we want to fix some terminology. We consider some constant coefficient linear partial differential operator $p(D)$ with real principal part p_m and consider $\xi^0 = (\xi^{0\prime}, \mu) \in R^n$ such that

$$p_m(\xi^0) = 0, \ \operatorname{grad} p_m(\xi^0) = 0, \ \operatorname{rank} \operatorname{Hess} p_m(\xi^0) = n - 1,$$

and such that for any $\sigma \neq 0$ which is orthogonal to ξ^0 we have the implication:

$$p_{m,1}(\sigma) = 0, \ \text{implies} \ (\partial/\partial\sigma_n)p_{m,1}(\sigma) \neq 0. \tag{8.2.1}$$

Here $p_{m,1}$ is the localization of p_m at ξ^0 and we recall that the rank of Hess $p_m(\xi^0)$ can be at most $n - 1$, since p_m is homogeneous. (Here the assumption that rank Hess $p_m(\xi^0)$ is maximal has no deep meaning. We could have considered more degenerate situations. Actually what we want to have is that $p_{m,1}$ is hyperbolic in the space of variables on which it effectively depends.) Further we assume that the signature of the quadratic form $p_{m,1}$ on the space of variables orthogonal to ξ^0 is either $+ - \cdots -$ or else $- + \cdots +$. We are then in a situation similar to the one encountered in section 1.3.

We shall now denote by G_μ^+, or sometimes by $G_{\xi^0}^+$, the cone

$$G_\mu^+ = \{\sigma \in R^n; \sigma_n > 0, \sigma \perp \xi^0, p_{m,1}(\sigma) > 0\},$$

and by G_μ^{+0} (or $G_{\xi^0}^{+0}$) the polar cone of G_μ^+ in the space of variables orthogonal to ξ^0:

$$G_\mu^{+0} = \{x \in R^n; x \perp \xi^0, \langle x, \sigma \rangle \geq 0, \forall \sigma \in G_\mu^+\}.$$

(In section 1.3 we have seen that G_μ^{+0} should be regarded as a symplectic polar, which also explains why G_μ^+ lives in a space of ξ-variables, whereas G_μ^{+0} lives in a space of x-variables. Cf. section 1.3 for a precise description.)

We recall from section 1.3 that if $p_{m,1}(\sigma) = 0, \sigma \perp \xi^0, \sigma \neq 0$, then we had grad $p_{m,1}(\sigma) \in (G_\mu^{+0} \cup -G_\mu^{+0})$ and that, conversely, the boundaries of G_μ^{+0} and $-G_\mu^{+0}$ can be obtained in this way. It is clear then from condition (8.2.1) that

$$(G_\mu^{+0} \cup -G_\mu^{+0}) \cap \{x; x_n = 0\} = \{0\}. \tag{8.2.2}$$

Definition 8.2.2. We call G_μ^{+0} and $-G_\mu^{+0}$ the *propagation cones associated with the triple* $(\mu, \xi^{0\prime}, x_n = 0)$. *The relation (8.2.2) shall be referred to as the "transversality condition".*

3. We shall now list the conditions on $p(D)$ needed in our main result. For this we denote by $I = \{\mu_1, \mu_2, ..., \mu_m\}$ the roots of $p_m(\xi^{\prime 0}, \mu) = 0$, counted with their respective

multiplicities, and consider a subset $J \subset I$ so that $\mu_i \neq \mu_j$ whenever $\mu_i \in J$ and $\mu_j \in I \setminus J$. Our assumptions on (p, ξ'^0, J) are as follows:

a) $x_n = 0$ is noncharacteristic for p and in fact the coefficient of ξ_n^m in p is one,

b) there is a conic neighborhood Γ of ξ'^0 and $c > 0$ so that $p(\xi', \tau) = 0$ implies $|Im\, \tau| \leq c$, if $\xi' \in \Gamma$,

c) the roots $\mu_i \in J$ of $p_m(\xi'^0, \mu) = 0$ of variable multiplicity are at most double,

d) if $\mu \in J$ and $\operatorname{grad} p_m(\xi'^0, \mu) \neq 0$, then $(\partial p_m / \partial \xi_n)(\xi'^0, \mu) \neq 0$,

e) if $\mu \in J$ but $\operatorname{grad} p_m(\xi'^0, \mu) = 0$, then the signature of $p_{m,1}$ on the space of variables orthogonal to (ξ'^0, μ) is either $+ - \cdots -$ or else $- + \cdots +$,

f) if $\mu \in J$, $\operatorname{grad} p_m(\xi'^0, \mu) = 0$, and if $\sigma \neq 0$ is orthogonal to (ξ'^0, μ) then (8.2.1) is valid.

Remark 8.2.3 *The condition d) is, in analogy with what we have already said for e), a transversality condition. Actually, the conditions c),d),e),f) are sometimes too restrictive, in that we would like to have replaced them at least by the same kind of conditions but in a situation of "constant multiplicity". We can in fact work with the following somewhat more general conditions g) and h):*

g) in a conic neighborhood of (ξ'^0, μ) we can write p_m in the form

$$p_m(\xi) = [Q(\xi)]^\nu Q'(\xi),$$

where Q and Q' are both real-analytic and homogeneous in a conic neighborhood of (ξ'^0, μ), ν is a natural number and $Q'(\xi'^0, \mu) \neq 0,$. Moreover, we have either that $(\partial / \partial \xi_n) Q(\xi'^0, \mu) \neq 0$ or else that

h) $\operatorname{grad} Q(\xi'^0, \mu) = 0$ and that the Hessian of the localization of Q at (ξ'^0, μ) is nondegenerate and has signature $+ - \cdots - $ or $- + \cdots +$. In the latter case we assume that (8.2.1) holds with $p_{m,1}$ replaced by the localization of Q at (ξ'^0, μ).

We shall give statements however only in the case when c),d),e),f) are valid, since they are then shorter. (The extension to the case g), h) is interesting for the case of crystal optics.)

4. Once J as above is fixed, it is interesting to factor p into a product of two factors, where the first factor is associated with the roots μ_i in J of $p_m(\xi'^0, \mu) = 0$ and the second

with the roots in $I \setminus J$. (A similar situation appears in Sjöstrand [2].) This is based on the remark that we can find a conic neighborhood Γ' of ξ'^0, a natural number k, $c' > 0$ and open cones $G \subset\subset G' \subset R^n$ such that for every $\xi' \in \Gamma'$ for which $|\xi'| > c'$, there are precisely k roots $\nu_i(\xi')$, $i = 1, ..., k$, of $p(\xi', \nu) = 0$ with $(\xi', \nu_i(\xi')) \in G$ and $m - k$ roots $\nu_i(\xi')$, $k + 1 \le i \le m$ of $p(\xi', \nu) = 0$ with $(\xi', \nu_i(\xi')) \in R^n \setminus G'$. We also denote by p_J and q_J the functions on $\{\xi; \xi' \in \Gamma', |\xi'| > c'\}$ defined by

$$p_J(\xi) = (\xi_n - \nu_1(\xi')) \cdots (\xi_n - \nu_k(\xi'))$$

$$q_J(\xi) = (\xi_n - \nu_{k+1}(\xi')) \cdots (\xi_n - \nu_m(\xi')),$$

so that

$$p = p_J q_J \text{ on the domain of definition of the } p_J, q_J.$$

It is also clear that if we write q_J in the form

$$q_J(\xi) = \xi_n^{m-k} + \sum_{i=0}^{m-k-1} a_i(\xi') \xi_n^i,$$

with $a_i : \{\xi' \in R^{n-1}; \xi' \in \Gamma', |\xi'| > c'\} \to C$, then the a_i are real-analytic. In fact, more is true, in that we can find a complex conic neighborhood $\Gamma_C \subset C^{n-1}$ of ξ'^0 and c'' so that the a_i have analytic extensions to the set

$$U = \{\zeta' \in C^{n-1}; \zeta' \in \Gamma_C, |\zeta'| > c''\}$$

and satisfy an estimate of the form

$$|a_i(\zeta')| \le c(1 + |\zeta'|)^{m-k-i} \text{ on } U.$$

We conclude that the a_i are symbols of analytic pseudodifferential operators in a conic neighborhood of ξ'^0.

Our next move is now to write ξ_n^i for $i \ge m - k$ in the form

$$\xi_n^i = q_J Q_i + \sum_{j=0}^{m-k-1} a_{ij} \xi_n^j, \tag{8.2.3}$$

where the a_{ij} are once more analytic symbols in a conic neighborhood of ξ'^0 at infinity. In fact, $a_{m-k,j} = -a_j$ and the a_{ij} for $i > m - k$ can be computed recurrently by an obvious division algorithm. When $I = J$, $q_J = 1$ and the sum in (8.2.3) is considered void, i.e. no a_{ij} are considered. We need to associate analytic pseudodifferential operators with the a_{ij}. Actually, the symbols a_{ij} are in $n - 1$ variables, whereas the operators will act on distributions in n variables. To avoid misunderstandings, we shall therefore define briefly what we need.

Definition 8.2.4. *Let $\Gamma_C \subset C^{n-1}$ be a complex conic neighborhood of $\xi'^0 \in \dot{R}^{n-1}$ and let*

$$a : \{\zeta' \in \Gamma_C; |\zeta'| > c\} \to C$$

be a holomorphic function such that

$$|a(\zeta')| \leq c'(1 + |\zeta'|)^b \text{ for some } c' \text{ and } b.$$

Denote $\Gamma_C \cap R^{n-1}$ by Γ. Also fix $\varphi \in C_0^\infty(R^n)$, identically one in a neighborhood of 0 and $\chi \in C^\infty(R^{n-1})$ such that the support of χ is contained in $\{\xi' \in \Gamma'; |\xi'| > 2c'\}$ where $\Gamma' \subset\subset \Gamma$ is a conic neighborhood in R^{n-1} of ξ'^0. Assume that χ is identically one on a set of form $\{\zeta' \in \Gamma''; |\zeta'| > 3c'\}$ for some other conic neighborhood Γ'' of ξ'^0. We shall then call $a(D')$ defined by

$$a(D')v = \mathcal{F}^{-1}(a(\xi')\chi(\xi')\mathcal{F}(\varphi v)(\xi)) \tag{8.2.4}$$

the pseudodifferential operator associated with a.

Note that the $a(D')$ thus depends on the choices of φ and χ, but we shall not make this explicit in the notation.

We shall need later on the following result on pseudolocality:

Lemma 8.2.5. *a) Assume that $(0, (0, ..., 0, 1)) \notin WF_A v$. Then it follows that*

$$(0, (0, ..., 0, 1)) \notin WF_A a(D')v.$$

b) Moreover, if $(0, (\xi'^0, \nu)) \notin WF_A v$ for some $\nu \in R$, then it follows that $(0, (\xi'^0, \nu)) \notin WF_A a(D')v$.

Note that the second part of the lemma is a consequence of the results on the pseudolocality of analytic pseudodifferential operators (cf. e.g. chapter IV), since $a(D')$ is a well-defined analytic pseudodifferential operator on sets of form $\{(x, \xi); \xi \in G_1\}$, when G_1 is a suitably small conic neighborhood in R^n of (ξ'^0, ν). (Note here that the estimates for derivatives of a are of type

$$|(\partial/\partial\xi)^\alpha a(\xi')| \leq c^{|\alpha|+1}\alpha!(1 + |\xi'|)^{b-|\alpha|}$$

rather than $|(\partial/\partial\xi)^\alpha a(\xi')| \leq c^{|\alpha|+1}\alpha!(1 + |\xi|)^{b-|\alpha|}$. On small cones of type G_1 these two estimates are however equivalent.) The only problem is therefore with part a) of the lemma, which is however very simple since $a(D')$ has constant coefficients. For the convenience of the reader, we shall sketch a proof of this result at the end of this section.

5. We can now state the main result of the section.

Proposition 8.2.6. *Assume that the above assumptions are satisfied for $(p(D), \xi'^0, J)$. Also let $a_{ij}(D')$ be pseudodifferential operators associated as in (8.2.4) (for fixed φ, χ) with the symbols a_{ij} from (8.2.3). Consider $u \in D'(R^n)$ a solution of $p(D)u = 0$ on the set $x_n > 0$ and assume that when $\mu \in J$ and $grad\, p_m(\xi'^0, \mu) \neq 0$, then*

$$\{(s\ grad\, p_m(\xi'^0, \mu), (\xi'^0, \mu)); s \neq 0\} \cap \{x; x_n > 0\} \cap WF_A u = \emptyset, \qquad (8.2.5)$$

whereas when $\mu \in J$ and $grad\, p_m(\xi'^0, \mu) = 0$, then

$$[(G_\mu^{+0} \cup -G_\mu^{+0}) \times (\xi'^0, \mu)] \cap WF_A u \cap \{x; x_n > 0\} = \emptyset. \qquad (8.2.6)$$

Under these assumptions the conclusion is that

$$(0, \xi'^0) \notin WF_A\{[(\partial/\partial x_n)^i - \sum_{j=0}^{m-k-1} a_{ij}(D')(\partial/\partial x_n)^j]u\}_{|x_n = 0_+} \qquad (8.2.7)$$

for any $i \geq m - k$.

The "local" version of this result is also true, in that it suffices to assume that u is only defined in a neighborhood of 0. For notational reasons we prefer here to work with a globally defined u.

Remark 8.2.7. *The most important particular case is perhaps when $J = I$. The conclusion is then that*

$$(0, \xi'^0) \notin WF_A(\partial/\partial x_n)^i u_{|x_n = 0_+}, \forall i.$$

(The proof of this particular case is actually somewhat simpler and does not depend on lemma 8.2.5. In the following section we shall come back to this result, but we shall assume there that the problem is local since the boundary is real analytic rather than flat.)

6. **Proof of proposition 8.2.6.** We consider some solution v of $p(D)v = 0$ which is defined in a neighborhood V of 0 and satisfies

$$(0, \xi'^0) \notin WF_A((\partial/\partial x_n)^i(u - v)_{|x_n = 0_+}), \forall i. \qquad (8.2.8)$$

From (8.2.8) and theorem 7.1.1 we can conclude that the regularity assumptions (8.2.5) and (8.2.6) remain valid for small x_n if we replace u by v there. We claim that

$$(0, (\xi^{0'}, \mu)) \notin WF_A v, \forall \mu \in J. \qquad (8.2.9)$$

In fact, when μ is a simple root of $p_m(\xi^{0\prime}, \mu) = 0$ this follows from (8.2.5) and standard results on propagation of singularites for operators of principal type and when μ is a double root of that equation, from (8.2.6) and theorem 1.1.6. (It is at this moment when we take full advantage of the fact that we have transformed our one-sided problem into a problem in free space.)

Since $p(D)v = 0$ it will now follow from (8.2.9) that

$$(0, (\xi^{\prime 0}, \nu)) \notin WF_A(\partial/\partial x_n)^i q_J(D)v, \forall i, \forall \nu \in R. \tag{8.2.10}$$

Here of course

$$(\partial/\partial x_n)^i q_J(D) = (\partial/\partial x_n)^{i+m-k} + \sum_{j=0}^{m-k-1} a_j(D')(\partial/\partial x_n)^{j+i},$$

the $a_j(D')$ being defined as in (8.2.4). (When $J = I$, (8.2.10) is trivial.)

It suffices to prove (8.2.10) for $i = 0$. To do so, we denote $q_J(D)v$ by w. Since $p(D)v = 0$ it follows that

$$(0, (\xi^{0\prime}, \nu)) \notin WF_A p_J(D)w, \forall \nu. \tag{8.2.11}$$

$p_J(D)$ is here a pseudodifferential operator associated with p_J in the same way in which $q_J(D)$ is associated with q_J, and it is in particular a classical analytic pseudodifferential operator in a conic neighborhood of $(0, (\xi^{0\prime}, \nu))$. From the Hörmander-Sato regularity theorem it follows then that (8.2.11) implies (8.2.10), unless $(0, (\xi^{0\prime}, \nu))$ is characteristic for $p_J(D)$. The only characteristic vectors of form $(0, (\xi^{0\prime}, \nu))$ for $p_J(D)$ are however those when $\nu \in J$. For such vectors, (8.2.10) follows from (8.2.9).

In addition to (8.2.10), we now also observe that we have

$$(0, (0, ..., 0, 1)) \notin WF_A q_J(D)v. \tag{8.2.12}$$

In fact, $(0, (0, ...0, 1)) \notin WF_A v$ since $p(D)v = 0$, so (8.2.12) is a consequence of lemma 8.2.5. From (8.2.12) and the analoguous relation for $(\partial/\partial x_n)^i q_J(D)v$, it follows now that it is possible to consider the traces to $x_n = 0$ of the $(\partial/\partial x_n)^i q_J(D)v$. We conclude then from (8.2.10) that

$$(0, \xi^{0\prime}) \notin WF_A[(\partial/\partial x_n)^i q_J(D)v]_{|x_n=0}, \forall i.$$

Actually, we can write the last relation, as is immediately seen, in the form

$$(0, \xi^{\prime 0}) \notin WF_A[(\partial/\partial x_n)^i - \sum_{j=0}^{m-k-1} a_{ij}(D')(\partial/\partial x_n)^j]v_{|x_n=0}.$$

This gives (8.2.7), if we also take into account (8.2.8).

Remark 8.2.8. *Later on we shall only use (8.2.7) for* $i = m - k, ..., m - 1$. *(The other relations follow anyhow from these if we also use* $p(D)u = 0$.) *(8.2.7) can then be written in the form*

$$(0, \xi^{0\prime}) \notin WF_A XU,$$

where X *is an explicitly computable* $k \times m$ *matrix of analytic pseudodifferential operators with entries of type* $a(D')$ *and* $U = (u_j^+)_{j=0}^{m-1}$ *is the vector of Cauchy data* $u_j^+ = (\partial/\partial x_n)^j u_{|x_n=0_+}$ *of* u.

7. It remains to prove lemma 8.2.5 a) and we shall start with some preliminary remarks. The assumption on v is that we can find $d' > 0$, $\varepsilon' > 0$, $b' \in R$, a conic neighborhood G of $(0, ..., 0, 1)$ and $c > 0$ so that $|v(g)| < c$ for any $g \in E'(R^n)$ for which

$$|\hat{g}(\zeta)| \leq \exp\ [d'|Re\ \zeta| + \varepsilon'|Im\ \zeta| + b'ln(1 + |\zeta|)], \text{ if } Re\ \zeta \in -G,$$

$$|\hat{g}(\zeta)| \leq \exp\ [\varepsilon'|Im\ \zeta| + b'ln(1 + |\zeta|)], \text{ if } Re\ \zeta \notin -G.$$

For later need, we shrink d', ε' so that $\varphi(x) = 1$ for $|x| < d' + \varepsilon'$ and for b' we assume that

$$\int |\mathcal{F}(\varphi v)(\xi)|(1 + |\xi|)^{b'}\ d\xi \leq c_1.$$

Moreover, we may assume that b' is such that $f \in E'(R^n)$ and $|\hat{f}(\xi)| \leq (1 + |\xi|)^{b'}$ implies

$$[a(D)v](f) = (2\pi)^{-n} \int \hat{f}(-\xi)a(\xi')\chi(\xi')\mathcal{F}(\varphi v)(\xi)\ d\xi,$$

where φ and χ are as in (8.2.4). We shall now prove the following

Lemma 8.2.9. *Consider* a, φ, χ, v *and let* d', ε', b' *and* G *be fixed as before. Then we can find a conic neighborhood* G' *of* $(0, ..., 0, 1)$, $d, \varepsilon, b \leq b', c'$ *with the following property:*

for any $h \in A(C^n)$ *which satisfies*

$$|h(\zeta)| \leq exp\ [d|Re\ \zeta| + \varepsilon|Im\ \zeta| + bln(1 + |\zeta|)], \text{ if } Re\ \zeta \in -G', \tag{8.2.13}$$

$$|h(\zeta)| \leq exp\ [\varepsilon|Im\ \zeta| + bln(1 + |\zeta|)], \text{ if } Re\ \zeta \notin -G', \tag{8.2.14}$$

we can find $h' \in A(C^n)$ *and* $h'' \in C^\infty(R^n)$ *such that*

$$a(-\xi')\chi(-\xi')h(\xi) = h'(\xi) + h''(\xi), \text{ if } \xi \in R^n,$$

$$|h'(\zeta)| \leq c'exp\ [d'|Re\ \zeta| + \varepsilon'|Im\ \zeta| + b'ln(1 + |\zeta|)], \text{ if } Re\ \zeta \in -G, \tag{8.2.15}$$

$$|h'(\zeta)| \leq c'exp\ [\varepsilon'|Im\ \zeta| + b'ln(1 + |\zeta|)], \text{ if } Re\ \zeta \notin -G \tag{8.2.16}$$

$$|h''(\zeta)| \leq c'(1 + |\xi|)^{b'}, \text{ if } \xi \in R^n. \tag{8.2.17}$$

Proof of lemma 8.2.9.(Sketch) We fix a conic neighborhood $G' \subset\subset G$ of $(0,...,0,1)$, $c_i > 0$, $i = 1,2,3,4$, and choose a $C^\infty(C^n)$ function ψ such that

i) $\psi(\zeta) \neq 0$ only when $a(-\zeta')$ is defined,

ii) $\psi(\zeta) = 0$ if $|\zeta| < c_1$, or $Re\,\zeta \notin G'$, or $|Im\,\zeta| > c_2|Re\,\zeta|$,

iii) $\psi(\zeta) \equiv 1$ if ζ is in a complex conic neighborhood of $(0,...,0,1)$ and $|\zeta| \geq c_3$,

iv) $|\bar{\partial}\psi(\zeta)| \leq c_4$.

We next define a function f' by $f'(\zeta) = a(-\zeta')\chi(-Re\,\zeta')h(\zeta)\psi(\zeta)$ if $\psi(\zeta) \neq 0$ and by $f'(\zeta) = 0$ if $\psi(\zeta) = 0$. It follows that

$$|\bar{\partial}f'(\zeta)| \leq c_5\exp\left[d|Re\,\zeta| + \varepsilon|Im\,\zeta| + b\ln(1 + |\zeta|)\right], \text{ if } Re\,\zeta \in -G',$$

$$|\bar{\partial}f'(\zeta)| \leq c_5\exp\left[(dc_6 + \varepsilon)|Im\,\zeta| + b\ln(1 + |\zeta|)\right], \text{ if } Re\,\zeta \notin -G'.$$

Starting from sufficiently small d, ε, b, we can then find g and c_7, for which $\bar{\partial}g = \bar{\partial}f'$ and so that

$$|g(\zeta)| \leq c_7\exp\left[d'|Re\,\zeta| + \varepsilon'|Im\,\zeta| + b'\ln(1 + |\zeta|)\right], \text{ if } Re\,\zeta \in -G,$$

$$|g(\zeta)| \leq c_7\exp\left[\varepsilon'|Im\,\zeta| + b'\ln(1 + |\zeta|)\right], \text{ if } Re\,\zeta \notin -G.$$

Functions as desired are then $h' = f' - g$, $h'' = a(-\zeta')\chi(-Re\,\zeta')h(\zeta) - f' + g$.

8. We can now return to the proof of lemma 8.2.3 and note at first that (8.2.15), (8.2.16), (8.2.17) show in particular that

$$|h'(\xi)| \leq c_8(1 + |\xi|)^{b'} \text{ if } \xi \in R^n,$$

$$|h'(\zeta)| \leq c_9(1 + |\zeta|)^{b'} \exp\left[(d + \varepsilon)|\zeta|\right].$$

It follows from these two inequalities and the Phragmén-Lindelöf principle that

$$|h'(\zeta)| \leq c_{10}(1 + |\zeta|)^{b'} \exp\left[(d + \varepsilon)|Im\,\zeta|\right],$$

so in particular, $\mathcal{F}^{-1}(h') \in E'(x; |x| \leq d + \varepsilon)$. It is also no loss of generality to assume that $\varphi = 1$ if $|x| < d + \varepsilon$, since we can always shrink d and ε until this is achieved.

Let us now choose $w \in E'(R^n)$ so that $h = \hat{w}$ satisfies the inequalities (8.2.13) and (8.2.14). If h', h'' are associated with this h as in lemma 8.2.9, it follows that

$$[a(D')v](w) = (2\pi)^{-n} \int h'(-\xi)\mathcal{F}(\varphi v)(\xi)\,d\xi + (2\pi)^{-n} \int h''(-\xi)\mathcal{F}(\varphi v)(\xi)\,d\xi.$$

Here

$$\int h'(-\xi)\mathcal{F}(\varphi v)(\xi)\, d\xi = (\varphi v)(\mathcal{F}^{-1}h') = v(\mathcal{F}^{-1}h').$$

It is then clear that

$$|\int h'(-\xi')\mathcal{F}(\varphi v)(\xi)\, d\xi| \le c_9, \int |h''(-\xi')\mathcal{F}(\varphi v)(\xi)|\, d\xi \le c_9,$$

for some constant c_9 which does not depend explicitly on w. We obtain $|[a(D')v](w)| \le 2c_9$ and can conclude that $(0,(0,...,0,1)) \notin WF_A a(D')v$.

8.3 Regularity of traces as a consequence of interior regularity, II.

1. In this section we come back to remark 8.2.7. What we want to do is to prove a result similar to that from the respective remark, but for a real-analytic boundary. We shall in fact assume that U is an open domain in R^n with real-analytic boundary ∂U and consider $x^0 \in \partial U$. We also consider a constant coefficient linear partial differential operator $p(D)$ and let $u \in D'(R^n)$ be a distribution which satisfies $p(D)u = 0$ on U. It is no loss of generality to assume that locally near x^0, ∂U and U are given by the equation $x_n = g(x')$, respectively by $x_n > g(x')$, for some real-analytic function g for which $\text{grad}(0) = 0$ and that $x^0 = (0,g(0)) = 0$. Since the problem is local, we may in fact assume that

$$U = \{x; |x'| < c', g(x') < x_n < c_1\},$$

for some constants c', c_1, $c_1 > g(x')$ if $|x'| < c'$. Finally we fix $\xi^{0\prime} \in R^{n-1}$ and assume that the conditions a), b), c), d), e), f) from section 8.2 hold for $p(D)$, $\xi^{0\prime}$ and Γ. Of course we replace here the condition (8.2.2) (notations are as in section 8.2) by

there is a neighborhood V of x^0 so that $V \cap (G_\mu^{+0} \cup -G_\mu^{+0}) \cap \{x_n = g(x')\} = \{0\}$.

(The condition for the bicharacteristic curves L_μ is formulated in a similar way.) Our main result is the following:

Theorem 8.3.1. *Assume that when* $p_m(\xi^{0\prime}, \mu) = 0$ *and* $\text{grad}\, p_m(\xi^{0\prime}, \mu) \neq 0$, *then*

$$\{(s\ \text{grad}\, p_m(\xi'^0, \mu), (\xi'^0, \mu)); s \neq 0\} \cap \{x; |x'| < c', g(x') < x_n < c_1\} \cap WF_A u = \emptyset, \quad (8.3.1)$$

whereas when $\text{grad}\, p_m(\xi'^0, \mu) = 0$, *then*

$$[(G_\mu^{+0} \cup -G_\mu^{+0}) \times (\xi'^0, \mu)] \cap WF_A u \cap \{x; |x'| < c', g(x') < x_n < c_1\} = \emptyset. \quad (8.3.2)$$

Then it follows that

$$(x^0, \xi^{0\prime}) \notin WF_A(\partial/\partial n)^k u_{|x_n = g(x')}, \forall k. \quad (8.3.3)$$

2. The proof of this result is very close to that of proposition 8.2.6, so we shall only describe the basic changes. As in section 8.2 the main idea is to transform the boundary regularity problem into a problem of propagation of regularity in free space to which we can apply the results from the first chapter. Since the boundary is not flat, we cannot extend u directly to the other side of the boundary as a solution (at least this is not possible by using only the results from these notes), but we can apply proposition 7.8.1 for some hyperplane of form $x_n = \delta$ sufficiently close to $x_n = g(0) = 0$. To take advantage of this, we fix $\Gamma' \subset\subset \Gamma$ so that Γ' is still a conic neighborhood of $\xi^{0\prime}$ and consider a solution v of $p(D)v = 0$ associated with u as in proposition 7.8.1. From theorem 7.1.1 it follows that

$$((x', x_n), (\xi', \xi_n) \notin WF_A(u - v), \forall \xi' \in \Gamma', \forall \xi_n, \forall x_n \text{ s.t. } |x_n - \delta| < \epsilon$$

so the relations (8.3.1), (8.3.2) remain valid when u is replaced by v, if we also intersect with $x_n > \delta$. When δ was sufficiently close to 0, we can now essentially argue as in the proof of proposition 8.2.6 and obtain

$$(x^0, \xi^{0\prime}) \notin (\partial/\partial x_n)^k v_{|x_n=0}.$$

Theorem 7.2.5 gives then

$$(x^0, \xi^{0\prime}) \notin (\partial/\partial n)^k v_{|x_n=g(x')},$$

which gives the desired conclusion immediately.

8.4   Incoming and outgoing bicharacteristics and propagation cones

1. In the present section we assume as in the first section of the chapter, that in our base space we have a well-distinguished time variable. We shall assume in fact that $p(D)$ is a constant coefficient linear partial differential operator in $n + 1$ variables (x,t) which is hyperbolic with respect to the surface $t = 0$ and for which the surface $\{(x, t) \in R_x^n \times R_t; x_n = 0\}$ is noncharacteristic. (The interface is thus still of form $x_n = 0$ and not $t = 0$.) One of our goals is to discuss the cones G_μ^+, G_μ^{+0} from section 8.2 in terms of slowness surfaces. We shall make assumptions on p_m so that the propagation cones can only be of two types: either they carry singularities from the region $x_n > 0$ to $x_n = 0$ with increasing time, or else singularities are carried away from $x_n = 0$ into the region $x_n > 0$ when time increases. Before we go into details, we discuss the analoguous question when μ is a simple root of $p_m(\xi'^0, \mu, 1) = 0$: Actually then, wave front set

singularites at $(\xi'^0, \mu, 1)$ propagate along the bicharacteristic

$$L = \{(s \text{ grad } p_m(\xi^0, 1), (\xi^0, 1)); s \in R\}, \quad \xi^0 = (\xi'^0, \mu).$$

The main point is here to assume $(\partial/\partial\tau)p_m(\xi^0, 1) \neq 0$. Note that if we assume that ξ^0 is a regular point of the slowness surface S (of p_m), then we will have $\text{grad}_\xi p_m(\xi^0, 1) \neq 0$. To assume that also $(\partial/\partial\tau)p_m(\xi^0, 1) \neq 0$ was then a standard assumption in this context. We also recall that from $(\partial/\partial\tau)p_m(\xi^0, 1) \neq 0$ it will follow that

$$\langle \xi^0, \text{ grad}_\xi p_m(\xi^0, 1) \rangle \neq 0. \tag{8.4.1}$$

If we assume $\xi'^0 = 0$, this means that

$$(\partial/\partial\xi_n)p_m(\lambda^0) \neq 0, \text{ where } \lambda^0 = (\xi^0, 1), \tag{8.4.2}$$

so x_n will not be constant on the bicharacteristic. The assumption $\xi'^0 = 0$ means here of course that the normal to $x_n = 0$ points in the direction of ξ^0, but clearly (8.4.2) will remain valid when ξ^0 is close to this normal. It follows whenever (8.4.2) holds that when time increases, then x_n will either monotonically increase or monotonically decrease on the bicharacteristic. We shall call L then, in analogy to the terminology in Sommerfeld's radiation condition, outgoing (with respect to $x_n > 0$) in the first case and incoming in the second. The terms "outgoing" and "incoming" have of course to be interchanged, when we are interested in the part of some bicharacteristic which lives in the region $x_n < 0$. Of course, L is incoming, respectively outgoing, with respect to $x_n > 0$ precisely if $(\partial/\partial\tau)p_m(\lambda^0)(\partial/\partial\xi_n)p_m(\lambda^0)$ is < 0, respectively > 0.

Remark 8.4.1.

Since $R_x^n \setminus \{x; x_n = 0\}$ has no natural "inner" respectively "outer" region, the terminology is perhaps misleading. An alternative could be "past-oriented" for incoming, respectively "future-oriented". (Sommerfeld would perhaps have said "wegstrahlend", i.e. "radiating away" for outgoing.)

2. We want to perform a similar discussion for the case when μ is a double root of $p_m(\xi'^0, \mu, 1) = 0$. We shall then assume that $F(\xi, \tau) = p_m(\xi, \tau)$ satisfies the assumptions I,II, III, from section 6.6 in a neighborhood of the point λ^0. (Recall that $\lambda^0 = (\xi'^0, \mu, 1)$.) As for n, we assume $n \geq 3$ (i.e. $n + 1 \geq 4$) in all this discussion. In particular our assumptions are that

$$\text{grad } p_m(\lambda^0) = 0 \tag{8.4.3}$$

and that $(\partial/\partial\tau)^2 p_m(\lambda^0) \neq 0$. Without loss of generality we may assume then that

$$(\partial/\partial\tau)^2 p_m(\lambda^0) > 0. \tag{8.4.4}$$

Moreover we assume rank Hess $p_m(\lambda^0) = n$. (Recall that we are in $n + 1$ variables, so this is maximal.)

Also let $p_{m,1}$ (or $p_{m,1,\xi^0}$, when we want to specify ξ^0) be the localization polynomial of p_m at λ^0. Our first concern is then to discuss the validity of condition e), section 8.2. With the notations from the present situation this means (notice also that here we are in $R_{x,t}^{n+1}$ rather than in R_x^n) that we want to check wether

$$(\partial/\partial\theta_n)p_{m,1}(\theta) \neq 0 \text{ for any } \theta \in W \tag{8.4.5}$$

where $W = \{\theta \in \dot{R}^{n+1}; p_{m,1}(\theta) = 0, \langle\theta, \lambda^0\rangle = 0\}$. Here we have of course

$$p_{m,1}(\theta) = \langle\text{Hess } p_m(\lambda^0)\theta, \theta\rangle/2.$$

To check (8.4.5), let us observe at first that $(\partial/\partial\theta_n)p_{m,1}(\theta) = 0$ for some point $\theta \in W$ would mean that the projection of $\theta^0 = (0, ..., 0, 1, 0) \in R^{n+1}$ to the orthogonal of λ^0 is tangent to W. In view of lemma 6.6.2 it follows (and it is here where we need $n \geq 3$) that

$$(\partial/\partial\theta_n)p_{m,1}(\theta) \neq 0 \text{ for all } \theta \in W, \text{ provided } p_{m,1}(\theta^0) > 0. \tag{8.4.6}$$

An interesting case for which $p_{m,1}(\theta^0) > 0$ holds is when θ^0 is proportional to $(\xi^0, 0)$. Indeed, we have seen in section 6.6 that it follows from (8.4.4) that

$$\langle\nabla_\xi, \xi^0\rangle^2 p_m(\lambda^0) > 0,$$

which gives $p_{m,1}(\theta^0) > 0$ immediately. It follows from this that (8.4.5) will hold if ξ^0 points in the direction of the normal to $x_n = 0$. It will then also hold if we require that this normal is sufficiently close to ξ^0. Actually what we want to say by this, is the following: for fixed $\xi^0 \in S$ which satisfies the condition above, we have

$$\langle\eta, \text{ grad}_\xi\, p_{m,1}(\theta)\rangle \neq 0 \text{ for any } \theta \in W, \tag{8.4.7}$$

provided $\eta \in \dot{R}^n$ is close enough to ξ^0. To come to our initial situation, we can then change coordinates orthogonally in R_x^n to make sure that η is proportional to $(0, ..., 0, 1)$. We continue our discussion of the case when μ is a double root, assuming that $p_{m,1}(\theta^0) > 0$, $n \geq 3$. In analogy to the case of bicharacteristics, the cones G_μ^{+0},

$$G_\mu^{+0} = \{(x,t); \langle x,\xi\rangle + t\tau > 0, \forall(\xi,\tau) \in G_\mu^+, \langle x,\xi^0\rangle + t = 0\}$$

can then be of two distinguished types: either x_n increases with increasing t, or x_n decreases with increasing t. Indeed, since G_μ^{+0} is the convex hull of vectors of form

$$\{y = s \text{ grad } p_{m,1}(\theta); \theta \in W, \theta_n > 0, s \neq 0\},$$

what we have to show is that

$$(\partial/\partial\theta_n)p_{m,1}(\theta)(\partial/\partial\tau)p_{m,1}(\theta) \tag{8.4.8}$$

does not change sign when $n \geq 3$ and θ varies in $W^+ = \{\theta \in W, \theta_n > 0\}$. In fact, since W^+ is connected when $n \geq 3$, we need only observe that the expression from (8.4.8) never vanishes on W^+: that $(\partial/\partial\theta_n)p_{m,1}(\theta)$ does not vanish is what we have discussed just before, and the same thing for $(\partial/\partial\tau)p_{m,1}(\theta)$ had been checked in section 6.6. Actually, more can be said about the sign of x_n. Indeed, since this sign does not change when (x,t) varies in G_μ^{+0}, we may compute it by just checking it on some particular vector there. The cones here are now second order, so they have a symmetry axis, which is proportional to the symmetry axis of G_μ^+, if we identify $R_{x,t}^{n+1}$ and $R_{\xi,\tau}^{n+1}$. By definition, we had however that $\xi_n > 0$ for vectors in G_μ^+, so we can conclude that $x_n > 0$ for $(x,t) \in G_\mu^{+0}$.

We are led to the following definition:

Definition 8.4.2. G_μ^{+0} *is called "incoming" with respect to $x_n > 0$ if $t < 0$ for $(x,t) \in G_\mu^{+0}$. It is called "outgoing" in the opposite case.*

Remark 8.4.3. *A similar definition makes sense with respect to $x_n < 0$. Rather than working with cones in the halfspace $\xi_n > 0$, $x_n > 0$, we have then to work with cones in $\xi_n < 0$, $x_n < 0$. Thus, e.g., we shall now consider the cones*

$$G_\mu^- = \{(\xi,\tau); \langle\xi,\xi^0\rangle + \tau > 0, \xi_n < 0, p_{m,1}(\xi,\tau) > 0\},$$

and define G_μ^{-0} to be the polar cone of G_μ^-.

3. Starting from some point $\xi^0 \in S$, we now return to the situation from theorem 8.3.1 where, together with the root ξ_n^0 of $p_m(\xi'^0,\mu,1) = 0$, we have to consider all other roots of that equation. (We assume again that all these roots are real.) A case of particular interest is when the normal to $x_n = 0$ points in the direction of ξ^0, i.e. when $\xi'^0 = 0$. To formulate conditions in their most natural form, it is however useful to leave coordinates as they were initially, i.e., work with halfspaces of form $\langle x,\xi^0\rangle > 0$ rather than with the halfspace $x_n > 0$. In particular, we define the cone G_μ^+ by the condition

$$G_\mu^+ = \{(v,w); \langle v,\xi^0\rangle + w > 0, \langle v,\xi^0\rangle > 0, p_{m,1}(v,w) > 0\},$$

and we shall say that it is associated with $(p_m, \xi^0, \langle x,\xi^0\rangle > 0)$, for short. If now by chance, $-\xi^0$ lies also in S, then together with G_μ^+ we have also to compute the cone $G_{-\xi^0}^+$ for

$-\xi^0$ and the same halfspace, i.e. the cone associated with $(p_m, -\xi^0, \langle x, \xi^0\rangle > 0)$. Indeed, had we used coordinates in which ξ^0 had been of form $(0, ..., 0, \xi_n^0)$, then the assumption $p_m(-\xi^0, 1) = 0$ would say precisely that $-\xi_n^0$ is also a root of $p_m(\xi^{0\prime}, \mu, 1) = 0$, where $\xi^{0\prime} = 0$. This explains our interest in the following result:

Proposition 8.4.4. *Assume that* f, $f(\xi) = p_m(\xi, 1)$, *satisfies* $f(\xi) = f(-\xi)$ *and consider* ξ^0 *such that* $f(\xi^0) = 0$, $grad f(\xi^0) = 0$, $Hess_{\xi\xi} f(\xi^0) \neq 0$. *Denote by* $f_{\xi^0}^T$, *respectively by* $f_{-\xi^0}^T$ *the localization polynomials of* f *at* ξ^0, *respectively at* $-\xi^0$. *Similarly, let* $p_{m,1,\xi^0}$ *and* $p_{m,1,-\xi^0}$ *be the localization of* p_m *at* $(\xi^0, 1)$, *respectively* $(-\xi^0, 1)$. *Finally, let* $G_{\xi^0}^+$, $G_{-\xi^0}^+$, $G_{\xi^0}^{+0}$, $G_{-\xi^0}^{+0}$, *be the cones associated with* $(p_m, \xi^0, \langle x, \xi^0\rangle > 0)$, *respectively* $(p_m, -\xi^0, \langle x, \xi^0\rangle > 0)$. *Then we have*

a) $f_{-\xi^0}^T(\xi) = f_{\xi^0}^T(-\xi)$,

b) $p_{m,1,-\xi^0}(v, w) = p_{m,1,\xi^0}(-v, w)$,

c) $G_{-\xi^0}^+ = \{(v, w); \ (v, -w) \in G_{\xi^0}^+\}$,

d) $G_{-\xi^0}^{+0} = \{(x, t); \ (x, -t) \in G_{\xi^0}^{+0}\}$.

It follows that of the two cones $G_{\xi^0}^{+0}$, $G_{-\xi^0}^{+0}$, by necessity one will have to be outgoing and the other incoming.

Remark 8.4.5. *The main result is here d). The proof of proposition 8.4.4 will be by straightforward verification and passage through a) is of course not necessary, in that we could directly use that the assumption implies* $p_m(\xi, \tau) = p_m(-\xi, \tau)$. *If we mention a) explicitly, it is for the fact that we want to refer as often as possible to conditions expressed explicitly in terms of the slowness surface.*

Proof of proposition 8.4.4. $f_{\xi^0}^T$ is uniquely defined by the relation

$$f(\xi) = f_{\xi^0}^T(\xi) + O(|\xi - \xi^0|^3) \text{ for } |\xi - \xi^0| \text{ small.}$$

It follows that

$$f(\xi) = f(-\xi) = f_{\xi^0}^T(-\xi) + O(|-\xi - \xi^0|^3) = f_{\xi^0}^T(-\xi) + O(|\xi + \xi^0|^3)$$

if $|\xi - \xi^0|$ is small. The last relation shows that $f_{\xi^0}^T(-\xi)$ can play the role of $f_{-\xi^0}^T(\xi)$, so we obtain a). b) is an immediate consequence. (Also cf. remark 8.4.5.) In fact,

$$p_{m,1,-\xi^0}(v, w) = f_{-\xi^0}^T(v + w\xi^0) = f_{\xi^0}^T(-v - w\xi^0) = p_{m,1,\xi^0}(-v, w).$$

We now read off c) from the definitions of $G_{\xi^0}^+$, $G_{-\xi^0}^+$:

$$G_{\xi^0}^+ = \{(v,w); \langle v, \xi^0 \rangle + w > 0, \langle v, \xi^0 \rangle > 0, p_{m,1,\xi^0}(v,w) \geq 0\},$$

$$G_{-\xi^0}^+ = \{(v,w); \langle v, -\xi^0 \rangle + w > 0, \langle v, \xi^0 \rangle > 0, p_{m,1,-\xi^0}(v,w) \geq 0\}.$$

d) follows from c) taking polars.

We have discussed in detail the situation when the normal to the interface points into the direction of some singular point ξ^0. We want to discuss also briefly what happens when $f(\xi) = f(-\xi)$ and when this normal is only close to $\pm \xi^0/|\xi^0|$. If we are close enough, we have seen that $G_{\xi^0}^{+0}$ will be either incoming or outgoing. However, if we choose orthogonal coordinates so that $H = \{x, x_n = 0\}$, then we will only have that $|\xi^{0\prime}|$ is small compared to ξ_n^0 rather than that $\xi^{0\prime} = 0$. In particular, none of the roots $\nu \neq \xi_n^0$ of the equation $p_{m,1}(\xi^{0\prime}, \nu, 1) = 0$ will correspond to the point $-\xi^0$ in the old coordinates. Two of these roots ν will however be so that $(\xi^{0\prime}, \nu, 1)$ is close to $(-\xi^0, 1)$ and we shall assume here that grad $p_m(\xi^{0\prime}, \nu, 1) \neq 0$ for them, so in particular the two roots will be distinct. Let us denote then by L_ν the bicharacteristics of form

$$L_\nu = \{(s \text{ grad } p_m(\xi^{0\prime}, \nu, 1), (\xi^{0\prime}, \nu, 1)); s \in R\}.$$

The interesting thing is now that these two bicharacteristics will have the same type, incoming or outgoing, with $G_{-\xi^0}^{+0}$, if $(\xi^{0\prime}, \nu, 1)$ is close enough to $(-\xi^0, 1)$. Indeed, the type of $G_{-\xi^0}^{+0}$ is the same with that of any of the relative bicharacteristics

$$\{(sp_{m,1,-\xi^0}(v,w), (-\xi^0, 1), (v,w)); s \in R, p_{m,1,-\xi^0}(v,w) = 0, \langle v, -\xi^0 \rangle + w = 0,$$

$$\langle v, \xi^0 \rangle > 0\},$$

so we can use proposition 1.1.12. (The type of a relative bicharacteristic is defined in analogy with the type of a bicharacteristic.)

4. We conclude the section with a discussion of a relation between microlocal hyperbolicity and the transversality of the cones G_+^{+0} to $\{(x,t); x_n = 0\}$. That there should be such a relation is easy to believe if one thinks for a moment at the related but simpler case of simple solutions of $p_m(\lambda^0) = 0$, $\lambda^0 \in R^{n+1}$, $(\xi^{0\prime}, \tau^0) \neq 0$. Let us assume more precisely that p_m is real-valued and that

i) $p_m(\lambda^0) = 0$,

ii) $(\partial/\partial \xi_n) p_m(\lambda^0) \neq 0$.

The solutions ξ_n of $\xi_n \to p_m(\lambda) = 0$ in a complex conic neighborhood of λ^0 are then of form $\xi_n = \xi_n(\xi', \tau)$ for some real analytic function $\xi_n(\xi', \tau)$ which is real-valued for real arguments. These solutions are then in particular real if (ξ', τ) is in a real conic neighborhood of $(\xi^{0\prime}, \tau^0)$.

We shall now obtain basically the same result also in the case of double roots. Choosing once more $\lambda^0 \in R^{n+1}$, $(\xi^{0\prime}, \tau^0) \neq 0$, we assume that the following conditions are satisfied for some real valued p_m:

i)$'$ $p_m(\lambda^0) = 0$,

ii)$'$ $\mathrm{grad}_\lambda p_m(\lambda^0) = 0$,

iii)$'$ rank $H_{\lambda\lambda} p_m(\lambda^0) = n$,

iv)$'$ $G_{\lambda^0}^{+0}$ is transversal to $x_n = 0$,

v)$'$ p_m is hyperbolic in the direction t.

A first consequence of the assumptions is that

$$(\partial/\partial\xi_n)^2 p_m(\lambda^0) \neq 0.$$

Indeed $(\partial/\partial\xi_n)^2 p_m(\lambda^0)$ is the coefficient of v_n in the localization polynomial $v \to p_{m,1}(v)$ of p_m at λ^0 and in view of iv)$'$, $(0, ..., 0, 1, 0)$ is not characteristic for $p_{m,1}$. Also recall that as a consequence of v)$'$, $p_{m,1}$ is hyperbolic in the variable v_{n+1}. On the other hand, $p_{m,1}$ is a quadratic form which is nondegenerate in the variables orthogonal to λ^0. From iv)$'$ it is then clear that $p_{m,1}$ is also hyperbolic in the v_n variable. Let us denote by D the discriminant of $p_{m,1}$ in this variable. It follows that $D \geq 0$ as a quadratic form. Clearly $D(\xi^{0\prime}, \tau^0) = 0$. With the sign condition this gives that $\mathrm{grad}\, D(\xi^{0\prime}, \tau^0) = 0$, so D only depends on the variables orthogonal to $(\xi^{0\prime}, \tau^0)$ effectively. Moreover, in these variables D is nondegenerate since we had rank $H_{\lambda\lambda} p_m(\lambda^0) = n$. We can now prove the following result:

Lemma 8.4.6. *We can find a real conic neighborhood* Γ' *of* $(\xi^{0\prime}, \tau^0)$ *and a complex conic neighborhood* Γ *of* λ^0 *so that any root* $\nu \in C$ *of* $p_m(\xi', \nu, \tau) = 0$ *is real if* $(\xi', \nu, \tau) \in \Gamma$, $(\xi', \tau) \in \Gamma'$.

Proof. Since p_m is homogeneous, it suffices to prove the assertion for λ in a small, not necessarily conic, neighborhood of λ^0. We apply the Weierstrass preparation theorem in a complex neighborhood of λ^0 to write that

$$p_m(\lambda) = ((\xi_n - \xi_n^0)^2 + a(\xi', \tau)(\xi_n - \xi_n^0) + b(\xi', \tau))g(\lambda) \tag{8.4.9}$$

where

a) a, b and g are real valued real-analytic functions,

b) $a(\xi^{0\prime}, \tau^0) = 0$, $b(\xi^{0\prime}, \tau^0) = 0$, $\operatorname{grad} b(\xi^{0\prime}, \tau^0) = 0$,

c) $g(\lambda^0) \neq 0$.

Let also $\Delta = a^2 - 4b$ be the discriminant of the polynomial $t \to t^2 + a(\xi', \tau)t + b(\xi', \tau)$. Although in (8.4.9) we have not kept track of homogeneities, it is immediate that Δ is homogeneous of degree two. Indeed, if $\xi_n^i(\xi', \tau)$, $i = 1, 2$, are the two solutions of $p_m(\xi', \xi_n, \tau) = 0$ with (ξ', ξ_n^i, τ) in a conic neighborhood of λ^0, it follows that $\Delta = (\xi_n^1(\xi', \tau) - \xi_n^2(\xi', \tau))^2$. We will have here real solutions ξ_n^i for real (ξ', τ) precisely if $\Delta(\xi', \tau) \geq 0$. Next we observe that $\Delta(\xi^{0\prime}, \tau^0) = 0$ and that $\operatorname{grad} \Delta(\xi^{0\prime}, \tau^0) = 0$. (We use b).) Since Δ is homogeneous, it suffices to show that the Hessian $H\Delta$ of Δ at $(\xi^{0\prime}, \tau^0)$ is a positive quadratic form in the variables orthogonal to $(\xi^{0\prime}, \tau^0)$. The latter is the case since $H\Delta$ is proportional to D and has the same sign.

We have now proved lemma 8.4.6. Combining the lemma and the discussion before it, we also obtain

Lemma 8.4.7. *Assume that for some given* $(\xi^{0\prime}, \tau^0)$ *the roots* ν *of* $p_m(\xi^{0\prime}, \nu, \tau^0) = 0$ *are all real and at most double. When* $\operatorname{grad} p_m(\lambda^0) \neq 0$ *assume that in fact already* $(\partial/\partial\xi_n)p_m(\xi^{0\prime}, \nu, \tau^0) \neq 0$ *and when* $\operatorname{grad} p_m(\xi^{0\prime}, \nu, \tau^0) = 0$, *then assume that the conditions iii)', iv)',v)' hold. Then* p_m *is microlocally hyperbolic with respect to* x_n *at* $(\xi^{0\prime}, \tau^0)$.

8.5 Conical refraction in transmission problems

1. In this section we show how the results from the preceding sections can be put together to lead to results on conical refraction for constant coefficient transmission problems when the interface is planar. Actually, all preparations have been made to consider real-analytic interfaces as well, so the reason why we mainly consider the planar case is notational. We shall assume then that $H \subset R_x^n$ is of form $\{x; x_n = 0\}$ and write

$$Y^0 = \{(x,t) \in R^{n+1}; x_n = 0,\ t_1 < t < t_2\},$$

$$Y^+ = \{(x,t) \in R^{n+1}; x_n > 0,\ t_1 < t < t_2\},$$

$$Y^- = \{(x,t) \in R^{n+1}; x_n < 0,\ t_1 < t < t_2\},$$

$$Y = \{(x,t) \in R^{n+1};\ t_1 < t < t_2\},$$

for some $t_1 < t^0 = 0 < t_2$. Further we consider two constant coefficient linear partial differential operators p^+ and p^- of order m on R^{n+1} which are hyperbolic with respect to $t = 0$ and for which Y^0 is noncharacteristic. We shall then consider a distribution $u \in D'(Y \setminus Y^0)$ such that the restrictions $u^\pm$ of u to $Y^\pm$ are extendible to distributions on Y and satisfy the equations

$$p^+ u^+ = 0 \text{ on } Y^+, p^- u^- = 0 \text{ on } Y^-. \tag{8.5.1}$$

Further, we denote by u_j^+ and u_j^- the traces to $x_n = 0$ of $(\partial/\partial x_n)^j u^+$, respectively $(\partial/\partial x_n)^j u^-$. Finally, we write

$$U^+ = (u_j^+)_{j=0}^{m-1}, U^- = (u_j^-)_{j=0}^{m-1}.$$

The transmission condition is

$$U^- = BU^+, \tag{8.5.2}$$

where B is a $(m \times m)$ matrix of analytic pseudodifferential operators, $B = (b_{ij})$, $i, j = 0, ..., m - 1$. Sometimes it is necessary to ask for specific conditions on B in order to obtain reasonable results and we shall discuss one type of such conditions later on. To "determine" the transmission problem (we do not make precise in what sense) we shall assume that we dispose of "enough information" for u^+, u^- when $t < 0$. Explicitly we may e.g. assume that u^+ is real analytic for $t_1 < t < 0$ and that the WF_A-singularities of u^- are known (and of suitable form) for $t_1 < t < 0$.

In order to say something about the singularities of u^+, u^- for the future, we shall now make a number of assumptions concerning p^+ and p^-. We shall also discuss then what kind of matrices B seem natural from the point of view of geometric optics. All our conditions refer to some fixed $\xi'^0 \in \dot{R}^{n-1}$.

I) Conditions on p^-. In order to obtain results which can be nicely formulated, we shall basically put conditions so that the analytic wave front set of u^- propagates along lines in $Y^- \times R^{n+1}$ which hit the interface transversally. Actually, we want (for the sake of formulation of results) make sure that conical refraction occurs only in association with p^+. If p^- is then e.g. of the type of the standard wave equation, much more than what we use here is known in the literature. In particular, we could formulate results for cases of nonflat tangential interfaces, provided we would fall into a case which we can treat for p^+. We shall not do so, but assume, for the flat case above, the following conditions:

i) the coefficient of ξ_n^m in p^- is one,

ii) there is a conic neighborhood $\Gamma \subset R_{\xi'}^{n-1} \times R_\tau$ of $(\xi'^0, 1)$ and a constant $c > 0$ so that for any fixed $(\xi', \tau) \in \Gamma$ it follows from $p^-(\xi', \nu, \tau) = 0$ that $|Im\, \nu| \leq c$.

iii) for any fixed $\mu^0 \in R$ which satisfies $p_m^-(\xi'^0, \mu^0, 1) = 0$ we can find a conic neighborhood Γ of $(\xi'^0, 1)$, a natural number k and a real analytic function $\lambda : \Gamma \to R$, which is positively homogeneous of degree one in (ξ', τ), such that

$$p_m^-(\xi, \tau) = (\xi_n - \lambda(\xi', \tau))^k q^-(\xi, \tau), \lambda(\xi'^0, 1) = \mu^0, q^-(\xi'^0, \mu^0, 1) \neq 0.$$

Moreover, we assume that $(\partial/\partial\tau)\lambda(\xi'^0, 1) \neq 0$.

iv) at least one of the bicharacteristics L_μ^- associated with the factor $\xi_n - \lambda(\xi', \tau)$ is incoming.

Note that the bicharacteristic $L_{\mu^0}^-$ of the factor $\xi_n - \lambda(\xi', \tau)$ which passes through $(0, (\xi'^0, \mu^0, 1))$ is

$$\{((s, -s\, \text{grad}_{\xi'}\lambda(\xi'^0, 1), -s(\partial/\partial\tau)\lambda(\xi'^0, 1)), (\xi'^0, \mu^0, 1)); \; s \in R\}$$

and it hits $x_n = 0$ transversally. Moreover, our assumption $(\partial/\partial\tau)\lambda(\xi'^0, 1) \neq 0$ shows that there is a nontrivial time evolution on $L_{\mu^0}^-$. We shall denote by J^- the set of those roots of $p_m^-(\xi'^0, \mu, 1) = 0$, counted with multiplicities, for which the bicharacteristic L_μ^- constructed as above is incoming with respect to $x_n < 0$, i.e. for which x_n increases when time increases on L_μ^-. We can then conclude from section 8.2 (cf. here remark 8.2.3) that

$$\{((s\, \text{grad}_{(\xi', \tau)}(\xi_n - \lambda(\xi', \tau))_{|(\xi'^0, \mu, 1)}, (\xi'^0, \mu, 1)); s \neq 0\} \cap WF_A u^-_{|Y^-} = \emptyset,$$

for all $\mu \in J^-$ implies

$$(0, (\xi'^0, 1)) \notin WF_A X^- U^-,$$

where X^- is associated with $p = p^-$ and $J = J^-$ as in section 8.2.

II. Conditions for p^+. The conditions for p^+ are closely related to the conditions a), b), c), d), e) from section 8.2. We shall in fact assume the following:

a) the coefficient of ξ_n^m in p^+ is one,

b) there is a conic neighborhood $\Gamma \subset R_{\xi'}^{n-1} \times R$ of $(\xi'^0, 1)$ and a constant $c > 0$ so that for any fixed $(\xi', \tau) \in \Gamma$ it follows from $p^+(\xi', \nu, \tau) = 0$ that $|Im\, \nu| \leq c$,

c) the roots μ of $p_m^+(\xi'^0, \mu) = 0$ are at most double,

d) if μ is a simple root of $p_m^+(\xi'^0, \mu, 1) = 0$, then $(\partial/\partial\xi_n)p_m^+(\xi'^0, \mu, 1) \neq 0$,

e) if μ is a double root of $p_m^+(\xi'^0, \mu, 1) = 0$, then the signature of $p_{m,1}^+$ or of $-p_{m,1}^+$, as a quadratic form on the space of variables orthogonal to $(\xi'^0, \mu, 1)$, is $+ - \cdots -$,

f) if μ is a double root of $p_m^+(\xi'^0, \mu, 1) = 0$, and if $\sigma \neq 0$ is a solution of $p_{m,1}^+(\sigma) = 0$ which is orthogonal to $(\xi'^0, \mu, 1)$, then $(\partial/\partial\xi_n)p_{m,1}^+(\sigma) \neq 0$.

(Here $p_{m,1}^+$ is the localization polynomial of p_m^+ at $(\xi'^0, \mu, 1)$. It thus in particular depends on μ.)

Also denote by J^+ the set of those roots μ of $p_m^+(\xi'^0, \mu, 1) = 0$ for which the associated bicharacteristics or propagation cones, are incoming. In analogy with the above for p^-, we obtain then from section 8.2 a well defined explicitly computable matrix X^+ of analytic pseudodifferential operators associated with $p = p^+$ and $J = J^+$ for which we have the following implication:

assume that u^+ is an extendable solution of $p^+u^+ = 0$ in Y^+ and assume that for all simple roots $\mu \in J^+$ of $p_m^+(\xi'^0, \mu, 1) = 0$

$$L_\mu^+ \cap WF_A u^+ \cap \{(x, t, \xi, \tau); x_n > 0\} = \emptyset, \qquad (8.5.3)$$

where L_μ^+ is the bicharacteristic of p_m^+ which passes through $(0, (\xi'^0, \mu, 1))$,

respectively that for every double root $\mu \in J^+$ of $p_m^+(\xi'^0, \mu, 1) = 0$

$$(G_\mu^{+0} \times (\xi'^0, \mu, 1)) \cap WF_A u^+ = \emptyset \qquad (8.5.4)$$

where G_μ^{+0} is the propagation cone associated with μ and p^+.

Then it folows that

$$(0, (\xi'^0, 1)) \notin WF_A X^+ U^+.$$

It is now natural to assume that the three conditions

$$(0, (\xi'^0, 1)) \notin WF_A X^- U^-, (0, (\xi'^0, 1)) \notin WF_A X^+ U^+ \text{ and } U^- = BU^+$$

imply

$$(0, (\xi'^0, 1)) \notin WF_A U^-, (0, (\xi'^0, 1)) \notin WF_A U^+.$$

Using obvious notations it will then suffice to assume that the system

$$\begin{pmatrix} I, & -B \\ X^+, & 0 \\ 0, & X^- \end{pmatrix}$$

is elliptic at $(0, (\xi'^0, 1))$ on vectors of type $\begin{pmatrix} U^+ \\ U^- \end{pmatrix}$.

Note that this is possible only if the system is at least determined. The meaning of these conditions is of course that if u^+ and u^- are microlocally smooth in the past for frequencies of form $(\xi'^0, \mu, 1)$, then they are also microlocally smooth in the future. We do not pursue this here any further but mention the following result in which only the microlocality of B is needed.

Proposition 8.5.1. *(On conical refraction) We convene in this statement that "incoming" and "outgoing" is with respect to $x_n < 0$ when the bicharacteristics or propagation cone is associated with p^- and with respect to $x_n > 0$ for such entities, when they are computed for p^+. Assume that p^- and p^+ satisfy the conditions above for ξ'^0 and let u^+, u^- be an extendible solution of the transmission problem (8.5.1), (8.5.2). Assume further that (8.5.3) and (8.5.4) hold for all incoming bicharacteristics, respectively incoming propagation cones, for p^+. If then there is some incoming bicharacteristic L_μ^- of p^- (of the type from I) such that*

$$L_\mu^- \cap WF_A u^- \neq \emptyset, \tag{8.5.5}$$

then we can find some outgoing bicharacteristic L_ν^+ of p_m^+ such that

$$L_\nu^+ \cap WF_A u^+ \neq \emptyset,$$

or some outgoing propagation cone G_μ^{+0} of p_m^+ such that

$$(G_\mu^{+0} \times (\xi'^0, \mu, 1)) \cap WF_A u^+ \neq \emptyset.$$

Proof of proposition 8.5.1 The proof is by now very simple. In fact, we can argue by contradiction. If the conclusion were not true, it would follow from section 8.2 that $(0, (\xi'^0, 1)) \notin WF_A U^+$. From the transmission condition we could conclude that $(0, (\xi'^0, 1)) \notin WF_A U^-$. In view of theorem 7.1.1 we would obtain

$$((x, t), (\xi, \tau)) \notin WF_A u^-,$$

provided $x_n < 0$, $|x| + |t| < \varepsilon$, $|\xi' - \xi'^0| + |\tau - 1| < \varepsilon$. This would contradict (8.5.5).

8.6 Conical reflection at the boundary

1. We describe here a result on transversal reflection of singularities which is closely related to the results on conical refraction from the preceding section. The main thing is

that also in the case of reflection we can have situations when an incoming bicharacteristic brakes up into a full propagation cone when reflected back into the interior. This is quite obvious after a moments thought, so the only point is here to give some explicit conditions when such a phenomenon occurs. We also mention that "conical reflection" has been observed, in a mathematically more delicate context, in Ivrii [3]. (There the operator had characteristics of constant multiplicity, but the bicharacteristics were tangent to the boundary. In the present section the main thing is that reflection will be transversal, although the operator will have characteristics of variable multiplicity.)

2. The assumptions here are similar to those for the operator p^+ in the preceding section. With the notation Y^+ from there we assume in fact that p is a constant coefficient linear partial differential operator for which $x_n = 0$ is noncharacteristic and which satisfies for fixed ξ'^0 the conditions a), b), c), d), e), f) from the preceding section. We denote by L_μ, respectively G_μ^{+0}, the bicharacteristics and propagation cones of p_m, associated with simple, respectively double, roots μ of $p_m(\xi'^0, \mu, 1) = 0$. Under the present conditions bicharacteristics and propagation cones are either incoming or outgoing with respect to $x_n > 0$. We shall accordingly split the set I of distinct roots μ of $p_m(\xi'^0, \mu, 1) = 0$ into four subsets:

$$I = I_{L,i} \cup I_{L,o} \cup I_{G,i} \cup I_{G,o},$$

where $I_{L,i}$ contains the simple roots for which the associated bicharacteristic L_μ is incoming, $I_{L,o}$ contains the remaining simple roots, $I_{G,i}$ contains the double roots for which the associated propagation cone G_μ^{+0} is incoming and $I_{G,o}$ contains the remaining double roots.

We have then the following result:

Proposition 8.6.1. *Let p and ξ'^0 be as above and assume (for simplicity) that $I_{G,i}$ is void. Denote by k the number of elements in $I_{L,i}$. Consider next an extendible solution u of $p(D)u = 0$ on Y^+ such that*

$$(0, (\xi'^0, 1)) \notin WF_A(\partial/\partial x_n))^i u_{|x_n=0_+}, i = 0, ..., k - 1, \tag{8.6.1}$$

$$L_\mu \cap WF_A u \cap \{x_n > 0\} = \emptyset \text{ if } \mu \in I_{L,o}, \tag{8.6.2}$$

$$(G_\mu^{+0} \times (\xi'^0, \mu, 1)) \cap WF_A u = \emptyset \text{ if } \mu \in I_{G,o}. \tag{8.6.3}$$

Then it follows that

$$L_\mu \cap WF_A u \cap \{x_n > 0\} = \emptyset \text{ for all } \mu \in I_{L,i}. \tag{8.6.4}$$

A similar result holds if instead of considering solutions defined on Y^+ we consider local solutions defined on sets of form $\{(x,t); \rho(x) > 0\} \cap U\}$ where U is a neighborhood of the origin and where ρ is a real analytic function such that $\mathrm{grad}\,\rho(0) = (0,...,0,1)$ and such that $\rho(x) > 0$ implies $x_n > 0$.

In particular, if $I_{L,o}$ is void but $I_{L,i}$ is not, then singularities coming in along the L_μ, $\mu \in I_{L,i}$ will be reflected along at least one of the propagation cones. If there is only one such propagation cone, we will have a case of clear cut conical reflection.

Remark 8.6.2. *a) Instead of the Dirichlet conditions from (8.6.1) one may consider more general boundary condition, provided some appropriate Lopatinski conditions are satisfied. (Cf. Agmon-Douglis-Nirenberg [1].)*

b) With the preparations above, one may formulate of course also similar results for cases where there is no distinguished time variable, or when the boundary is of type $\rho(x,t) > 0$, or, finally, when $\rho(x) > 0$ does not imply $x_n > 0$.

3. **Proof of proposition 8.6.1.(Sketch)** Only the flat case needs to be considered. The proof consists of two steps. In the first step we show that if the assumptions are satisfied, then

$$(0, \xi'^0) \notin WF_A(\partial/\partial x_n)^i u_{|x_n=0_+}, \forall i \in \{0,...,m-1\}. \tag{8.6.5}$$

That this implies (8.6.4) is then seen as in the last part of the proof of proposition 8.5.1.

To prove (8.6.5), we denote by $J = I_{L,o} \cup I_{G,o}$ and by a_{ij} the symbols associated with p and J in section 8.2. It follows from there that

$$(0, \xi'^0) \notin WF_A[(\partial/\partial x_n)^i u_{|x_n=0_+} - \sum_{j=0}^{m-k-1} a_{ij}(D')(\partial/\partial x_n)^j u_{|x_n=0_+}] = \emptyset, \forall i.$$

This gives (8.6.5) in view of the assumption.

8.7 The case of crystal optics

1. We can finally turn our attention to the transmission problem in crystal optics. As in the case of propagation of singularities in free space we have to deal with a system rather than with a scalar equation, but here it is important to consider the full overdetermined system. The problem, as we consider it, is of local nature, but for notational simplicity we shall assume that we are working in all physical space R^3. More precisely, we assume that X^+, X^- and Z are given in R^3 so that X^+, X^- are open domains, that Z is a

smooth analytic interface, that $R^3 = X^+ \cup Z \cup X^-$ (disjoint union) and that X^+ and X^- are separated precisely by Z. On Z we can choose a smooth normal, which is supposed to point, to make a choice, into X^+. We assume that X^+ is filled with some homogeneous optically biaxial but magnetically isotropic medium (examples of such media are biaxial crystals), whereas X^- is filled (for simplicity) with some homogeneous isotropic medium (examples are air, vacuum, but also crystals from the cubic class.) For the components $E^\pm$ and $H^\pm$ in $X^\pm \times R$ of the electromagnetic field (E, H) (E is the electric and H the magnetic component) we have then, with an obvious notation for the dielectric tensor ε and the magnetic permeability tensor μ the equations, assuming that the time varies in the interval (t_1, t_2):

$$(\partial/\partial t)(\varepsilon^\pm E^\pm) - \mathrm{curl} H^\pm = 0 \text{ in } X^\pm \times (t_1, t_2), \tag{8.7.1}$$

$$(\partial/\partial t)(\mu^\pm H^\pm) + \mathrm{curl} E^\pm = 0 \text{ in } X^\pm \times (t_1, t_2), \tag{8.7.2}$$

$$\mathrm{div}(\varepsilon^\pm E^\pm) = 0 \text{ and } \mathrm{div}\,(\mu^\pm H^\pm) = 0 \text{ in } X^\pm \times (t_1, t_2). \tag{8.7.3}$$

Along the space-time interface $S = Z \times (t_1, t_2)$, (E^+, H^+) and (E^-, H^-) are related by the transmission conditions. Explicitly these conditions are

$$[n \times E] = 0, [n \times H] = 0, \tag{8.7.4}$$

$$[\langle n, \varepsilon E\rangle] = 0, [\langle n, \mu H\rangle] = 0, \tag{8.7.5}$$

where $[A]$ is the "jump" of the quantity A on S, i.e.,

$$[A] = A_{|S}^+ - A_{|S}^-.$$

(Here $A^\pm$ is the restriction of A to $X^\pm \times (t_1, t_2)$. The fact that the jumps of the quantities $[n \times E]$, etc. make sense will be established later on.)

The transmission conditions (8.7.4), (8.7.5) are established e.g. in Born-Wolf [1]. For the convenience of the reader we explain here briefly how they are deduced from Maxwell's equations, putting more emphasis on distribution-theoretical aspects than in Born-Wolf [1]. For simplicity we shall assume that $t_1 = -\infty$ and that $t_2 = +\infty$. Starting point in the argument is that we may regard our two-media configuration as a unique medium with discontinuous (E, H, ε, μ). Maxwell's equations will then still hold, but in weak form. Explicitly we shall in fact have

$$\int \{\langle \varepsilon E, (\partial/\partial t)\varphi\rangle - \langle H, \mathrm{curl}\,\varphi\rangle\} dx dt = 0 \tag{8.7.6}$$

$$\int \{\langle \mu H, (\partial/\partial t)\psi\rangle + \langle E, \mathrm{curl}\,\psi\rangle\} dx dt = 0, \tag{8.7.7}$$

$$\int_{R^4} \langle \varepsilon E, \text{ grad } \rho \rangle dx dt = 0, \int_{R^4} \langle \mu H, \text{ grad } \lambda \rangle dx dt = 0, \tag{8.7.8}$$

for any $\varphi \in [C_o^\infty(R^4)]^3$, $\psi \in [C_o^\infty(R^4)]^3$, $\rho \in C_o^\infty(R^4)$, and $\lambda \in C_o^\infty(R^4)$.

To prove (8.7.4) we start from the calculus relation

$$\langle E^\pm, \text{ curl } \psi \rangle = \text{ div } (E^\pm \times \psi) + \langle \text{ curl } E^\pm, \psi \rangle \text{ on } X^\pm \times R.$$

Combining this with (8.7.2) and (8.7.7) we obtain (when $E^\pm$, $H^\pm$ are smooth on $X^\pm \times R$) that

$$\int_{R^4} \text{ div } (E^\pm \times \psi) dx dt = 0, \forall \psi \in [C_o^\infty(R^4)]^3.$$

From the Gauss theorem we can therefore conclude that

$$\int_S \langle n, E^+ \times \psi \rangle_{|S} - \langle n, E^- \times \psi \rangle_{|S} d\sigma = 0, \ \forall \psi \in [C_o^\infty(R^4)]^3,$$

where $d\sigma$ is the surface element on S. The first relation in (8.7.4) is a consequence, if we also observe that

$$\langle n, (E^+ - E^-) \times \psi \rangle_{|S} = \langle n \times (E^+ - E^-), \psi \rangle_{|S},$$

and the second relation in (8.7.4) is proved in a similar way. As for the relations (8.7.5), we start from the remark that

$$\text{div } (\rho \varepsilon^\pm E^\pm) = \rho \text{ div } (\varepsilon^\pm E^\pm) + \langle \varepsilon^\pm E^\pm, \text{ grad } \rho \rangle \text{ on } X^\pm \times R.$$

It follows from (8.7.3) and (8.7.8), using once more the Gauss theorem, that

$$\int_S \langle n, \rho \varepsilon^+ E^+ \rangle_{|S} - \langle n, \rho \varepsilon^- E^- \rangle_{|S} d\sigma = 0, \ \forall \rho \in C_o^\infty(R^4).$$

This gives the first relation in (8.7.5). The second relation there is of course obtained in a similar way.

We have now justified the relations (8.7.4) and (8.7.5), assuming that we already know that the restrictions of $E^\pm$ and $H^\pm$ to S exist. This will be often the case, and we discuss it later on, if $E^\pm, H^\pm$ can be extended as distributions accross S. To simplify the notations we shall in the sequel assume that $S = \{(x,t); x_3 = 0\}$, and normalize to the situation when ε^- and μ^- are both the identity 3×3 matrices. It is then also convenient to denote ε^+ and μ^+ just by ε and μ. Explicitly, the conditions mentioned so far are then to find a distribution

$$u = (E, H) \in [D'((X^+ \cup X^-) \times (t_1, t_2))]^6$$

such that:

a) the restrictions u^+ and u^- of u to $X^+ \times (t_1, t_2)$, respectively $X^- \times (t_1, t_2)$, are extendible to distributions on $R^3 \times (t_1, t_2)$,

b) $(\partial/\partial t)(\varepsilon E) - \text{curl} H = 0$, for $x_3 > 0$, $(\partial/\partial t)(\mu H) + \text{curl} E = 0$, for $x_3 > 0$,

c) $\text{div} (\varepsilon E) = 0$, $\text{div} (\mu H) = 0$ for $x_3 > 0$,

d) $(\partial/\partial t)E - \text{curl} H = 0$, for $x_3 < 0$, $(\partial/\partial t)H + \text{curl} E = 0$, for $x_3 < 0$,

e) $\text{div} E = 0$, $\text{div} H = 0$ for $x_3 < 0$,

f) $[n \times E] = [n \times H] = 0$, $[\langle n, \varepsilon E \rangle] = [\langle n, \mu H \rangle] = 0$, where $n = (0, 0, 1)$ and $[v] = v_{|x_3 = 0_+} - v_{|x_3 = 0_-}$.

Finally, to complete the setting of the transmission problem, one should assume that E and H are given for $t = t^0$, for some fixed t^0, or that "enough" information is known about E and H in some time interval $t_1 < t < t_2$, $t^0 \in (t_1, t_2)$. We shall in fact assume, to simplify the situation, that E^+ and H^+ vanish identically for $t \leq t^0$ whereas for E^-, H^- we assume that we have specific information on the analytic wave front set for $t^1 < t < t^0$. (Situations when we had information on $E(t^0), H(t^0)$, etc., could often be handled with only minor modifications.) Physically this could come from the following situation: X^+ has been kept in the dark in the past, but some light has been lit on in X^- in the period $t^3 < t < t^1$ and has been spent at $t = t^1$. Moreover, no light source is used after time t^1, so the homogeneous Maxwell system is valid after that time.

2. In the computations which follow it is now convenient to assume temporarily that orthogonal coordinates have been chosen so that

$$\varepsilon = \begin{pmatrix} \varepsilon_1 & & 0 \\ & \varepsilon_2 & \\ 0 & & \varepsilon_3 \end{pmatrix}, \quad \mu = \begin{pmatrix} \nu & & 0 \\ & \nu & \\ 0 & & \nu \end{pmatrix}.$$

We shall then denote as in section 6.7 by $\psi(\xi) = (d_2 + d_3)\xi_1^2 + (d_3 + d_1)\xi_2^2 + (d_1 + d_2)\xi_3^2$, and by $\varphi(\xi) = d_2 d_3 \xi_1^2 + d_3 d_1 \xi_2^2 + d_1 d_2 \xi_3^2$, d_i as in section 6.7. It follows then that the generic components $v^\pm$ of $u^\pm$ satisfy the scalar equations

$$Q^+ v^+ = 0, Q^- v^- = 0, \qquad (8.7.9)$$

where

$$Q^+(\xi, \tau) = \tau^2(\tau^4 - \psi(\xi)\tau^2 + \varphi(\xi)|\xi|^2)$$

and

$$Q^-(\xi, \tau) = \tau^2(\tau^4 - 2|\xi|^2\tau^2 + |\xi|^4) = \tau^2(\tau^2 - |\xi|^2)^2.$$

The trouble with the equations (8.7.9) is that $x_3 = 0$ is characteristic for Q^+ and Q^-. It is essentially for this reason, that we will have to rely also on the divergence equations from c) and e). Actually, we have the following result

Lemma 8.7.1. *Denote by* $p^+ = \tau^4 - \psi(\xi)\tau^2 + \varphi(\xi)|\xi|^2$ *and let* $p^+(D)$ *be the associated constant coefficient linear partial differential operator. Assume that* $u^+ \in [D'(X^+ \times (t_1, t_2))]^6$, $u^+ = \begin{pmatrix} E^+ \\ H^+ \end{pmatrix}$, *satisfies the equations b) and c) above in* $x_3 > 0$ *and assume that for some* t^0, *supp* $u^+(t^0)$ *is compact as a distribution in* $\{x; x_3 > 0\}$. *Then the components* v_i *of* u^+ *satisfy the equation* $p^+(D)v = 0$.

$(u^+(t^0) = u^+(\cdot, t^0)$ is the restriction of u^+ to $t = t^0$. It exists since $t = t^0$ is noncharacteristic for the system b).)

Remark 8.7.2. *A similar result is valid of course for solutions of d), e), provided we replace* p^+ *by* $p^- = (\tau^2 - |\xi|^2)^2$.

Proof of lemma 8.7.1. The proof is elementary. Denote $p(D)u^+$ by f, so that $D_t^2 f = 0$. We may then write $f(x,t) = h(x) + tg(x)$ for some (vector-valued) distributions h and g which have compact support in X^+ for each fixed t since f has compact support in X^+ for t close to t^0. (The system in b) is hyperbolic in t, so we have finite propagation speed.) What we must show then is that $h = g \equiv 0$. To do so, let us write

$$h = \begin{pmatrix} h' \\ h'' \end{pmatrix}, \quad g = \begin{pmatrix} g' \\ g'' \end{pmatrix},$$

where h', respectively g', and h'', respectively g'', denote the first and last 3 components of h respectively g. The assumptions on u^+ now imply

$$curl\ g' = 0,\ div\ \varepsilon g' = 0, \tag{8.7.10}$$

$$curl\ g'' = 0,\ div\ g'' = 0, \tag{8.7.11}$$

$$curl\ h'' = \varepsilon g',\ div\ h'' = 0, \tag{8.7.12}$$

$$curl\ h' = -\mu g'', div\ \varepsilon h' = 0. \tag{8.7.13}$$

Next we note that the systems $\begin{pmatrix} curl \\ div \end{pmatrix}$, $\begin{pmatrix} curl \\ div\varepsilon \end{pmatrix}$ are elliptic and cannot therefore have nontrivial solutions with compact support in x. The equations (8.7.10),(8.7.11) thus imply that $g' = g'' \equiv 0$. Inserting this into (8.7.12), (8.7.13), we can moreover repeat the last argument for the equations there to deduce that also $h' = h'' \equiv 0$.

3. We have now concluded that the components v_i of E and H satisfy the fourth order equations $p^+(D)v_i = 0$, $p^-(D)v_i = 0$ on the parts $Y^+ = \{(x,t); x_3 > 0\}$ and $Y^- = \{(x,t); x_3 < 0\}$ of their domains of definition. (We have returned here of course to coordinates in which the interface is of form $x_3 = 0$.) The advantage of having reduced ourselves to these fourth order equations is that for them $x_3 = 0$ is noncharacteristic. In particular, restrictions to $x_3 = 0$ for extendible solutions with compact support in x will exist as a consequence of Peetre's theorem. Moreover, $p^-(D)$ is just the square of the standard wave equation. This makes p^- an operator with characteristics of constant multiplicity and propagation of WF_A singularities for solutions of $p^-v = 0$ is very simple in free space and has been analyzed in great detail for boundary problems. As for p^+, propagation of WF_A-singularities in free space, has been analyzed in section 6.7.

We want here next to see how such singularities are transmitted across $x = x_3$ by the transmission conditions. Since we have already seen that the components of our solutions satisfy fourth order equations, we could in principle now try to rewrite the conditions f) for each component v_i as a condition which involves only the Cauchy traces of that component. Actually this is not necessary if we only want to obtain results similar to the one in proposition 8.5.1. Indeed, looking into the argument in section 8.5 we see that in the present situation it suffices to show the following:

assume that (E^+, H^+, E^-, H^-) is a solution of the above transmission problem such that

$$(0, (\xi'^0, 1)) \notin WF_A \begin{pmatrix} E^+ \\ H^+ \end{pmatrix}_{|x_3=0_+}$$

Then it follows for any component v^- of $\begin{pmatrix} E^- \\ H^- \end{pmatrix}$ that

$$(0, (\xi'^0, 1)) \notin WF_A(\partial/\partial x_3)^i v^-_{|x_3=0_+}, \forall i. \qquad (8.7.14)$$

Actually, for $i = 0$ this is trivial, so it will also follow that

$$(0, (\xi'^0, 1)) \notin WF_A(\partial/\partial x)^\alpha(\partial/\partial t)^j \begin{pmatrix} E^- \\ H^- \end{pmatrix}_{|x_3=0_+}$$

for all multiindices α of form $(\alpha_1, \alpha_2, 0)$ and all j. We conclude from the equations for E^-, H^- that (8.7.14) holds for $i = 1$. (Also here we need the full system of Maxwell's equations.) To continue, we can now argue in a similar way, after having derived the Maxwell system in x_3, etc.

4. All the preceding shows that we can argue for the system of crystal optics precisely as in the scalar case. What we need next is then to check under what conditions the assumptions on microlocal hyperbolicity and the transversality conditions for bicharacteristics and propagation cones are satisfied. We discuss this for p^+. The number of real roots μ of the equation $p^+(\xi'^0, \mu, 1) = 0$ can be 0, 2 or 4 (when multiplicities are counted.) In fact this equation has for fixed ξ'^0 precisely 4 complex roots and nonreal roots come in pairs: if μ is a root, so is its complex conjugate. Actually the real roots μ of $p^+(\xi'^0, \mu, 1) = 0$ are precisely the ξ_3- coordinates of the points of intersection of the line $\xi' = \xi'^0$ with Fresnel's surface $\tilde{S} = \{\xi \in R^3; p^+(\xi, 1) = 0\}$, so one can get a feeling of how many real roots one will have in a given situation by looking at Fresnel's surface. (For pictures of Fresnel's surface, see the book of Sommerfeld [1] or Fladt-Baur [1].) Thus for example it is clear that we will have 4 real roots whenever $\xi' = \xi'^0$ intersects the innermost component of $R^3 \setminus \tilde{S}$ nontrivially. This will be for example the case when the ξ_3 axis points in the direction of some singular point $\tilde{\xi}$ of Fresnel's surface, or to some point nearby. More precisely, we have the situation described in the following remark:

Remark 8.7.3. *Assume that $\tilde{\xi} \in \tilde{S}$ is a singular point on the Fresnel surface of p^+. Then there is $\varepsilon > 0$ so that if $\eta \in R^3$, $|\eta| = 1$, is chosen with $|\eta - \tilde{\xi}/|\tilde{\xi}|| < \varepsilon$ and if orthogonal coordinates are chosen so that $\eta = (0, 0, 1)$, then we can find a conic neighborhood Γ of $(\tilde{\xi}_1, \tilde{\xi}_2, 1)$ ($\tilde{\xi}$ is of course written in the new coordinates and $\tilde{\xi} = (\tilde{\xi}_3, \tilde{\xi}_2, \tilde{\xi}_3)$) so that the equation $p^+(\xi_1, \xi_2, \mu, 1) = 0$ has 4 real roots if $(\xi_1, \xi_2, \tau) \in \Gamma$.*

While this gives a situation with 4 real roots, we should also mention that from the point of view of the results which we want to obtain, 4 nonreal roots are not very interesting in that no point of form $(x, (\xi'^0, \nu, 1))$, $x_3 > 0$, will then ever be in $WF_A u^+$.

5. We have now discussed microlocal hyperbolicity and recall from section 6.7 that we know already that the multiplicity of the roots of $p^+(\xi'^0, \mu, 1) = 0$ is at most two and we have already observed a couple of times that it will be two when ξ^0 points in the direction of a singular point on Fresnel's surface. We shall assume that this is the case in the remainder of this section and that ξ^0 is close to the normal at $x_3 = 0$ and that $t^0 = 0$. The propagation cone $G_{\xi^0}^{+0}$ associated with (p^+, ξ^0) is then transversal to $x_3 = 0$ and we assume that it is future oriented. We have then two more roots of $p^+(\xi'^0, \mu, 1) = 0$ and the bicharacteristics or propagation cone associated with these roots will also be transversal to $x_3 = 0$ if the direction of ξ^0 and of the normal to $x_3 = 0$ are sufficiently close. Moreover, these bicharacteristics, or the propagation cone, are past oriented. All

this has been discussed in section 8.4. The conditions for p^+ are then all satisfied and we also note (see section 8.2) that it follows from the fact that u^+ vanishes for negative times that we can have

$$(0, (\xi'^0, 1)) \in WF_A u^+_{|x_3=0_+}$$

only if

$$(G_{\xi^0}^{+0} \times (\xi^0, 1)) \cap WF_A u^+ \neq \emptyset. \tag{8.7.15}$$

To obtain (8.7.15), and that means conical refraction, it suffices then to make sure that we cannot have (8.7.14). If we assume that the equation $p^-(\xi'^0, \mu, 1) = 0$ has two real nonvanishing roots (and this means that we must assume $|\xi'^0| < 1$), then this is quite easy to achieve. Indeed, we will have two bicharacteristics of type L_μ^- associated with our two roots, one of which will be outgoing with respect to $x_3 < 0$ and the other incoming.(As usual we assume that bicharacteristics pass through the point $(0, (\xi'^0, \mu, 1))$.) Let us denote by L' the incoming one: it is this one on which we can act by some experiment in some period $t^3 < t < t^1$. It suffices then to assume that

$$L' \cap WF_A u^- \cap \{(x, t, \xi, \tau); x_3 < 0, t^1 < t < 0\} \neq \emptyset.$$

When ξ^0 points precisely in the direction of the normal to $x_3 = 0$, this coresponds to the classical case of conical refraction in crystal optics.

8.8   Other phenomena: External conical refraction

1. The type of conical refraction which we have considered in section 8.7 is called traditionally "internal". The reason is that it is a phenomenon which appears when light passes from the outside into the inside of some crystal. There is another type of conical refraction, called "external" and which appears when light comes out from the crystal. Although external conical refraction is a much simpler phenomenon, we want to analyze it here also briefly to explain how it can be understood from the microlocal point of view. Actually in doing so we shall remain within the frame set by the results on propagation of singularities for operators of principal type, so strictly speaking, it is not of the type of the results considered in these notes. The question which we shall study is roughly speaking this. We assume that the region $x_3 \leq 0$ is filled by some biaxial crystal, whereas in the region $x_3 > 0$ we have air. Also consider a solution u of Maxwell's equations for which the analytic singular support contains a ray of form

$$L = \{(x, t); x = \nu x^0, t = \nu t^0, \nu \in R, x_3 < 0\}.$$

From the Sato-Hörmander regularity theorem, we will then be able to find for each fixed $(x, t) \in L$ some (ξ^0, τ^0) with $(x, t, \xi^0, \tau^0) \in WF_A u$ and

$$\tau^{02} p(\lambda^0) = 0, \text{ where } p(\lambda) = \tau^4 - \psi(\xi)\tau^2 + \varphi(\xi)|\xi|^2.$$

(The notations are as in section 6.7.) Assuming that (ξ^0, τ^0) is in the smooth part of the characteristic variety of Maxwell's system and that $\tau^0 \neq 0$, it is also natural to assume that (x^0, t^0) and (ξ^0, τ^0) are related by

$$(x^0, t^0) \text{ is proportional to } \operatorname{grad} p(\xi^0, \tau^0), \tag{8.8.1}$$

such that in fact the whole ray L is obtained from propagation of singularities, starting e.g. from $(x^0, t^0, \xi^0, \tau^0)$. (Even when we assume this, there might exist of course other points $(\tilde{\xi}, \tilde{\tau})$ with the property that $(x^0, \tau^0, \tilde{\xi}, \tilde{\tau}) \in WF_A u$.) Also note that once (ξ^0, τ^0) has been fixed, the future fate of L is easy to establish: the transmission conditions will relate the Cauchy traces from the part $x_3 < 0$ to the part $x_3 > 0$ and will lead to a well-defined ray of form

$$(\mu(y^0, v^0), (\eta^0, \theta^0) ; \ \mu \in R, x_3 > 0\}$$

in the analytic wave front set of u in the region $x_3 > 0$. An observer which observes the light ray moving in direction x^0 inside the crystal, will see it continued outside the crystal in direction y^0. The problem is then if (ξ^0, τ^0) is uniquely determined from (x^0, t^0) and relation (8.8.1). We shall see in a moment that for four exceptional directions (x^0, t^0) there are complete circular cones of (ξ^0, τ^0) associated with (x^0, t^0), so for these directions the observer will see a cone of light coming out from the crystal: external conical refraction has occured. It is convenient to argue here in terms of the wave surface for Maxwell's system, which was called Fresnel's surface. We denote it by S. After some calculations already performed above we see then that what we would like to know is if a point on the surface is uniquely determined from the direction of the normal to the surface at that point: in other words, we want to study wether or not the Gauss map which sends $x \in S$ into the normal $n(x)$ at x can be used as a coordinatization of S. A very beautiful remark of R.W.Hamilton in this context is now the following: there are four planar circles imbedded in Fresnel's surface so that all points on a fixed such circle have a common tangent plane. (Cf. e.g. Courant-Hilbert [1].) It is clear then that conical refraction must be expected in general for rays propagationg in these directions. It should be observed however that this type of conical refraction is not of the quality of the conical refraction associated with the four singular points of Fresnel's surface and external conical refraction is not visible at the level of the first analytic wave front set.

(The precise meaning of this is the following: if in our situation L corresponds to one single (ξ^0, τ^0) in the past, then no conical refraction occurs.) The only problem is that a continuum of bicharacteristics for Maxwell's system in biaxial crystals may have the same projection into the physical space. It is also interesting to note that apart from the points on the four exceptional circles, the total curvature is nonvanishing at all points in the smooth part of S. In particular at least locally and outside the four exceptional circles, the normal determines the point. We omit further details. At this moment, we should perhaps also stress a point, related to conical refraction which comes from singular points in the characteristic variety. At least as far as phenomena of this type considered in the present notes are concerned, conical refraction disappears when wave front sets of sufficiently high order are considered. It seems therefore likely that, one can produce solutions for which the first wave front set is spreading on just one ray lying on the cone of conical refraction. The point is then that it is not possible to decide in terms of first microlocalization alone which ray this will be.

2. In these notes we have mainly insisted on conical refraction for the first analytic wave front set. One could likewise study conical refraction in the C^∞- category or for higher analytic wave front sets. While part of the results from these notes might remain true for the C^∞- case, we should explicitly say that stronger assumptions will in general be needed. In fact, one of the pleasant things in the analytic category is that in general the results do not depend on lower order terms. In the C^∞- category we must on the other hand expect that some Levi-type conditions will be needed to obtain nice results. (Cf. e.g. Lascar [1].) Actually, if no such conditions are imposed, conical refraction can come from lower order terms in situations when there is no such refraction if the appropriate Levi conditions hold. More seems to be true however: we have had in our results multihomogeneous principal parts and whatever remained in the principal part of the operator once the multihomogeneous principal part was taken away did not matter for propagation phenomena at the level of higher analytic wave front sets. (Here we needed the $WF_{A,s}^k$ wave front sets.) No such phenomena can be expected in the C^∞ case, so we must expect that we need Levi-type conditions inside the principal part if we want to have nice results. (Actually, in an earlier version of the results from these notes, where only the WF_A^k were used, such conditions were needed to prove a result of the type of theorem 1.1.6.) The second question, that of results for higher order wave front sets, poses a different kind of problem. In fact, it is difficult to imagine that higher order microlocalization is of the same physical relevance than is first microlocalization. Indeed, in light propagation, it might for example be difficult to produce rays for which the energy is concentrated on small bi- or multineighborhoods. It seems therefore that

while higher order microlocalization is a very useful tool to understand and structuralize arguments in conical refraction (and perhaps in other questions) it is not very interesting to study propagation of higher order wave front sets of the type considered here for their own sake.

3. In concluding this section we should observe that all our study was stated in terms of wave front sets. As is well known, an important phenomenon in light propagation is polarization. Polarization has also been studied from the point of view of microlocalization. The relevant notion is here the polarization set of vector valued distributions introduced by N.Dencker [1]. Also cf. Esser [2], Gerard [1,2], Martinez [1]. A study of propagation of polarization sets for operators with characteristics of variable multiplicity has perhaps not yet been done.

Chapter 9

Partial analyticity, higher microlocalization and sheaves

1. Standard microlocalization constitutes a convenient frame to localize arguments to conic sets in the cotangent space of some given manifold. In the C^∞-category such localizations can be achieved by using partitions of unity built up from pseudodifferential operators. In the analytic category this is not possible within the class of classical analytic pseudodifferential operators but an alternative approach has been developped by stressing the sheaf-theoretic aspects underlying microlocalization. Indeed, as we should perhaps recall here, it were precisely such sheaf-theoretical constructions which induced M.Sato to consider standard microlocalization in the first place. The situation is similar for second microlocalization, in that a number of sheaves of microfunctions are staying behind the constructions from these notes and have been effectively used in the theory of Kashiwara-Laurent. The most interesting of these sheaves is certainly the sheaf of partially analytic microfunctions. Actually, the fact that partial analyticity cannot be characterized appropriately in terms of the first analytic wave front set alone, may be considered as one of the initial justifications for a theory of higher microlocalization. (Cf. the comments made on this in section 9.2.) A problem which we have with partial analyticity is that, in the frame set for these notes, there seems to be no unique candidate for this notion which could be given preference over all others. Here the reader may recall that a related problem appears with partial C-infinity: also there we have to distinguish between, at least, two different notions of that sort (sometimes called "weak" and "strong" C-infinity: cf. e.g. Palamodov [1]). We shall discuss in this chapter two concepts of partial analyticity and also mention briefly a third one. The reason for going into all this trouble is that different variants of partial analyticity are associated with different variants of higher order wave front sets. In particular, this discussion sheds

new light on one of the reasons why different variants of higher order wave front sets appeared at all. (The other reason, which has led to more significant differences between definitions, was mentioned in section 3.1 and was that on the set of multiindices we had different order relations which were useful in higher microlocalization.)

9.1   The G_φ- class of a partially analytic distribution

1. Roughly speaking a distribution $u \in D'(U)$, U open in R^n, should be called partially analytic in the variables $x' = (x_1, \ldots, x_d)$, $d \leq n$, if we can find an open domain $\Omega \subset R^{n+d}$ in which the variables shall be denoted by (x, y'), $x \in R^n$, $y' \in R^d$, and an extension v of u to Ω with the following properties

a) $U = \{x; (x, 0) \in \Omega\}$,

b) $[(\partial/\partial x_j) + i(\partial/\partial y_j)]v = 0$, for $j \leq d$.

The problem with these conditions is that we have not specified what kind of extension of u, v should be. Since we have assumed that u is a distribution on U, a natural choice could be $v \in D'(\Omega)$, but since we are working in the analytic category, we may as well ask that v be a hyperfunction: $v \in B(\Omega)$. Accordingly, we shall obtain different notions of partial analyticity which are interesting for us, since they are the background against which some of the choices in the definition of WF_A^k are made.

2. It seems convenient here to characterize the various notions of partial analyticity in terms of G_φ-classes, i.e. we want to find φ so that u is partially analytic in some specified sense, precisely when $u \in G_\varphi(U)$ for that particular weight function. Actually this is not possible in all cases, unless we drop one of the requirements on the weight functions φ made in section 2.2. In fact, the basic weight function in this context is $\psi(\xi) = |\xi'|$, and it will turn out (cf. proposition 9.1.2 below) that when we ask for v to belong to $D'(\Omega)$, then u will lie, roughly speaking, in G_φ with $\varphi \sim \psi$ on a set of type $|\xi'| \geq b\ln(2 + |\xi|)$ for some $b \geq 0$. The problem with our requirements in section 2.2 is that it is not true, of course, that

$$\psi(\xi) = |\xi'| \geq c(1 + |\xi|^{c'}) \quad \text{with} \ c > 0, 1 > c' > 0$$

on a set of type $|\xi'| \geq b\ln(2 + |\xi|)$.

Here we may recall that in Liess-Rodino [1] the condition $\varphi(\xi) \geq c(1 + |\xi|^{c'})$ was mainly introduced to make factors φ^{-j}, $j > 0$ large, useful to obtain integrability in integrals after some partial integrations. This referred to the theory of pseudodifferential or Fourier integral operators, but the definition of G_φ-classes makes of course sense without this condition, and indeed a number of results concerning G_φ-classes are valid without it. (Actually in fact, conditions of type $\varphi(\xi) \geq c(1 + |\xi|^{c'})$ were not considered in the initial definition of G_φ-classes in Liess [1].) We mention here the following variant of proposition 2.4.1, which is a consequence of the results in Liess [1], as was proposition 2.4.1 itself.

Proposition 9.1.1. *Let $\varphi : R^n \to R_+$ be a Lipschitzian function such that*

$$\varphi(\xi) \sim |\xi'| \quad on \ |\xi'| \geq b \ln(2 + |\xi|),$$

for some constant b. Let also $w \in D'(U)$ and x^0 be given. Then there are equivalent:

i) There are c, c', b' and a bounded sequence of distributions u_j in $E'(R^n)$ so that

$$u = u_j \quad for \ |x - x^0| < c,$$

$$|\hat{u}_j(\xi)| \leq c'(c'j/|\xi'|)^j, \ for \ |\xi'| \geq b' \ln(2 + |\xi|).$$

ii) There are $c, \varepsilon, d, b', b''$ so that $|u(g)| \leq c$ for any $g \in C_0^\infty(R^n)$ which satisfies

$$|\hat{g}(\zeta)| \leq \exp\left[d\varphi(-Re\,\zeta) + \langle x^0, Im\,\zeta \rangle + \varepsilon |Im\,\zeta| + b' \ln(1 + |\zeta|)\right],$$
$$for \ |Re\,\zeta'| \geq b'' \ln(2 + |Re\,\zeta|),$$

$$|\hat{g}(\zeta)| \leq \exp\left[\varepsilon |Im\,\zeta| + \langle x^0, Im\,\zeta \rangle + b' \ln(1 + |\zeta|)\right], \ for \ |Re\,\zeta'| \leq b'' \ln(2 + |Re\,\zeta|).$$

We have now clarified the sense in which we shall use the notation G_φ in proposition 9.1.2 below. When the extension v of u is considered in hyperfunctions, then actually we will remain within the range of the definitions used in the previous chapters. In fact partial analyticity of u comes then to

$$(x^0 \times A) \cap WF_{|\xi'|}u = \emptyset$$

for A a set of type $A = \{\xi; |\xi'| > f(|\xi|)\}$ for some suitable sublinear function f, which we may of course assume to be larger than $c|\xi|^{c'}$ for some positive constants $c, c', c' < 1$. To simplify notation we shall express this by saying that $\varphi \sim |\xi'|$ on a set of type $|\xi'| > f(|\xi|)$, etc. Finally we mention, without going into details (details are closer to the "distribution" than to the "hyperfunction" case) that the situation when we ask for $|\xi'| > c|\xi|^\delta$ for some $\delta < 1$, corresponds to the assumption that v is chosen to be a Gevrey ultradistribution. (As long as we do not ask for an optimal δ, it does not matter if ultradistributions are taken of Roumieu or Beurling type.)

The precise statements of what we shall prove are given in the following two propositions.

Proposition 9.1.2. *Let $u \in D'(U)$. Then there are equivalent:*

i) There is $\Omega \subset R^{n+d}$ and an extension $v \in D'(\Omega)$ of u such that a) and b) from the beginning of this section are valid.

ii) For every $x^0 \in U$ there is b and a Lipschitzian function φ so that $u \in G_{\varphi,x^0}$, and $\varphi \sim |\xi'|$ on $|\xi'| \geq b \ln(2 + |\xi|)$.

Proposition 9.1.3. *Let $u \in D'(U)$. Then there are equivalent:*

i) There is $\Omega \subset R^{n+d}$ and an extension $v \in B(\Omega)$ of u such that a) and b) above are valid.

ii) For every $x^0 \in U$ there is a sublinear function f and a Lipschitzian function φ so that $u \in G_{\varphi,x^0}$, $\varphi \sim |\xi'|$ on $|\xi'| \geq f(|\xi|)$.

The proof of proposition 9.1.2 is quite elementary and will be given in section 9.3. For the proof of proposition 9.1.3 it seems advisable to review the theory of hyperfunctions in a form which is most convenient here. We shall do so in the sections 9.4, 9.5, 9.6 and prove proposition 9.1.3 in section 9.7. Before we consider these proofs at all, we shall discuss what kind of corollaries we can obtain from them on the relation between partial analyticity and wave front sets.

9.2   The first two wave front sets and partial analyticity

1. Let Ω be some open neighborhood of $U \subset R^n$ in R^{n+d} and let $v \in B(\Omega)$ be a solution of

$$[(\partial/\partial x_j) + i(\partial/\partial y_j)]v = 0, \text{ for } j \leq d.$$

When $((x^0, y^0), (\xi^0, \eta^0)) \in WF_A v$ it follows that $\xi^{0\prime} = \eta^0 = 0$. ($(\xi' = (\xi_1, ..., \xi_d)$, of course.) We also observe that v can be restricted to $y = 0$ since the subspaces $y_j = 0$ are noncharacteristic for the system

$$\left(\frac{\partial}{\partial x_j} + i\frac{\partial}{\partial y_j}\right) v = 0, \ j = 1, ..., d.$$

Denote by $u = v_{|y=0}$. From general results on the wave front set of a restriction we can conclude that

$$(x, \xi^0) \in WF_A u \text{ implies } \xi^{0\prime} = 0, \forall x \in U. \tag{9.2.1}$$

Thus we obtain complete information on the first wave front set of u when $\xi^{0\prime} \neq 0$, but in fact nothing can be said on the first wave front set of u when $\xi^{0\prime} = 0$. From proposition 9.1.3 we obtain however the following information on the second wave front set then:

Proposition 9.2.1. *Assume that $u \in D'(U)$ satisfies condition i) from proposition 9.1.3 and consider $x^0 \in U$, $\xi^0 \in R^n$ with $\xi^{0\prime} = 0$ and $\sigma^0 \in \dot{R}^d$. (As always, we regard σ^0 also as an element in $R^n = R^d \times R^{n-d}$.) Then it follows that*

$$(x^0, \xi^0, \sigma^0) \notin W F_A^2 u.$$

Remark 9.2.2. *A similar result is of course also valid for the situation from proposition 9.1.2, if we replace $W F_A^2$ with its temperated version.*

A converse to the statements made in the above is also true. In fact we have

Proposition 9.2.3. *Assume that*

a) $(x^0, \xi^0, \sigma^0) \notin W F_A^2 u, \forall \xi^0 \in \{0\} \times \dot{R}^{n-d}, \forall \sigma^0 \in \dot{R}^d$,

b) $(x^0, \xi^0) \notin W F_A u$, whatever $\xi^0 \in R^n$ with $\xi^{0\prime} \neq 0$ is.

Then it follows that u is partially analytic (in the hyperfunction sense) near x^0.

The proposition follows from a combination of the propositions 2.2.8 and 9.1.3.

Remark 9.2.4. *A related result is discussed in proposition 9.8.2 below.*

9.3 Proof of proposition 9.1.2

1. i) $\Rightarrow$ ii). The statement being local, we may assume that $\Omega = \{(x, y'); |x| < \epsilon, |y'| < \epsilon\}$ and prove then that $u \in G_{\varphi,0}$. Let us first note that a particular solution of

$$\left(\frac{\partial}{\partial x_j} + i \frac{\partial}{\partial y_j} \right) w = 0, \ j \leq d, \ w \in D'(\Omega) \tag{9.3.1}$$

is given by

$$e^{i\langle x, \zeta \rangle + i \langle y', \eta' \rangle}, \ \zeta \in C^n, \ \eta' \in C^d, \tag{9.3.2}$$

provided that $\zeta' + i\eta' = 0$. Since we assumed that Ω is convex it follows from the general theory of linear constant coefficient partial differential equations that the general solution of (9.3.1) is a superposition of exponentials as in (9.3.2). This is closely related to theorem 3.11.2 and is in fact the "representation form" of the fundamental principle of Ehrenpreis

[1] and Palamodov [1]. (Note that we use this principle here for an overdetermined system of equations rather than for the scalar case. This is not strictly speaking necessary, as we shall see in a related situation in section 9.6. On the other hand this system is extremely simple, so the result is actually not difficult to prove.) What this means for the particular situation at hand and for fixed $\varepsilon' < \varepsilon$ is, that we can find $b \in R$ and a Radon measure μ on $V = \{(\zeta, \eta') \in C^{n+d}; \zeta' + i\eta' = 0\}$ so that

$$\int_V d|\mu(\zeta, \eta')| < \infty$$

and so that

$$v(f) = \int_V \hat{f}(-\zeta, -\eta')(1 + |(\zeta, \eta')|)^b e^{-\varepsilon'|Im\,\zeta| - \varepsilon'|Im\,\eta'|} d\,\mu(\zeta, \eta), \qquad (9.3.3)$$

if $f \in C_0^\infty((x, y') \in R^{n+d}; |(x, y')| < \varepsilon')$. Thus, formally

$$v(x) = (2\pi)^{-n} \int e^{i\langle x, \zeta\rangle}(1 + |(\zeta, \eta')|)^b e^{-\varepsilon'|Im\,\zeta| - \varepsilon'|Im\,\eta'|} d\,\mu(\zeta, \eta).$$

It is here convenient to parametrize V by C^n and to write

$$(1 + |(\zeta, \eta')|)^b d\mu(\zeta, \eta') = (1 + |\zeta|)^b d\nu(\zeta),$$

for some Radon measure ν on C^n. We then still have $\int_{C^n} d\,|\nu| < \infty$ and (9.3.3) has transformed to

$$v(f) = \int_{C^n} \hat{f}(-\zeta, i\zeta')(1 + |\zeta|)^b e^{-\varepsilon'|Im\,\zeta| - \varepsilon'|Re\,\zeta'|} d\,\nu(\zeta).$$

As for the restriction u of v to $y' = 0$, we obtain

$$u(g) = \int_{C^n} \hat{g}(-\zeta)(1 + |\zeta|)^b e^{-\varepsilon'|Im\,\zeta| - \varepsilon'|Re\,\zeta'|} d\,\nu(\zeta),$$

if $g \in C_0^\infty(|x| < \varepsilon')$. It is now easily seen that if

$$|\hat{g}(\zeta)| \le (1 + |\zeta|)^{-b} e^{\varepsilon'|Im\,\zeta| + \varepsilon'|Re\,\zeta'|}$$

then $|u(g)| \le c$. This gives ii) in view of proposition 9.1.1.

ii) $\Rightarrow$ i). By Holmgren's theorem, we may define the extension locally. Assume then that $x^0 = 0$ and that we can find φ and b with $\varphi \sim |\xi'|$ on $|\xi'| \ge b\ln(2 + |\xi|)$ so that

$$|\hat{g}(\zeta)| \le (1 + |\zeta|)^{b'} e^{\varepsilon\varphi(-Re\,\zeta) + \varepsilon|Im\,\zeta|}$$

implies

$$|u(g)| \le c$$

for some ε, c, b'. It follows then from the Hahn-Banach theorem that we can find a Radon measure ν on C^n with $\int d|\nu| < \infty$ so that

$$u(g) = \int_{C^n} \hat{g}(-\zeta)(1 + |\zeta|)^{-b'} e^{-\varepsilon|Im\,\zeta| - \varepsilon\varphi(Re\,\zeta')} \, d\,\nu(\zeta).$$

We can now define a distribution v in a neighborhood of the origin in R^{n+d} which extends u locally by

$$v(f) = \int_{C^n} \hat{f}(-\zeta, i\zeta')(1 + |\zeta|)^{-b'} e^{-\varepsilon|Im\,\zeta| - \varepsilon\varphi(Re\,\zeta')} \, d\,\nu(\zeta).$$

The distribution so defined is a solution of (9.3.1). Indeed, if

$$f = {}^t\left(\partial/\partial x_j + i\partial/\partial y_j\right)g, \; g \in C_0^\infty(R^{n+d})$$

then $\hat{f}(-\zeta, i\zeta') = 0$.

9.4   Review of hyperfunctions

1. For the convenience of the reader we collect in this section some elementary facts about hyperfunctions which we shall need later on. Standard texts on microlocalization in hyperfunctions are Sato-Kawai-Kashiwara [1] or Kawai-Kashiwara-Kimura [1]. There are several complementary points of view to look at hyperfunctions. Indeed, one may regard hyperfunctions as equivalence classes of formal sums of holomorphic functions defined on wedges, as locally finite sums of real analytic functionals or as (formal) infinite order derivatives of distributions or functions. The last point of view is here particularily revealing, since it shows from where all the sublinear functions above come. Moreover, each of these viewpoints proves advantageous for some specific applications, so we shall describe them briefly in this section. (For yet another point of view, cf. Hörmander [5] and Komatsu [1], [2].)

Let us then consider at first U open in R^n. We call "wedge over U", or simply "wedge", a set of type

$$\omega = (U + i\Gamma) \cap \tilde{U} \subset C^n$$

where Γ is an open connected cone in R^n and $\tilde{U}$ is a complex neighborhood of U in C^n. (Sets of this form are sometimes called "tuboids" in the literature: cf. e.g. Esser-Laubin, [2]. The set U is often called the "edge" of ω.)

Let us now consider a finite number of wedges ω_j over U and let h_j be holomorphic functions on ω_j. The wedges will later on be (without any further specification) over the same U. With the h_j we associate the formal sum $\sum h_j$. It is then clear that in the space of such formal sums we have a natural commutative addition. (In other words, we consider the free group with generators in $O(\omega)$.) Let us denote the space of all such formal sums by $\mathcal{M}$.

2. The next thing is to observe that if $p(D)$ is a constant coefficient linear partial differential operator on R^n, then $p(D)$ acts in a natural way on $\sum h_j$ by the formula

$$p(D) \sum h_j = \sum p(D_z) h_j.$$

(Here $p(D_z)$ is the operator obtained by replacing $\partial/\partial x$ in $p(D_x)$ formally by $\partial/\partial z$.) Actually, more is true: if we consider a formal infinite order constant coefficient linear partial differential operator of form

$$J(D_z) = \sum_{|\alpha| \geq 0} a_\alpha (-i\partial/\partial x)^\alpha \quad , a_\alpha \in C,$$

such that

$$\lim_{|\alpha| \to \infty} (\alpha! |a_\alpha|)^{1/|\alpha|} = 0 \,, \tag{9.4.1}$$

then $f \in O(\omega)$ implies that

$$z \to \sum_{|\alpha| \geq 0} a_\alpha (-i\partial/\partial z)^\alpha f(z) \in O(\omega).$$

This is seen from Cauchy's inequalities. It makes then sense to write that

$$J(D_z) \sum_j h_j = \sum_j \left[\sum_{|\alpha| \geq 0} a_\alpha D_z^\alpha h_j \right].$$

We thus have an action of a class of infinite order constant coefficient linear partial differential operators on formal sums of holomorphic functions defined on wedges. Let us also note here incidentally that (9.4.1) is equivalent with the fact that the function

$$J(\zeta) = \sum_{|\alpha| \geq 0} a_\alpha \zeta^\alpha \,, \zeta \in C^n,$$

which is called the symbol of $J(D)$, is an entire function of infraexponential type. For this reason, we shall call henceforth $J(D_z)$ an infinite order linear partial differential operator of infraexponential type (with constant coefficients). We also mention that if $J(D_z)$ and $J'(D_z)$ are two such operators associated with the functions J and J', then the composition $J(D_z) \circ J'(D_z)$ is the operator associated with the symbol $J''(\zeta) = J(\zeta) J'(\zeta)$.

3. In $\mathcal{M}$ we shall now introduce an equivalence relation by saying that

$$\sum h_j \sim \sum h'_k \, ,$$

for $h_j \in O((U + i\Gamma_j) \cap \tilde{U}_j)$, $h'_k \in O((U + i\Gamma'_k) \cap \tilde{U}'_k)$, if we can find

$$f_{ls} \in O((U + i \text{ co } (\Gamma_l \cup \Gamma'_s)) \cap \tilde{U}_{ls}), \text{ such that}$$

$$h_j = \sum_s f_{js}, \qquad\qquad (9.4.2)$$

$$h'_k = \sum_l f_{lk}. \qquad\qquad (9.4.3)$$

("co" is here, as elsewhere in these notes, "convex hull".)

It is not difficult to see that

$$\sum h'_j \sim \sum h''_k \, , \; \sum f'_l \sim \sum f''_s$$

implies

$$\sum h'_j + \sum f'_l \sim \sum h''_k + \sum f''_s.$$

Further, if h_1, h_2 are in $O(\omega)$ for the same wedge ω, then the pointwise sum $h_3 \in O(\omega)$ defined by $h_3(z) = h_1(z) + h_2(z)$ is easily seen to be equivalent with $h_1 + h_2$, the latter sum being performed in $\mathcal{M}$. Also note that a convenient way to characterize

$$\sum_{j=1}^m h_j \sim 0 \, , \; h_j \in O((U + i\Gamma_j) \cap \tilde{U}_j)$$

is, that we can find $f_{jk} \in O((U + i \text{ co } (\Gamma_j \cup \Gamma_k)) \cap \tilde{U}_{jk})$ with

$$f_{jk} = -f_{kj} \; \text{ and } \; h_j = \sum_k f_{jk}.$$

Indeed, this is obvious if $m = 2$ and for $m > 2$ it suffices to write down that

$$-h_1 \sim \sum_{j=2}^m h_j,$$

if $\sum h_j \sim 0$, etc.

We shall denote the factor space $\mathcal{M}/\sim$ by $B(U)$. The projection $\mathcal{M} \to \mathcal{M}/\sim$ is called "hyperfunctional boundary value". In fact, the above is one of the standard ways to define the space $B(U)$ of hyperfunctions on U. $B(U)$ is a commutative group on which we have a natural action of infinite order constant coefficient linear partial operators of infraexponential type. Next we observe that if $f \in O((U + i\Gamma) \cap \tilde{U})$ and if $\tilde{U}'$ is some other complex neighborhood of U, then

$$f \sim f_{|\tilde{U} \cap \tilde{U}'} , \quad \text{as elements in } \mathcal{M}.$$

It follows easily that in $B(U)$ we can introduce a multiplication with real-analytic functions on U. Moreover, hyperfunctions, when considered for all $U \subset R^n$, form a (pre)-sheaf: restrictions are defined in an obvious way and linear partial differential operators (with analytic coefficients), respectively infinite order partial differential operators of infraexponential type, are local operators in $B(U)$. It is often useful (although not trivial) that this sheaf is flabby: if $U \subset R^n$ is given and if $u \in B(U)$, then we can find $v \in B(R^n)$ so that u is the restriction of v to U. We shall use this result implicitly, in that it is used in the proof of a theorem of Kawai-Kaneko, which we need and state later on as theorem 9.4.2. (Actually, instead of applying the theorem of Kawai-Kaneko, we could do with a slightly less powerful result from Liess [4], which does not depend on the flabbyness of the sheaf of hyperfunctions.)

4. Our next concern is to analyze how distributions are imbedded into hyperfunctions. Actually we shall describe two ways to obtain this imbedding. Since both, distributions and hyperfunctions, are sheaves, it suffices to show (in both procedures) that for any fixed $U' \subset\subset U$ there is a natural imbedding $i : E'(U') \to B(U')$. To construct i we may now at first consider a collection of sharp open convex cones $\Gamma^j \subset \dot{R}^n$ so that

$$\cup \Gamma^j = \dot{R}^n,$$

and continuous functions $f_j : R^n \to R$ with $0 \leq f_j(\xi) \leq 1$, $\sum f_j \equiv 1$, $f_j(\xi) \neq 0 \Rightarrow \xi \in \Gamma^j$ if $|\xi| \geq 1$. If $v \in E'(U')$ is fixed, we then define

$$h_j(z) = (2\pi)^{-n} \int_{R^n} e^{i\langle z, \xi \rangle} \hat{v}(\xi) f_j(\xi) d\xi, \text{ if } z \in \{t \in C^n; Im\, t \in (\Gamma^j)^0\}. \qquad (9.4.4)$$

Here we have denoted by $(\Gamma^j)^0$ the polar of Γ^j. It is important that the integral in (9.4.4) converges and defines a holomorphic function. Moreover, it is not difficult to see that for fixed j and fixed $G^j \subset\subset (\Gamma^j)^0$ the limit

$$\lim_{y \to 0, y \in G^j} h_j(\cdot + iy) = v_j \text{ exists in } D'(U') \qquad (9.4.5)$$

and that

$$v = \sum v_j \text{ in } D'(U'). \qquad (9.4.6)$$

Thus v is the sum of distributional boundary values of some holomorphic functions defined on wedges. It is then natural to set

$$i(v) = \sum b(h_j) \text{ in } B(U'),$$

boundary values being taken in hyperfunctions. Thus, to repeat it shortly, $i(v)$ is the class of the formal sum $\sum h_j$, provided v is related to some h_j by (9.4.5) and (9.4.6). It is interesting to note - and this is in fact the true reason why all this is useful- that $i(v)$ does not depend on the particular way in which we did arrive at the representation from (9.4.5), (9.4.6): if $\sum h'_k$ is another formal sum of holomorphic functions for which

$$\lim_{y \to 0,\, y \in G^{k\prime} \subset\subset (\Gamma^{k\prime})^0} h'_j(\cdot + iy) = v'_k \text{ exists in } D'(U')$$

and if $\sum v'_k = v$, then $\sum h'_k \sim \sum h_j$ in $\mathcal{M}$ (over U'). This is a consequence of Martineau's version of the edge-of-the-wedge theorem. (We may here recall that the distributional version of the edge-of-the-wedge theorem is somewhat more precise than $\sum h'_k \sim \sum h_j$, in that it establishes that the functions f_{kj} which relate the h'_k to the h_j have distributional boundary values.) It follows in particular from this discussion, that if h is a holomorphic function on a wedge, which admits a distributional boundary value u, then $i(u) = b(h)$. (i is here considered for distributions in $D'(U)$ and not only for distributions with compact support.)

(In this context it might be useful to recall that a holomorphic function h defined on a wedge $\omega = (U + i\Gamma) \cap \tilde{U}$ admits a boundary value $u \in D'(U)$ on the edge U if and only if it is of temperated growth near the edge. By this we mean that for each compact $K \subset U$ and each cone $\Gamma' \subset\subset \Gamma$ there are constants c and k so that

$$|h(x + iy)| \le c|y|^{-k}$$

if $x \in K$, $y \in \Gamma'$ and $(x + iy) \in \tilde{U}$.)

5. The second way to embedd $E'(U')$ into $B(U')$ has the advantage that it works equally well to embedd real analytic functionals into $B(U')$. To be more precise, we shall denote by $\mathcal{A}'(K)$ the space of real analytic functionals on R^n which are supported by K and recall that analytic functionals can be identified with hyperfunctions with compact support and that, more generally, the space of hyperfunctions on a bounded open set U can be identified with the set $\mathcal{A}'(\bar{U})/\mathcal{A}'(\partial U)$. ($\partial U$ is the boundary of U.) For more details on the relation between hyperfunctions and real analytic functionals, cf. Schapira [1] and Hörmander [5]. Let us also recall that if v is a real analytic functional, then we can associate with v a hyperfunction $\tilde{i}(v)$ by the following prescription

$$\tilde{i}(v) = \sum_\sigma b(h_\sigma), \tag{9.4.7}$$

where σ runs through the set of all $\sigma = (\sigma_1, ..., \sigma_n)$ with $\sigma_j \in \{-1, 1\}$ and $h_\sigma \in A(z \in C^n; \sigma_j \operatorname{Im} z_j > 0, j = 1, ..., n)$ is defined by

$$h_\sigma(z) = (-1)^{\sigma_1 + \cdots + \sigma_n}(2\pi i)^{-n} v(\frac{1}{(t_1 - z_1)\cdots(t_n - z_n)}). \tag{9.4.8}$$

(Thus, v acts in the variables t.)

Lemma 9.4.1.. *When $v \in E'(R^n)$, then $i(v) = \tilde{i}(v)$.*

Sketch of proof. For σ of form $\sigma = (\sigma_1, ..., \sigma_n)$, $\sigma_j = \pm 1$, we denote by

$$\Gamma_\sigma = \{y \in R^n; y_j \sigma_j > 0, j = 1, ..., n\}.$$

The lemma is now based on the remark that

$$\int_{\Gamma_\sigma} e^{i\langle z - t, \xi \rangle} d\xi = (-1)^{\sigma_1 + \cdots + \sigma_n} i^{-n} \frac{1}{(z_1 - t_1) \cdots (z_n - t_n)} \; , \text{ if Im } z \in \Gamma_\sigma.$$

Next we observe that we may write $i(v) = \sum_\sigma b(f_\sigma)$, with

$$f_\sigma(z) = (2\pi)^{-n} \int_{\Gamma_\sigma} e^{i\langle z, \xi \rangle} \hat{v}(\xi) d\xi.$$

Thus

$$f_\sigma(z) = (2\pi)^{-n} \int_{\Gamma_\sigma} e^{i\langle z, \xi \rangle} v(e^{-i\langle t, \xi \rangle}) d\xi =$$

$$= (2\pi i)^{-n}(-1)^{\sigma_1 + \cdots + \sigma_n} \; v\left(\frac{1}{(z_1 - t_1) \cdots (z_n - t_n)}\right).$$

In view of this lemma, we shall now also write i instead of $\tilde{i}$, whenever we want to imbedd analytic functionals into hyperfunctions.

6. To conclude the section, we mention the following result due to Kawai [1] and Kaneko [1] :

Theorem 9.4.2. *Consider $u \in B(U)$. Then we can find a constant coefficient infinite order infraexponential linear partial differential operator $J(D)$ and a C^∞ function f on U so that*

$$u = J(D)f.$$

It follows from this theorem that one may as well regard hyperfuntions as infinite order derivatives of C^∞- functions. This point of view shall be useful when we shall deal with hyperfunctions which are holomorphic in part of the variables.

9.5   Hyperfunctions and real analytic functionals

1. As in the preceding section we denote by $\mathcal{A}'(R^n)$ the space of real analytic functionals and, for K compact in R^n, by $\mathcal{A}'(K)$ the space of real analytic functionals supported

by K. As we have seen real analytic functionals may be identified locally with hyperfunctions using the relations (9.4.7) and (9.4.8). It is further interesting to note that in $\mathcal{A}'(R^n)$ we have a natural action of infinite order infraexponential linear partial differential operators with constant coefficients defined by

$$(J(D)v)(f) = v(J(-D)f),$$

if f is analytic in a complex neighborhood of R^n. We can now prove the following result:

Lemma 9.5.1. *Let u, v be real analytic functionals and suppose that*

$$i(u) = J(D)i(v) \qquad (9.5.1)$$

for some infraexponential infinite order constant coefficient linear partial differential operator.

Then it follows that

$$u = J(D)v \qquad (9.5.2)$$

in the sense of analytic functionals.

Proof. Since i is injective, it suffices to show that

$$i(u) = i(J(D)v). \qquad (9.5.3)$$

To prove the latter, we represent $i(v)$ in the form $i(v) = \sum_{\sigma} b(h_{\sigma})$, where the h_{σ} are given by (9.4.8) and the σ runs once more through the set of all $\sigma = (\sigma_1, ..., \sigma_n)$, $\sigma_j = \pm 1$. Thus

$$J(D)i(v) = \sum_{\sigma} b(J(D_z)h_{\sigma}),$$

where, with the notation $\varepsilon(\sigma) = \sigma_1 + \cdots + \sigma_n$,

$$J(D_z)h_{\sigma}(z) = (-1)^{\varepsilon(\sigma)}(2\pi i)^{-n} v \left(J(D_z) \frac{1}{(t_1 - z_1) \cdots (t_n - z_n)} \right)$$

$$= (-1)^{\varepsilon(\sigma)}(2\pi i)^{-n} v \left(J(-D_t) \frac{1}{(t_1 - z_1) \cdots (t_n - z_n)} \right)$$

$$= (-1)^{\varepsilon(\sigma)}(2\pi i)^{-n} J(D_t v) \left(\frac{1}{(t_1 - z_1) \cdots (t_n - z_n)} \right).$$

This is now precisely what is needed for relation (9.5.3).

2. Having identified hyperfunctions with compact support with real analytic functionals, it makes sense to consider FBI-transforms of hyperfunctions with compact support. We

shall here only look briefly into the case of the FBI²-transform. In fact when $u \in \mathcal{A}'(R^n)$, $t \in R^n$, $\xi \in R^n$, we define

$$FBI^2 u(t, \xi) = u(E(\cdot, t, \xi)),$$

where E is, as in section A.3, the function

$$E(x, \xi, t) = \exp[-i\langle x, \xi \rangle - |\xi'| \, |x' - t'|^2/2 - |\xi''| \, |x'' - t''|^2/2].$$

We would now like to read off regularity properties of u from estimates of the FBI²-transform, but we have not even introduced notions like e.g. WF_A^2 for hyperfunctions. (Technically this would not be difficult if we would base definitions on the FBI transform.) We shall therefore only consider a very elementary situation.

Lemma 9.5.2. *Let $1 \le d < n$ be a fixed integer and denote $\xi' = (\xi_1, ..., \xi_d)$, $\xi'' = (\xi_{d+1}, ..., \xi_n)$. Consider $u \in \mathcal{A}'(R^n)$ and assume that $0 \notin \ supp \ u$. Then we can find c, c', $\delta > 0$ and a sublinear function f such that*

$$|(FBI^2 u)(t, \xi)| \le c \, e^{-c'|\xi'|} \tag{9.5.4}$$

if $|\xi'| \le |\xi''|$, $|t| \le \delta$, and $|\xi'| \ge f(|\xi|)$.

Proof. Since $0 \notin \ supp \ u$ we can write u in the form $u = \sum u_j$, where the u_j are real analytic functionals such that $0 \notin \ co \ supp \ u_j = K_j$. Next we observe that if we consider $\varepsilon > 0$, then we can find C_ε such that

$$|(FBI^2 u_j)(t, \xi)| \le C_\varepsilon \sup_z |E(z, t, \xi)|,$$

where the supremum is for all complex z such that dist $(z, K_j) \le \varepsilon$. It is immediate that we can find $c'' > 0, \delta > 0$, which do not depend on ε, so that

$$\sup_z |E(z, t, \xi)| \le e^{\varepsilon |\xi| - c''|\xi'|},$$

if $|\xi'| \le |\xi''|$ and $|t| \le \delta$.

We conclude that

$$|(FBI^2 u)(t, \xi)| \le \inf_{\varepsilon > 0} C_\varepsilon e^{\varepsilon |\xi| - c''|\xi'|} \quad \text{for } |\xi'| \le |\xi|, |t| \le \delta. \tag{9.5.5}$$

Here we observe that

$$\inf_\varepsilon C_\varepsilon e^{\varepsilon t} = e^{\tilde{f}(t)},$$

with $\tilde{f}(t) = \inf_\varepsilon(\varepsilon t + \ln C_\varepsilon)$. Obviously, $\tilde{f}$ is sublinear and from (9.5.5) it follows that $|(FBI^2 u)(t, \xi)| \le \exp(-c''|\xi'|/2)$ if $|\xi'| \le |\xi|$, $|t| \le \delta$ and $(c''/2)|\xi'| \ge \tilde{f}(|\xi|)$. This is the desired conclusion, if we set $f = (2/c'')\tilde{f}$.

Remark 9.5.3. *It is also transparent from the argument that the estimate (9.5.4) is all that can be obtained from the assumption* $0 \notin supp\ u$. *Since* $0 \notin supp\ u$ *means maximal regularity of* u *near* 0, *this is another justification for the need of a condition of type* "$|\xi'| \geq f(|\xi|)$" *in the definition of* WF_A^2.

9.6 Partially analytic distributions and infinite order operators

1. The fundamental principle which we have used in the proof of proposition 9.1.2 is valid also in the frame of hyperfunctions. This has been established in Kaneko [1,2]. In the proof of proposition 9.1.3 we could therefore argue in a way similar to that used in the proof of proposition 9.1.2. The computations are however less straightforward, so we shall work in a slightly different way and base the computations more directly on an intermediate result considered by Kaneko in the proof of the fundamental principle, rather than on the principle itself. Actually Kaneko has considered only the case of scalar equations, but this is not really embarassing: indeed, if in the conditions from proposition 9.1.3, v is a hyperfunction solution of the overdetermined system from relation b), section 9.1, then the restriction v^l of v to $\{(x, y') \in \Omega; y_k = 0$ if $k \neq l\}$ exists for all $l \leq d$ (in view of the fact that v satisfies $(\partial/\partial x_k + i\partial/\partial y_k)v = 0$ and that $y_k = 0$ is noncharacteristic for the operator $(\partial/\partial x_k + i\partial/\partial y_k)$), and satisfies

$$(\partial/\partial x_l + i\partial/\partial y_l)v^l = 0.$$

If the conclusion of proposition 9.1.3, $i) \Rightarrow ii)$ is proved for $d = 1$, then it will follow in the case of a general $d \leq n$ that

$$v = \bigcap_{j \leq d} G_{\psi_j, x^0},$$

where $\psi_j(\xi) \sim |\xi_j|$ for $|\xi_j| > f_j(|\xi|)$, for some sublinear functions f_j. This will give the desired conclusion in view of proposition 2.2.8.

The result of Kaneko now gives in the particular case of interest to us:

Lemma 9.6.1. *Let* $\Omega \subset R^{n+1}$ *be a convex open neighborhood of* $U \subset R^n$ *in* R^{n+1}, *the variables in* R^{n+1} *being denoted by* (x, y_1). *Let further* $v \in B(\Omega)$ *satisfy* $(\partial/\partial x_1 + i\partial/\partial y_1)v = 0$. *Then we can find* $f \in C^\infty(\Omega)$ *and an infinite order constant coefficient partial differential operator* $\mathcal{K}$ *of infraexponential type, which acts in the variables* (x, y_1), *such that*

$$v = \mathcal{K}(D_x, D_{y_1})f \ , (\partial/\partial x_1 + i\partial/\partial y_1)f = 0. \tag{9.6.1}$$

We could now in principle add to this the Fourier representation which f inherits as a solution of $(\partial/\partial x_1 + i\partial/\partial y_1)f = 0$ and could obtain a representation of v in this way. We shall instead contend ourselves with a simplification of (9.6.1) and write then down what follows from that simplification for the situation encountered in proposition 9.1.3. The reason why we want to simplify (9.6.1) is that we want to restrict the representation from there to $y_1 = 0$. As long as $\mathcal{K}(D_x, D_{y_1})$ effectively operates in y_1 this gives a rather complicated expression concerning the higher order traces of v. We can avoid this by observing that

$$\mathcal{K}(\zeta, \eta_1) = \mathcal{K}(\zeta, -i\zeta_1) + \mathcal{K}'(\zeta, \eta_1)(\zeta_1 + i\eta_1).$$

Here it is immediate that $\mathcal{K}(\zeta, -i\zeta_1)$ and $\mathcal{K}'(\zeta, \eta_1)$ are entire analytic functions of infra-exponential type. Since

$$(\partial/\partial x_1 + i\partial/\partial y_1)f = 0$$

we can now conclude from (9.6.1) that

$$v = \mathcal{K}(D_x, -iD_{x_1})f.$$

Let us then return to the situation when v is a hyperfunction solution of equation b) from the beginning of section 9.1, in the case when $d = 1$. To simplify the situation further we shall assume that U is a neighborhood of the origin in R^n. It follows from the above that we can find $\varepsilon > 0$, an infinite order linear partial differential operator of infraexponential type $J(D_x)$ in the variables x and a C^∞-function f defined on $\{(x, y_1) \in R^{n+1}; |x| < \varepsilon, |y_1| < \varepsilon\}$ such that

$$u = J(D_x)f(x, 0) \text{ for } |x| \le \varepsilon,$$

and

$$(\partial/\partial x_1 + i\partial/\partial y_1)f = 0 \text{ for } |x| < \varepsilon, |y_1| < \varepsilon.$$

9.7   Proof of proposition 9.1.3

1. Proof of i) $\Rightarrow$ ii). We have seen in the preceding section that it is no loss in generality to assume $d = 1$. Moreover, to simplify notations, we shall assume that $x^0 = 0$. We must then show that we can find a sublinear function f and a weight function ψ for which

$$\psi \sim |\xi_1| \quad \text{on } |\xi_1| \ge f(|\xi|),$$

and for which $u \in G_{\psi,0}$. It is, moreover, no loss in generality to assume that $u \in E'(R^n)$.

To check that $u \in G_{\psi,0}$ we shall argue using the FBI^2- transform. In fact, from the proof of proposition 2.1.7 it is clear that it suffices to show that

$$u(e^{-i\langle x,\xi\rangle} - |\xi_1|\,|x_1 - t_1|^2/2 - |\xi''|\,|x'' - t''|^2/2) = O(e^{-c|\xi_1|}) \text{ if } |\xi_1| \geq f'(|\xi|) \quad (9.7.1)$$

for some sublinear function f' and for $|t| < c'$, c' small. Here we recall that u in fact satisfies the assumptions from i). We can therefore use the results from the preceding section to conclude that there is $v \in C^\infty(R_{x,y_1}^{n+1}; |x| < \varepsilon, |y_1| < \varepsilon)$ and an infinite order infraexponential operator $J(D_x)$ so that

$$u = J(D_x)v_{|y_1} = 0 \text{ on } |x| < \varepsilon,$$

$$\left(\frac{\partial}{\partial x_1} + i\frac{\partial}{\partial y_1}\right) v = 0.$$

Next, it is convenient to cut off v near 0. More precisely, we consider $g \in C_0^\infty(R^n)$ so that

$$g(x) = 0 \text{ if } |x| > 2\varepsilon/3$$

and

$$g(x) = 1 \text{ if } |x| < \varepsilon/2.$$

It follows then that we still have

$$u = J(D_x)(gv_{|y_1} = 0) \text{ if } |x| < \varepsilon/2,$$

so that

$$u = J(D)(gv_{|y_1} = 0) + w \quad (9.7.2)$$

for some $w \in B(R^n)$ which has compact support in R^n and for which $0 \notin \text{supp } w$. We want to compute FBI^2u starting from the representation 9.7.2. It is then convenient to regard (9.7.2) as a relation in real analytic functionals (cf. lemma 9.5.1 on this point) and to apply the FBI^2 transform in real-analytic functionals. In particular it follows from lemma 9.5.2 that we can find c, c', δ and a sublinear function f' so that

$$|(FBI^2w)(t,\xi)| \leq c\, e^{-c'|\xi_1|} \text{ if } |t| < \delta, |\xi_1| > f'(|\xi|),$$

so we are essentially left with a study of $I(t,\xi) = FBI^2(J(D)gv_{|y_1=0})$. It also follows from the definition of the action of $J(D)$ on real-analytic functionals that

$$I(t,\xi) = (gv_{|y_1=0})(J(-D_z)E(z,t,\xi)), \quad (9.7.3)$$

where the right hand side of (9.7.3) is the action of the analytic functional $g(v_{|y_1=0})$ on the analytic function $J(-D_x)E(z,t,\xi)$. Actually, $gv_{|y_1=0}$ is just a C^∞-function and $J(-D_z)E(z,t,\xi)$ is for Im $z = 0$

$$\sum a_\alpha(-D_x)^\alpha E(x,t,\xi),$$

if we had $J(D_z) = \sum a_\alpha (D_z)^\alpha$. It therefore follows from 9.7.3 that

$$I(t,\xi) = \int g(x)v(x,0)J(-D_x)E(x,t,\xi)\,dx, \qquad (9.7.4)$$

the integral being standard, once $J(-D_x)E(x,t,\xi)$ has been computed.

The main thing is now that the integral in (9.7.4) can be computed using contour deformations in x_1 for small $|x|$. Indeed, v has a holomorphic extension then and the holomorphic extension of $J(-D_x)E(x,t,\xi)$ is just $J(-D_x)E(z,t,\xi)$. Let us then fix $\xi \in R^n$ with $|\xi_1| \leq |\xi''|$ and assume, to make a choice, that

$$\xi_1 \geq 0.$$

We also denote by $\Lambda \subset C$ the contour

$$\Lambda = (-\infty, \ -\varepsilon/2) \cup \Lambda_1 \cup \Lambda_2 \cup \Lambda_3 \cup (\varepsilon/2, \ \infty),$$

where the Λ_i are the oriented segments with endpoints

$$-\varepsilon/2, \ -\varepsilon/4 - i\varepsilon/4 \ \text{in the case of} \ \Lambda_1$$

$$-\varepsilon/4 - i\varepsilon/4, \varepsilon/4 - i\varepsilon/4 \ \text{in the case of} \ \Lambda_2$$

$$\varepsilon/4 - i\varepsilon/4, \varepsilon/2 \ \text{in the case of} \ \Lambda_3.$$

It follows that

$$I(t,\xi) = \left[\int_{|x''|\geq \varepsilon/2} \int_{-\infty}^{\infty} + \int_{|x''|<\varepsilon/2} \int_{|x_1|>\varepsilon/2} \right] g(x)v(x,0)J(-D_x)E(x,t,\xi)\,dx_1\,dx''$$

$$+ \int_{|x''|<\varepsilon/2} \int_{\Lambda_1 \cup \Lambda_2 \cup \Lambda_3} v(z_1,x'')J(-D_z)E((z_1,x''),t,\xi)\,dz_1\,dx''.$$

Here we have of course used the notation $v(z_1,x'')$ for the function $v(x_1,x'',y_1)$. The fact that

$$\left[\int_{|x''|\geq \varepsilon/2} \int_{-\infty}^{\infty} + \int_{|x''|<\varepsilon/2} \int_{|x_1|>\varepsilon/2} \right] g(x)v(x,0)J(-D_x)E(x,t,\xi)\,dx_1\,dx''$$

can be estimated by $c \exp\,(-c'|\xi_1|)$ for $|t|$ small, $|\xi_1| \leq |\xi''|$, $|\xi| \geq f'(|\xi|)$, for some suitable sublinear function f', can be proved with arguments similar to the ones from the proof of lemma 9.5.2. To study the remaining part of I we note that

$$|E(z,t,\xi)| = \exp[\xi_1(y_1 - (x_1 - t_1)^2/2) + \xi_1 y_1^2/2 - |\xi''|((x'' - t'')^2 - |y''|^2)/2]$$

in the case $\xi_1 > 0$. We may here assume that $|y_1| \leq \varepsilon$ implies $y_1^2 < |y_1|$ so that $\xi_1 y_1^2/2$ will not produce difficulties. When z is in a ε' neighborhood of $\Lambda_1 \cup \Lambda_2 \cup \Lambda_3$, ε' is suitably small, $|t|$ is small and $\xi_1 \leq |\xi''|$, we conclude that

$$|E(z,t,\xi)| \leq e^{-\xi_1 d} + (\varepsilon')^2 |\xi''| \ \text{for some} \ d > 0.$$

We conclude that

$$|J(-D_z)E(z,t,\xi)| \le C_{\varepsilon'} e^{-d\xi_1} + (\varepsilon')^2 |\xi''|$$

for $z \in \Lambda_1 \cup \Lambda_2 \cup \Lambda_3$. The argument can now be continued as in the preceding section and we omit further details.

2. **ii) $\Rightarrow$ i).** It suffices again to work in a small neighborhood of $x^0 = 0$. We shall argue essentially as in the proof of the part ii) $\Rightarrow$ i) of proposition 9.1.2. It follows under the assumptions valid in proposition 9.1.3, ii), that we can find $\varepsilon > 0$, $b \in R$, a Radon measure ν on C^n with $\int d\,|\nu| < \infty$, and a sublinear function f so that

$$u(g) = \int_{C^n} \hat{g}(-\zeta) \exp[-\varepsilon|Im\,\zeta| - (\varepsilon|Re\,\zeta'| - f(|Re\,\zeta|))_- + b\ln(1 + |\zeta|)]\,d\nu(\zeta).$$

$(a_- = (|a| - a)/2$ is the negative part of a.) Here we use proposition 2.2.12. We want to extend u to a hyperfunction v in the variables (x, y') which satisfy the Cauchy-Riemann equations in z'. Formally we shall set

$$v(x, y') = \int_{C^n} \exp[i\langle x, \zeta\rangle - \langle y', \zeta'\rangle]\exp[-\varepsilon|Im\,\zeta| - (\varepsilon|Re\,\zeta'| - f(|Re\,\zeta|))_- \\ + b\ln(1 + |\zeta|)]\,d\nu(\zeta).$$

To give a meaning to this, we shall denote by (z, t') the variables from C^{n+d}, with the understanding that $Re\,(z, t') = (x, y')$. Furthermore, we choose a fininte number of sharp open convex cones Γ^l such that

$$\bigcup_l \Gamma^l = \dot{R}^n$$

and fix Radon measures ν_l so that

$$\sum \nu_l = \nu\,, \ \sum \int d|\nu_l| < \infty,$$

$$\text{supp } \nu_l \cap \{\zeta; |Re\,\zeta| \ge 1\} \subset \{\zeta\,;\,Re\,\zeta \in \Gamma^l\}.$$

Next, we fix $G^l \subset\subset (\Gamma^l)^0$ and define for (z, t') functions

$$v_l(z, t) = \int \exp[i\langle z, \zeta\rangle - \langle t', \zeta'\rangle - \varepsilon|Im\,\zeta| - (\varepsilon|Re\,\zeta'| - f(|Re\,\zeta|))_- + b\ln(1 + |\zeta|)] \\ d\nu_l(\zeta), \tag{9.7.5}$$

whenever the integral converges. We claim

a) that this happens when $Im\,z \in G^l$, provided that $|z| + |t'|$ is small,

b) that the functions v_l are analytic when (z, t) are as in a),

c) that

$$\left(\frac{\partial}{\partial z_j} + \frac{\partial}{\partial t_j}\right) v_l(z,t') = 0, \forall j \leq d, \text{ for } (z,t') \text{ as in a) },$$

d) that $\sum b(v_{l|t'=0}) = u.$

To prove a) , we observe that when $Im\, z \in \Gamma'$, $\zeta \in \text{supp } \nu_l$, $|\text{Re }\zeta| \geq 1$, then -
$Re\,\langle Imz\,,\, Im\zeta\rangle \leq -c|Im\, z|\,|Re\,\zeta|$. For the same ζ, z we can conclude that

$$Re\,(i\langle z,\zeta\rangle - \langle t',\zeta'\rangle) - \varepsilon|Im\,\zeta| - (\varepsilon|Re\,\zeta'| - f(|Re\,\zeta|))_- + b\ln(1+|\zeta|)$$

$$|z|\,|Im\,\zeta| - c|Im\, z|\,Re\,\zeta| + |Re\,t'|\,Re\,\zeta'| + |Im\,t'|\,|Im\,\zeta'| - \varepsilon|Im\,\zeta| -$$

$$-(\varepsilon|Re\,\zeta'| - f(|Re\,\zeta|))_- + b\ln(1+|\zeta|) \leq c(z)$$

if $|Re\, z| + |Im\, t'|$ is small.

This gives a). b) and c) follow immediately. As for d) we have that

$$v_l(z,0) = \int \exp[\langle z,\zeta\rangle - \varepsilon|Im\,\zeta| - (\varepsilon|Re\zeta'| - f(|Re\,\zeta|))_- + b\ln(1+|\zeta|)]\,d\nu_l(\zeta).$$

It is an easy matter to see that $\sum b(v_l(z,0)) = u$ and we shall omit the details.

9.8   The sheaf of partially analytic microfunctions

1. Let U be open in R^n. The basic ingredient in Sato's theory is the sheaf of microfunctions on T^*U. We recall that, starting from the sheaf (of hyperfunctions) B , M.Sato had introduced microfunctions as equivalence classes of hyperfunctions modulo the following equivalence relation (which depends on (x^0, ξ^0)):

$$v \sim u \text{ at } (x^0, \xi^0), \text{ if } (x^0, \xi^0) \notin WF_A^1(u-v). \tag{9.8.1}$$

Since in most of these notes we have worked with distributions rather than with hyperfunctions, we shall work in the sequel only with microfunctions which have representatives in distributions. We may then as well define, from the very beginning, the stalk of microfunctions at (x^0, ξ^0) by

$$\mathcal{C}_{(x^0,\xi^0)} = D'_{x^0}/\sim, \tag{9.8.2}$$

where D'_{x^0} denotes the space of germs of distributions at x^0 and $\sim$ is again the equivalence relation from (9.8.1). We may also equivalently set

$$\mathcal{C}_{(x^0,\xi^0)} = D'(U)/\sim, \tag{9.8.3}$$

where U is any open set which contains x^0. Note that (9.8.2) defines a projection

$$\Pi_{(x^0,\xi^0)} : D'_{x^0} \to \mathcal{C}_{(x^0,\xi^0)},$$

which sends $u \in D'_{x^0}$ in its equivalence class. Actually it is often simpler to work with representants in D'_{x^0} rather than with the classes from $\mathcal{C}_{(x^0,\xi^0)}$. The following definition corresponds to the notion of partially analytic microfunctions, if we use representants:

Definition 9.8.1. *Assume* $\Sigma = \{\xi; \xi_i = 0, i \leq d\}$ *for some* d. *We shall say that* u *is partially analytic in the directions parallel to* Σ *at* (x^0, ξ^0), *if we can find* $v \in D'_{x^0}$ *which is partially analytic in the directions parallel to* Σ *near* x^0 *so that*

$$(x^0, \xi^0) \notin WF^1_A(u - v).$$

(When we speak about partial analyticity in this section, we shall always think, to make a choice, on hyperfunction extensions of distributions. Cf. section 9.1.)

We can then prove the following result:

Proposition 9.8.2. *Consider* $(x^0, \xi^0) \in \Sigma$ *and* $u \in D'(U)$. *Then there are equivalent:*

i) u *is partially analytic in the variables parallel to* Σ *at* (x^0, ξ^0).

ii) There is no $\sigma \in N_{(x^0,\xi^0)}\Sigma$ *so that* $(x^0, \xi^0, \sigma) \in WF^2_A u$.

(The WF^2_A which we consider here is, in conformity with our choice of "partial analyticity", the one with the condition "$|\Pi_1\xi| \geq f(|\xi|)$" for some suitable sublinear function f.)

Proof of proposition 9.8.2. i) $\Rightarrow$ ii) is obvious from the preceding section. To prove that ii) $\Rightarrow$ i) we first apply proposition 2.2.12, assuming $x^0 = 0$. It follows in fact from the assumption ii) and that proposition that we can find

a) a Radon measure ν on C^n with $\int d|\nu| < \infty$,

b) an open cone $\Gamma \subset R^n$ which contains ξ^0, constants ε, c, b, a function $l : R^n \to R_+$ and a sublinear function g so that

$$u(f) = \int \hat{f}(\zeta) \exp[-l(-Re\,\zeta) - \varepsilon|Im\,\zeta| + b\ln(1 + |\zeta|)]\, d\nu(\zeta),$$

$$l(\xi) \geq c|\xi'|, \text{ if } \xi \in \Gamma, |\xi'| \geq g(|\xi|).$$

Let us then consider an open cone $\Gamma' \subset\subset \Gamma$ which contains ξ^0 and a continuous function $\rho : R^n \to R$ so that :

$$0 \leq \rho(\xi) \leq 1, \ \forall \xi \in R^n,$$

$$\rho(\xi) = 0, \text{ if } \xi \notin \Gamma,$$

$$\rho(\xi) = 1, \text{ if } \xi \in \Gamma', \ |\xi| \geq 1.$$

We also set

$$v(f) = \int \hat{f}(\zeta) \rho(-Re\,\zeta) \exp[-l(-Re\,\zeta) - \varepsilon|Im\,\zeta| + b\ln(1 + |\zeta|)]\, d\nu(\zeta).$$

It is clear that $(x^0, \xi^0) \notin WF_A^1(u - v)$, so u and v define the same microfunction at (x^0, ξ^0). Moreover, it is clear from section 9.1 that v is partially analytic in the variables parallel to Σ.

Remark 9.8.3. *Definition 9.8.1 is only interesting when* $(x^0, \xi^0) \in \Sigma$. *In fact otherwise* u *is partially analytic in the directions parallel to* Σ *at* (x^0, ξ^0) *precisely when* $(x^0, \xi^0) \notin WF_A^1 u$. *This can be seen arguing, e.g., as in the preceding proof.*

2. Since all the concepts used in the preceding definition and proposition have an invariant meaning, one can extend them for the case when Σ is an analytic homogeneous regular involutive submanifold from T^*U. This is in fact already indicated by our terminology. Definitions similar to the above ones also make sense for higher order wave front sets. For simplicity we have preferred here to state only the case of two-microlocalization explicitly. It is of course only in this case that we have given an invariant meaning to the underlying notions.

9.9   Second microlocalization and sheaves

1. We discuss in this section, remaining on a rather elementary level, which are the basic sheaves related to second microlocalization and describe some properties of these sheaves which seem to be inherent to any process of second microlocalization. In doing so we shall refer, to make a choice, to the wave front set WF_A^2 defined by the condition "$|\Pi_1\xi| \geq f(|\xi|)$" for some sublinear f, other variants of second wave front sets being analyzed in a similar way. In principle it should be possible to obtain related results also in the case of higher microlocalization. At the present stage of this work, it does not make much sense however to do so, since we have not studied the invariant meaning of our definitions.

2. Let U open in R^n (or a real-analytic manifold) be given. The three sheaves considered by Sato as a frame for first microlocalization, are:

a) the sheaf $\mathcal{A}$ of germs of real-analytic functions on U;

b) the sheaf B of germs of hyperfunctions on U,

c) the sheaf C of microfunctions on T^*U.

The fundamental condition which relates $\mathcal{A}$, B, C is that

$$u \in B_{x^0}, \Pi_{(x^0, \xi^0)} u = 0, \forall \xi^0 \in \dot{R}^n \text{ implies } u \in \mathcal{A}_{x^0}.$$

Here $\Pi_{(x^0, \xi^0)}$ is, as in the preceding section, the canonical projection $B_{x^0} \to C(x^0, \xi^0)$ which maps u in its equivalence class modulo the equivalence relation $\sim$ from (9.8.1).

In the case of second microlocalization, the "base space" will be some fixed analytic homogeneous regularly involutive manifold $\Sigma \subset T^*U$ and the role of T^*U will be taken by $N\Sigma$. It is also clear that the role of B will be played by the restriction of the sheaf C of microfunctions to Σ, since it is C which we want to microlocalize further. Since in most of these notes we worked with distributions, we shall restrict our attention, as in the preceding section, to microfunctions which have representatives in distributions, i.e., we set once more

$$C(x^0, \xi^0) = D'_{x^0} / \sim .$$

It is also easy to see which are the sheaves which we shall use to replace $\mathcal{A}$ and C from before : the role of $\mathcal{A}$ will be played by the sheaf $\mathcal{F}$ of microfunctions on Σ which are partially analytic in the directions parallel to Σ, and that of C will be played by the sheaf of two-microfunctions defined on $N\Sigma$ in the following way. If $(x^0, \xi^0, \sigma^0) \in N\Sigma$ is fixed, we introduce an equivalence relation $\sim^2$ (which depends on (x^0, ξ^0, σ^0)) :

if $u, v \in C(x^0, \xi^0)$, then we write $u \sim^2 v$, if $(x^0, \xi^0, \sigma^0) \notin WF_A^2(u - v)$.

(Here we shall say for $w \in C(x^0, \xi^0)$ that $(x^0, \xi^0, \sigma^0) \notin WF_A^2 w$, if we can find a representant f of w so that $(x^0, \xi^0, \sigma^0) \notin WF_A^2 f$. Of course, if this is true for some representant of w, then it is true for any other representant, so the second wave front set is well-defined for microfunctions.)

We then define the stalk of $\mathcal{E}$ at (x^0, ξ^0, σ^0) by

$$\mathcal{E}(x^0, \xi^0, \sigma^0) = C(x^0, \xi^0) / \sim^2 .$$

We thus have here, too, a natural projection

$$\Pi_{(x^0, \xi^0, \sigma^0)} : C(x^0, \xi^0) \to \mathcal{E}(x^0, \xi^0, \sigma^0).$$

and the "fundamental relation" between $\mathcal{E}$, C and F is :

$$\cap_{\sigma^0} \ker \Pi_{(x^0, \xi^0, \sigma^0)} = \mathcal{F}_{(x^0, \xi^0)}.$$

(The fact that this is true was proved in the preceding section.)

3. We have now introduced the basic sheaves on which one may build up the theory of second microlocalization. Since in analytic microlocalization one cannot localize with the aid of partitions of unity, the cohomological properties of these sheaves are fundamental for the success of the theory. We have not tried to work out which are the cohomological properties which are of greatest use in the applications. We shall however at least state two simple results of this type. (We give explicit rather than cohomological statements.) Let us first mention the following definition:

Definition 9.9.1. *Let u be a two-microfunction on $W \subset \Sigma$, i.e., $u \in C(W)/ \sim^2$. We define*

$$ess\ supp\,^2 u = \{(x^0, \xi^0, \sigma^0) \in N\Sigma; (x^0, \xi^0) \in W, (x^0, \xi^0, \sigma^0) \in WF_A^2 u.\}$$

The following result is now related to the edge-of-the-wedge theorem:

Proposition 9.9.2. *Let $(x^0, \xi^0) \in \Sigma$ be fixed and consider open cones $Y^{j'} \subset\subset Y^j$ in R^d. Assume u in $C(x^0, \xi^0)$ is given so that*

$$ess\ supp\,^2 u \subset \cup\{(x^0, \xi^0, \sigma); \sigma \in Y^{j'}\}.$$

Then we can find $u_j \in C(x^0, \xi^0)$ so that

$$u = \sum u_j \ and \quad ess\ supp\,^2 u_j \subset \{(x^0, \xi^0, \sigma); \sigma \in Y^j\}.$$

The proof of this is not difficult and can e.g. be performed with arguments used in Liess [4]. We omit the details.

Remark 9.9.3. *We believe of course that more is true than what is stated in the proposition, in that one should be able to obtain decompositions of u associated with decompositions of $U \subset N\Sigma$. One may expect that this problem can be treated with the methods from Bengel-Shapira [1], Laurent [1] or Laubin [3].)*

Finally we mention

Proposition 9.9.4. *Let $Y^{j'} \subset\subset Y^j$ be open cones in R^d and consider $u_j \in C(x^0, \xi^0)$ so that*

$$\{(x^0, \xi^0, Y^j \cap Y^k)\} \cap WF_A^2(u_j - u_k) = \emptyset, \forall j, \forall k.$$

Then there is $u \in C(x^0, \xi^0)$ so that

$$\{(x^0, \xi^0, Y^{j'})\} \cap WF_A^2(u_j - u) = \emptyset, \forall j.$$

The proof of this result is closely related to that of theorem 1.16 in Liess [5]. Also in this case we omit the details of the argument, and a remark similar to remark 9.9.3 applies.

9.10   The Fourier transform of measures on C^n with supports concentrated on polycones.

1. When u is a tempered distribution on R^n concentrated in some sharp convex cone $\Gamma \subset R^n$, then its Fourier transform can be extended analytically to the set $\{z \in C^n; -\operatorname{Im}\zeta \in \Gamma^0\}$, where Γ^0 is the polar of Γ. In fact, formally, the extension is just

$$\hat{u}(\zeta) = u(\exp -i\langle x, \zeta \rangle),$$

which makes sense for $\zeta \in R^n - i\Gamma^0$. The same simple idea is underlying the arguments from this section.

Let us in fact consider a polyhomogeneous structure $M_0, M_1, ..., M_k$ in R^n. For simplicity we assume that $M_j = \{\xi; \xi_j = 0 \text{ for } j \geq 1 + d_j\}$. Let us also denote, as in section 2.1, by

$$\dot{M}_i = M_i \ominus M_{i+1}, \; i = 0, ..., k-1,$$

and by

$$P_i = I - \Pi_i.$$

Next we consider $\xi^i \in \dot{M}_i, i = 0, ..., k-1$. A generic multineighborhood of $(\xi^0, \xi^1, ..., \xi^{k-1})$ is then of form

$$\Gamma = \{\xi; \; (\Pi_i - \Pi_{i+1})\xi \in V_i, |\Pi_{i+1}\xi| \leq \tilde{c}|\Pi_i\xi|, \; i = 0, ..., k-1\}, \tag{9.10.1}$$

where the V_i are open conic neighborhoods of ξ^i in $\dot{M}_i$ and $\tilde{c} > 0$. We may also assume later on that the V_i are sharp convex cones. Let now also μ be a Radon measure on C^n so that

$$\int_{C^n} d|\mu|(\zeta) < \infty, \; \text{supp } \mu \subset \{\zeta; \; Re\,\zeta \in \Gamma\}.$$

The purpose of this section is to discuss the analytic extendibility of the distribution $u \in D'(x \in R^n; |x| < \varepsilon)$ defined, for fixed $\varepsilon > 0, b \in R$, by

$$u(f) = \int \hat{f}(-\zeta) \exp[-\varepsilon|\Pi_{k-1} Re\,\zeta| - \varepsilon|Im\,\zeta| + b\,ln(1 + |\zeta|)]\,d\mu(\zeta), \tag{9.10.2}$$

if $f \in C_0^\infty(x \in R^n; |x| < \varepsilon)$. The following result is immediate:

Lemma 9.10.1. *Denote by S the interior of the set*

$$\tilde{S} = \{y \in R^n; -\langle y, \xi \rangle - \varepsilon|\Pi_{k-1}\xi| \leq 0, \forall \xi \in \Gamma\}.$$

Then u admits an analytic extension to the set

$$U = \{z; z = x + iy, |x| < \varepsilon, y \in S\}.$$

Remark 9.10.2. *If $y \in \tilde{S}$ and $0 < t \leq 1$, then $ty \in \tilde{S}$ and the same property is valid for S. Indeed, if already $\langle y, \xi \rangle \geq 0, \forall \xi \in \Gamma$, then $ty \in \tilde{S}$ is trivial and if $\langle y, \xi \rangle \leq 0$ for some $\xi \in \Gamma$, then we must have at least $|\langle y, \xi \rangle| \leq \varepsilon |\Pi_{k-1}\xi|$. But then we also have $|\langle ty, \xi \rangle| \leq \varepsilon |\Pi_{k-1}\xi|$ for $0 < t \leq 1$.*

Note that this remark gives sense to the statement in lemma 9.10.1, in that we may now speak of boundary values of holomorphic functions $h(z)$ on U when $\text{Im } z \to 0$, $z \in U$.

Proof of lemma 9.10.1. When $z = x + iy \in U$ is fixed we can find $c > 0$ so that

$$-\langle y, \xi \rangle - \varepsilon |\Pi_{k-1}\xi| \leq -c|\xi|, \ \forall \xi \in \Gamma.$$

It is then easy to see that $v : U \to C$ given by

$$v(z) = \int \exp[i \langle z, \zeta \rangle - \varepsilon |\Pi_{k-1} Re \, \zeta| - \varepsilon |Im \, \zeta| + b \, ln(1 + |\zeta|)] \, d\mu(\zeta) \qquad (9.10.3)$$

is well-defined and holomorphic. For fixed $y \in S$ we will also have

$$\int v(x + iy) f(x) dx = \int \hat{f}(-\zeta) \exp[i \langle y, \zeta \rangle - \varepsilon |\Pi_{k-1} Re \, \zeta| - \varepsilon |Im \, \zeta| + b \, ln(1 + |\zeta|)] \, d\mu(\zeta),$$

whenever $f \in C_0^\infty(x \in R^n; |x| < \varepsilon)$. It is then immediate that

$$\lim_{y \to 0, y \in S} v(\cdot + iy) = u, \text{ weakly in } D'(|x| < \varepsilon).$$

2. Let us also analyze the geometric shape of $\tilde{S}$ for a moment. If $y \in \tilde{S}$, then we obtain in particular

$$-\langle y, \mu_i(t)\xi \rangle - \varepsilon |\Pi_{k-1}\mu_i(t)\xi| \leq 0, \ \forall i, \forall \xi \in \Gamma, \forall t > 0.$$

For $t \to 0$ this gives

$$\langle P_i y, \ P_i \xi \rangle \geq 0, \forall \xi \in \Gamma, \forall 0 < i \leq k - 1.$$

Conversely, we have

Lemma 9.10.3. *Consider $0 < i \leq k - 1$ and denote, for fixed c, by*

$$T_i = \{ y \in R^n; |\Pi_{k-1}y| < \varepsilon, \langle P_i y, P_i \xi \rangle > c|P_i \xi| \, |P_i y|, \forall \xi \in \Gamma,$$
$$\langle (\Pi_j - \Pi_{j+1})y, \xi \rangle \geq 0, \text{ if } \Pi_j \xi \in V_j, j = i, ..., k - 1 \},$$

if $i \leq k - 2$, respectively,

$$T_{k-1} = \{ y \in R^n; |\Pi_{k-1}y| < \varepsilon, \langle P_{k-1}y, P_{k-1}\xi \rangle > c|P_{k-1}\xi| |P_{k-1}y|, \forall \xi \in \Gamma \}.$$

Then $T_i \subset S$.

Indeed, when $y \in T_i$ we have

$$-y = -[P_i + \sum_{j=i}^{k-2}(\Pi_j - \Pi_{j+1}) + \Pi_{k-1}]y,$$

so that $-\langle y, \xi \rangle - \varepsilon|\Pi_{k-1}\xi| \leq -c|P_i\xi|\,|P_iy|$, $\forall \xi \in \Gamma$. Taking into account that $|\xi| \leq c'|P_i\xi|$ for $\xi \in \Gamma$, this can be improved to

$$- \langle y, \xi \rangle - \varepsilon|\Pi_{k-1}\xi| \leq -c''|\xi|\,|P_iy|, \forall \xi \in \Gamma. \tag{9.10.4}$$

If therefore $|y - \tilde{y}| < c''|P_iy|/2$, we will obtain

$$-\langle \tilde{y}, \xi \rangle - \varepsilon|\Pi_{k-1}\xi| \leq -c''|\xi||P_iy|/2, \forall \xi \in \Gamma.$$

It follows that a full neighborhood of y is in $\tilde{S}$, so $y \in S$. (Note that $P_iy \neq 0$ since we could not have otherwise $\langle P_iy, P_i\xi \rangle > c|P_iy|\,|P_i\xi|$.)

Remark 9.10.4. *The situation is particularily simple when $k = 2$. In fact, only T_0 is then nontrivial and actually equal to*

$$T_1 = \{y \in R^n; |\Pi_1y| < \varepsilon, \langle P_1y, P_1\xi \rangle > c|P_1\xi|\,|P_1y|, \forall \xi \in \Gamma\}.$$

Since $\Gamma = \{\xi; P_1\xi \in V_0, \Pi_1\xi \in V_1, |\Pi_1\xi| < c|\xi|\}$, the second condition in the definition of T_1 therefore gives

$$\langle P_1y, \xi \rangle > c|\xi|\,|P_1y| \text{ if } \xi \in V_0,$$

i.e., P_1y must lie in the polar of V_0.

3. From the proof of lemma 9.10.1, or also, post factum from the conclusion of the lemma, and simple functional analytic considerations, it is clear that the extension v of u is of temperated growth when $y \to 0$. We prove a quantitatively more precise version of this:

Lemma 9.10.5. *There are c_1 and $k \geq 0$ so that v defined by (9.10.3) satisfies, for fixed i,*

$$|v(z)| \leq c_1|P_iy|^{-k}, \quad \text{if } z = x + iy, |x| < \varepsilon, y \in T_i. \tag{9.10.5}$$

Proof. Using the inequality (9.10.4) from the proof of lemma 9.10.3, we see that it suffices to check that

$$(1 + |\xi|)^b \exp[-c''|P_iy|\,|\xi|] \leq c_1|P_iy|^{-k},$$

if $\zeta \in R^n$ and c_1, respectively k is conveniently chosen. Actually this comes to

$$(1 + \beta)^b \exp[-c''\alpha\beta] \leq c_1\alpha^{-k}, \forall \alpha > 0, \forall \beta > 0, \qquad (9.10.6)$$

which is immeditate to check. (We may, e.g., study the maximum for $\beta > 0$ of

$$(1 + \beta)^b \exp[-c''\alpha\beta]$$

for fixed $\alpha > 0$.)

Remark 9.10.6. *An interesting feature of (9.10.5) is that it is associated with a multi-homogeneous structure of R^n which is complementary to that given by the M_i. In fact, the natural multiplications are now $\nu_i(t)\xi = \Pi_i\xi + tP_i\xi$, rather than $\mu_i(t)\xi = P_i\xi + t\Pi_i\xi$.*

4. A situation which is very similar to the one above appears when, rather than considering (9.10.2), we look at the distribution

$$u(f) = \int \hat{f}(-\zeta)\exp[-\varepsilon|Im\ \zeta| + b\,ln(1 + |\zeta|)]\,d\mu(\zeta), \qquad (9.10.7)$$

μ as before. This time, u can be extended analytically to the set $U' = \{|x| < \varepsilon\} + iS'$, where S' is the polar of Γ. When $k = 2$ this polar contains, provided Γ still has the form from (9.10.1), the sets

$$\{y; \langle P_1 y, \xi \rangle > 0, \xi \in V_0, \langle \Pi_1 y, \xi \rangle > 0, \xi \in V_1\}.$$

9.11 Higher order wave front sets and boundary values of holomorphic functions

1. If $G^1, ..., G^s$ is a collection of open convex cones in R^n so that $\cup G^j = \dot{R}^n$ and if, e.g., $u \in E'(R^n)$ is given, then we have seen that we can write that

$$u = \sum b(h_j), \qquad (9.11.1)$$

where the h_j are holomorphic in $R^n + iG^{j0}$. Similar statements are valid for general hyperfunctions, although they are then a little more complicated. (A local variant of the decomposition (9.11.1) can be achieved for hyperfunctions, using the argument from section 9.3 for analytic functionals.) Actually, for a given u and locally near some x^0, we do not always need a full range of cones G^j to arrive at a representation of type (9.11.1): it is this remark which is the starting point for Sato's theory of microlocalization. Let us in fact recall Sato's definition of the first analytic wave front set.

Definition 9.11.1. *Let* $u \in B(U)$ *and consider* $x^0 \in U$, $\xi^0 \in R^n$. *We shall say that*

$$(x^0, \xi^0) \notin WF_A u,$$

if we can find a collection of open cones $\Gamma^1, ..., \Gamma^l$ *in* $\dot{R}^n$, $\varepsilon > 0$, *a complex neighborhood* $\tilde{U}$ *of* U *in* C^n *and holomorphic functions*

$$h_j \in O(z \in \tilde{U}; |Re\, z - x^0| < \varepsilon,\, Im\, z \in \Gamma^j),$$

so that

$$u = \sum b(h_j) \quad for \ |x - x^0| < \varepsilon, \tag{9.11.2}$$

$$\xi^0 \notin \Gamma^{j0}, \quad for \ j = 1, ..., l. \tag{9.11.3}$$

Note in particular, that if $u \in D'(U)$ is of form $u = b(h)$, $h \in O(z \in \tilde{U}; Re\, z \in U,\, Im\, z \in \Gamma)$, then the wave front set of u (over U) is smaller for larger Γ. This corresponds to the intuitive feeling that u should be considered more regular if it is a boundary value of a holomorphic function defined on a larger wedge.

Definition 9.11.1 was, historically speaking, the first explicit definition of a concept of wave front set. Several other defintions of wave front sets have been formulated later on, which all turned out to be equivalent in the analytic case. In fact, it has been proved by Bony [3,4] that there is essentially only one way to introduce an analytic wave front set which possesses all the structural properties one wants an analytic wave front set to have. (For a more direct argument, see e.g. Liess [4].) Let us mention here, for comparision sake, a distributional version of definition 9.11.1

Definition 9.11.2 *Let* $u \in D'(U)$ *and consider* $x^0 \in U$, $\xi^0 \in \dot{R}^n$. *We shall write that*

$$(x^0, \xi^0) \notin wf_A u,$$

if we can find a collection of open cones $\Gamma^1, ..., \Gamma^l$ *in* $\dot{R}^n$, $\varepsilon > 0$, *distributions* $u_j \in D'(x \in R^n; |x - x^0| < \varepsilon)$, *a complex neighborhood* $\tilde{U}$ *of* U *in* C^n *and holomorphic functions*

$$h_j \in O(z \in \tilde{U}; |Re\, z - x^0| < \varepsilon,\, Im\, z \in \Gamma^j),$$

so that

$$u = \sum b(h_j) \quad for \ |x - x^0| < \varepsilon, u_j = \lim_{y \to 0, y \in \Gamma^j} h_j(\cdot + iy),$$

$$\xi^0 \notin \Gamma^{j0}, \quad for \ j = 1, ..., l.$$

3. It is part of Bony's theorem that if $u \in D'(U)$, then

$$(x^0, \xi^0) \notin WF_A u \text{ is equivalent to } (x^0, \xi^0) \notin wf_A u. \tag{9.11.4}$$

Remark 9.11.3. *The fact that definition 9.11.2 is equivalent with Hörmander's first definition of the standard analytic wave front set had been established independently of Bony's theorem by Nishiwada [1] and had also been realized to be a consequence of the work of Bros-Iagolnitzer, cf. Iagolnitzer [1]. (Also cf. Liess [6] where a characterization by duality similar to proposition 2.2.12 for Hörmander's wave front set had been given. Actually it is not difficult to see that that proposition immediately leeds to the result of Nishiwada. A related argument is given in Björk [1].)*

4. As a consequence of (9.11.4) the frame, distributional or hyperfunctional, in which we introduce the first analytic wave front set for distributions has no bearing on the results. The situation is slightly different for higher microlocalization: we have seen above that several variants of WF_A^k seem interesting, and in particular it is worthwhile sometimes to analyze if one wants to remain in distributions, or if one is better off by working in some larger frame.

5. It is natural to ask if one could not possibly characterize (absence of) higher order wave front sets with the aid of representations similar to those which appear in definition 9.11.2. This seems indeed possible and in fact for the case of two-microlocalization the problem has been solved in Okada-Tose [1]. Actually one of the relations between Sato-type definitions and definitions based on the FBI-transform in the case $k = 2$ had been discussed earlier already in Esser-Laubin [2]. We shall discuss here briefly the implication not considered in Esser-Laubin [2], for the case of general microlocalization. For simplicity we shall place ourselves in a purely distributional frame. What we shall do then is to discuss what kind of a representation of u as a sum of boundary values of holomorphic functions one can obtain if one knows that

$$(x^0, \xi^0, \xi^1, ..., \xi^{k-1}) \notin WF_A^k u. \tag{9.11.5}$$

The fact now, that we consider the "distributional" case is meant to say not only that $u \in D'(U)$ (actually we have not considered higher order wave front sets for anything else) but also that the conditions $|\Pi_i \xi| \geq f_i(|\Pi_{i-1}\xi|)$, f_i sublinear, are replaced by $|\Pi_i \xi| \geq m \ln(1 + |\xi|)$, $i = 1, ..., k - 1$. (The reason why we work here "completely" in distributions is that the result of Esser-Laubin [2], to which we want to relate our discussion, is more elaborate for distributions, than it is in the case of the hyperfunctional WF_A^2.)

To simplify the discussion we shall proceed inductively. It is then convenient to write - and how this is done effectively will be discussed in a moment-

$$u = \sum_{j=0}^{k-1} u_j, \tag{9.11.6}$$

where the u_j have the property that

a) $(x^0, \xi^0, \xi^1, ..., \xi^{i-1}) \notin WF_A^i u_i$, $i = 0, ..., k-1$,

b) $(x^0, \eta^0, \eta^1, ..., \eta^{s-1}) \in WF_A^s u_i$ implies for $s < i$ that η^r is in a small conic neighborhood of ξ^r if $r \leq s - 1$.

The intuitive interpretation of this step of the argument is of course that it is not all of u that matters, but only the k-microfunction class of u, which is obtained by throwing away all parts of u which are negligible modulo lower order wave front sets.

Let us show how one can obtain a decomposition of type (9.11.6), assuming for simplicity that $x^0 = 0$. In fact, from (9.11.5) it follows that we can find $d > 0$, $\varepsilon > 0$, $m \geq 0, b \in R$ and a multineighborhood Γ of $(\xi^0, \xi^1, ..., \xi^{k-1})$ so that

$$|u(f)| \leq c \tag{9.11.7}$$

for any $f \in C_0^\infty(|x| < \varepsilon)$ which satisfies

$$|\hat{f}(\zeta)| \leq \exp[d|\Pi_{k-1} Re\, \zeta| + \varepsilon|Im\, \zeta| - b\ln(1 + |\zeta|)], \tag{9.11.8}$$

if $Re\, \zeta \in -\Gamma$, $|\Pi_i Re\, \zeta| \geq m\ln(1 + |Re\, \zeta|)$, $i > 0$, respectively

$$|\hat{f}(\zeta)| \leq \exp[\varepsilon|Im\, \zeta| - b\ln(1 + |\zeta|)], \tag{9.11.9}$$

for the remaining ζ. When $|\Pi_i Re\, \zeta| \leq m\ln(1 + |Re\, \zeta|)$ for some i, then $d|\Pi_{k-1} Re\, \zeta| \leq m'\ln(1 + |Re\, \zeta|)$. It follows that (9.11.7) will also be valid if we replace (9.11.8), (9.11.9) by

$$|\hat{f}(\zeta)| \leq \exp[d|\Pi_{k-1} Re\, \zeta| + \varepsilon|Im\, \zeta| - b'\ln(1 + |\zeta|)], \text{ if } Re\, \zeta \in -\Gamma, \tag{9.11.10}$$

$$|\hat{f}(\zeta)| \leq \exp[\varepsilon|Im\, \zeta| - b'\ln(1 + |\zeta|)], \text{ if } Re\, \zeta \notin -\Gamma, \tag{9.11.11}$$

for suitable b'.

Let us now choose a continuous function $0 \leq r(\xi) \leq |\Pi_{k-1}\xi|$ on R^n so that $r(\xi) = |\Pi_{k-1}\xi|$ in some multineighborhood of $(\xi^0, \xi^1, ..., \xi^{k-1})$ and so that $|\xi| \geq 1, \xi \notin \Gamma$, implies $r(\xi) = 0$. We conclude that (9.11.7) will hold if

$$|\hat{f}(\zeta)| \leq \exp[dr(-Re\, \zeta) + \varepsilon|Im\, \zeta| - b\ln(1 + |\zeta|)].$$

It is now a consequence of Hahn-Banach's theorem that we can find a Radon measure μ on C^n with $\int d|\mu| < \infty$ and for which

$$u(f) = \int_{C^n} \hat{f}(-\zeta) \exp[-r(Re\,\zeta) - \varepsilon|Im\,\zeta| + b'\ln(1 + |\zeta|)]\,d\mu(\zeta),$$

if $f \in C_0^\infty(|x| < \varepsilon)$. (We write $u(f) = \int \hat{f}(-\zeta) \cdots$ rather than $u(f) = \int \hat{f}(\zeta) \cdots$ in order to be closer to Parsevals formula.)

To construct some u_0 as needed in (9.11.6) we shall now use a continuous function $\rho_0 : R^n \to R$, $0 \leq \rho_0(\xi) \leq 1$, such that

c) $\rho_0(\xi) = 0$ if $|\xi| \geq 1$ and ξ is outside some previously fixed small convex neighborhood of ξ^0,

d) $\rho_0(\xi) = 1$ if ξ is in some still smaller conic neighborhood of ξ^0.

We then set

$$u_0(f) = \int_{C^n} (1 - \rho_0(Re\,\zeta))\hat{f}(-\zeta) \exp[-r(-Re\,\zeta) - \varepsilon|Im\,\zeta| + b'\ln(1+|\zeta|)]\,d\mu(\zeta), \quad (9.11.12)$$

We are thus left with a distribution of type (9.11.12), where $1 - \rho_0$ is replaced by ρ_0. We can then continue to argue in this way and set inductively (and we do not make this here completely explicit), as long as $j < k - 1$,

$$u_j(f) = \int_{C^n} \hat{f}(-\zeta) \exp[-r(Re\,\zeta) - \varepsilon|Im\,\zeta| + b'\ln(1 + |\zeta|)]\,d\nu_j(\zeta),$$

for some Radon measures ν_j on C^n which satisfy $\int d|\nu_j| < \infty$ and which have the following properties e) and f):

e) $\zeta \in \text{supp } \nu_j$ and $|\Pi_l Re\,\zeta| \geq 1$ implies for $l < j - 1$ that $\Pi_l Re\,\zeta \in G^l$,

f) $\Pi_j Re\,\zeta \in G^j$ implies $\zeta \notin \text{supp } \nu_j$.

Here the G^l are some previously fixed sharp open convex cones in M_l which contain ξ^l.

For $j = k - 1$ the situation is similar, although now rather than asking for f), we shall ask for the condition

g) $\zeta \in \text{supp } \nu_{k-1}$ implies $r(Re\,\zeta) = |\Pi_{k-1} Re\,\zeta|$.

The fact that a) is valid is now easy to check using the test by duality and b) is true by construction.

6. We have now described how to obtain a decomposition of type (9.11.6) with a) and b). Since we may assume that we already know how to write the u_j, $j < k-1$, in a relevant way as sums of boundary values of holomorphic functions, it remains to see what can be said about u_{k-1}. To decompose u_{k-1} into pieces related to a situation similar to that in definition 9.11.2, we may now argue in a way similar to the above and write

$$u_{k-1} = \sum_{l=0}^{s} v_l$$

for some s, where the v_l have representations of form

$$v_l(f) = \int_{C^n} \hat{f}(-\zeta) \exp[-r(Re\,\zeta) - \varepsilon|Im\,\zeta| + b'\ln(1 + |\zeta|)]\,d\kappa_l(\zeta),$$

and where the κ_l are Radon measures on C^n with the following properties

h) $\int d|\kappa_l|(\zeta) < \infty$,

i) supp $\kappa_l \subset$ supp ν_{k-1}, $l = 0, ..., s$,

j) for $l > 1$, supp $\kappa_l \subset X_l + iR^n \subset C^n$, where X_l is a sharp convex multicone which avoids a multineighborhood of $(\xi^0, \xi^1, ..., \xi^{k-1})$.

k) for $l = 0$, $\zeta \in$ supp κ_0, $|Re\,\zeta| \geq 1$ implies $r(Re\,\zeta) = |\Pi_{k-1} Re\,\zeta|$.

With each v_l we are now in a situation of the type studied in section 9.10. It is clear from there that one can compute realistic domains to which the v_l can be extended holomorphically. We do not write down the result explicitly, in that the geometric shape of the domain of extensions is rather complicated. Moreover, lemma 9.10.5 shows that these extensions are of temperated growth near y=0 in several groups of variables. From Esser-Laubin [2] it follows that this is all there is in $(x^0, \xi^0, \xi^1) \notin WF_A^2 u_1$, but if this is also true for $k > 2$ should still be checked. We have not tried to do so, since it is not even clear that what is interesting for higher microlocalization are representations in terms of boundary values of holomorphic functions.

References

Agmon S.-Douglis A.-Nirenberg L. : [1] *Estimates near the boundary for solutions of elliptic partial differential equations satisfying general boundary conditions I.* C.P.A.M. 12 (1959), 623-727.

Alber H.D. : [1] *Untersuchungen von Gleichungen der mathematischen Physik mit der Methode der geometrischen Optik am Beispiel der konischen Refraktion.* Z.A.M.M. 68:4 (1988), T5-T12.

Arisumi T. : [1] *Propagation of Gevrey singularities for a class of microdifferential operators.* J.Fac. Sci. Univ. Tokyo, Sect. IA, 38 (1991), 339-350.

Arnold V.I. /Gusein-Zade S.M./ Varchenko A.N. : [1] *Singularities of differentiable maps*, vol.I. Birkhäuser Verlag, Boston- Basel- Stuttgart, 1985.

Artin M. : [1] *On the solutions of analytic equations.* Inventiones Math. 5 (1968), 277-291.

Atiyah M.-Bott R.-Gårding L. : [1] *Lacunas for hyperbolic operators with constant coefficients.* Acta Math., 129 (1972), 109-189.

Beals R. : [1] *A general calculus of pseudodifferential operators.* Duke Math.J., 42 (1975), 1-42.

Bengel G.-Schapira P. : [1] *Décomposition microlocale analytique des distributions.* Ann. Inst. Fourier Grenoble, 29 (1979), 101-124.

Bernardi E. : [1] *Propagation of singularities for hyperbolic operators with multiple involutive characteristics.* Osaka J. Math. 25 (1988), 19-31.

Bernardi E.-Bove A. : [1] *Propagation of Gevrey singularities for a class of operators with triple characteristics* I,II. Duke Math.J., 60 (1989, 1990), 187-205, 207-220.

Bernardi E.-Bove A.-Parenti C. : [1] *Propagation of C^∞ singularities for a class of operators with double characteristics.* In Pitman Research Notes in Mathematics, (1988), 255-265.

Björk J.E. : [1] *Rings of differential operators.* North- Holland Publ. Comp., Amsterdam/New York, 1979.

Bony J.M. : [1] *Extension du théorème de Holmgren.* Séminaire Goulaouic-Schwartz 1975-1976, exposé 17.

[2] *Analyse microlocale des équations aux dérivées partielles non linéaires.* Lectures at the C.I.M.E. Summer School, Montecatini, 1989. In *"Microlocal Analysis and Applications"*, Lecture Notes in Mathematics of the Springer Verlag Nr. 1495, ed. by L.Cattabriga and L.Rodino, (1992), 1-46.

[3] *Propagation des singularités différentiables pour une class d'opérateurs à coefficients analytiques.* Astérisque, 34-35, (1976), 186-220.

[4] *Equivalence des diverses notions de spectre singulière analytique.* Sém. Goulaouic-Schwartz, Ecole Polytechnique, 1976-1977, exp. n° 3.

Bony J.M.-Schapira P. : [1] *Propagation des singularités analytiques pour les solutions des équations aux dérivées partielles.* Ann. Inst. Fourier Grenoble, 26 (1976), 81-140.

Boutet de Monvel L. : [1] *Propagation des singularités des solutions d'équations analogues à l'équation de Schrödinger.* In "Fourier Integral Operators and Partial Differential Equations", Lecture Notes in Math. of the Springer Verlag, vol. 459, ed. by J. Chazarain, (1975), 1-15.

Bronstein M.D. : [1] *On the smoothness of roots of polynomials, which depend on parameters.* Sibirskii Mat. Journ. t.XX:6 (1979), 493-501.

Courant R.-Hilbert D. : [1] *Methods of mathematical physics.* Interscience Publ., 1962.

Darboux G. : [1] *Leçons sur la théorie générale des surfaces* ,vol.IV. Gauthier-Villars, 1896.

Delort J-M. : [1] *F.B.I. Trasformation. Second Microlocalization and Semilinear Caustics.* Springer Lecture Notes in Mathematics, vol. 1522 (1992).

Dencker N. : [1] *The propagation of singularities for pseudodifferential operators with self-tangential characteristics.* Ark. Mat. 27:1 (1989), 65-88.
[2] *On the propagation of polarization sets for systems of real principal type.* J.Funct. Anal. 46:3 (1982), 351-372.

Duistermaat J.J. : [1] *Fourier integral operators.* Courant Institute Lecture Notes, New York, 1974.

Ehrenpreis L. : [1] *Fourier analysis in several complex variables.* Interscience Publ., New York, 1970.

Encyklopädie der mathematischen Wissenschaften : [1] Bd.III, 2.Teil, 2.Hälfte, Teubner Verlag, Leipzig.

Esser P. : [1] *Second analytic wave front set in crystal optics.* Appl. Analysis, vol. 24 (1987), 189-213.
[2] *Polarisation analytiques dans les problems aux limites non-diffractifs de multiplicité constante.* Bull. Soc. Royale Liège, 55:3 (1988), 371-465.

Esser P.-Laubin P. : [1] *Second microlocalization on involutive submanifolds.* Séminaire d'analyse supérieure, Univ. de Liège, Institut de Mathematique, Liege 1987.
[2] *Second analytic wave front set and boundary values of holomorphic functions.* Applicable Analysis, vol. 25 (1987), 1-27.

Fefferman C. : [1] *The uncertainty principle.* Bull.A.M.S. 9:2 (1983), 129-206.

Fefferman C.- Phong D.H. : [1] *On the asymptotic eigenvalue distribution of a pseudo-differential operator.* Proc. Nat. Acad. Sci. U.S.A., vol. 77:10 (1980). 5622-5625.

Fladt K.-Baur H. : [1] *Analytische Geometrie spezieller Flächen und Raumkurven.* Vieweg Verlag, Braunschweig, 1975.

Gårding L. : [1] *History of mathematics of double refraction.* Archive for the History of exact sciences. Vol. 40:4 (1990), 355-387.

Gerard C. : [1] *Réflexion du front d'onde polarisé des solutions des systèmes d'équations aux dérivées partielles.* C.R. Acad. Sci. Paris, 297 (1983), 409-412.
[2] *Propagation de la polarisation pour des problèmes aux limites.* Thèse, Orsay, 1984.

Grigis A.-Lascar R. : [1] *Equations locales d'un système de sous-varietés involutives.* C.R. Acad. Sci. Paris, 283 (1976), 504-506.

Grigis A.-Schapira P.-Sjöstrand J.. : [1] *Propagation des singularités analytiqyes pour les opérateurs à caractéristiques multiples.* C.R. Acad. Sci. Paris, 293 (1981), 397-400.

von Grudzinski O. : *Quasihomogeneous distribution.* North Holland Mathematical Studies, vol.165, North Holland, New York Amsterdam, 1991.

Hanges N. : [1] *Propagation of analyticity along real bicharacteristics.* Duke Math. J., 48:1 (1981), 269-279.

Hanges N.-Sjöstrand J. : [1] *Propagation of analyticity for a class of non-micro-characteristic operators.* Ann. of Math., 116 (1986), 559-577.

Hansen S. : [1] *Singularities of transmission problems.* Math. Ann., 268 (1984). 233-253.

Herglotz G. : [1] *Über die Integration linearer partieller Differentialgleichungen mit konstanten Koeffizienten* I-III, Berichte Sächs. Akad. Wiss., 78 (1926), 93-126. 287-318, 80 (1928). 69-114.

Hörmander L. : [1] *Fourier integral operators I.* Acta Math., 127 (1971), 79-183.
[2] *Uniqueness theorems and wave front sets for solutions of linear differential equations with analytic coefficients.* C.P.A.M., 24 (1971), 671-704.

[3] *On the existence and regularity of solutions of linear pseudodifferential operators.* L'enseignement math., 17:2 (1971), 99-163.

[4] *Convolution equations in convex domains.* Inventiones Math., 4:6 (1968), 306-318.

[5] *The analysis of linear partial differential operators* I,II,III. Springer Verlag, Grundlehren der mathematischen Wissenschaften, vol.251, 252, 274, Berlin-New York, 1983, 1985.

[6] *An introduction to complex analysis in several variables.* D. van Nostrand Comp., Princeton, 1966.

[7] *On the existence of real analytic solutions of partial differential equations with constant coefficients.* Inv. Math. 21 (1973), 151-182.

[8] *The Weyl calculus of pseudodifferential operators.* C.P.A.M. XXXII (1979), 359-443.

Iagolnitzer,D. : [1] *Microlocal essential support of a distribution and decomposition theorems-an introduction.* In "Hyperfunctions and Theoretical Physics". Lecture Notes in Math. vol. 449, 121-133, ed. F.Pham, Springer Verlag, 1975.

Ivrii V.Ia. : [1] *On wave fronts of solutions of the system of crystal optics.* Sov.Math. Dokl., 18 (1977), 139-141.

[2] *Wave fronts of solutions of symmetric pseudodifferential systems.* Sibirskii Mat. Zhurn., 20:3 (1979), 557-578.

[3] *Wave fronts of solutions of boundary-value problems for symmetric hyperbolic systems II. Systems with characteristics of constant multiplicity.* Sibirskii Mat. Zhurn., 20:5, (1979), 1022-1038.

Kajitani K. : [1] *Sur la condition necessaire du probleme mixed bien posé pour les systémes hyperboliques à coefficients variables.* Publ. R.I.M.S., Kyoto Univ., 9 (1974), 261-284.

[2] *A necessary condition for the well-posed hyperbolic mixed problem with variable coefficients.* J. Math. Kyoto Univ. 14 (1974), 231- 242.

Kajitani K.- Wakabayashi S. : [1] *Microhyperbolic operators in Gevrey classes.* Publ. R.I.M.S. Kyoto Univ., 25 (1989), 169-221.

[2] *The hyperbolic mixed problem in Gevrey classes.* Japan J. Math. 15 (1989), 315-383.

Kaneko K. : [1] *Representation of hyperfunctions by measures and some of its applications.* J.Fac. Sc. Tokyo IA, 19:3, 1972, 321- 353.

Kashiwara M.-Kawai T. : [1] *Deuxième microlocalisation* Proc. Conf. Les Houches 1976, Lecture Notes in Physics Nr. 126, Springer Verlag, Berlin Heidelberg New York.

[2] *Microhyperbolic pseudodifferential operators I.* J. Math. Soc. Jap. 27 (1975), 359-404.

Kashiwara M.- Kawai T.- Kimura T. : [1] *Foundations of algebraic Analysis.* Princeton Mathematica Series, vol. 37, Princeton University Press, 1986, Princeton.

Kashiwara M.-Laurent Y. : [1] *Théorème d'annulation et deuxième microlocalisation.*

K.Kataoka : [1] *A microlocal approach to general boundary value problems,* Publ. R.I.M.S., Kyoto Univ. 12, Suppl. (1977), 147-153.

[2] *Microlocal theory of boundary value problems I.* J. Fac. Sc. Tokyo IA, 27 (1980), 355-399.

[3] *Microlocal theory of boundary value problems II.* J. Fac. Sc. Tokyo IA, 28 (1981), 31-56.

Kawai T. : [1] *On the theory of hyperfunctions and its applications to partial differential operators with constant coefficients.* J.Fac. Sc. Tokyo IA, 17:3, 1970, 467- 519.

Kessab A. : [1] *Propagation du front d'onde Gevrey des solutions d'èquations à caractéristiques multiples involutives.* C.R. Acad. Sc. Paris, t.299 I (1984), 977-978.

Kiselman C.O. : [1] *Existence and approximation theorems for solutions of complex analogues of boundary problems.* Ark. Mat. 6:3 (1966), 193-207.

[2] *Regularity classes for operators in convexity theory.* Typescript, Lund, July 1991.

Kohn J.J.- Nirenberg L. : [1] *On the algebra of pseudodifferential operators.* C.P.A.M. XVIII (1965), 269-305.

Komatsu H. : [1] *An elementary theory of hyperfunctions and microfunctions.* Banach Center Publ. vol. 27 (1992), 233-256.

[2] *Microlocal analysis in Gevrey classes and in complex domains.* In Springer Lecture Notes in Math., vol. 1495 (edited by L.Cattabriga and L.Rodino) (1991), 161-237.

Lascar R. : [1] *Propagation des singularités des solutions d'équations pseudo differentielles a caracteristiques de multiplicités variables* Lecture Notes in Math., vol. 856, Springer Verlag, Berlin Heidelberg New York, 1981.

Laubin P. : [1] *Refraction conique et propagation des singularités analytiques.* J. Math. pures et appl., 63 (1984), 149-168.

[2] *Propagation of the second analytic wave front set in conical refraction.* Proc. Conf. on hyperbolic equations and related topics, Padova, 1985.

[3] *Front d'onde analytique et décomposition microlocale des distributions.* Annals Inst. Fourier, 33:3, 1983, 179-200.

[4] *Etude 2-microlocale de la diffraction.* Bull. Soc. Royale de Science de Liège, 56:4 (1987), 296-416.

Laubin P.- Esser P. : [1] *Second analytic wave front set of the fundamental solution of hyperbolic operators.* Comm. in P.D.E., 11:5 (1986), 459-482.

Laubin P.- Willems B. : [1] *Second front d'onde analytique le long du conormal à l'origin.* Bull. Soc. Royale de Science de Liège, 59:3-4, (1990), 289-300.

Laurent Y. : [1] *Theorie de la deuxième microlocalisation dans le domaine complexe.* Birkhäuser Verlag, Basel, Progress in Math., vol.53, 1985.

Lebeau G. : [1] *Deuxième microlocalisation sur les sous- varietés isotropes.* Ann. Inst. Fourier Grenoble, XXXV:2 (1985), 145-217.

[2] *Deuxième microlocalisation à croissance* Seminaire Goulaouic-Meyer-Scwartz, 1982/1983. École Polytecn., Paris 1983.

Leray J. : [1] *Hyperbolic Differential Equations.* Princeton Univ. Press, Princeton, 1952.

Liess O. : [1] *Intersection properties of weak analytically uniform classes of functions.* Ark.Mat., 14:1 (1976), 93-111.

[2] *Necessary and sufficient conditions for propagation of singularities for systems of linear partial differential operators with constant coefficients.* C. P.D.E., 8:2 (1983), 89-198.

[3] *Localization and propagation of analytic singularities for operators with constant coefficients.* C. P.D.E., 13:6 (1988), 729-767.

[4] *Boundary behaviour of analytic functions, analytic wave front sets and the intersection problem of Ehrenpreis.* Rend. Sem. Mat. Torino, 42:3 (1984), 103-152.

[5] *The Fourier-Bros-Iagolnitzer transform in inhomogeneous Gevrey classes.* Simon-Stevin Journal, vol. 60:2 (1986), 105-121.

[6] *The microlocal Cauchy problem for constant coefficient linear partial differential operators.* Revue Roum. Math., 21:9, (1976), 1221-1239.

[7] *The Cauchy problem in inhomogeneous Gevrey classes.* C.P.D.E., 11 (1986), 1379-1439.

[8] *Conical refraction and higher microlocalization.* Proc. Seminars in Complex Analysis and Geometry 1988. Edited by J.Guenot and D.C.Struppa. Editel Publ. Company, Cosenza 1990, 89-115.

Liess O.-Rodino L. : [1] *Inhomogeneous Gevrey classes and related pseudodifferential operators.* Boll. Un. Mat. Ital., 3-C (1984), 233-323.

[2] *Fourier integral operators and inhomogeneous Gevrey classes.* Annali Mat. Pura ed Appl., (IV) vol. CL (1988), 167-262.

Liess,O.- Rosu.R. : [1] *Propagation of singularities for constant coefficient linear partial differential equations in a situation when the direction of propagation depends on the solution.*

Boll. U.M.I. (6) 3-B (1984), 351-382.

Loria G. : [1] *Il passato e il presente delle principali teorie geometriche.* Torino,Carlo Clausen editore, 1896.

Ludwig D. : [1] *Conical refraction in crystal optics and hydrodynamics.* C.P.A.M., 14 (1961), 113-124.

Martinez A. : [1] *Propagation du front d'onde polarisé analytique.* C.R. Acad. Sci. Paris, 298 (1984), 513-516.

Melrose R. : [1] *Transformation of boundary problems.* Acta Math., 147 (1981), 149-236.

Melrose R.- Uhlmann G. : [1] *Microlocal structure of involutive conical refraction.* Duke Math. J., 46 (1979), 571-582.

Musgrave M.J.P. : [1] *Crystal acoustics.* Holden Day, San Francisco, 1979.

Nishiwada K. : [1] *On local characterizations of wave front sets in terms of local boundary values of holomorphic functions.* Publ. R.I.M.S. 14, 1978, 309-321.

Nosmas J.C. : [1] *Parametrix du problème de transmission pour l'équation des ondes.* C.R. Acad. Sci. Paris, Série A, 280 (1975), 1213- 1216.

Okada Y. : [1] *Second microlocal singularities of tempered and Gevrey classes.* Journal Fac. Sc. Tokyo, Sec. IA, 39:3, (1992), 475-505.

Okada Y. - Tosé N. : [1] *FBI-transformation and microlocalization- equivalence of the second analytic wave front sets and the second singular spectrum.* Journal de Math. Pures et Appl., t. 70:4 (1991), 427-455.

Palamodov,V. : [1] *Linear partial differential operators with constant coefficients.* Mir Publ., Moscow 1967. Also available as vol. 168 in the Grundlehren series of the Springer Verlag.

Parenti C. : [1] *Microiperbolicità.* Seminario di Analisi Mat. Dip. Mat. Bologna, 1985-1986, V.1- V.22.

Petrini M.-Sordoni V. : [1] *Propagation of singularities for a class of hyperbolic operators with triple characteristics.* C.P.D.E. 16: 4-5, (1991), 683-705.
[2] *Propagation of singularities for hyperbolic pseudodifferential operators with multiple involutive characteristics.* To appear.

Rodino L. : [1] *Microlocal analysis for spatially inhomogeneous pseudodifferential operators.* Ann. Scuola Norm. Sup. Pisa, ser. IV:9 (1982), 211-253.
[2] *On the Gevrey wave front set of the solutions of a quasielliptic equation.* Rend. Sem. Mat. Univ. Polit. Torino. Conference on "Linear partial and pseudodifferential operators". Torino **1982** (1983), 221-234.

Sakamoto R. : [1] L^2 *well-posedness for hyperbolic mixed problems.* Publ. R.I.M.S. Kyoto Univ. 8 (1972/73), 265-293.

Sato M.-Kawai T.-Kashiwara M. : [1] *Hyperfunctions and pseudodifferential operators.* Lecture Notes in Math., vol. 287, Springer Verlag,Berlin Heidelberg New York, 1973, 265-529.

Schapira P. : [1] *Théorie des Hyperfonctions.* Lecture Notes in Mathematics, vol.126, Springer Verlag New York, Heidelberg.
[2] *Propagation au bord et réflexion des singularités analytiques des solutions des équations aux dérivées partielles I.* Publ. R.I.M.S. 1976 Suppl. (Also cf. Sem. Goulaouic-Schwartz 1975-1976, exposé 6, Ecole Polit. Paris), II. Sem. Goulaouic-Schwartz 1976-1977, exposé 9, Ecole Polit. Paris.
[3] *Microfunctions for boundary value problems.* In " Algebraic Analysis", vol.II, volumes in honour of M.Sato, ed. by M.Kashiwara and T.Kawai, Academic Press, 1989, 809-819.

Sjöstrand J. : [1] *Propagation of singularities for operators with multiple involutive characteristics.* Ann. Inst. Fourier Grenoble, 26:1, (1976), 141-155.

[2] *Propagation of analytic singularities for second order Dirichlet problems I.* Comm.P.D.E., 5:1 (1980), 41-94.

[3] *Analytic singularities and microhyperbolic boundary values problems.* Math. Ann. 254 (1980), 211-256.

[4] *Singularités analytiques microlocales.* Astérisque vol.95, 1982, Soc.Math. France.

Sommerfeld A. : [1] *Vorlesungen über theoretische Physik.* Bde III u. IV, Akademische Verlagsgesellschaft, 1964.

Sommerville, D.M.Y. : [1] *Analytical Geometry of three Dimensions.* Cambridge, at the University Press, 1934.

Taylor M. : [1] *Reflection of singularities of solutions to systems of differential equations.* C.P.A.M. 28 (1975), 457-478.

[2] *Grazing rays and reflection of singularities of solutions to wave equations. Part II (Systems).* C.P.A.M. 29 (1976), 463-481.

Tose N. : [1] *On a class of microdifferential operators with involutory double characteristics-as an application of second microlocalization.* J. Fac. Sci. Univ. Tokyo, Sect. IA, Math. 33 (1986), 619-634.

[2] *The 2-microlocal canonical form for a class of microdifferential equations and propagation of singularities.* Publ. R.I.M.S. Kyoto, 23-1 (1987), 101-116.

[3] *Second microlocalisation and conical refraction.* Ann. Inst. Fourier Grenoble, 37:2 (1987), 239-260.

[4] *Second microlocalisation and conical refraction, II.* "Algebraic analysis", Vol.II. Volumes in honour of Prof. M.Sato, edited by T.Kawai and M.Kashiwara, Academic Press, 1989,867-881.

[5] *On a class of nonhyperbolic microdifferential equations with involutory double characteristics.* Proc. Jap. Acad., 63 Ser. A (1987), 258-261.

[6] *On a class of 2-microhyperbolic systems.* J. Math. Pures et Appl. 67 (1988), 1-15.

Treves F. : [1] *Introduction to pseudodifferential and Fourier integral operators,volume 1: pseudodifferential operators.* Plenum Press, New York and London, 1980.

Uchida M. : [1] *Un théorème d'injectivité du morphisme de restriction pour les microfonctions.* C.R. Acad. Sci. Paris Ser.I, 308, (1989), 83-85.

Uhlmann G. : [1] *Light intensity distribution in conical refraction.* C.P.A.M. 35 (1982), 69-80.

Wakabayashi S. : [1] *A necessary condition for the mixed problem to be C^∞ well posed.* C.P.D.E. 5 (1980), 1031-1064.

[2] *The mixed problem for hyperbolic systems.* In "Singularities for Boundary Value Problems", Nato Summer School in Maratea, ed. by H.G.Garnir, D.Reidel Publ. Comp. (1981), 327- 370.

[3] *Singularities of solutions of the Cauchy problem for symmetric hyperbolic systems.* C.P.D.E., 9 (1984), 1147- 1177.

[4] *Generalized Hamilton Flows and Singularities of Solutions of the Hyperbolic Cauchy Problem.* Proc. Hyperbolic Equations and Related Topics, Taniguchi Symposium 1984, Kinokuniya, Tokyo, 415-423.

[5] *Singularities of solutions of the Cauchy problem for hyperbolic systems in Gevrey classes.* Japan.J.Math. 11:1 (1985), 157-201.

Notations

The following is a (partial) list of notations which are used in the paper.

$\mathcal{A}$ denotes the sheaf of real analytic functions.

$A(C^n)$, $A(U)$ denotes the space of analytic functions on C^n respectively on U.

$\mathcal{A}'(R^n)$, $\mathcal{A}'(U)$, $\mathcal{A}(K)$, space of real-analytic functionals on R^n and U, respectively with support in K.

$A_{\Omega,b,b'}$, space of entire analytic functions, introduced in section 7.5.

$|\ |_{\Omega,b,b'}$ is the norm in $A_{\Omega,b,b'}$.

$a_+ = max(a,0)$ and $a_- = (-a)_+$ denote the positive, respectively the negative, part of $a \in R$.

$A^\perp$ is the orthogonal complement of A with respect to the symplectic two-form.

α, β, γ etc. are often used to denote multiindices. Multiindex notation is used in standard fashion.

$\mathcal{B}$ is the sheaf of hyperfunctions.

$B_{c\varphi}$ denotes a φ-neighborhood of B.(Cf. section 2.2.)

$\mathcal{C}$ is the standard sheaf of microfunctions.

$\mathcal{C}(x^0, \xi^0)$ is the stalk of $\mathcal{C}$ at (x^0, ξ^0).

$C^\infty(U)$ are the C^∞ functions on U.

$D'(U)$ are the distributions on U.

df is the differential of f.

δ_{ij} is the Kronecker Delta.

∂A is the boundary of A.

$\mathcal{E}$ is a sheaf of microfunctions. (Cf. section 9.9.)

ess supp$^2 u$ essential support of u in two-microlocalization. (Cf. section 9.9.)

$E'(U)$ are the distributions on U which have compact support.

$E(x, \xi, t), E(x, \xi, t, \alpha, \beta)$ denote the functions defined in section A.3 by

$$E(x, \xi, t) = \exp[-i\langle x, \xi\rangle - |\xi'|\,|x' - t'|^2/2 - |\xi''|\,|x'' - t''|^2/2],$$

$$E(x, \xi, t, \alpha, \beta) = \exp[-i\langle x, \xi\rangle - \alpha\,|\xi'|\,|x' - t'|^2/2 - \beta\,|\xi''|\,|x'' - t''|^2/2].$$

$\mathcal{F}$, applied to some distribution, denotes usually the Fourier transform.

(Our convention is, written formally, $\mathcal{F}u(\xi) = \int \exp\langle x, \xi\rangle u(x)\,dx$.)

When it stands alone it is a sheaf of microfunctions. (Cf. chapter IX.)

$\mathcal{F}^{-1}$ is the inverse Fourier transform.

F_x denotes for a, possibly vector-valued function F, the partial derivatives in x. Similar notation for derivatives in other variables.

$F_{(x)}$ denotes the x-components of a vectorial function with values in (e.g.) (x, ξ) space. Similar notations are used in related situations.

$\hat{f}$ or $\mathcal{F}(f)$ is the Fourier or Fourier-Borel (when we consider complex arguments) transform of f.

g_Γ is for some function g and some set Γ the function $g_\Gamma(\xi) = g(\xi)$ if $\xi \in \Gamma$ and $g_\Gamma(\xi) = 0$ if $\xi \notin \Gamma$.

G^0 is the polar of G (in the euclidean structure of R^n and if the 0 is not an index !).

$G' \subset\subset G$, for multineighborhoods. (Cf. section 3.1.)

$G(\xi^0, \xi^1, ..., \xi^{k-1}, \delta, \beta)$, family of weight functions introduced in section 2.4.

$G^\perp$, $G^{\perp_r}$, symplectic dual cones. ("$\perp$" refers to the standard symplectic structure, "$\perp_r$" to the relative symplectic structure.)

H_f is the Hamiltonian field of f.

$[H_f, H_g]$ is the commutator of H_f and H_g.

H_K is the support function of K: $H_K(\xi) = \sup_{x \in K}\langle x, \xi\rangle$.

I is the identity operator, $I : R^n \to R^n$.

I_α denotes various auxiliary expressions in proofs.

Id denotes the identity $n \times n$ matrix.

$J(D_z) = \sum_\alpha a_\alpha D_z^\alpha$ is when $\lim_{|\alpha|\to\infty}(\alpha!|a_\alpha|)^{1/|\alpha|} = 0$ an infraexponential infinite order partial differential operator.

$\{0\} = M_k \subset M_{k-1} \subset \cdots \subset M_0 = R^n$, $M_i \neq M_{i-1}$, is called a polyhomogeneous structure on R^n. (The M_j are linear subspaces.)

$\dot{M}_j = M_j \ominus M_{j+1}$ is the orthogonal complement of M_{j+1} in M_j.

$\mathcal{M}$ denotes, in chapter IX, the set of formal sums of holomorphic functions defined on wedges.

$\mu_i(t)$ are the multiplications associated with a polyhomogeneous structure. Explicitly $\mu_i(t)\xi = t\Pi_i(\xi) + (I - \Pi_i)(\xi)$.

p_m is the principal part of p. It is of order m.

$p_m^T, p_{m,1}, p_{m,k}$ are localizations of p_m. Cf. section 1.1.

$\Pi_j : R^n \to M_j$ is the orthogonal projection from R^n onto M_j.

$\Pi_{(x^0,\xi^0)}$ is the projection from distributions (or hyperfunctions) to microfunctions.

$\Pi_{(x^0,\xi^0,\sigma^0)}$ is the projection from microfunctions to two-microfunctions.

R^n: n-dimensional euclidean space. When $d < n$, R^d is often identified with $\hat{R}^d = \{x \in R^n; x_i = 0 \text{ for } i \leq d+1\}$.

$\dot{R}^n = R^n \setminus \{0\}$.

$\mathcal{R}(\xi^0, \xi^1, ..., \xi^{k-2}, \delta, \beta)$. Family of weight functions introduced in section 2.4.

S^{n-1} is the unit sphere in R^n.

S_φ^m, SF_φ^m, $\tilde{S}_\varphi^m$, $\tilde{S}F_\varphi^m$- various symbol spaces. (Cf. section 3.3.)

$\sum'$ is usually a sum over some subset of indices.

TY tangent space to the differential variety Y.

$T_\Sigma\Gamma$ is the tangent space to Γ over the points of $\Sigma \subset \Gamma$.

T^*X cotangent space of X.

$\dot{T}^*X$ is the cotangent space with the zero-section removed.

$T_{rel}\Sigma$ relative tangent space of Σ. Section 4.6.

WF_A: standard analytic wave front set. The same thing is denoted occasionally by wf_A.

WF_A^k: k-th order analytic wave front set: cf. section 2.1. Is associated with a given polyhomogeneous structure which may not be specified explicitly if the $(\xi^0, \xi^1, ..., \xi^{k-1})$ appear in some natural way. In that case M_j will be the orthogonal complement of span $\{\xi^0, \xi^1, ..., \xi^{j-1}\}$. WF_A^k comes in different variants.

$WF_{A,s}^k$: k-th order semiisotropic analytic wave front set. Cf. section 2.1. It is one of the variants of higher order analytic wave front sets.

$WF_{A,\Sigma}^2$: second analytic wave front set with respect to the regular analytic involutive homogeneous subvariety $\Sigma \subset T^*X$.

$WF_\varphi u$ is the G_φ- wave front set of u. (Cf. section 2.2.)

$\{f, g\}$ is the Poisson bracket of f, g.

$\{f, g\}_r$ is the relative Poisson bracket of f, g. (Cf. section 4.6.)

Δ is the Laplace operator on on R^n.

ω is the standard fundamental two-form $\Sigma dx_j \wedge d\xi_j$.

ω_r is the relative fundamental two-form.(Cf. section 4.6.)

$\sim$ equivalence relations for weight functions or hyperfunctions.

$\sim_A$ equivalence of two weight functions on the set A. (Cf. 2.2.)

$\sim^2$ equivalence relation for microfunctions. (Cf. section 9.9.)

Subject index

Printing: Weihert-Druck GmbH, Darmstadt
Binding: Buchbinderei Schäffer, Grünstadt

Vol. 1412: V.V. Kalashnikov, V.M. Zolotarev (Eds.), Stability Problems for Stochastic Models. Proceedings, 1987. X, 380 pages. 1989.

Vol. 1413: S. Wright, Uniqueness of the Injective III₁Factor. III, 108 pages. 1989.

Vol. 1414: E. Ramirez de Arellano (Ed.), Algebraic Geometry and Complex Analysis. Proceedings, 1987. VI, 180 pages. 1989.

Vol. 1415: M. Langevin, M. Waldschmidt (Eds.), Cinquante Ans de Polynômes. Fifty Years of Polynomials. Proceedings, 1988. IX, 235 pages.1990.

Vol. 1416: C. Albert (Ed.), Géométrie Symplectique et Mécanique. Proceedings, 1988. V, 289 pages. 1990.

Vol. 1417: A.J. Sommese, A. Biancofiore, E.L. Livorni (Eds.), Algebraic Geometry. Proceedings, 1988. V, 320 pages. 1990.

Vol. 1418: M. Mimura (Ed.), Homotopy Theory and Related Topics. Proceedings, 1988. V, 241 pages. 1990.

Vol. 1419: P.S. Bullen, P.Y. Lee, J.L. Mawhin, P. Muldowney, W.F. Pfeffer (Eds.), New Integrals. Proceedings, 1988. V, 202 pages. 1990.

Vol. 1420: M. Galbiati, A. Tognoli (Eds.), Real Analytic Geometry. Proceedings, 1988. IV, 366 pages. 1990.

Vol. 1421: H.A. Biagioni, A Nonlinear Theory of Generalized Functions, XII, 214 pages. 1990.

Vol. 1422: V. Villani (Ed.), Complex Geometry and Analysis. Proceedings, 1988. V, 109 pages. 1990.

Vol. 1423: S.O. Kochman, Stable Homotopy Groups of Spheres: A Computer-Assisted Approach. VIII, 330 pages. 1990.

Vol. 1424: F.E. Burstall, J.H. Rawnsley, Twistor Theory for Riemannian Symmetric Spaces. III, 112 pages. 1990.

Vol. 1425: R.A. Piccinini (Ed.), Groups of Self-Equivalences and Related Topics. Proceedings, 1988. V, 214 pages. 1990.

Vol. 1426: J. Azéma, P.A. Meyer, M. Yor (Eds.), Séminaire de Probabilités XXIV, 1988/89. V, 490 pages. 1990.

Vol. 1427: A. Ancona, D. Geman, N. Ikeda, École d'Eté de Probabilités de Saint Flour XVIII, 1988. Ed.: P.L. Hennequin. VII, 330 pages. 1990.

Vol. 1428: K. Erdmann, Blocks of Tame Representation Type and Related Algebras. XV. 312 pages. 1990.

Vol. 1429: S. Homer, A. Nerode, R.A. Platek, G.E. Sacks, A. Scedrov, Logic and Computer Science. Seminar, 1988. Editor: P. Odifreddi. V, 162 pages. 1990.

Vol. 1430: W. Bruns, A. Simis (Eds.), Commutative Algebra. Proceedings. 1988. V, 160 pages. 1990.

Vol. 1431: J.G. Heywood, K. Masuda, R. Rautmann, V.A. Solonnikov (Eds.), The Navier-Stokes Equations – Theory and Numerical Methods. Proceedings, 1988. VII, 238 pages. 1990.

Vol. 1432: K. Ambos-Spies, G.H. Müller, G.E. Sacks (Eds.), Recursion Theory Week. Proceedings, 1989. VI, 393 pages. 1990.

Vol. 1433: S. Lang, W. Cherry, Topics in Nevanlinna Theory. II, 174 pages.1990.

Vol. 1434: K. Nagasaka, E. Fouvry (Eds.), Analytic Number Theory. Proceedings, 1988. VI, 218 pages. 1990.

Vol. 1435: St. Ruscheweyh, E.B. Saff, L.C. Salinas, R.S. Varga (Eds.), Computational Methods and Function Theory. Proceedings, 1989. VI, 211 pages. 1990.

Vol. 1436: S. Xambó-Descamps (Ed.), Enumerative Geometry. Proceedings, 1987. V, 303 pages. 1990.

Vol. 1437: H. Inassaridze (Ed.), K-theory and Homological Algebra. Seminar, 1987–88. V, 313 pages. 1990.

Vol. 1438: P.G. Lemarié (Ed.) Les Ondelettes en 1989. Seminar. IV, 212 pages. 1990.

Vol. 1439: E. Bujalance, J.J. Etayo, J.M. Gamboa, G. Gromadzki. Automorphism Groups of Compact Bordered Klein Surfaces: A Combinatorial Approach. XIII, 201 pages. 1990.

Vol. 1440: P. Latiolais (Ed.), Topology and Combinatorial Groups Theory. Seminar, 1985–1988. VI, 207 pages. 1990.

Vol. 1441: M. Coornaert, T. Delzant, A. Papadopoulos. Géométrie et théorie des groupes. X, 165 pages. 1990.

Vol. 1442: L. Accardi, M. von Waldenfels (Eds.), Quantum Probability and Applications V. Proceedings, 1988. VI, 413 pages. 1990.

Vol. 1443: K.H. Dovermann. R. Schultz, Equivariant Surgery Theories and Their Periodicity Properties. VI, 227 pages. 1990.

Vol. 1444: H. Korezlioglu, A.S. Ustunel (Eds.), Stochastic Analysis and Related Topics VI. Proceedings, 1988. V, 268 pages. 1990.

Vol. 1445: F. Schulz, Regularity Theory for Quasilinear Elliptic Systems and – Monge Ampère Equations in Two Dimensions. XV, 123 pages. 1990.

Vol. 1446: Methods of Nonconvex Analysis. Seminar, 1989. Editor: A. Cellina. V, 206 pages. 1990.

Vol. 1447: J.-G. Labesse. J. Schwermer (Eds), Cohomology of Arithmetic Groups and Automorphic Forms. Proceedings, 1989. V, 358 pages. 1990.

Vol. 1448: S.K. Jain, S.R. López-Permouth (Eds.), Non-Commutative Ring Theory. Proceedings, 1989. V, 166 pages. 1990.

Vol. 1449: W. Odyniec, G. Lewicki, Minimal Projections in Banach Spaces. VIII, 168 pages. 1990.

Vol. 1450: H. Fujita, T. Ikebe, S.T. Kuroda (Eds.), Functional-Analytic Methods for Partial Differential Equations. Proceedings, 1989. VII, 252 pages. 1990.

Vol. 1451: L. Alvarez-Gaumé, E. Arbarello, C. De Concini, N.J. Hitchin, Global Geometry and Mathematical Physics. Montecatini Terme 1988. Seminar. Editors: M. Francaviglia, F. Gherardelli. IX, 197 pages. 1990.

Vol. 1452: E. Hlawka, R.F. Tichy (Eds.), Number-Theoretic Analysis. Seminar, 1988–89. V, 220 pages. 1990.

Vol. 1453: Yu.G. Borisovich, Yu.E. Gliklikh (Eds.), Global Analysis – Studies and Applications IV. V, 320 pages. 1990.

Vol. 1454: F. Baldassari, S. Bosch. B. Dwork (Eds.), p-adic Analysis. Proceedings, 1989. V, 382 pages. 1990.

Vol. 1455: J.-P. Françoise. R. Roussarie (Eds.), Bifurcations of Planar Vector Fields. Proceedings, 1989. VI, 396 pages. 1990.

Vol. 1456: L.G. Kovács (Ed.), Groups – Canberra 1989. Proceedings. XII, 198 pages. 1990.

Vol. 1457: O. Axelsson, L.Yu. Kolotilina (Eds.), Preconditioned Conjugate Gradient Methods. Proceedings. 1989. V, 196 pages. 1990.

Vol. 1458: R. Schaaf, Global Solution Branches of Two Point Boundary Value Problems. XIX, 141 pages. 1990.

Vol. 1459: D. Tiba, Optimal Control of Nonsmooth Distributed Parameter Systems. VII, 159 pages. 1990.

Vol. 1508: M. Vuorinen (Ed.), Quasiconformal Space Mappings. A Collection of Surveys 1960-1990. IX, 148 pages. 1992.

Vol. 1509: J. Aguadé, M. Castellet, F. R. Cohen (Eds.), Algebraic Topology - Homotopy and Group Cohomology. Proceedings, 1990. X, 330 pages. 1992.

Vol. 1510: P. P. Kulish (Ed.), Quantum Groups. Proceedings, 1990. XII, 398 pages. 1992.

Vol. 1511: B. S. Yadav, D. Singh (Eds.), Functional Analysis and Operator Theory. Proceedings, 1990. VIII, 223 pages. 1992.

Vol. 1512: L. M. Adleman, M.-D. A. Huang, Primality Testing and Abelian Varieties Over Finite Fields. VII, 142 pages. 1992.

Vol. 1513: L. S. Block, W. A. Coppel, Dynamics in One Dimension. VIII, 249 pages. 1992.

Vol. 1514: U. Krengel, K. Richter, V. Warstat (Eds.), Ergodic Theory and Related Topics III, Proceedings, 1990. VIII, 236 pages. 1992.

Vol. 1515: E. Ballico, F. Catanese, C. Ciliberto (Eds.), Classification of Irregular Varieties. Proceedings, 1990. VII, 149 pages. 1992.

Vol. 1516: R. A. Lorentz, Multivariate Birkhoff Interpolation. IX, 192 pages. 1992.

Vol. 1517: K. Keimel, W. Roth, Ordered Cones and Approximation. VI, 134 pages. 1992.

Vol. 1518: H. Stichtenoth, M. A. Tsfasman (Eds.), Coding Theory and Algebraic Geometry. Proceedings, 1991. VIII, 223 pages. 1992.

Vol. 1519: M. W. Short, The Primitive Soluble Permutation Groups of Degree less than 256. IX, 145 pages. 1992.

Vol. 1520: Yu. G. Borisovich, Yu. E. Gliklikh (Eds.), Global Analysis – Studies and Applications V. VII, 284 pages. 1992.

Vol. 1521: S. Busenberg, B. Forte, H. K. Kuiken, Mathematical Modelling of Industrial Process. Bari, 1990. Editors: V. Capasso, A. Fasano. VII, 162 pages. 1992.

Vol. 1522: J.-M. Delort, F. B. I. Transformation. VII, 101 pages. 1992.

Vol. 1523: W. Xue, Rings with Morita Duality. X, 168 pages. 1992.

Vol. 1524: M. Coste, L. Mahé, M.-F. Roy (Eds.), Real Algebraic Geometry. Proceedings, 1991. VIII, 418 pages. 1992.

Vol. 1525: C. Casacuberta, M. Castellet (Eds.), Mathematical Research Today and Tomorrow. VII, 112 pages. 1992.

Vol. 1526: J. Azéma, P. A. Meyer, M. Yor (Eds.), Séminaire de Probabilités XXVI. X, 633 pages. 1992.

Vol. 1527: M. I. Freidlin, J.-F. Le Gall, Ecole d'Eté de Probabilités de Saint-Flour XX – 1990. Editor: P. L. Hennequin. VIII, 244 pages. 1992.

Vol. 1528: G. Isac, Complementarity Problems. VI, 297 pages. 1992.

Vol. 1529: J. van Neerven, The Adjoint of a Semigroup of Linear Operators. X, 195 pages. 1992.

Vol. 1530: J. G. Heywood, K. Masuda, R. Rautmann, S. A. Solonnikov (Eds.), The Navier-Stokes Equations II – Theory and Numerical Methods. IX, 322 pages. 1992.

Vol. 1531: M. Stoer, Design of Survivable Networks. IV, 206 pages. 1992.

Vol. 1532: J. F. Colombeau, Multiplication of Distributions. X, 184 pages. 1992.

Vol. 1533: P. Jipsen, H. Rose, Varieties of Lattices. X, 162 pages. 1992.

Vol. 1534: C. Greither, Cyclic Galois Extensions of Commutative Rings. X, 145 pages. 1992.

Vol. 1535: A. B. Evans, Orthomorphism Graphs of Groups. VIII, 114 pages. 1992.

Vol. 1536: M. K. Kwong, A. Zettl, Norm Inequalities for Derivatives and Differences. VII, 150 pages. 1992.

Vol. 1537: P. Fitzpatrick, M. Martelli, J. Mawhin, R. Nussbaum, Topological Methods for Ordinary Differential Equations. Montecatini Terme, 1991. Editors: M. Furi, P. Zecca. VII, 218 pages. 1993.

Vol. 1538: P.-A. Meyer, Quantum Probability for Probabilists. X, 287 pages. 1993.

Vol. 1539: M. Coornaert, A. Papadopoulos, Symbolic Dynamics and Hyperbolic Groups. VIII, 138 pages. 1993.

Vol. 1540: H. Komatsu (Ed.), Functional Analysis and Related Topics, 1991. Proceedings. XXI, 413 pages. 1993.

Vol. 1541: D. A. Dawson, B. Maisonneuve, J. Spencer, Ecole d' Eté de Probabilités de Saint-Flour XXI - 1991. Editor: P. L. Hennequin. VIII, 356 pages. 1993.

Vol. 1542: J.Fröhlich, Th.Kerler, Quantum Groups, Quantum Categories and Quantum Field Theory. VII. 431 pages. 1993.

Vol. 1543: A. L. Dontchev, T. Zolezzi, Well-Posed Optimization Problems. XII, 421 pages. 1993.

Vol. 1544: M.Schürmann, White Noise on Bialgebras. VII, 146 pages. 1993.

Vol. 1545: J. Morgan, K. O'Grady, Differential Topology of Complex Surfaces. VIII, 224 pages. 1993.

Vol. 1546: V. V. Kalashnikov, V. M. Zolotarev (Eds.), Stability Problems for Stochastic Models. Proceedings, 1991. VIII, 229 pages. 1993.

Vol. 1547: P. Harmand, D. Werner, W. Werner, M-ideals in Banach Spaces and Banach Algebras. VIII, 387 pages. 1993.

Vol. 1548: T. Urabe, Dynkin Graphs and Quadrilateral Singularities. VI, 233 pages. 1993.

Vol. 1549: G. Vainikko, Multidimensional Weakly Singular Integral Equations. XI, 159 pages. 1993.

Vol. 1550: A. A. Gonchar, E. B. Saff (Eds.), Approximation Theory in Complex Analysis and Mathematical Physics. IV, 222 pages. 1983.

Vol. 1551: L. Arkeryd, P. L. Lions, P.A. Markowich, S.R. S. Varadhan. Nonequilibrium Problems in Many-Particle Systems. Montecatini, 1992. Editors: C. Cercignani, M. Pulvirenti. VII, 158 pages 1993.

Vol. 1552: J. Hilgert, K.-H. Neeb, Lie Semigroups and their Applications. XII, 315 pages. 1993.

Vol. 1553: J.-L- Colliot-Thélène, K. Kato, P. Vojta. Arithmetic Algebraic Geometry. Editor: E. Ballico. VII, 223 pages. 1993.

Vol. 1554: A. K. Lenstra, H. W. Lenstra, Jr. (Eds.), The development of the number field sieve. VIII, 131 pages. 1993.

Vol. 1555: O. Liess, Conical Refraction and Higher Microlocalization. X, 389 pages. 1993.